Advances in Sustainability Science and Technology

The book series aims at bringing together valuable and novel scientific contributions that address the critical issues of renewable energy, sustainable building, sustainable manufacturing, and other sustainability science and technology topics that have an impact in this diverse and fast-changing research community in academia and industry.

The areas to be covered are

- Climate change and mitigation, atmospheric carbon reduction, global warming
- Sustainability science, sustainability technologies
- Sustainable building technologies
- Intelligent buildings
- Sustainable energy generation
- Combined heat and power and district heating systems
- Control and optimization of renewable energy systems
- Smart grids and micro grids, local energy markets
- Smart cities, smart buildings, smart districts, smart countryside
- Energy and environmental assessment in buildings and cities
- Sustainable design, innovation and services
- Sustainable manufacturing processes and technology
- Sustainable manufacturing systems and enterprises
- Decision support for sustainability
- Micro/nanomachining, microelectromechanical machines (MEMS)
- Sustainable transport, smart vehicles and smart roads
- Information technology and artificial intelligence applied to sustainability
- Big data and data analytics applied to sustainability
- Sustainable food production, sustainable horticulture and agriculture
- Sustainability of air, water and other natural resources
- Sustainability policy, shaping the future, the triple bottom line, the circular economy

High quality content is an essential feature for all book proposals accepted for the series. It is expected that editors of all accepted volumes will ensure that contributions are subjected to an appropriate level of reviewing process and adhere to KES quality principles.

The series will include monographs, edited volumes, and selected proceedings.

Jorge Filipe Leal Costa Semião ·
Nelson Manuel Santos Sousa ·
Rui Mariano Sousa da Cruz ·
Gonçalo Nuno Delgado Prates
Editors

INCREaSE 2023

Proceedings of the 3rd INternational CongRess on Engineering and Sustainability in the XXI CEntury

Springer

Editors
Jorge Filipe Leal Costa Semião
Instituto Superior de Engenharia
Universidade do Algarve
Faro, Portugal

Nelson Manuel Santos Sousa
Instituto Superior de Engenharia
Universidade do Algarve
Faro, Portugal

Rui Mariano Sousa da Cruz
Instituto Superior de Engenharia
Universidade do Algarve
Faro, Portugal

Gonçalo Nuno Delgado Prates
Instituto Superior de Engenharia
Universidade do Algarve
Faro, Portugal

ISSN 2662-6829 ISSN 2662-6837 (electronic)
Advances in Sustainability Science and Technology
ISBN 978-3-031-44005-2 ISBN 978-3-031-44006-9 (eBook)
https://doi.org/10.1007/978-3-031-44006-9

This Springer imprint is published by the registered company Springer Nature Switzerland AG
The registered company address is: Gewerbestrasse 11, 6330 Cham, Switzerland

Paper in this product is recyclable.

Preface

It is our pleasure to present the proceedings of the 3rd International Congress on Engineering and Sustainability in the XXI Century (INCREaSE 2023). This book of proceedings aims to bring together valuable and novel scientific contributions to the sustainable development in a multidisciplinary way, that have an impact in diverse and fast-changing research areas, both in academia and industry, reflected in the fields of civil, electronics, food, and mechanical engineering. This book presents 31 works from authors from different countries in several transversal areas, such as big data and data analytics, climate change and mitigation, carbon reduction, sustainable food processing and safety, sustainability in water management, sustainable energy generation and management, construction sustainability, and other subjects related to the sustainable development.

This year's INCREaSE was organized by the Institute of Engineering and hosted by the University of Algarve during July 5–7, 2023, in Faro, Portugal.

All members of the organizing committee, authors, and reviewers played a key role with their dedicated work and efforts.

INCREaSE 2023 had an excellent group of keynote speakers: Enrique Cabrera-Rochera—University of Valencia, Spain, Paulo Sérgio Duque de Brito—Polytechnic of Portalegre, Portugal, and Jorge A. Saraiva—University of Aveiro, Portugal. We are thankful to these leading experts for their participation in INCREaSE 2023.

We wish to express our gratitude to all the above participants who contributed to the success of the third edition of INCREaSE.

July 2023

Jorge Filipe Leal Costa Semião
Nelson Manuel Santos Sousa
Rui Mariano Sousa da Cruz
Gonçalo Nuno Delgado Prates

Organization

INCREaSE is an international congress organized by the Institute of Engineering from the University of Algarve, Faro, Portugal.

Organizing Committee

Damian Bęben	Opole University of Technology, Poland
Gonçalo Prates	University of Algarve and IGOT—University of Lisbon, Portugal
Irene Albertos	University of Valladolid, Spain
J.David Bienvenido-Huertas	University of Granada, Spain
Jooyeoun Jung	Oregon State University, USA
Juan Gutierrez-Estrada	University of Huelva, Spain
Jorge Semião	University of Algarve and INESC-ID, Portugal
Nelson Sousa	University of Algarve, Portugal
Pedro Coelho	IST - University of Lisbon, Portugal
Raúl Martin-Garcia	University of Cádiz, Spain
Rui Cruz	University of Algarve and MED, Portugal
Víctor Muñoz-Martinez	University of Malaga, Spain

Scientific Committee

Alexander Martín-Garín	University of País Vasco, Spain
Ana Carreira	University of Algarve, Portugal
Ana Filipa Ferreira	IST - University of Lisbon, Portugal
António Fernando Sousa	University of Algarve, Portugal
Antonio Illana	University of Cádiz, Spain
António Saraiva Lopes	IGOT—University of Lisbon and International Association for Urban Climate, Portugal
Carlos Álvarez	Teagasc Food Research Center, Ireland
Carlos Otero Silva	University of Algarve, Portugal
Carlos Rubio-Bellido	University of Seville, Spain
Clauciana Schmidt Bueno de Moraes	UNESP, Brazil
Cristiano Cabrita	University of Algarve, Portugal
Cristina Torrecillas-Lozano	University of Seville, Spain
Damla Dag	Oregon State University, USA

Daniel Sánchez-García	University Carlos III de Madrid, Spain
Fátima Farinha	University of Algarve, Portugal
Gil Fraqueza	University of Algarve, Portugal
Inmaculada Pulido-Calvo	University of Huelva, Spain
Ivone Alejandra Czerwinski	Instituto Español de Oceanografía, Centro Ocean-ográfico de Cádiz, Spain
Jânio Monteiro	University of Algarve and INESC-ID, Portugal
João Gomes	University of Algarve, Portugal
Joaquín Gómez-Estaca	Institute of Food Science Technology and Nutrition, Spain
Jorge Isidoro	University of Algarve, Portugal
José Enrique Díaz	University of Cádiz, Spain
Krzysztof Drozdzol	Opole University of Technology, Poland
Long Chen	Cornell University, USA
Manuela Moreira da Silva	University of Algarve and CIMA-ARNET and CEiiA—COLAB Smart and Sustainable Living, Portugal
Maria Rossella R. Massimino	University of Catania, Italy
Miguel Oliveira	University of Algarve, Portugal
Miguel Suffo Pino	University of Cádiz, Spain
Muriel Iten	ISQ, Portugal
Pedro J. S. Cardoso	University of Algarve and LARSyS, Portugal
Qingyang Wang	Oregon State University, USA
Remedios Cabrera Castro	University of Cádiz, Spain
Roberto Laranja	University of Algarve, Portugal
Rossana Villa Rojas	University of Nebraska–Lincoln, USA
Tomasz Maleska	Opole University of Technology, Poland

Invited Speakers

Enrique Cabrera-Rochera	University of Valencia, Spain
Paulo Sérgio Duque de Brito	Polytechnic of Portalegre, Portugal
Jorge A. Saraiva	University of Aveiro, Portugal

Institutional Support

UNIVERSIDADE DO ALGARVE

Sponsors

CÂMARA MUNICIPAL DE LOULÉ
CÂMARA MUNICIPAL DE FARO
RTA-VISIT ALGARVE
SMART CAMPUS
STAP-Reparação, Consolidação e Modificação de Estruturas, S.A.
HUBEL GRUPO
ÁGUAS DE MONCHIQUE

Contents

Special Session on Water Management Challenges

Rainwater Harvesting in University Buildings. Feasibility Analyses in Mediterranean Climate

Armando Silva-Afonso[1,2] and Carla Pimentel-Rodrigues[1,2](✉)

[1] RISCO – Universidade de Aveiro, Aveiro, Portugal
anqip@anqip.pt

[2] ANQIP – Associação Nacional para a Qualidade nas Instalações Prediais, Aveiro, Portugal

Abstract. Situations of water stress or even water scarcity are increasing in different regions of the planet, due to exponential population growth, the economic development model and/or climate change. One of the regions on the planet where climate change is causing increasingly recurrent situations of drought is southern Europe, in the Mediterranean basin. In parallel with this effect, an increase in the intensity and frequency of heavy rains is also observed, generating, for example, very significant impacts on urban environments. Rainwater harvesting systems (RWHS) in buildings are a sustainable solution that allows tackling these two effects of climate change, promoting a retention of water, which can be used for non-potable purposes and mitigating urban flooding. For this reason, RWHS are being implemented in several countries affected by climate change, such as Portugal, where the University of Aveiro promoted a study in two pilot buildings, described in this article, with the aim of assessing the feasibility installation of this system in new university buildings. The conclusions show high benefits not only from the point of view of adaptation to climate change, but also from the economic point of view, with payback periods between 10 and 12 years.

Keywords: Rainwater Harvesting in Buildings · University Buildings · Sustainable Buildings

1 Introduction

Fresh water is an essential resource for sustainable development and is fundamental for socioeconomic development, for maintaining terrestrial ecosystems and for human survival itself [1–5]. However, situations of water stress or even water scarcity are increasing in different regions of the planet, due to exponential population growth, the economic development model and/or climate change [6–14].

One of the regions on the planet where climate change is causing increasingly recurrent situations of drought or even water scarcity is southern Europe, in the Mediterranean basin [15–18]. In parallel with this effect, an increase in the intensity and frequency of heavy rains is also observed in this region, generating, for example, very significant impacts on urban environments, imposing the widespread adoption of measures to improve adaptation to more extreme climatic conditions [19–25].

J. F. L. C. Semião et al. (Eds.): INCREaSE 2023, ASST, pp. 3–12, 2023.
https://doi.org/10.1007/978-3-031-44006-9_1

Rainwater harvesting systems (RWHS) in buildings are a sustainable solution that allows tackling these two effects of climate change, promoting a retention of water, which can be used for non-potable purposes, and mitigating urban flooding [26–29]. For this reason, RWHS is being promoted in several countries affected by climate change, even in Mediterranean climates, although in this case the reduction in total precipitation, especially in summer, may call into question the technical-economic interest of the solution.

This article presents a study carried out by the University of Aveiro, Portugal, aiming to evaluate the feasibility of installing RWHS in new university buildings, based on the results obtained in two pilot buildings. In addition to the obvious environmental benefits, in terms of constituting a reserve of water for non-potable purposes in periods of drought, reducing the consumption of traditional sources.

The use of rainwater in a Mediterranean climate is sometimes hampered by the fact that, in this climate, the periods of greatest consumption generally occur in the summer, which corresponds to a period of less or even zero precipitation. There is therefore an obvious mismatch between availability and needs.

However, school buildings, in general, constitute an exception to this rule, as school holidays occur in the summer, implying a significantly reduced use of buildings in this period. University buildings are no exception, although research activities are sometimes maintained throughout the year, regardless of the interruption of classes in the summer.

2 Materials and Methods

2.1 Characteristics of the Buildings

The application of RWHS was studied in two new buildings at the University of Aveiro: the building of the Interdisciplinary Complex of Sciences Applied to Nanotechnology and Oceanography, which will be called Building 1, and the Building of Optical Communications, Radio Communications and Robotics - Telematics, which will be designated by Building 2. These two buildings are essentially dedicated to scientific research, although they also hold classes.

Building 1 has several sanitary facilities, with a total of 21 toilets with flushing cisterns and two wash taps mainly used for cleaning floors. The use of rainwater was considered only for these uses, excluding showers and washbasin taps for reasons of health safety.

Figure 1 shows Building 1 already completed and Fig. 2 shows the types of sanitary devices that will be fed with rainwater. Building 2 is relatively similar in terms of sanitary fixtures, having a total of 15 toilets with flushing cisterns and five wash taps. As in Building 1, the use of rainwater was considered only for flushing cisterns and wash taps, excluding sinks, showers and basin taps.

Figure 3 shows the Building 2. The devices to be supplied with rainwater are similar to those in Building 1.

2.2 Design Bases and Sizing Methods

In establishing the design bases and sizing methods, the applicable technical documents were observed. In Portugal, there is a Technical Specification for RWHS in buildings, the

Fig. 1. Building 1.

Fig. 2. Type of devices to be fed with rainwater.

Fig. 3. Building 2.

ETA ANQIP 0701:2007 [30], which was the main reference used, although the European Standard EN 16941–1:2018 (*On-site non-potable water systems - Part 1: Systems for the use of rainwater*) [31] was also consulted. For the dimensioning of the piping networks, the European Standards EN 806–3:2006 [32], EN 12056–2:2000 [33] and EN 12056–3:2000 [34] were also considered, as well as other applicable bibliographies in the building hydraulic installations sector.

According to ETA ANQIP 0701, the usable annual volume of rainwater can be determined by the expression:

$$V_a = C \times P \times A \times \eta_f \tag{1}$$

where V_a corresponds to the annual volume of rainwater that can be used, C is assumed as a runoff coefficient that considers water retention, absorption and diversion on the collection surface (80% for flat or low slope waterproof roof, according to ETA ANQIP 0701 and EN 16941–1:2018), P refers to the average annual value of accumulated precipitation in the city of Aveiro (≈1800 mm), A is equivalent to the coverage catchment area measured in horizontal projection and η_f considers the hydraulic filtration efficiency (in general close to 90%). For the sizing of storage tanks, expression (1) can be used on a monthly basis, if monthly average precipitation values are available locally.

In the case of Building 1, the roof essentially consists of sandwich panels, but there is also a small accessible terrace area. For reasons of water quality, the area corresponding to the accessible terrace was not considered as a catchment area, considering that the total area covered with panels will be fully sufficient to meet the required consumption needs. The usable area considered totals 540 m^2, from which an annual volume of rainwater of about 700 m^3 is obtained by applying expression (1).

In the case of Building 2, only the utilization of an area of 460 m^2 was considered for the RWHS, as the full utilization of the roof was difficult for architectural reasons. The annual volume of usable rainwater obtained is around 635 m^3 in this case.

At several weather stations near Aveiro, monthly precipitation values are available. The average values calculated for the Castelo Burgães station (40.853 °N; –8.379 °W), which has data recorded since 1930/31, available on the website of the National System of Information on Water Resources (SNIRH), were adopted in this study. Table 1 summarizes the average values determined from the records available from 1930 to the date of the study, which was adopted in the calculations.

Table 1. Average monthly rainfall in Aveiro (mm)

Month	Jan	Feb	Mar	Apr	May	Jun	Jul	Aug	Sep	Oct	Nov	Dec
Runoff	249	233	185	142	143	68	26	33	83	173	217	247

It is likely that, due to climate change, there is a tendency for the values indicated in Table 1 to decrease over time (15% to 20% in 2100, according to IPMA - Portuguese Institute of Sea and Atmosphere), but this reduction is difficult to estimate in the present. However, it was considered that, during the useful life of the installation, the possible reductions in precipitation do not compromise the conclusions of the study.

3 Results and Discussion

Storage tank sizing is generally an important aspect of a RWHS design, especially as it is the most expensive component. Technical Specification ETA ANQIP 0701 presents a simplified sizing procedure for estimating the volume of tanks, although it recommends that, in large installations, an optimization methodology be applied, based on the average monthly precipitation in the location and on the monthly consumption diagram.

In the present study, the simplified method of ETA ANQIP 0701 was adopted, obtaining, in both buildings, tanks with a volume of 6.5 m^3 for a maximum storage period of 30 days. It should be noted that the volume has been rounded according to traditional commercial values for tanks of this type.

Regarding non-potable water demand in buildings, the values suggested in Annex 2 of ETA ANQIP 0701 were adopted, which are transcribed in Table 2. The indicated values assume the installation of efficient devices, labeled in category A or higher in the ANQIP water efficiency labeling scheme for products, as is the case of the devices installed in these buildings.

Table 2. Non-potable water demand [30]

Device	Consumption
Flushing cistern (category A) in service buildings	4400 L/ ind and year
Wash tap (dominant use for floor washing)	5 L/ m^2 of floor

In Building 1, it is estimated that there are 100 daily users, weighting weekdays and weekends. For the area covered by the washing taps, the value of 150 m^2 was considered according to the project, estimating that the washes have a monthly frequency. Based on these assumptions, the following annual demand for non-potable water was estimated in Building 1 (D_1):

$$D_1 = 4400 \times 100 + 5 \times 12 \times 150 = 449.000\, L/year = 449.0\, m^3/year \quad (2)$$

This value corresponds to 37.40 m^3/month. However, in the months of July, August and September, half of this value (18.70 m^3/month) was considered, bearing in mind that, in these months, some research activities are continued, despite being the usual holiday months in Portugal.

In Building 2, it is estimated that there are 125 weighted daily users. In this building, the area considered for the coverage of the washing taps was 616 m^2, according to the project, and a quarterly frequency was estimated for washing, given the characteristics of the building. Based on these assumptions, the following annual demand for non-potable water was estimated in Building 2 (D_2):

$$D_2 = 4400 \times 125 + 5 \times 4 \times 616 = 562,320\, L/year \approx 562.3\, m^3/year \quad (3)$$

This value corresponds to 46.86 m^3/month. However, as in Building 1, in the months of July, August and September only half of this value was considered (23.43 m^3/month).

Tables 3 and 4 present summarized maps with the simulation of the operation of the RWHS in the two buildings. On the security side, it is assumed that the storage tank is empty at the beginning of the simulation cycle, which was made to correspond to the hydrological year in Mediterranean climates (October 1st to September 30th).

Table 3. Simulation of the operation of the RWHS in Building 1

Month	Average monthly rainfall	Monthly demand	Available vol. of the monthly rainfall	Demand-available. Difference	Storage tank volume	Volume of water in the tank		Public network supply require
						Beg.	End	
	(mm)	(m^3)	(m^3)	(m^3)	(m^3)	(m^3)	(m^3)	(m^3)
October	173	37.40	67.26	29.86	6.50	0.00	6.50	0.00
November	217	37.40	84.37	46.97		6.50	6.50	0.00
December	247	37.40	96.03	58.63		6.50	6.50	0.00
January	249	37.40	96.81	59.41		6.50	6.50	0.00
February	233	37.40	90.59	53.19		6.50	6.50	0.00
March	185	37.40	71.93	34.53		6.50	6.50	0.00
April	142	37.40	55.21	17.81		6.50	6.50	0.00
May	143	37.40	55.60	18.20		6.50	6.50	0.00
June	68	37.40	26.22	−10.96		6.50	0.00	4.46
July	26	18.70	10.11	−8.59		0.00	0.00	8.59
August	33	18.70	12.83	−5.87		0.00	0.00	5.87
September	83	18.70	32.27	13.57		0.00	6.50	0.00
TOTAL	1799	392.70						18.92

The estimated cost for the RWHS in Building 1 is € 8100.00, including the 6500 L tank, the leaf filter, the valves and connection pipes, the technical and pumping pack and the additional cost of water supply and rainwater drainage networks in the building. The RHWS for Building 2 was estimated at a similar value.

To size the pumping group (booster type), a peak flow of 1 L/s was adopted, considering the EN 806–3 standard and a pump head of 22 m was set, a value in accordance with Portuguese regulations [35] for buildings with up to three floors. The commercial power obtained for these values was 0.8 kW. With regard to Building 2, the calculation parameters were the same, whereby the same result was obtained.

With regard to the price of water in the region of Aveiro, the reference value for the University is 1.88 €/m^3. For electricity, the reference value is 0.203 €/kWh.

It can be seen that the energy consumption in the two RHWS will be very low. For the flow rate of 1 L/s (3.6 m^3/h), it can be estimated:

$$\textit{Annual hours of operation}\ (\textit{Building}\ 1) = 392.70/3.6 \approx 109\, h/year \tag{4}$$

Table 4. Simulation of the operation of the RWHS in Building 2

Month	Average monthly rainfall	Monthly demand	Available vol. of the monthly rainfall	Demand-available. Difference	Storage tank volume	Volume of water in the tank		Public network supply required
						Beg.	End	
	(mm)	(m^3)	(m^3)	(m^3)	(m^3)	(m^3)	(m^3)	(m^3)
October	173	46.86	57.30	10.44	6.50	0.00	6.50	0.00
November	217	46.86	71.87	25.01		6.50	6.50	0.00
December	247	46.86	81.81	34.95		6.50	6.50	0.00
January	249	46.86	82.47	35.61		6.50	6.50	0.00
February	233	46.86	77.17	30.31		6.50	6.50	0.00
March	185	46.86	61.27	14.41		6.50	6.50	0.00
April	142	46.86	47.03	0.17		6.50	6.50	0.00
May	143	46.86	47.36	0,50		6.50	6.50	0.00
June	68	46.86	22.52	−24.34		6.50	0.00	17.84
July	26	23.43	8.61	−14.82		0.00	0.00	14.82
August	33	23.43	10.93	−12.50		0.00	0.00	12.50
September	83	23.43	27.49	4.06		0.00	6.50	0.00
TOTAL	1799	492.03						45.16

$$Energy\ cost = 109 \times 0.8 \times 0.203 \approx 18\ €/year \tag{5}$$

$$Annual\ hours\ of\ operation\ (Building\ 2) = 492.03/3.6 \approx 137\ h/year \tag{6}$$

$$Energy\ cost = 137 \times 0.8 \times 0.203 \approx 22\ €/year \tag{7}$$

In economic terms, water savings at current prices, to an average year, correspond to the following values:

$$Savings\ in\ Building\ 1 = 373.78 \times 1.88 \approx 702.71\ €/year \tag{8}$$

$$Savings\ in\ Building\ 2 = 446.87 \times 1.88 \approx 840.11\ €/year \tag{9}$$

In a simplified way, the payback periods of these investments can be estimated in:

$$Payback\ in\ Building\ 1 = 8,100.00/702.71 \approx 11.5\ years \tag{10}$$

$$Payback\ in\ Building\ 2 = 8,100.00/840.11 \approx 9.6\ years \tag{11}$$

In fact, these payback times will be slightly longer, considering monetary updating as well as annual energy and maintenance costs [36]. However, as all these factors are of low value and the payback period is short, the results will not be significantly altered.

4 Conclusions

The analysis of Table 3 shows that, in an average year, 373.78 m^3 of rainwater is used in Building 1 (392.70 – 18.92 = 373.78), but it is necessary to resort to the public network in the summer months to complete the demand, with a supply of 18.92 m^3. These numbers mean that 95% of non-potable water needs in Building 1 can be met by rainwater (373.78/392.78 = 0.95).

In Building 2, 446.87 m^3 of rainwater is used (492.03 – 45.16 = 446.87), but it is also necessary to resort to the public network in the summer months to complete the demand, with a supply of 45.16 m^3. These numbers mean that, in Building 2, 91% of non-potable water needs can be met by rainwater (446.87/492.03 = 0.91).

As for the payback periods for these investments, values between 10 and 12 years can be considered. Considering that the average useful life of these installations can reach 40 years [35], these payback periods are very interesting from an economic point of view.

Although the installation of RWHS in Mediterranean climates is sometimes questioned in terms of economic viability, these two case studies from the University of Aveiro show that, in university buildings (and, in general, in school buildings), these systems are interesting not only from the point of view of environmental sustainability and adaptation to climate change but also from an economic point of view. This finding results from the fact that school holidays coincide with the dry season, nullifying the usual mismatch in the Mediterranean climate between water needs and availability.

References

1. European Commission: COM 718 Communication from the Commission to the Council and the European Parliament on Thematic Strategy on the Urban Environment; Commission of the European Communities—SEC: Brussels, Belgium (2006)
2. United Nations: The Sustainable Development Goals Report 2022. 2022. Available online: https://unstats.un.org/sdgs/report/2022/, accessed on 12 February 2023
3. United Nations. *World Urbanization Prospects: The 2014 Revision (Highlights)*; United Nations—Department of Economic and Social Affairs: New York, NY, US (2014)
4. UNEP: UN Environment Programme. Global Forum on Cities Highlights Need for Sustainable Development. 2022. Available online: https://www.unep.org/news-and-stories/story/global-forum-cities-highlights-need-sustainable-development, accessed on 12 February 2023
5. United Nations: The future we want. New York (2012)
6. Trenberth, K.: Changes in precipitation with climate change. Climate Res. **47**, 123–138 (2011)
7. Chalmers, P.: Climate Change. Implications for Buildings. Key Findings from the Intergovernmental Panel on Climate Change, Fifth Assessment Report; European Climate Foundation, Building Performance Institute Europe, Global Buildings Performance Network, World Business Council for Sustainable Development, University of Cambridge's Judge Business School, Institute for Sustainability Leadership. BPIE: Brussels, Belgium (2014)
8. Wilk, J., Wittgren, H.: Adapting Water Management to Climate Change; Swedish Water House Policy Brief 2009, Nr. 7. SIWI: Stockholm, Sweden (2009)
9. Haines, A., Kovats, R., Campbell-Lendrum, D., Corvalan, C.: Climate change and human health: Impacts, vulnerability and public health. Public Health **120**, 585–596 (2006)
10. Wilby, R.: A review of climate change impacts on the built environment. Built Environment Journal **33**, 31–45 (2007)

11. World Health Organization. Summary and Policy Implications Vision 2013: The Resilience of Water Supply and Sanitation in the Face of Climate Change; WHO—Department for International Development: Geneva, Switzerland (2009)
12. United Nations. Buildings and Climate Change: Summary for Decision-Makers. United Nations Environment Program: Paris, France (2009)
13. Ekström, M., Fowler, H., Kilsby, C., Jones, P.: New estimates of future changes in extreme rainfall across the UK using regional climate model integrations. 2. Future estimates and use in impact studies. Journal of Hydrology **300**, 234–251 (2005)
14. Giorgi, F., Lionello, P.: Climate change projections for the Mediterranean region. Glob. Planet. Chang. **63**, 90–104 (2007)
15. Pausas, J.G.: Changes in fire and climate in the Eastern Iberian Peninsula (Mediterranean Basin). Clim. Chang. **63**, 337–350 (2004)
16. Spinoni, J., Vogt, J.V., Naumann, G., Barbosa, P., Dosio, A.: Will drought events become more frequent and severe in Europe? Int. J. Climatol. **38**, 1718–1736 (2018)
17. Monteiro, C.A., Calheiros, C., Pimentel-Rodrigues, C., Silva-Afonso, A., Castro, P.M.L.: Contributions to the design of rainwater harvesting systems in buildings with green roofs in a Mediterranean climate. Water Sci. Technol. **73**, 1842–1847 (2016)
18. Silva-Afonso, A., Pimentel-Rodrigues, C.: The importance of water efficiency in buildings in Mediterranean countries; The Portuguese experience. Int. J. Sys. Appli. Eng. Develop. **2**(5), 17–24. Cambridge (2011)
19. Pimentel-Rodrigues, C., Silva-Afonso, A.: Contributions of water-related building installations to urban strategies for mitigation and adaptation to face climate change. Appl. Sci. **9**, 3575 (2019)
20. UNEP: Technologies for Climate Change Mitigation: Building Sector; UNEP Risø Centre on Energy. Climate and Sustainable Development: Copenhagen, Denmark (2012)
21. Freni, G., Liuzzo, L.: Effectiveness of rainwater harvesting systems for flood reduction in residential urban areas. Water **11**, 1839 (2019)
22. Akter, A., Tanim, A., Islam, M.: Possibilities of urban flood reduction through distributed-scale rainwater harvesting. Water Sci. Eng. **13**, 95–105 (2020)
23. Campisano, A., Modica, S.: Rainwater harvesting as source control option to reduce roof runoff peaks to downstream drainage systems. Hydroinformatics **18**, 23–32 (2016)
24. Palla, A., Gnecco, I., La Barbera, P.: The impact of domestic rainwater harvesting systems in storm water runoff mitigation at the urban block scale. Environ. Manage. **15**, 297–305 (2017)
25. Campisano, A., et al.: Urban rainwater harvesting systems: Research, implementation and future perspectives. Water Resour. **115**, 195–209 (2017)
26. Behzadian, K., Kapelan, Z., Mousavi, S.J., Alani, A.: Can smart rainwater harvesting schemes result in the improved performance of integrated urban water systems? Environ. Sci. Pollu. Res. **25**, 19271–19282 (2018)
27. Bocanegra-Martinez, A., Ponce-Ortega, J., Nápoles-Rivera, F., Serna-González, M., Castro-Montoya, A.J., El-Halwagi, M.M.: Optimal design of rainwater collecting systems for domestic use into a residential development. Resou. Conser. Recy. J. **84**, 44–56 (2014)
28. Norman, P., Porporato, A.: Sizing a rainwater harvesting cistern by minimizing costs. J. Hydrol. **541**, 1340–1347 (2016)
29. AWWA: Climate Change and Water Resources: A Primer for Municipal Water Providers; AWWA Research Foundation, University Corporation for Atmospheric Research, American Water Works Association; 1P-5C-91120–05/06-NH. IWA Publishing: Denver, CO, USA (2006)
30. ANQIP: Technical Specification ETA 0701—Rainwater Harvesting Systems in Buildings (Version 11). ANQIP: Aveiro, Portugal (2022). (In Portuguese)

31. European Committee for Standardization: European Standard EN 16941-1—On-Site Non-Potable Water Systems—Part 1: Systems for the Use of Rainwater. CEN, Brussels, Belgium (2018)
32. European Committee for Standardization: European Standard EN 806-3—Specifications for installations inside buildings conveying water for human consumption - Part 3: Pipe sizing - Simplified method. CEN, Brussels, Belgium (2006)
33. European Committee for Standardization: European Standard EN 12056-2—Gravity Drainage Systems Inside Buildings - Part 2: Sanitary Pipework, Layout and Calculation. CEN, Brussels, Belgium (2000)
34. European Committee for Standardization: European Standard EN 12056-3—Gravity Drainage Systems Inside Buildings - Part 3: Roof Drainage, Layout and Calculation. CEN, Brussels, Belgium (2000)
35. PORTUGAL: Regulamento Geral dos Sistemas Públicos e Prediais de Distribuição de Água e de Drenagem de Águas Residuais – Decreto Regulamentar n.º 23/95. Imprensa Nacional, Lisbon (portuguese) (1995)
36. Hajani, E., Rahman, A.: Reliability and cost analysis of a rainwater harvesting system in peri-urban regions of greater Sydney. Australia. Water **6**, 945–960 (2014)

Stormwater Attenuation and Enhanced Infiltration System to Mitigate Flood and Drought Conditions

Stephen D. Thomas[1] (✉), George R. French[1], Joseph A. O'Meara[2], and Matthew P. Dale[2]

[1] OGI Groundwater Specialists, Durham, UK
stephen.thomas@ogi.co.uk
[2] Groundwater Dynamics Ltd, Leamington Spa, UK

Abstract. This paper describes a stormwater attenuation & enhanced infiltration system, comprising the construction of attenuation trenches, perforated pipes with gravel surround, integrated with large numbers of drilled vertical infiltrators installed through the base of the trenches.

Combining the pressure head of the collected stormwater in the attenuation trenches, with the negative suction pressure within the unsaturated vadose zone, provides the differential pressure to force the water into the ground via these infiltrators. Furthermore, by installing vertical infiltrators, higher permeable ground strata are often encountered, particularly in anisotropic ground conditions where the horizontal permeability is much greater than the vertical permeability.

Stormwater Attenuation & Enhanced Infiltration Systems (SAEIS) have been used regularly in the UK for many years, both to mitigate flooding, and to provide recharge to groundwater resources. With groundwater resources similarly in decline in Portugal, together with more frequent flooding conditions occurring during intense rainfall, such collection and enhanced infiltration systems would also be most beneficial to the hydrogeology and weather conditions of Portugal.

Mathematical modelling is presented which simulates the water flow through the infiltrator system, and into the unsaturated soil strata. Combining the storage capacity of the attenuation trenches with a large number of infiltrators, produces a highly efficient attenuation and infiltration system that not only mitigates the impact of flooding but has no requirement for ongoing energy consumption.

With stormwater collected, filtered, attenuated, and then infiltrated to the vadose zone, collected stormwater will eventually find its way down to the water table, and so mitigate drought conditions by supplementing the aquifer water resources.

The stormwater attenuation & enhanced infiltration system (SAEIS) exemplifies true low carbon sustainability, and meets with the United Nations Department of Economic and Social Affairs for Sustainable Development - Goal 6: *Ensure availability and sustainable management of water and sanitation for all.*

Keywords: Stormwater · Attenuation · Enhanced Groundwater Infiltration

J. F. L. C. Semião et al. (Eds.): INCREaSE 2023, ASST, pp. 13–24, 2023.
https://doi.org/10.1007/978-3-031-44006-9_2

1 Introduction

Climate change and global warming are affecting all countries of the world, with more extreme weather events happening with greater frequency [8]. In the UK, one major impact of global warming has been the increase in hotter periods combined with an increasing frequency of high-intensity rainfall events [6, 7]. With more of the UK being covered with urban developments, roads and industrial parks, the available open land which enables natural infiltration to the ground is receding (Charlesworth and Booth, 2016).

Over time this means less water is being recharged back into the ground and into the major aquifers. As a consequence, the UK Government has enforced where possible the need for a developer to construct a Sustainable Drainage System (SuDS) to mitigate the peak flooding that inevitably occurs from rapid runoff from hard impermeable surfaces such as roofs, roads, car parks or other paved areas [3]. In 2010 the UK Government published Building Regulations with Part H (Drainage and Waste Disposal, 2015). Within these Regulations, Part H3 (3) describes the design of systems to manage the drainage of rainwater with the following priority:

a) An adequate soakaway or some other adequate infiltration system; or where that is not reasonably practicable,
b) A watercourse; or where that is not reasonably practicable,
c) A stormwater sewer; or where that is not reasonably practicable,
d) A foul sewer.

The building regulations guidance presented above clearly states that the priority is to discharge collected groundwater to an adequate soakaway or some other adequate infiltration system. In response to these regulations a Stormwater Attenuation and Enhanced Infiltration System (SAEIS) has been developed to provide an efficient method to collect, attenuate and infiltrate stored water into soils of variable permeability at depth. This system has been designed to infiltrate water at greater depths (typically in the range of 3 to 12 m deep) than conventional soakaway systems which means the system is more suitable in areas of the UK where low permeability anisotropic soils such as laminated sandy and silty clays are prominent, (Jarvis et al., 1984).

By enabling infiltration at depth into laminated silt or sand lenses within clay soils, an enhanced infiltration system helps to prevent local surface water flooding by minimizing the flow of water to watercourses and sewers. This in turn helps to mitigate the peak water levels in local rivers following storm events, as shown by the hydrograph in Fig. 1. This paper presents mathematical modelling undertaken to demonstrate how the enhanced infiltration system works in practice by enabling infiltration of stormwater into anisotropic low permeability soils which are most common in the United Kingdom.

With storm flood conditions occurring more frequently in areas of Portugal over recent decades, enhanced infiltration techniques have the potential of mitigating the worst impact of such flooding events. Furthermore, the resulting enhanced infiltration of storm rainwater back to the unsaturated vadose zone, will also serve to provide a contribution to the mitigation of drought conditions, as the infiltrated rainwater will eventually find its way to below the water table.

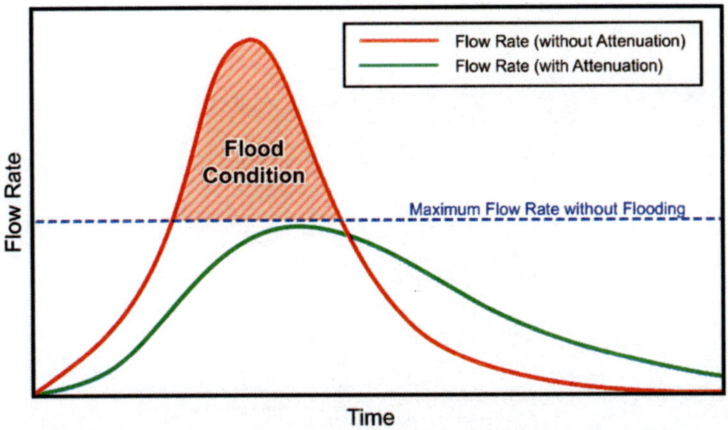

Fig. 1. Hydrograph showing the difference between non-attenuated and attenuated flow rate following a storm event.

2 Stormwater Attenuation and Enhanced Infiltration System

The Enhanced Infiltration System has been designed to attenuate and infiltrate large volumes of rainwater following storm events. The schematic in Fig. 2 shows how the system is set up and how rainwater passes through the system and into the soil. Rainwater that lands on buildings and paved surfaces such as car parks enters the drainage system through a series of gutters and drains. Water then flows through underground drains towards a silt trap where fine particles are removed from the water to prevent clogging of the infiltration system. Water then flows into a series of attenuation trenches set a minimum of 5.0 m away from any buildings. The trenches are typically in the order of 1.8 m wide and 1.7 m deep.

Each trench can contain one or two perforated pipes, designed to provide a large storage volume to attenuate rainwater following a storm event. The trench is backfilled around the perforated pipes with gravel or angular stone. At the base of the trench the pipes typically sit on a layer of 10 mm pea gravel. Gravel or stone is then backfilled around the pipes and typically to 300 mm above the crown of the pipes. Above the crown of the pipes, graded stone is often placed to provide greater bearing capacity if the land above is for vehicular use. The attenuation trench is surrounded by a specialist geotextile designed to prevent the intermixing of granular layers, thus stabilising the sub-base construction.

Beneath the attenuation trenches are a series of vertical infiltrators (Figs. 2 and 3) which are specially designed plastic pipes which aid the infiltration of water into the soil. Prior to installing the attenuation trench, each infiltrator is installed by placing it inside a 90 mm diameter drilled borehole. The top of each infiltrator is set normally 300 mm below the base of the trench (Fig. 3). The length of infiltrator normally varies between 3 m and 12 m, depending on the geology and volumes of water that need to be infiltrated, together with the ambient water table level.

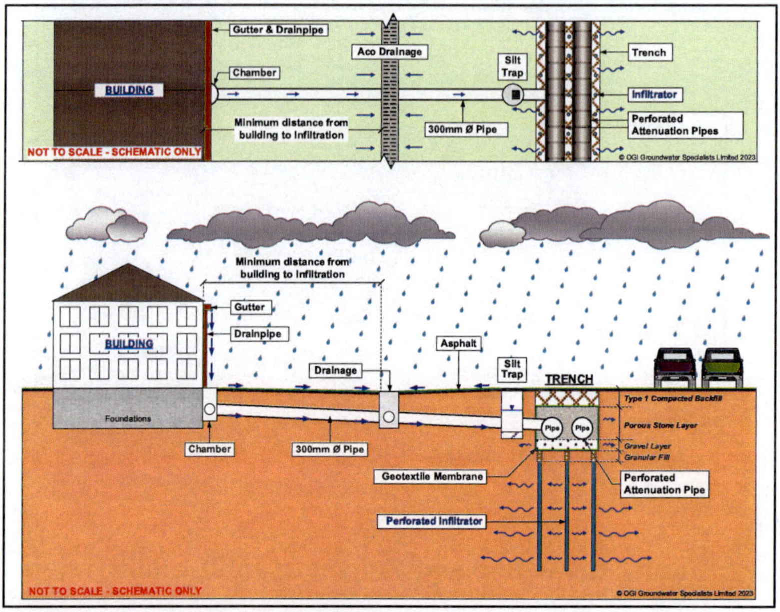

Fig. 2. Schematic of a Typical Attenuation and Enhanced Infiltration System.

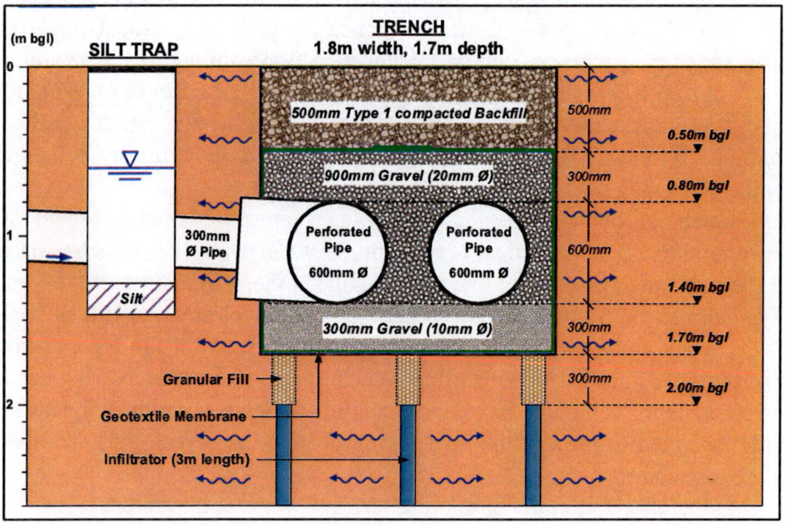

Fig. 3. Schematic of the Attenuation and Enhanced Infiltration system in section

The above enhanced infiltration system has been successfully implemented to discharge collected water at over 500 sites in the UK. The system provides an elegant and energy efficient solution to the flooding challenges of each particular site (Fig. 4).

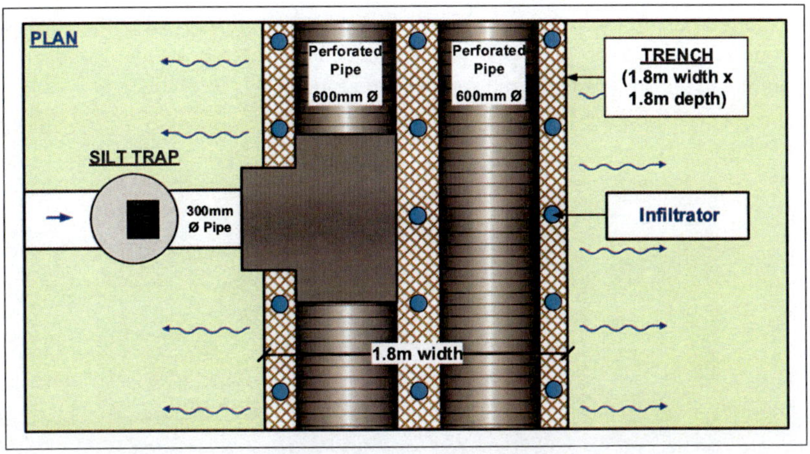

Fig. 4. Schematic of the Attenuation and Enhanced Infiltration system in plan.

The system has proved particularly successful to drain waterlogged sports fields, including rugby fields, cricket pitches and horse racing tracks. However, the main area of growth is for new build housing and commercial construction, where city councils are legislating for 100% of the collected rainfall to be put back to ground.

To illustrate the application of this attenuation and enhanced infiltration system, Figs. 5, 6 and 7 present the application at an industrial unit and car park in the northeast of England. The silty-clay ground was initially considered to be unsuitable to soakaway stormwater; but after installation, the 180 m of trench, together with c. 500 infiltrators, was sufficient to manage the rainwater over a 4 Ha area.

Fig. 5. Construction of perforated pipes within attenuation trenches.

Horizontal perforated pipes are then installed and backfilled with angular stone or gravel, typically to 300 mm above the pipe crown. The geotextile wrapping of the stone is then completed before graded stone is infilled on top of the geotextile to provide greater bearing capacity when the land above is for vehicular use.

Fig. 6. Connection of attenuation pipework to maintenance chamber.

To mitigate the potential of silts from entering the trench attenuation system, there are many silt traps and chambers constructed to intercept the silts, and these can be regularly cleaned as part of a planned maintenance program.

To degrade any possible hydrocarbon collected within the attenuation trenches, when excavated to the chosen depth, the trench is fully wrapped with a specialist geotextile (Fig. 7) before adding a 300 mm bedding layer of 10 mm pea gravel.

The combination of the geotextile of a unique fiber structure, together with the layers of pea gravel, adds a level of treatment for hydrocarbon contamination by promoting growth of a microbic biofilm which digests the hydrocarbon.

3 Rainwater Infiltration Theory into Unsaturated Ground

The theory of rainwater infiltration into unsaturated ground above the water-table is complex. When stormwater fills the underground attenuation storage volume within the trenches, this also fills the infiltrators directly beneath the trenches. This will result in high differential pore water pressure between the water filled infiltrator, and the natural suction pressure in the surrounding unsaturated ground.

Fig. 7. Final backfilling of attenuation trench and wrapping with biofilm geotextile

(i) The pressure in the infiltrators will be governed by the elevation of the water head in the attenuation storage (Fig. 5). For example, if the water level fills the trenches to 600 mm above the base of the trench, and the infiltrators are installed to 3 m below the base of the trench, there will be a pressure of circa 35 kPa at the tip of the infiltrator.

(ii) The initial pore water pressure in the unsaturated ground surrounding the infiltrator is negative, i.e. less that atmospheric pressure. Negative pore water pressure is also commonly referred to as soil suction. In fine-grained soils, suction pressure can be high, with values of −50 kPa regularly produced.

(iii) When an infiltrator fills and under a water pressure, combining these two conditions, can result in circa 85 kPa in differential pore-water pressure between the inside of the infiltrator and the surrounding unsaturated soil. This results in both the pressure in the infiltrator "pushing" the water into the surrounding ground, and with the unsaturated ground also effectively "pulling" the water from the water filled infiltrator.

Experience from the UK has demonstrated that the performance of an attenuation and enhanced infiltration system far exceeds the rate of infiltration from a normal & traditional soakaway trench, i.e. without the addition of drilled infiltrators.

This performance is regularly met with surprise by those who have traditionally viewed that a soakaway does not work sufficiently in a silt or clay soil. However, when the concept of high anisotropy that occurs in a laminated soil, together with the fact that an unsaturated soil "pulls" the water from the infiltrator, especially when mathematical models are used to simulate and validate the performance, this innovative system is now becoming accepted in the construction industry (Fig. 8).

Nevertheless, although over 500 enhanced infiltration systems have been installed in the UK, there is still always resistance to a new technique. The main opposition

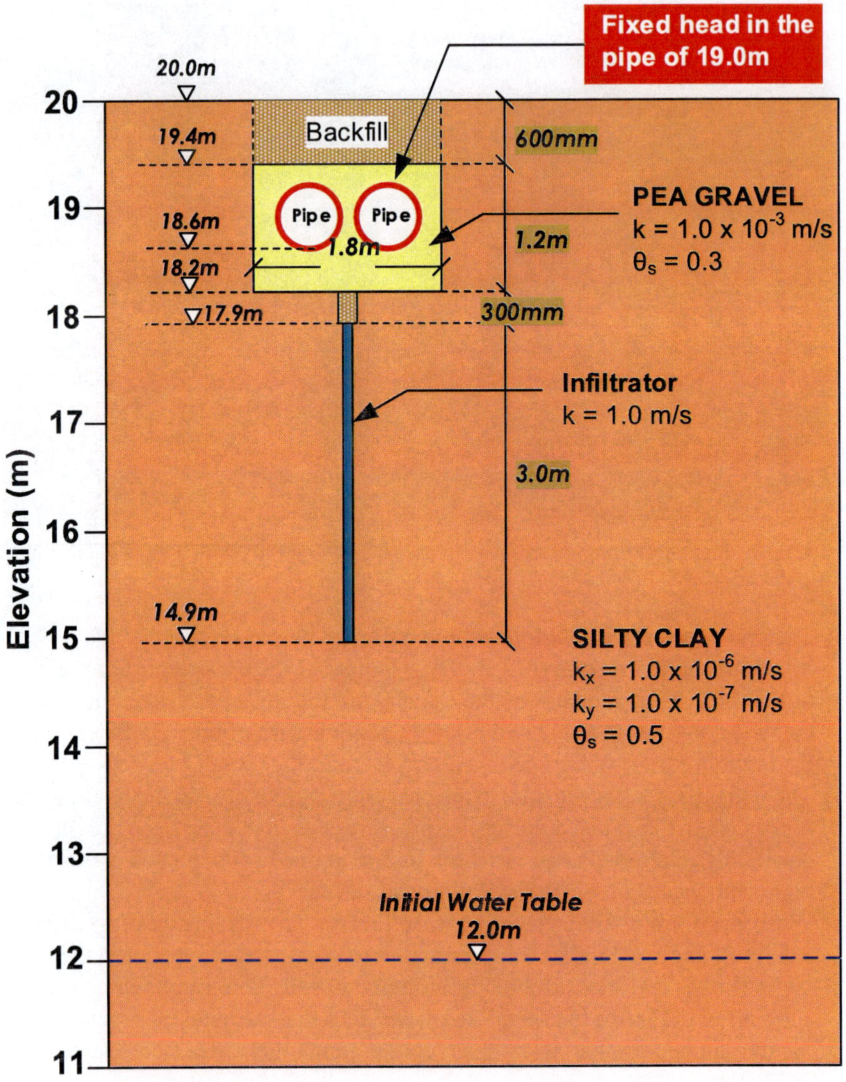

Fig. 8. Conceptual model of a single infiltrator with ground properties.

to the use of the system to dispose of stormwater, is the view by some that this high water infiltration rate into low permeability ground is considered not feasible, despite the system having been tested and demonstrated to work in practice.

For this reason, more "science" has been required to demonstrate how it is indeed feasible to infiltrate high water flow rates into low to medium permeability soils.

A finite element model was subsequently developed by the authors to demonstrate that a carefully constructed and tested system can be proved to work scientifically.

To illustrate the movement of the collected stormwater through the infiltrator and into the ground, a finite element model has been developed using GeoStudio's finite element model SEEP/W [4]. The finite element modelling requires the "discretization" of the ground surrounding the infiltrators into tens of thousands of smaller zones called "elements" to geometrically model the ground.

The model then enables different properties to be allocated to each element for which the governing groundwater flow equation is solved. Finite elements are connected with common "nodes" for which the dependent variables (such as pore-pressure or hydraulic head) are calculated. The finite element model is solved on a computer with millions of calculations taking place in a few seconds to solve for the value of head at each node. From these outputs, other information can be calculated such as pore pressure, degree of saturation and groundwater velocity.

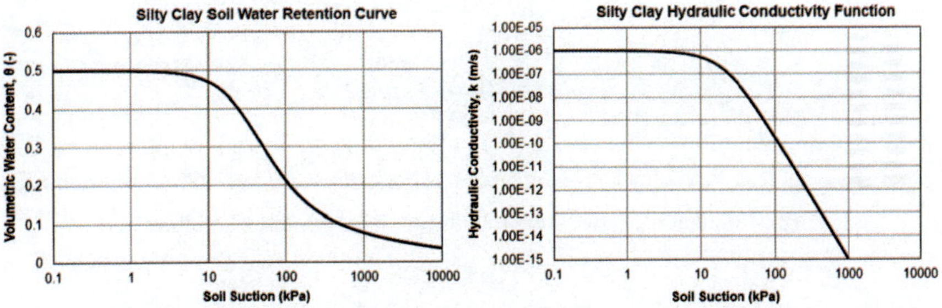

Fig. 9. Typical soil-water retention curve & hydraulic conductivity function for silty clay

An axisymmetric finite element model has been developed to simulate steady state two-dimensional radial flow of collected water into the unsaturated vadose soil zone surrounding the infiltrator. For this scenario, the backfilled trench, perforated pipes and infiltrator have been simulated in SEEP/W. The simulated soil is based on a typical anisotropic silty clay which is common to the northeast of England.

Horizontal permeability is simulated as 1.0×10^{-6} m/s and the vertical permeability as 1.0×10^{-7} m/s. To simulate the unsaturated behavior of the soil, a sample soil water retention curve (SWRC) along with a sample hydraulic relative conductivity function have been selected based on typical functions for silty clays, as shown in Fig. 9. Fixed head is applied to the perforated pipe to simulate the effect of a storm event where the pipe and trench have become flooded with rainwater.

Finite element simulation outputs are presented in Figs. 10 and 11. Fig. 10 depicts the simulated steady state groundwater head contours (ranging from 19 m to 12 m) and the simulated flow lines (shown by the green lines with arrows). The dashed blue line represents the position of the water table (which is the position where pore water pressure equals 0.0 kPa and saturation equals 100%).

Groundwater flow velocity is driven by a gradient of hydraulic head; so explaining the pattern of groundwater flow that can be seen in Fig. 10. In the system, the head changes from 19.0 m at the trench to 15.0 m at the bottom of the infiltrator. This 4.0 m head differential drives the water from the attenuation trench, and down through the vertical infiltrator, as depicted by the green arrows.

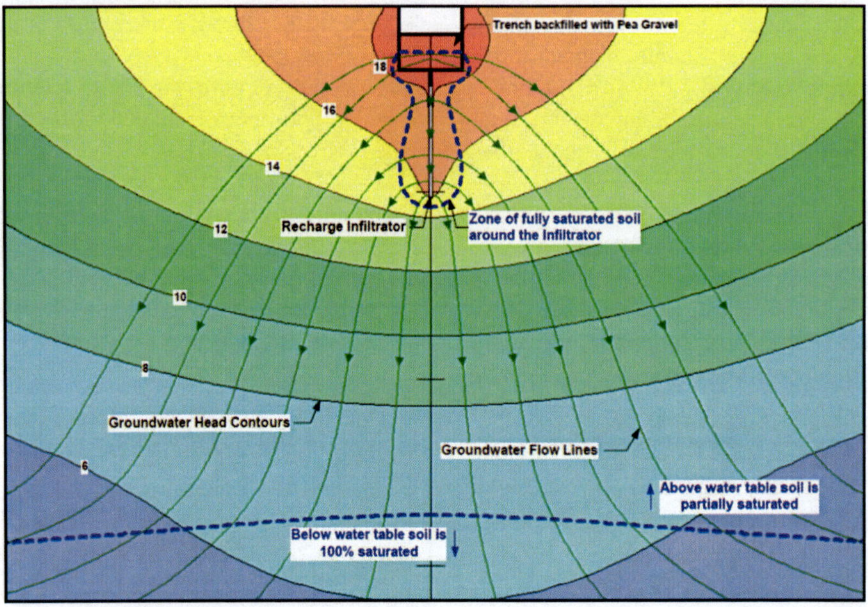

Fig. 10. Simulated steady state groundwater head contours and flow lines.

However, another driver is that the pore pressure in the unsaturated vadose zone above the water table will be negative, i.e. under suction. This means there is a greater differential head produced between the positive pressure in the infiltrator, and the negative pore-water pressure in the unsaturated vadose zone.

The water then enters the soil at depth, and then flows into the horizontal layers of silt or sand which have higher permeability than the clay layers. This horizonal flow into the surrounding soil results in a zone of fully saturated soil immediately surrounding the infiltrator, as depicted by the blue water table line in Fig. 11.

The enhanced saturation enables greater flow of groundwater into the surrounding soil as this increases the permeability of the soil. Moving horizontally away from the infiltrator, and out of the saturated zone, saturation decreases resulting in an increase in soil suction, together with an associated reduction in soil permeability.

Although a reduction in the permeability of the soil occurs in the unsaturated zone, the soil suction pressure gradient, combined with the elevation head gradient within the infiltrator, is sufficient to migrate the collected water into the vadose zone surrounding the infiltrator.

Over time, this water will slowly infiltrate through the soil until it reaches the water table at depth, as depicted by the groundwater flow lines in Fig. 11.

This has the effect of raising the water table at some time following a storm event; however, it is unlikely it will stay in this position over time as the water-table naturally fluctuates due to changing weather patterns throughout the year.

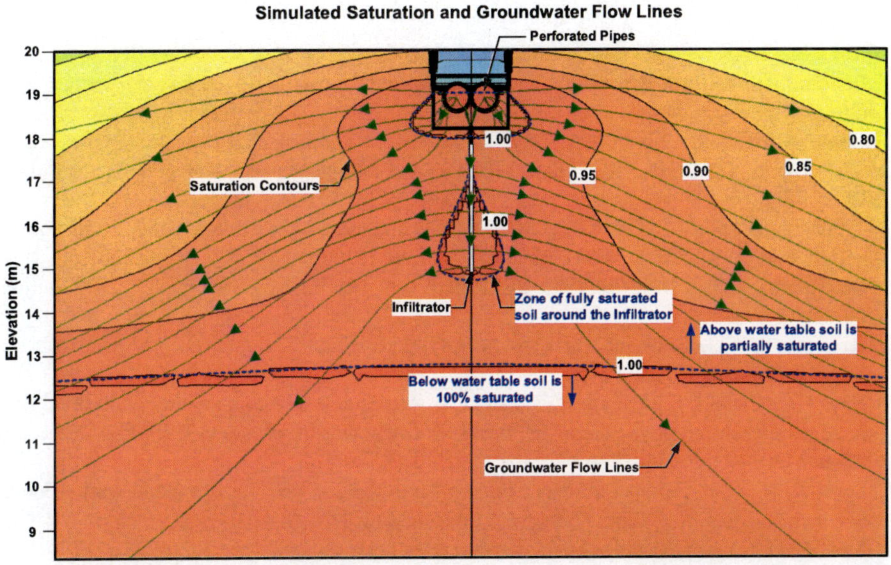

Fig. 11. Simulated steady state saturation contours and groundwater flow lines

4 Conclusions

A Stormwater Attenuation and Enhanced Infiltration System (SAEIS) enables the collection, attenuation and infiltration of stormwater back to ground. This paper explains how the system functions in practice; and demonstrates by mathematical modelling how water flows through the system and infiltrates into the soil at a far greater rate than can be achieved using a conventional soakaway system. The soil characteristics used in this study are based on a typical laminated anisotropic silty clay commonly encountered in the UK, and so would most likely be suitable for Portugal, where anisotropic geological conditions also exist in some regions.

Mathematical modelling demonstrates that infiltration systems can be effective in both medium and low permeability ground, benefitting from the anisotropy of the ground where horizontal silt and sand bands exist within otherwise clay soil, as it enables the infiltration of stormwater at depth into more permeability soil layers.

Where storm water has traditionally been sent to watercourses and sewers, the application of effective infiltration systems has a significant benefit by mitigating the risk of localised flooding, together with replenishing water resources.

This opens up the option of using enhanced infiltration systems anywhere in the world where shallow soakaway systems were not originally thought suitable.

Stormwater attenuation & enhanced infiltration system (SAEIS) exemplifies true low carbon sustainability, achieving the United Nations Department of Economic and Social Affairs for Sustainable Development - Goal 6: *Ensure availability and sustainable management of water and sanitation for all.*

References

1. Building Regulations: Part H - Drainage and Waste Disposal, 2015 Edition. Planning Portal (2015)
2. Charlesworth, S.M., Booth, CA.: Sustainable surface water management. In: Charlesworth, S.M., Booth, C.A. (eds.) Sustainable Surface Water Management: A Handbook for SuDS. Wiley (2016)
3. Fletcher, T.D., et al.: SuDS, LID, BMPs, WSUD and more – the evolution and application of terminology surrounding urban drainage. Urban Water Journal **12**(7), 525–542 (2015)
4. GeoStudio: Seepage Modelling with SEEP/W: An engineering methodology, July 2012 Edition (2012)
5. Jarvis, R.A., et al.: Soils and Their Use in Northern England. Soils and Their Use in Northern England. Rothamsted Research (1984)
6. Kay, A.L., Crooks, S.M., Pall, P., Stone, D.A.: Attribution of autumn/winter 2000 flood risk in england to anthropogenic climate change: a catchment-based study. J. Hydrol. **406**(1–2), 97–112 (2011)
7. Schaller, N., et al.: Human influence on climate in the 2014 southern england winter floods and their impacts. Nat. Clim. Chang. **6**(6), 627–634 (2016)
8. Stott, P.: How climate change affects extreme weather events. Science **352**(6293), 1517–1518 (2016)

Microplastics in Portuguese Effluents: Extraction and Characterization

Solange Magalhães[1(✉)], Luís Alves[1], Anabela Romano[2], Maria da Graça Rasteiro[1], and Bruno Medronho[2,3(✉)]

[1] CIEPQPF, Department of Chemical Engineering, University of Coimbra, Pólo II – R. Silvio Lima, 3030-790 Coimbra, Portugal
{solangemagalhaes,mgr}@eq.uc.pt, luisalves@ci.uc.pt

[2] MED—Mediterranean Institute for Agriculture, Environment and Development & CHANGE – Global Change and Sustainability Institute, Faculdade de Ciências e Tecnologia, Campus de Gambelas, Universidade do Algarve, Ed. 8, 8005-139 Faro, Portugal
{aromano,bfmedronho}@ualg.pt

[3] Fibre Science and Communication Network (FSCN), Mid Sweden University, 851 70 Sundsvall, Sweden

Abstract. Microplastics (MPs) awareness has been growing particularly after several alarming reports about "garbage patches" in the world. Plastics do not biodegrade in any meaningful way and, up to now, only a small percentage of plastic waste is recycled, being all the rest dumped in landfills, incinerated or simply not collected. The distribution of MPs within the water ecosystem depends on particle density and size and environmental characteristics, such as winds and currents. In the present study, different Portuguese industrial effluents were analysed and characterised to determine which MPs in the treated water released from wastewater treatment plants (WWTP), predominate and contribute the most to the environmental contamination of aquifers which, eventually, will end up in the coast of Continental Portugal. Overall, this work suggests strategies for MPs analysis in WWTP, thus allowing mapping of the different types of MPs prevalent in Portugal. The establishment of such database will enable the creation of reliable laboratory models to test new and green removal processes, based on the flocculation by, for instance, bio-based flocculants.

Keywords: Microplastics · Portuguese effluents · Wastewater · Polyethylene

1 Plastics and Microplastics

Since their discovery, plastics own different applications in plural areas due to several favourable properties, such as being durable, inert, lightweight and resistant to corrosion [1, 2]. The massified production and use of plastics engages several sectors, such as packaging, construction, automotive, bottles, households, agriculture, high-technology, clothes, and cosmetics. In other words, plastics are virtually everywhere [1, 3].

Depending on the polymer type, plastics find different applications, being polyethylene terephthalate (PET) the most widely (Fig. 1).

J. F. L. C. Semião et al. (Eds.): INCREaSE 2023, ASST, pp. 25–36, 2023.
https://doi.org/10.1007/978-3-031-44006-9_3

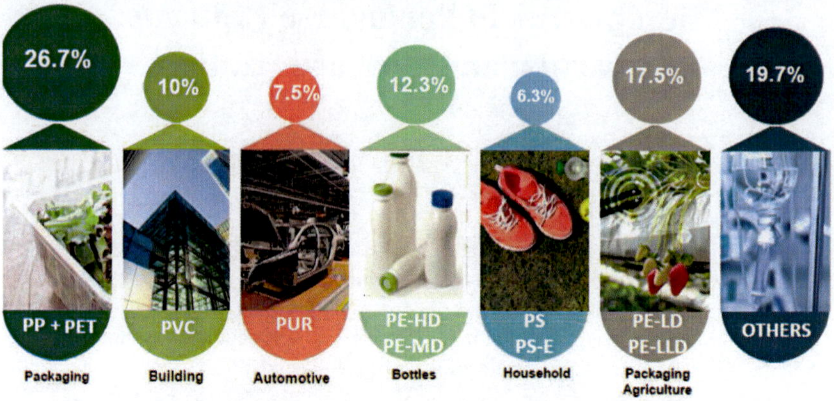

Fig. 1. Worldwide plastic demand by polymer type [1].

Despite the convenience and versatility of plastics, their non-biodegradable nature poses a significant environmental problem [4]. These materials can persist in the environment for hundreds of years and, currently, only a small percentage of plastic waste is effectively recycled. The remaining ends up in landfills, is incinerated, or is not properly collected and disposed. The progressive and slow deterioration of plastics may result in smaller particles that Thompson et al. (2004) initially coined as microplastics (MPs) to describe the accumulation of microscopic pieces of plastic in marine sediments and in the water column of European waters [5]. Later, Frias and Nash proposed an upper size limit to the initial term and MPs were then defined as particles smaller than 5 mm [6]. In 2011, Cole et al. distinguished MPs according to their origin, into primary and secondary [7]. In brief, primary MPs have a microscopic size between 2–5 mm in diameter and may be used, on purpose, in different formulations and applications, such as air-basting and cosmetic products. On the other hand, secondary MPs are the above mentioned tiny plastic fragments derived from the slow and progressive degradation of larger plastic debris, both at sea and on land [2, 3]. This phenomenon typically comprises the slow deterioration of macroplastics via three different mechanisms: biodeterioration, biofragmentation and assimilation [1, 8]. Over time, a combination of physical, biological and chemical processes can reduce the structural integrity of plastic debris, resulting in their fragmentation, ending in a very heterogeneous assemblage of pieces that vary in size, shape, colour, specific density, chemical composition, among other characteristics [9–11]. MPs can also suffer photo-degradation and some loss of structural integrity via abrasion, wave-action and turbulence phenomena.

2 Environmental Impact of the MPs

Despite the extensive use of plastics for decades, MPs have only recently become a central discussion theme around the world, not because of their societal benefits but mainly due to their threating potential to different ecosystems. MP particles have been detected in multiple environmental settings, including fresh- and seawater, atmosphere, sediments, soils, sewage sludge-tested effluents, biota, food and even breast milk. Their presence may cause a negative impact, particularly on the marine environment and biota, leading to increasing environmental awareness [7, 12].

Although plastics are generally regarded as inert materials, with low/negligible chemical reactivity [12], there have been questions about whether their dissemination in the body and organs may not originate different health issues [13]. Additionally, depending on the source, plastics may also contain different types of chemical additives which, due to their nature, may be also absorbed contributing to unexpected adverse health problems [14]. Different studies have reported on the potential threats upon MPs intake, such as the reduction in photosynthesis of plants and negative impacts on the feeding activity of zooplankton and marine animals (adverse effects to gill, stomach and alterations in histology) [15–19]. Due to their capacity to move among different tissues of plants and animals, along the food chain, and strong indications of associated health issues, their "inert" features should be seriously reconsidered.

Although debatable, the scientific community still considers that the current levels of MPs and nanoplastics (i.e., MPs smaller than 1 μm) in the environment are factually too low to affect human health [20]. Nevertheless, their amount is constantly rising and, by 2040, it is estimated that ca. 10 million tons of MPs can be released to the environment (without including the particles that are continuously being degraded from existing plastic waste!). The scientific community urges for reliable data, but it is fair to say that the potential threat of MPs should not be neglected.

3 Detection and Extraction of Microplastics: Approach and Results

It has been estimated that ca. 80 % of aquatic litter is delivered into aquatic systems by land-based sources, i.e., public littering, improper waste disposal, waste dump run-offs, tourism and industrial activities, and combined sewer systems are regarded as the main contributors to the pollution of the aquatic media with plastic debris [21, 22].

Although there are no current standard norms in EU regulating the amount of MPs in aquatic and terrestrial environments, it is important to understand their potential impact in such environments and develop feasible removal strategies. A fairly complete review on the current available methods for removing MPs from wastewater and sludge (and estimated costs associated) was provided by Zhang and Chen [23]. The authors concluded that with the current available technologies, the MPs recovery rate is still poor, and the associated costs are high. To address this challenge, it is mandatory to develop simple and cost-effective methods, which allow to save time and labor, while delivering reliable detection and high recovery rates of MPs. Other authors suggested a sort of trade-off paradox between aqueous media and soil as the recipients of MPs; the more efficient the process is in removing MPs, the more particles are generated and deposit into the sludge, thus enhancing its pollution threat [24, 25]. Recently, Vuori and Ollikainen [26] have shown that removing MPs from wastewater can be technically viable and cost-effective. The authors have considered three wastewater treatments (i.e., activated sludge, rapid sand filtering and membrane bioreactor) and two sludge management approaches (i.e., anaerobic digestion and incineration). The membrane bioreactor approach combined with sludge incineration was found to be the most cost-effective solution and prevents MPs from accumulating in soils.

The scenario is complex since although technically and economically feasible, the majority of the removed MPs can have different fates and transport pathways back into the environment via, for instance, the sludge, disposed membranes, or reject streams [27].

Although the Portuguese is not as robust as other European or global economies, it is relevant to evaluate the release of MPs in the treated effluents from wastewater treatment plants (WWTPs) and elucidate on the role of different industries as potential sources of MPs contamination to ecosystems. Therefore, we have evaluated different effluents from Portuguese paint, resin, pharmaceutical, textile and PVC industries located in a central area of Portugal (Coimbra) and the MPs were extracted and analysed regarding their concentration and polymer type. Effluents are strongly dependent on the source and thus differ from each other, since these are complex mixtures of several compounds, including synthetic chemicals, hydrocarbons, acrylic polymers, inorganic materials, and even heavy metals. Effluents from the resin industry are usually composed of acrylic polymers and inorganic compounds, such as calcium carbonate, plasticizers, etc., while PVC effluents are generally composed only of polymers and additives that are incorporated during the polymerization process for the control of PVC properties [28]. On the other hand, effluents from the pharmaceutical industry are typically heterogeneous mixtures of polymers, surfactants, antibiotics, and organic contaminants, among others [29]. Similarly, textile-based effluents can contain numerous toxic compounds, such as nonylphenol ethoxylates, benzothiazole, dyes, etc., and are usually composed of nylon and polyethylene polymers [1, 30]. Similarly in complexity, a municipal WWTP effluent was also evaluated. These effluents are often composed of heterogeneous mixtures with high content of organic and biological material.

These different effluents require generic but flexible extraction and characterization methods to adapt the specificities of each effluent. As alluded, effluents of different nature were used in this work, but their composition is expected to contain organic, inorganic and biological compounds. Therefore, in order to isolate the MPs in a reliable way, a step-by-step method was developed to deal with such types of compounds: a caustic treatment was used to remove the organic matter, an acid treatment was employed to remove the inorganic material and a peroxide-based agent was used to remove the biological matter. The "cleaning" procedure was as follows:

1- Digestion with 10 wt.% of the alkali agent during 12 h at 50 °C
2- Filtration through 1–2 µm filter
3- Digestion with an acidic agent until stop fizzing
4- Oxidation with H_2O_2
5- Washing with water
6- Washing with ethanol

In the first step, an aqueous alkaline solution containing KOH or NaOH is often used due to its high ionic conductivity and low cost [31–34]. Its effect was analyzed in a textile effluent, since this is known to be rich in textile fibers (often synthetic). Both alkali-based solvents (50 ml of 10 wt. % alkaline agent) were used in 400 ml of the effluent sample during 12 h at 50 °C with a constant mixing rate of 200 rpm. As can be seen in Fig. 2, no perceptible differences were observed in the textile fibers upon KOH or NaOH treatment.

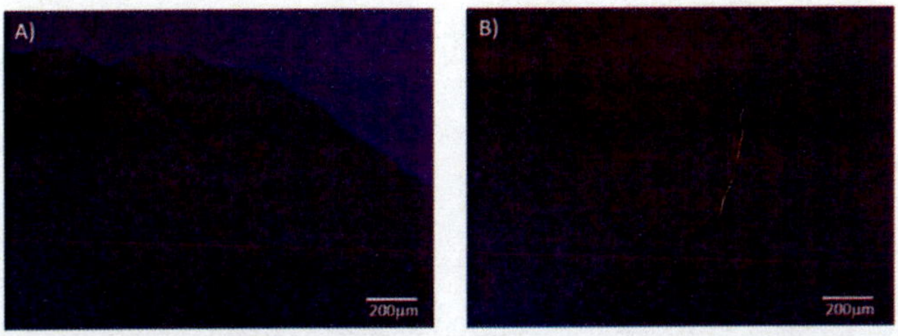

Fig. 2. Optical micrographs of a textile effluent treated with alkaline solvent. A) Sample treated with NaOH(aq.) and b) sample treated with KOH(aq.).

Both alkali-based solvents present good capacity to clean the effluents and do not manifest any perceptible effect on the shape and size of the MPs. The use of a digestion step is crucial when the MPs identification is mainly done by visual inspection. However, several authors consider that samples generally own low organic matter content and underestimate this step [35]. In our opinion, this initial procedure should not be neglected.

In order to remove the inorganic matter, a diluted HCl(aq.) solution was used. This acid has been suggested in literature for the digestion of inorganic material without damaging MPs [35, 36]. Later, the H_2O_2 solution was used to remove biological matter, and ethanol was finally added to sterilize the MPs.

Following this step-by-step method, we ensure that all contaminants (i.e., organic, inorganic and biological material) are sequentially removed in a reliable way.

The pharmaceutical effluent was trickier to analyze since after the sequential cleaning approach previously discussed, a viscoelastic structure was still present in the filter hindering the proper cleaning effluent (Fig. 3A). Polymers, such as hydroxypropylcellulose, hydroxyethylcellulose, carboxymethyl cellulose, hydroxypropyl methylcellulose, polyvinylpyrrolidone, pectin, carrageenan and guar gum are widely used in pharmaceutical industry and thus the observed gel-like material may be due to their gelation/precipitation [37]. Therefore, the generic cleaning method was slightly changed, and an anionic surfactant (sodium dodecyl sulfate) was used together with the alkali. As it can be observed in Fig. 3B, the effluent looks "cleaner". Overall, it is important to highlight that depending on the nature and effluent source, the amount of organic and inorganic compounds and/or other contaminants changes and thus the cleaning protocol may vary. Proper adjustments are required to ensure the reliable isolation, identification and quantification of MPs.

All industries analysed in this work have their own in-house WWTP. After the cleaning process previously discussed, it was possible to understand that the applied in-house WWTP decrease the percentage of MPs released (Table 1). Nevertheless, significant amounts of MPs are still liberated and potentially contaminating different ecosystems. Among the different sectors analyzed, the textile industry is surprisingly observed to contribute the least. Yet, some studies have shown that synthetic fibres are the dominant type of polyester MPs detected in aqueous media, sediments, and in various organisms [5, 30, 38]. This can be reasoned by the massified and extensive domestic and industrial washing cycles continuously releasing MPs into wastewaters. It should be highlighted that the textile WWTP shows a lesser ability to retain MPs particularly due to their smaller size. Table 1 summarizes the quantification of the MPs release of the main effluents studied in this work.

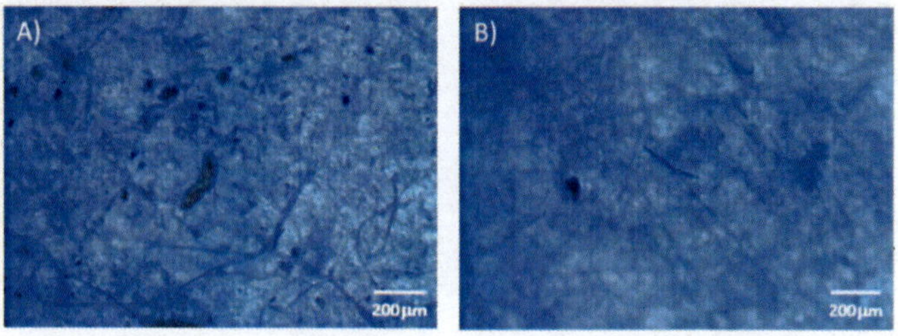

Fig. 3. Optical micrographs of a pharmaceutical effluent. A) Sample treated with NaOH(aq.) and b) sample treated with NaOH(aq.) plus anionic surfactant.

Table 1. Gravimetric quantification of MPs before and after each company's in-house WWTP. An estimative of the MPs amount released per ton of effluent is also presented.

	Before in-House WWTP (g MPs/100 mL Effluent)	After in-House WWTP (g MPs/100 mL Effluent)	MPs Released after in-House WWTP (g MPs/Ton Effluent)
Resins	0.044 ± 0.020	0.004 ± 0.002	41.66
Paint	0.172 ± 0.095	0.009 ± 0.004	89.01
Pharmaceutical	0.004 ± 0.003	0.002 ± 0.002	24.69
PVC	0.029 ± 0.001	0.007 ± 0.006	70.58
Textile	0.002 ± 0.002	0.002 ± 0.002	15.65
Domestic WWTP	0.019 ± 0.003	0.005 ± 0.005	54.32

The shape of the MPs recovered from the different effluent samples was mainly elongated fibres and fragments. Fibres constitute the majority of the MPs analysed as this shape allows easier diffusion through the in-house WWTP filters. Apart from being easily retained in the WWTP filters, fragments can form larger aggregates with the flocculating agents used and thus become more accessible for trapping in the in-house WWTP. It was reported that ca. 80 % of the MPs fall within a size range of 125–500 μm [39]. However, smaller MPs have been detected, which can be either due to the specific nature of the effluents tested and/or partial MPs degradation during the cleaning procedure. The sequential alkaline, acidic, and oxidizing treatments used may further deteriorate the MPs [2, 40]. To infer the effect of our developed "cleaning" treatment on the MPs structure, samples were analysed before and after its application. Although the trends are not fully clear, three main conclusions can be drawn: (i) the lowest size observed (i.e., 10–50 μm) tends to decrease upon "cleaning" the samples (this is particularly striking in the resins effluent); (ii) the bigger size reported (i.e., 500–600 um) decreases after "cleaning" the effluents; (iii) the intermediate sizes (i.e., 50–500 um) tend to increase their number density, most likely due to the fractionation of the bigger fibres. Therefore, this analysis showed that the cleaning procedure (applied to remove unwanted inorganic and organic compounds) might not be completely innocuous and can, in fact, partially affect the MPs shape and size.

A useful technique to infer on the chemical composition of the MPs is fluorescence microscopy [35, 41]. The fundamentals are fairly easy to understand; samples are stained with suitable chromopher, such as pyrene, Nile red or 4-dimethylamino-4′-nitrostilbene [28, 30], and the interaction of the MPs with a chromophore may induce the appearance of one colour depending on the material properties, such as the dielectric constant of the MPs. As observed in Fig. 4, polymers with different polarity have different dielectric constant and display different fluorescence behaviour.

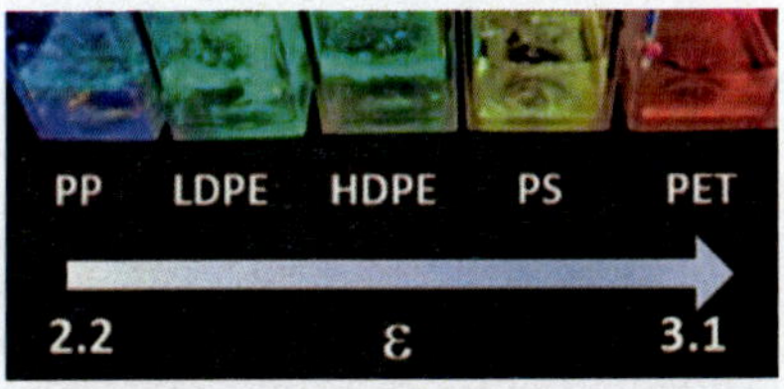

Fig. 4. Representative fluorescence of the different plastic materials stained with 4-dimethylamino-4′-nitrostilbene [30].

Pyrene was used in our samples as chromophore, because its more accessible and does not interact with organic impurities. The spectral emission of pyrene shifts to higher wavelengths, ranging from blue and green, for more hydrophobic plastics such as PP and PE, and to red when interacting with more polar plastics such as nylon or PET [28, 30, 42]. An example of MPs stained with pyrene is shown in Fig. 5. Data was compared with the literature to determine the main type of polymer in the MPs from the different effluents.

As can be seen in Fig. 5, the steady-state fluorescence emission of pyrene-stained samples mostly present MPs with green, blue, and red colours, which can be assigned to high-density polyethylene (HDPE), polyethylene (PE), and polyethylene terephthalate (PET), respectively [41]. In the textile effluent, apart from PET, polyamide (particles with red colour) were also observed, which is in agreement with the results previously reported [42, 43]. Overall, the presence of PET can be observed in the majority of the effluents, which is reasonable considering that this is the most used thermoplastic polymer in the world. The worldwide production of plastics, according to their polymer composition, is as follows: 36% polyethylene (PE), 21% polypropylene (PP), < 10% polyethylene terephthalate (PET), < 10% polyurethane (PUR), and < 10% polystyrene (PS) [39].

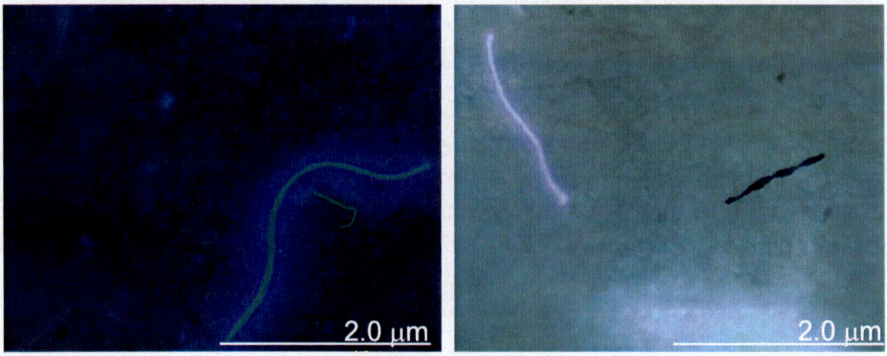

Fig. 5. Fluorescence microscopy images of the MPs before (A) and after (B) applying the cleaning procedure developed for the pharmaceutical effluent. Samples were stained with a pyrene solution during 15min prior analysis in the fluorescence microscope. Samples were imaged in a fluorescence microscope (Olympus BX51M), equipped with an objective lens of 100 ×, a filter set type U-MNU2 (360–370 nm excitation and 400 nm dichromatic mirror).

Apart from the fluorescence microscopy assays, FTIR-ATR was also used to complement the identification and to infer the MPs composition. In Fig. 6, typical FTIR spectra are shown for the MPs extracted from the pharmaceutical effluent, before and after being submitted to the in-house WWTP treatment. This effluent was selected as example due to the presence of different types of MPs in its composition. The spectrum of neat PE is also shown as reference. The pharmaceutical effluent shows vibrational modes that are characteristic of model PE, such as the rocking deformation, wagging deformation and asymmetric stretching of CH_2 groups. It is also possible to identify vibrational bands that can suggest the presence of PP. In particular, the band at 600 cm^{-1} can be attributed to the CH wagging mode, the band at 844 cm^{-1} is assigned to C-CH stretching, the band at 961 cm^{-1} is assigned to the trans -CH wagging, while the band at 1240 – 1252 cm^{-1} is assigned to CH rocking vibration and the band at 2890 – 2950 cm^{-1} can be attributed to the CH_2 asymmetric stretching of PP. In general, similar assignments can be done for the remaining effluents. Another striking observation for all samples tested is that after each in-house WWTP treatment, the intensity of the FTIR bands decreases. Although FTIR is not a truly quantitative method, these results suggest that the in-house WWTP reduce the number of MPs in the treated effluent, which agrees with the gravimetric quantifications presented in Table 1. Overall, FTIR complements and supports the fluorescence microscopy data showing that the main compounds present in the MPs are PET and PP.

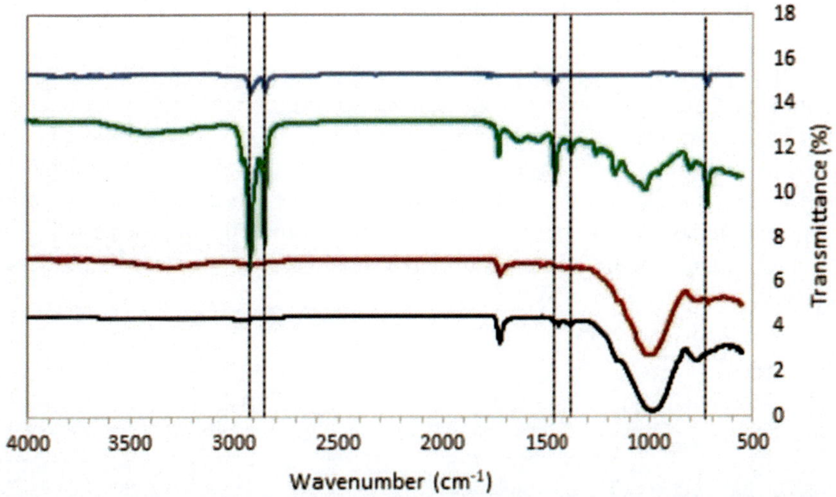

Fig. 6. FTIR spectra of the MPs from the pharmaceutical effluent before (red curve) and after (green curve) the in-house WWTP. The FTIR spectra from the filter (black curve) and neat PE (blue curve) are also shown. The main vibrational modes are highlighted with dashed vertical lines and their assignment is discussed in the main text [41, 44].

4 Conclusions

MPs are a well-distributed contaminant, whose potential threatening impact on the different ecosystems is still debatable and hard to evaluate. Robust data is clearly lacking not only regarding the extraction, quantification and identification but also focussing on evaluating the potential effect of MPs on different aquatic and terrestrial ecosystems.

In this work, we have analysed effluents from different Portuguese industrial sectors. Although we are aware of the limitations of this brief analysis, some important aspects can be outlined. Data suggests that the dominant types of MP in wastewaters from industrial areas in Portugal are PE, PP, and PET. Our study also suggests that the textile industry is the sector that releases the least number of MPs, while the PVC and paint industries were found to release the most. Furthermore, the domestic WWTP presents a significant contribution with some large particles, due to the possible formation of polymeric aggregates during the treatment. It is also clear that for a reliable and robust characterization of the MPs, extra care is need during their isolation. A cleaning protocol has been successfully established which might need to be adjusted depending on the specificities of each effluent. Fluorescence and FTIR microscopy were successfully employed to identify the MPs. The WWTP treatments used nowadays, at least in the industries analysed, are not sufficiently robust to remove/retain all MPs. Thus, it is urgent to develop methods that allow improved efficiency and higher removal yields of these contaminants from industrial effluents.

As mentioned, the contamination with MPs is a potential threat whose dimensions and real long-term impact are not consensual among the scientific community but needs to be urgently addressed. This study aims to raise awareness about the potential risk involving MPs and we thus suggest that similar analysis should be conducted in other areas of Portugal and in other countries to understand the global trend and the real extent of the MP problem.

Although there is no current robust knowledge on the real impact MPs cause in aquatic and terrestrial ecosystems, precaution measures should be taken and eventually EU should revise the Urban Waste Water directives and consider including mandatory removal requirements of MPs. Moreover, there is a need to understand how mitigation efforts and costs should be allocated to companies that supply goods causing MPs dissemination and how to eventually use associated tax instruments to finance robust approaches to remove MPs in WWTPs.

References

1. Magalhães, S., Alves, L., Medronho, B., Romano, A., Rasteiro, M.D.G.: Microplastics in ecosystems: from current trends to bio-based removal strategies. Molecules **25**, 3954 (2020). https://doi.org/10.3390/molecules25173954
2. Wagner, M., Lambert, S.: Freshwater Microplastics: Emerging Environmental Contaminants? (2017). https://doi.org/10.1007/978-3-319-61615-5
3. Cocca, M., et al.: Proceedings of the 2nd International Conference on Microplastic Pollution in the Mediterranean Sea (2020). https://doi.org/10.1007/978-3-319-71279-6
4. Muzafar, M.: Environmental Biotechnology: For Sustainable Future. 241–258 (2020). ISBN 978-3-030-40332-4
5. Thompson, R.: et al.: Lost at Sea: Where Is All the Plastic? Science (New York, N.Y.) **304**, 838 (2004). https://doi.org/10.1126/science.1094559
6. Frias, J.P.G.L., Nash, R.: Microplastics: finding a consensus on the definition. Mar. Pollut. Bull. **138**, 145–147 (2019). https://doi.org/10.1016/j.marpolbul.2018.11
7. Cole, M., Lindeque, P., Halsband, C., Galloway, T.S.: Microplastics as contaminants in the marine environment: A review. Mar. Pollut. Bull. **62**, 2588–2597 (2011). https://doi.org/10.1016/j.marpolbul.2011.09.025
8. Emadian, S.M., Onay, T.T., Demirel, B.: Biodegradation of bioplastics in natural environments. Waste Manage. **59**, 526–536 (2017). https://doi.org/10.1016/j.wasman.2016.10.006
9. Chubarenko, I., Bagaev, A., Zobkov, M., Esiukova, E.: On some physical and dynamical properties of microplastic particles in marine environment. Mar. Pollut. Bull. **108**, 105–112 (2016). https://doi.org/10.1016/j.marpolbul.2016.04.048
10. Andrady, A.L., et al.: Oxidation and fragmentation of plastics in a changing environment; from UV-radiation to biological degradation. Sci. Total Environ. **851**, 158022 (2022). https://doi.org/10.1016/j.scitotenv.2022.158022
11. Dimassi, S.N., et al.: Degradation-fragmentation of marine plastic waste and their environmental implications: a critical review. Arab. J. Chem. **15**, 104262 (2022). https://doi.org/10.1016/j.arabjc.2022.104262
12. Botterell, Z.L.R., et al.: Bioavailability and effects of microplastics on marine zooplankton: a review. Environ. Pollut. **245**, 98–110 (2019). https://doi.org/10.1016/j.envpol.2018.10.065
13. Prinz, N., Korez, Š.: Understanding how microplastics affect marine biota on the cellular level is important for assessing ecosystem function: a review. In: Jungblut, S., Liebich, V., Bode-Dalby, M. (eds.) YOUMARES 9 - The Oceans: Our Research, Our Future. Springer, Cham (2020). https://doi.org/10.1007/978-3-030-20389-4_6
14. Wright, S.L., Rowe, D., Thompson, R.C., Galloway, T.S.: Microplastic ingestion decreases energy reserves in marine worms. Curr. Biol. **23**, R1031–R1033 (2013). https://doi.org/10.1016/j.cub.2013.10.068

15. Michels, J., Stippkugel, A., Lenz, M., Wirtz, K., Engel, A.: Rapid aggregation of biofilm-covered microplastics with marine biogenic particles. Proceedings of the Royal Society B: Biological Sciences **285**, 20181203 (2018). https://doi.org/10.1098/rspb.2018.1203
16. Van Cauwenberghe, L., Janssen, C.R.: Microplastics in bivalves cultured for human consumption. Environ. Pollut. **193**, 65–70 (2014). https://doi.org/10.1016/j.envpol.2014.06.010
17. Sendra, M., Sparaventi, E., Novoa, B., Figueras, A.: An overview of the internalization and effects of microplastics and nanoplastics as pollutants of emerging concern in bivalves. Sci. Total Environ. **753**, 142024 (2021). https://doi.org/10.1016/j.scitotenv.2020.142024
18. Zhang, F., Man, Y.B., Mo, W.Y., Man, K.Y., Wong, M.H.: Direct and indirect effects of microplastics on bivalves, with a focus on edible species: a mini-review. Crit. Rev. Environ. Sci. Technol. **50**, 2109–2143 (2020). https://doi.org/10.1080/10643389.2019.1700752
19. Cozzolino, L., de los Santos, C.B., Zardi, G.I., Repetto, L., Nicastro, K.R.: Microplastics in commercial bivalves harvested from intertidal seagrasses and sandbanks in the Ria Formosa lagoon, Portugal. Marine and Freshwater Research **72**, 1092–1099 (2021). https://doi.org/10.1071/MF20202
20. Lim, X.: Microplastics are everywhere — but are they harmful? Nature, News Feature, 22–25 (2021). https://doi.org/10.1038/d41586-021-01143-3
21. Murshed, M.F., Mokhtar Kamal, N.H., Fagbenro, O., Wang, L., Wang, M.-H.: 305–346 (2022)
22. Shen, M., et al.: Removal of microplastics via drinking water treatment: current knowledge and future directions. Chemosphere **251**, 126612 (2020). https://doi.org/10.1016/j.chemosphere.2020.126612
23. Zhang, Z., Chen, Y.: Effects of microplastics on wastewater and sewage sludge treatment and their removal: a review. Chem. Eng. J. **382**, 122955 (2020). https://doi.org/10.1016/j.cej.2019.122955
24. Lares, M., Ncibi, M.C., Sillanpää, M., Sillanpää, M.: Occurrence, identification and removal of microplastic particles and fibers in conventional activated sludge process and advanced MBR technology. Water Res. **133**, 236–246 (2018). https://doi.org/10.1016/j.watres.2018.01.049
25. Lv, X., et al.: Microplastics in a municipal wastewater treatment plant: fate, dynamic distribution, removal efficiencies, and control strategies. J. Clean. Prod. **225**, 579–586 (2019). https://doi.org/10.1016/j.jclepro.2019.03.321
26. Vuori, L., Ollikainen, M.: How to remove microplastics in wastewater? A cost-effectiveness analysis. Ecological Economics **192**, 107246 (2022). https://doi.org/10.1016/j.ecolecon.2021.107246
27. Tang, K.H.D., Hadibarata, T.: Microplastics removal through water treatment plants: Its feasibility, efficiency, future prospects and enhancement by proper waste management. Environmental Challenges **5**, 100264 (2021). https://doi.org/10.1016/j.envc.2021.100264
28. Prata, J.C., et al.: A new approach for routine quantification of microplastics using Nile Red and automated software (MP-VAT). Sci. Total Environ. **690**, 1277–1283 (2019). https://doi.org/10.1016/j.scitotenv.2019.07.060
29. Rodríguez-Narvaez, O.M., Goonetilleke, A., Perez, L., Bandala, E.R.: Engineered technologies for the separation and degradation of microplastics in water: A review. Chem. Eng. J. **414**, 128692 (2021). https://doi.org/10.1016/j.cej.2021.128692
30. Sancataldo, G., et al.: Identification of microplastics using 4-dimethylamino-4′-nitrostilbene solvatochromic fluorescence. Microsc. Res. Tech. **84**, 2820–2831 (2021). https://doi.org/10.1002/jemt.23841
31. Gulizia, A.M., et al.: Evaluating the effect of chemical digestion treatments on polystyrene microplastics: recommended updates to chemical digestion protocols. Macromol. Chem. Phys. **223**, 2100485 (2022). https://doi.org/10.1002/macp.202100485

32. Tirkey, A., Upadhyay, L.S.B.: Microplastics: An overview on separation, identification and characterization of microplastics. Mar. Pollut. Bull. **170**, 112604 (2021). https://doi.org/10.1016/j.marpolbul.2021.112604
33. Akyildiz, S.H., Sezgin, H., Yalcin, B., Yalcin-Enis, I.: Optimization of the textile wastewater pretreatment process in terms of organics removal and microplastic detection. J. Clean. Prod. **384**, 135637 (2023). https://doi.org/10.1016/j.jclepro.2022.135637
34. Habibi, P., et al.: A new force field for OH– for computing thermodynamic and transport properties of H2 and O2 in aqueous NaOH and KOH solutions. J. Phys. Chem. B **126**, 9376–9387 (2022). https://doi.org/10.1021/acs.jpcb.2c06381
35. Prata, J.C., da Costa, J.P., Duarte, A.C., Rocha-Santos, T.: Methods for sampling and detection of microplastics in water and sediment: a critical review. TrAC, Trends Anal. Chem. **110**, 150–159 (2019). https://doi.org/10.1016/j.trac.2018.10.029
36. Nuelle, M.-T., Dekiff, J.H., Remy, D., Fries, E.: A new analytical approach for monitoring microplastics in marine sediments. Environ. Pollut. **184**, 161–169 (2014). https://doi.org/10.1016/j.envpol.2013.07.027
37. Guo, J.-H., Skinner, G.W., Harcum, W.W., Barnum, P.E.: Pharmaceutical applications of naturally occurring water-soluble polymers. Pharm. Sci. Technol. Today **1**, 254–261 (1998). https://doi.org/10.1016/S1461-5347(98)00072-8
38. Uddin, S., Fowler, S.W., Saeed, T., Naji, A., Al-Jandal, N.: Standardized protocols for microplastics determinations in environmental samples from the Gulf and marginal seas. Mar. Pollut. Bull. **158**, 111374 (2020). https://doi.org/10.1016/j.marpolbul.2020.111374
39. Ugwu, K., Herrera, A., Gómez, M.: Microplastics in marine biota: a review. Mar. Pollut. Bull. **169**, 112540 (2021). https://doi.org/10.1016/j.marpolbul.2021.112540
40. Wei, S., et al.: Characteristics and removal of microplastics in rural domestic wastewater treatment facilities of China. Sci. Total Environ. **739**, 139935 (2020). https://doi.org/10.1016/j.scitotenv.2020.139935
41. Gulmine, J.V., Janissek, P.R., Heise, H.M., Akcelrud, L.: Polyethylene characterization by FTIR. Polym. Testing **21**, 557–563 (2002). https://doi.org/10.1016/S0142-9418(01)00124-6
42. Sancataldo, G., Avellone, G., Vetri, V.: Nile Red lifetime reveals microplastic identity. Environ Sci Process Impacts **22**, 2266–2275 (2020). https://doi.org/10.1039/D0EM00348D
43. Magalhães, S., Alves, L., Romano, A., Medronho, B., Rasteiro, M.D.G.: Extraction and characterization of microplastics from portuguese industrial effluents. Polymers **14**, 2902 (2022). https://doi.org/10.3390/polym14142902
44. Pandey, M., Joshi, G.M., Mukherjee, A., Thomas, P.: Electrical properties and thermal degradation of poly(vinyl chloride)/polyvinylidene fluoride/ZnO polymer nanocomposites. Polym. Int. **65**, 1098–1106 (2016). https://doi.org/10.1002/pi.5161

Sorraia's Valley Restoration Project – Strategies, Problems and Experiences

Carla Rolo Antunes[1](✉) and Miguel Azevedo Coutinho[2]

[1] Faculdade de Ciências e Tecnologia, Universidade do Algarve and MED, Faro, Portugal
cmantunes@ualg.pt

[2] Universidade de Lisboa, Instituto Superior Técnico, Lisbon, Portugal
m.azevedo.coutinho@tecnico.ulisboa.pt

Abstract. The application of efficient measures, in the Sorraia's river valley, to mitigate the problems related to water resources management must be performed through preventive and structural measures, resulting from an integrated approach of methodologies, namely, hydrological, hydraulic, biophysical, ecological, landscape and social.

Aiming to an integrated management of the land, a series of measures, mainly concerning the restoration of the hydrographic network, of the Sorraia's valley, and the appropriate performance of the stream channels were proposed, applying mainly bioengineering techniques.

In Sorraia's river catchment changes in the dynamics of the thalwegs of the watercourses have been observed, with incisions of over 2.0 m in the last 10 years, being responsible for the lowering of the groundwater level, near the margins of watercourses, with very strong influence on river morphology, stability of the riverbanks and on the riparian vegetation.

This work focuses on the preliminary phase of this restoration project, having as main objective the identification of major issues and experiences perceived by project team members and the lessons learned during the construction and to select the most adequate bioengineering restoration typologies for the Mediterranean climatic region, particularly for nonpermanent water courses.

Keywords: Mediterranean watercourses · requalification · bioengineering

1 Introduction

In a Mediterranean climatic region, with nonpermanent water courses but also with sub humid and semiarid characteristics, is fundamental to establish principles and rules to ensure an effective response of the dominant hydrological and hydraulic conditions of river systems to the pressures of land use and occupation [2].

The effective application of actions to mitigate the problems related to water resources management must be done by preventive and structural measures, resulting from the application of integrated methodologies, namely, hydrological, hydraulic, biophysical, ecological, landscape, and social, in agreement with the land use planning objectives [10, 15, 20, 22].

J. F. L. C. Semião et al. (Eds.): INCREaSE 2023, ASST, pp. 37–48, 2023.
https://doi.org/10.1007/978-3-031-44006-9_4

The conservation of soil and water is a key priority for the proper management of watersheds and for the landuse. Controlling runoff/infiltration and combating soil erosion are the two major aspects to take into account for the related management strategies [7].

The novel approaches used by engineers with nature-based solutions (NBS) necessitate soil and water bioengineering (SWB) concepts and techniques. NBS is a unifying concept, embracing many aspects of SWB applications [16, 17, 25]. It involves techniques that use plants as living building materials, for: (i) natural hazard control (e.g., soil erosion, torrential floods and landslides) and (ii) ecological restoration or nature-based re-introduction of species on degraded lands, river embankments, and disturbed environments [6, 19]. Bioengineering combines engineering principles with a knowledge of vegetation and its interaction with soil, water and climate [4]. The needs of the soil bioengineering practitioners, the associated construction professionals, and the scientific/academic community have been reviewed and investigated in some depth in the last decade [11, 19, 21].

Plants and parts of plants are used as living building materials, through knowledge of their mechanical and/or biological properties [21, 24] in such a way that, through their development in combination with soil and rock, they ensure a significant contribution to the long-term protection against all forms of erosion. Live vegetation has been used for a very long time, to reduce soil erosion, for stream bank and bed stabilization, or to protect seawalls or sand dunes from the force of water [9]. In the initial phase, they often have to be combined with non-living building materials, which may, in some cases, ensure more or less temporarily, most of the supporting functions. The use of organic materials is preferred, because parallel to the development of the vegetation and its increasing stabilization ability, these materials will rot and be reincorporated in the natural biogeochemical cycles. Also preferred are indigenous (autochthonous) and site-specific plants, as they promote a biodiversity suited to the landscape [8].

The areas of application of soil and water bioengineering are various, for example stabilisation of embankments, slopes, river banks [13]. Soil and water bioengineering techniques contribute to the protection of erosion prone riverbanks, channel realignment, revitalization of non-natural watercourses and channels as well as increasing flood-retention in floodplains and the improvement of flood control always in accordance with the promotion of the ecological efficiency of the watercourses [5, 8].

Bioengineering solutions should provide a combination of the benefits of immediate hazard control, comprising techniques such as: (i) brush layers (that provide deep-seated protection), (ii) drain fascines or live pole drains (which drain excess water to allow vegetation establishment), (iii) vegetated crib walls (that immediately protect stream banks), (iv) brush mattresses (providing roughness from establishment against flow), and the long-term stabilization due to plant reinforcement effects. As with any stabilization technique, there is a stress (or load) transfer between the soil and the structure, but, in contrast to other solutions, this initial response is modified by the evolving role of the living material used in the bioengineering structure [18, 23].

In this framework and with the aim of an integrated management of the land, a series of interventions, aimed to the restoration of the hydrographic network of the Sorraia's valley were proposed by the application of bioengineering technologies which seek for interdisciplinary integration and coordination in promoting, stimulating and accelerating the restoration process, and taking in account the particular nature of each stream course.

In Sorraia's river catchment, major changes in the dynamics of the thalwegs of the watercourses have been observed, with incisions of over 2.0 m in the last 10 years, being responsible for the lowering of the groundwater level, near the margins of watercourses, with very strong influence on river morphology, stability of the riverbanks and on the riparian vegetation.

The proposed type of solutions, in this study context, besides promoting the improvement of aquatic and riparian ecosystems, create conditions to the establishment of an ecologic river "continuum", contribute to the restoration of the drainage system and sets of a framework of conditions, namely, the reduction of flooding discharges, restoration of riverbeds, delimitation of flooding areas, bank conservation and stability, besides other environmental improvements.

The study area is in the center of Portugal, particularly in the intervention area of the Associação de Regantes e Beneficiários do Vale do Sorraia – ARBVS (Irrigation and Beneficiary Association of the Sorraia Valley), with approximately 120 km of main river's corridor, covering the municipalities of Ponte de Sor, Avis, Mora, Coruche and Benavente (see Fig. 1.).

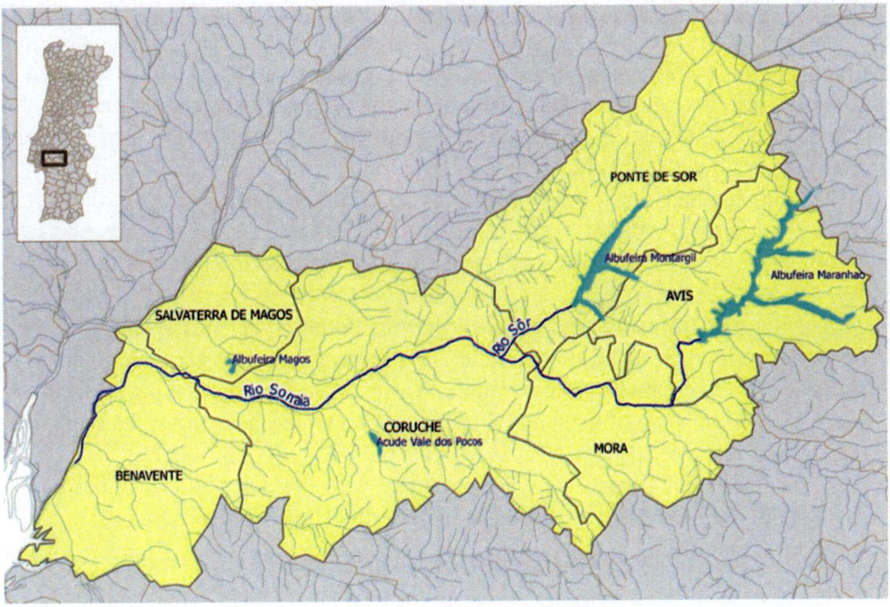

Fig. 1. Sorraia's valley case study area.

The Sorraia River is the tributary of the Tagus River Sorraia's watershed area is approximately 7730 km^2, the medium annual precipitation is 650 mm and the medium annual temperatue is 15,4 °C. Two reservoirs, Montargil and Maranhão, were built upstream in the watershed during the second half of the twentieth century. Sorraia Valley is one of the largest irrigation areas in Portugal, with 16 000 ha, in which corn (*Zea mays* L.), rice (*Oryza sativa* L.) and tomato (*Solanum lycopersicum* L.) predominate. The land use in the remaining area of the watershed is characterized by holm oak forest, rainfed cereals and pasture.

Agricultural practices have affected the hydromorphological characteristics of river systems in the wake of climate change. According to the Intergovernmental Panel on Climate Change (IPCC)'s Fifth Assessment Report [12], the Mediterranean climate, which is comprised of two contrasting seasons, i.e., the wet season with mild temperatures and the dry season with high temperatures, will show extreme variations due to climate change. This region will be highly affected by extreme events like droughts, floods and heat waves [1].

Climate change will deeply influence the Sorraia area: there will be less water available for crops and the ecoflows, in a short time span. This will drive changes in land use [14].

2 General Plan for Intervention in the Sorraia's Valley

2.1 Methodological Approach

Fieldwork in the project area was carried out in the first phase of the study, to characterize the conditions of the hydrographic network and interfluvial space. For the characterization of the different reaches of intervention an exhaustive photographic survey, 130 observation points was performed. Also, the recognition of the riparian vegetation, the biophysical characterization of the territory and the evaluation of the infrastructures conservation status were made, as well with hydrologic and hydraulic studies.

The field reconnaissance work of was made to establish the diagnosis of the system and to define the main types of problems in the study area, which allowed to support solutions and priorities and to establish a comprehensive intervention plan for the Sorraia's valley [3].

The solutions, to restore and enhance the functionality and sustainability of these ecological corridors adopted to stimulate and accelerate the processes in a natural as possible way, were defined in an integrated and articulated approach, taking into account, namely, the current state of the study area, the different needed actions, the implement costs, the guarantee of success.

Several Technical Datasheets were produced with the description and design criteria of works, grouped by type, which include the items and recommendations for construction, monitoring and maintenance.

2.2 Diagnostic

General hydrological modifications are reflected in changes on the average values of runoff and on its temporal distribution. Consequently, the sediment transport regime changed; this may be related to various aspects of the hydrological behavior of river catchments, due to human activity and to the extraction of sand from the river' bed.

During fieldwork the major type of problems diagnosed in the Sorraia's valley (see Fig. 2.) Among others, one can refer the dynamics of the thalwegs of the riverbeds, with general degradation of over 2.0 m, surpassing 2.5 m in several places.

It is considered that the incision of the riverbed must be essentially linked to decrease of the sediment transportation, and a significant part should result from the retention at Montargil and Maranhão dams. Another factor which contributes significantly is the extraction of sand, from the riverbed.

Fig. 2. Field recognition - Diagnostic of the main problems.

Locally, for the case of extraction of sand, we may see the generalized effect of high levels of bed degradation, with very strong influence on river morphology, stability of the banks and on the riparian vegetation.

Also, in recent years, the dams were responsible for modification of flow regimes, particularly during floods, by decreasing discharges, reducing the frequency and intensity of floods, with smaller magnitude (recurrence interval from 2 to 20 years), which are responsible for over 80% of the sediment transport [3]. These variations of hydro-sedimentological regimes results the general incisement of the river channel and the reduction riverbed's slopes which are more noticeable in upstream reaches.

From these changes in the system, the river channel width increased, and the thalweg is wandering all across the riverbed with a width larger than the corresponding for the reduced dominant discharge. Also connected relevant constrictions of the riverbed can be observed in several locations associated with the degradation the riparian vegetation (tree and shrubs), namely, willows (*Salix* sp.) and narrow-leafed ash (*Fraxinus angustifolia*), with the invasion of newly established vegetation. Inside the river channel, infestation patches are dominated by willows sprouts and other infesting varieties, mainly blackberries' bushes (brambles) (*Rubus fruticosus*) and giant reeds (canes) (*Arundo donax*) can be found, strongly competing for water in the ecosystem.

By one hand the riverbed incision allows for increased flow capacity, but on the other, the vegetation, increasing the roughness of the bed and clogging the cross section, is responsible for the opposite effect. In the margins and banks, the farmers put pressure on the width of the river channel, which leads to migration and destabilization of the banks and to the strangulation of the channel.

These behaviors and lowering of the water table induce migration of vegetation to the middle of the riverbed and, in the medium and long term, are responsible for the riverbed shift in the opposite direction. The most critical situations observed are located in places where the thalweg is meandering more or, where the lower riverbed crosses from one margin to the other.

2.3 Solution Types and Design of Works

Given the diagnostic the different types of woks and interventions the master plan of the Sorraia's valley was established (see Fig. 3) and the cost of works and proposed measures was estimated.

The developed interventions have as main objective to fix the plan morphology of the normal channel and the stabilization and repair of degraded banks and areas. In this framework, works and structural interventions are grouped in the following typologies.

Sills in the main channel, with "continuum" ramp: They are used to promote conditions for the establishment of an ecologic river "continuum". These sills, established in the normal channel perpendicular to the flow, are used to set the bed elevation levels, to reduce the speed of flow and the transport capacity and to promote a general equilibrium of the river system. They promote the retention of sediments, reduce the bedload, fix the base level of the thalweg of the lower bed, establish the equilibrium slope and stabilize the banks. The ramps, when applied alone, usually in areas of rocky bottoms or of coarser material, perform functions similar to those of the sills.

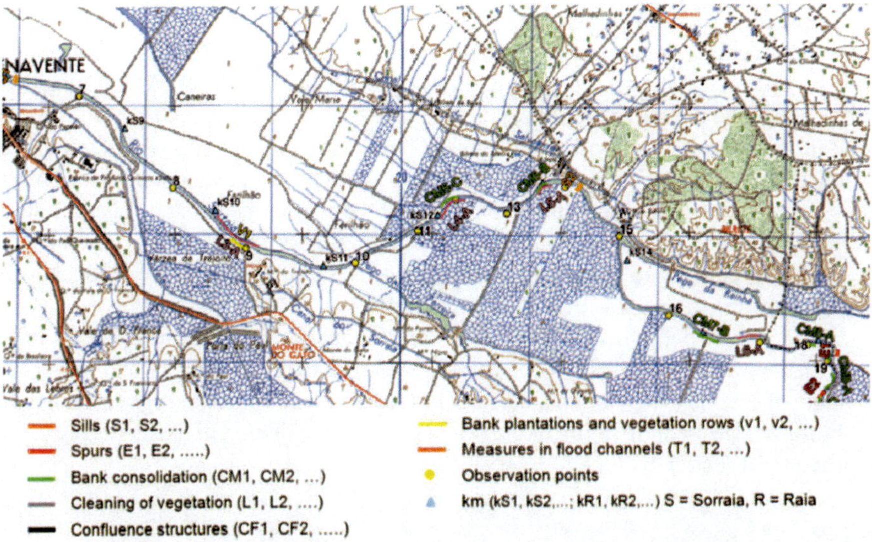

Fig. 3. Initial reach of the Master plan for the Sorraia's valley works.

The ramps, to be located in the base of the sill's spillway or separately at the entire width of the bed, are intended to ensure the "river continuum", to promote continuity of the fluvial corridor habitats and to allow the flow of aquatic fauna.

Spurs protruding from the banks: The spurs, to be constructed in the banks of the channel, are intended to deflect and guide the vein of the current and to protect the banks from erosion and base scouring. The spurs, according to their adopted geometry, are also intend, the make retention of sediments to the establishment of habitats, fixing the bed level at the base of the banks and deflecting the thalweg of the central area of the channel.

Bank consolidation: The bank protection to be established in the channels, are intended to protect the margins from erosion and scouring. For this type of works three different situations were considered: Type A - leveling at the top of bank levees, with precision leveling and smoothing of the surfaces to avoid bank breaching; Type B - global intervention to stabilize the slope of the bank, with protective rip-rap covering up to about half the height; and, Type C - setting up a rip-rap at the base of the bank, with the smoothing of the surface, filled with coarse bed material and with rocky revetment, where needed.

Cleaning of vegetation, on the bank (Type I) and removal of Willows and other type of vegetation from sand bars in the middle of the channels (Type II): These measures have as main objective to remove invasive vegetation in the bottom of the channels to improve flow conditions.

Bridge's protection: The bridge's protection works are similar to bank protection interventions (consolidation) but have the particular aim of protect the bridge embankments, both upstream and downstream.

Confluence structures: The confluence structures consist of works that comprise margin protections at the place of watercourses confluences, which, simultaneously, promote right orientation of currents to allow their adequate insertion, with the smallest turbulence in the downstream flow.

Bank plantations and vegetation rows: The plantations and vegetation rows to be establish on the channel's banks are interventions aimed at consolidating the tops and slopes of the banks, marking the transition from the normal bed channel and the flood channel. This transition allows the reduction of the velocity of the currents flowing from the bank full channel to the flood channel, and in the reverse situation, reducing the possibility of creating zones of concentrated erosion.

Measures in flood channels: The sills, to establish in flood channels, are intended to regularize the flow over flood channels, decreasing the flow velocity and promoting retention of sediments. Over time, the restored areas will allow limited agricultural used.

In Table 1 the quantification of the priority works considered in the first phase of the project are presented, by type of intervention.

Table 1. Quantification of priority works.

Type of intervention	Quantities
Sills	20
Ramps for an ecologic river "continuum"	9
Spurs protruding from the banks	15
Bank consolidation: Type A - Type B - Type C	3083 m - 1460 m - 1525 m
Cleaning of vegetation, on the bank	12 080 m
Bridge's protection (Rebolo, Amieira, Gravinha, Escusa, Santa Justa, Ponte Canal – Figueiras Gambas; Ponte Canal)	7
Bank plantations and vegetation rows	800 m
Measures in flood channels	9

3 Implementation of the Proposed Works

Implementation of the measures was the fundamental phase in the stream corridor restoration process. The revelation of the major issues and experiences perceived by project team members and the lessons learned during the construction are of key importance to the assessment of different type of solutions and for their application in future bioengineering restoration typologies, namely in the Mediterranean climatic region, for nonpermanent water courses.

To document some of the constructed works the following figures show some measures built in 2013, namely the sills for base level control. It is important to point out that these sills have performed accordingly to the goals defined in the project; in most cases, they supported already the occurrence of relevant floods (even in duration) which occurred in the winters of 2013/2014, 2014/2015 and 2015/2016.

In the following figures (see Figs. 4, 5, 6, 7 and 8) the comparisons between previous situations and the most actual ones recorded in the spring of 2016 are presented.

Fig. 4. Sill located close to Benavente: left) march, 14th 2014; right) june, 30th 2016.

Fig. 5. Sill located downstream Canal do Peso: left) march, 18th 2014; right) june, 30th 2016.

4 Discussion and Final Considerations

Although it is fundamentally in urban areas that the most critical issues associated with the presence of water, are present and are shown as natural hazards such as flooding, threatening the populations quality of life, in rural areas it is essential to consider measures that can conciliate the rational use and protection of the resources, in a global perspective for the watershed, and to mitigate the conflicts between sectors of activity and all the other stake holders.

Monitoring, evaluation, adjustment and maintenance of works are essential components of the adaptive management which must be undertaken to ensure the success of bioengineering solutions.

Fig. 6. Sill located downstream Ponte da Torrinha: left) march, 18th 2014; right) june, 30th 2016.

Fig. 7. Scouring at Ponte da Gravinha: left) february, 9th 2012; right) june, 30th 2016.

Fig. 8. Spur E4, close to km17 of the initial reach; left) november, 11th 2014; right) june, 30th 2016.

In this project it was considered very important to control rigorously the specifications and quality of the adopted materials as well as the construction phase. The rock components which constitute a significant part of the construction must be chosen carefully, in what concerns the type of rock material, namely the limestone applied, that must be of good quality and of strong physical resistance; the diameter distribution, should present a of large spectrum of sizes, along the described median diameter, in the range of sizes (all diameters). These characteristics are fundamental, because they determine the behavior of the structure, its permeability and their ability to become rapidly naturalized.

It is important to evaluate the performance of the constructed works and to prompt adjustments as soon as possible to leave them closer to the wanted performance. There are interventions that should be made after each hydrological year and mainly after the occurrence of major flooding, such as repairing any broken geometry and to fix zones were the permeably increased largely.

Finally, it should be understood that the intervention measures in these dynamic systems –stream corridors (including the river channels and marginal vegetation) - which encompasses several environmental functionalities and services, must be subject to close monitoring, to ensure the observation of the behavior and performance of the system's components and, also, the overall performance of the implemented measures. The lessons extracted, both from the construction phase and the monitoring, are fundamental to allow the establishment of necessary adjustments since the status of the systems is permanently changing in time.

References

1. Almeida, C., Ramos, T.B., Segurado, P., Branco, P., Neves, R., Oliveira, R.: Water quantity and quality under future climate and societal scenarios: a basin-wide approach applied to the Sorraia river, Portugal. Water (10), 1186 (2018)
2. Antunes, C., Coutinho, M.A.: Práticas de Bioengenharia na Reabilitação de Sistemas Fluviais em Clima Mediterrânico. Caso de Estudo: Ribeira de Barcarena, 11° Congresso da Água, Porto, Portugal (2012)
3. Antunes, C.R., Coutinho, M.A., Coutinho, M.A., Sousa, G.: Bioengineering technology for the restoration of river systems in the Sorraia's valley. In: Proceedings of World Fórum on Soil Bioengineering and Land Management New Challenges, pp. 239–247. Cascais (2014). E-BOOK / ISBN: 978-989-20-4788-1
4. Bischetti, G.B., Di Fi Dio, M., Florineth, F.: On the origin of soil bioengineering. Landsc. Res. 1–13 (2014)
5. Clewell, A.F., Aronson, J.: Ecological restoration: principles, values, and structure of an emerging profession. Society for Ecological Restoration 336 (2013)
6. Dhital, Y.P., Kayastha, R.B., Shi, J.: Soil bioengineering application and practices in Nepal. Environ. Manag. **51**, 354–364 (2013)
7. Dhital, Y.P., Tang, Q.: Soil bioengineering application for flood hazard minimization in the foothills of Siwaliks. Nepal. Ecol. Eng. **74**, 458–462 (2015)
8. European Federation of Soil Bioengineering (EFIB): European Guidelines for Soil and Water Bioengineering (2015)
9. Evette, A., Labonne, S., Rey, F., Liebault, F., Jancke, O., Girel, J.: History of bioengineering techniques for erosion control in rivers in western Europe. Environ. Manage. **43**, 972–984 (2009)
10. Federal Interagency Stream Restoration Working Group (FISRWG): Stream Corridor Restoration: Principles, Processes and Practices. Natural Resources Conservation Service, USA (1998)

11. Fernandes, J.P., Guiomar, N.: Nature-based solutions: the need to increase the knowledge on their potentialities and limits. Land Degrad. Dev. **29**(6), 1925–1939 (2018)
12. Intergovernmental Panel on Climate Change (IPCC): Climate Change 2013. The Physical Science Basis. Contribution of Working Group I to the Fifth Assessment Report of the Intergovernmental Panel on Climate Change. Cambridge University Press: Cambridge, UK, New York, NY, USA (2013)
13. Li, X., Zhang, L., Zhang, Z.: Soil bioengineering and the ecological restoration of riverbanks at the Airport Town, Shanghai. China. Ecol. Eng. **26**, 304–314 (2006)
14. MERLIN project, https://project-merlin.eu, last accessed 07 March 2023
15. Mickovski, S.B.: Re-thinking soil bioengineering to address climate change challenges. Sustainability **13**(6), 3338 (2021)
16. Moreau, C., Cottet, M., Rivière-Honegger, A., François, A., Evette, A.: Nature-based solutions (NbS): a management paradigm shift in practitioners' perspectives on riverbank soil bioengineering. J. Environ. Manage. (308), 114638 (2022)
17. Preti, F., Capobianco, V., Sangalli, P.: Soil and Water Bioengineering (SWB) is and has always been a nature-based solution (NBS): a reasoned comparison of terms and definitions. Ecological Engineering (181), 106687 (2022)
18. Preti, F., Giadrossich, F.: Root reinforcement and slope bioengineering stabilization by Spanish Broom (*Spartium* junceum L.). Hydrol. Earth Syst. Sci. (13), 1713–1726 (2009)
19. Rey, F., et al.: Soil and water bioengineering: practice and research needs for reconciling natural hazard control and ecological restoration. Science of the Total Environment (648), 1210–1218 (2019)
20. Sauli, G., Cornelini, P., Preti, F.: Manuale di ingegneria naturalistica applicabile al settore idraulico. Regione Lazio, Roma (2002)
21. Stokes, A., et al.: Ecological mitigation of hillslope instability: ten key issues facing practitioners and researchers. Plant Soil **377**, 1–23 (2014)
22. Tánago, M.G., Jalón, D.G.: Restauración de Rios e Riberas", Escuela Técnica Superior de Ingenieros de Montes. Fundación Colde del Valle de Salazar e Ediciones Mundi-prensa. Madrid. Espanha (1998)
23. Tardio, G., Mickovski, S.B.: Implementation of eco-engineering design into existing slope stability design practices. Ecol. Eng. **92**, 138–147 (2016)
24. Verstraeten, G., Poesen, J., Gillijns, K., Govers, G.: The use of riparian vegetated filter strips to reduce river sediment loads: an overestimated control measure? Hydrol. Process. **20**, 4259–4267 (2006)
25. Zaimes. G.N., Tardio G., Iakovoglou, V., Martin Gimenez, M., Garcia-Rodriguez, J.L., Sangalli, P.: New tools and approaches to promote soil and water bioengineering in the Mediterranean. Science of the Total Environment (693) (2019)

Water Use Efficiency in School Environment - The School as a Living Lab for Sustainability

Anabela Cordeiro[1](✉), Nadir Almeida[2], and Manuela Moreira da Silva[3,4]

[1] Universidade do Algarve, Instituto Superior de Engenharia, Campus Penha, Faro, Portugal
anabela.assuncao@gmail.com

[2] Universidade do Algarve, Faculdade de Economia, Campus Gambelas, Faro, Portugal
sv19.0084@iscee.edu.cv

[3] CIMA-ARNET, Universidade do Algarve, Instituto Superior de Engenharia, Campus Penha, Faro, Portugal
msanti@ualg.pt

[4] CEiiA, Centre of Engineering and Product Development, Matosinhos, Portugal

Abstract. The population growth and current consumption patterns, are increasing the pressure on the planet´s natural resources, including on water, that requires practices of more efficient management for sustaining the planet's biocapacity. Education is a very powerful tool for transforming current and future behaviors. This study was carried out in Algarve during 2021, to engage young people in improving the eco-efficiency of their school, family, and city. For this, the environment of a Basic School was used as a living lab, with students between 12 and 18 years old, and considering the consumption of water, energy and propane gas, the use of plastic water bottles and the calculation of the carbon balance in the school environment. Several activities were promoted using ICT with stakeholder's collaboration to explore the importance of the efficient use of water, to reduce the Water Footprint (WF), to promote the consumption of tap water and reduce the plastics use. The external green spaces of the school building were characterized identifying and counting trees and shrubs. The carbon emissions related to the consumption of water, energy and propane gas in the school building were quantified and the carbon sequestration by plants was estimated. The WF of students was about 279 L. Several water losses were detected, and the proposed measures will save 270 L/ min of water in the taps and 16 L/ min in the showers. This school was responsible for the emission of 31 t CO_2/year, and vegetation sequestered just 16% of those emissions.

Keywords: Environmental Education · Water Scarcity · Plastic Waste · Carbon Balance · Urban Nature

1 Introduction

The planet's capacity to support humanity's needs is not limitless and is widely considered to have been exceeded [1–3], largely due to the fact that 55% of the world's population already lives in cities [3] where 75% of greenhouse gases are emitted [4, 5].

J. F. L. C. Semião et al. (Eds.): INCREaSE 2023, ASST, pp. 49–62, 2023.
https://doi.org/10.1007/978-3-031-44006-9_5

In the Agenda 2030 and the 17 Sustainable Development Goals, it is very clear that efficiency in natural resource management, particularly water, requires not only scientific and technological solutions in the current digital world. It is also fundamental a transformation in citizens' behavior, who must be empowered to reverse current practices of excessive consumerism. Environmental Education, centered on the balance between people's quality of life and the preservation of natural ecosystems, represents a strategic tool that can involve young people in an inclusive way, in the definition of behaviors to ensure the sustainable use of fundamental resources for life on Earth. This participatory education can serve as the foundation for more empowered, capable, sustainable, inclusive, and peaceful societies.

The urban water cycle must therefore recognize that water, essential to life as it is, holds an intangible value in its environmental, social, and economic dimensions. The interaction of cities with the natural water cycle is based on a balanced logic of prosperity, promoting the use of various sources (rainwater, seawater desalination, water reuse, etc.), while protecting natural habitats and the biodiversity. Urban green spaces are essential because they collect, purify, and use rainwater, ensuring diverse ecosystem services to communities, in addition to promoting the beauty and harmony of the urban landscape [10]. In the current scenario, with increasingly prolonged and severe drought periods [11], there is a need to improve the efficiency of water use, that is, to optimize water consumption and its sources, ensuring that the quantity and quality requirements are met. The efficient use of water is the consumption of the minimum amount of water necessary to effectively fulfill its function(s), whether they are tasks, processes, or services [12]. This concept focuses on reducing water waste rather than restricting its use. Small to large technological and behavioral changes contribute to efficient use, especially behavioral changes, on the part of consumers, reducing water waste and/or choosing to purchase more efficient products/services. The concept of WF [13] represents the volume of water consumed per day, per individual or organization, or the volume of water required to produce a good or guarantee a service. In recent years, WF has been used as a tool to raise awareness among the population, especially the younger generations, of the need for efficient water use [14, 15].

In addition to consuming natural resources, cities are responsible for producing large amounts of waste (solid, liquid and gaseous) that gradually deplete the carrying capacity of natural ecosystems [1] and threaten public health [16]. Among various urban solid wastes, plastic waste is considered an emerging threat to habitats, wildlife, and humanity [10, 17]. In 2020, Portugal produced 5.3 t of such solid waste, and the high consumption of bottled water in plastic packaging made them one of the main types of waste produced [18]. In Portugal 99% of water treated for human consumption and distributed to the consumer's tap is considered safe [19]. However, many people prefer to consume mineral water sold in plastic packaging, claiming that the presence of chlorine (residual disinfectant present in water) alters its organoleptic properties, mainly its taste. In terms of gas emissions, the various anthropogenic activities associated with technological development, urbanization and mobility, supported by high consumption of energy and fossil fuels, have been responsible for the emission of different greenhouse gases, namely, CO, SO_2, O_3, NOx, CH_4, C_6H_6, among others. In addition, on average cities have the capacity to sequester less than 30% of their emissions through urban vegetation [20]. As cities become more densely populated, the area available for nature and its ecosystem services tends to decrease [21]. The number of parks, public gardens, urban forests, and other green spaces within the city have decreased with the increase in population density. This fact is leading to the loss of urban ecosystem services, namely regulatory ones, which contribute to the local mitigation of the effects of climate change. Specifically, with the existence of less urban vegetation, mainly trees and shrubs, water retention in the city is reduced, the ability to mitigate extreme temperature peaks is lost, and the potential for carbon sequestration in vegetation biomass is reduced [22].

This study was conducted at Secondary School João da Rosa – Olhão, and its main objective was to use this school environment as a living laboratory, where young students and teachers worked as a team with other stakeholders (University of Algarve, Municipality of Olhão, Portuguese Environmental Agency, and Águas do Algarve, S.A.) to improve its eco-efficiency. The main resources consumed (water, energy, and propane gas) for the normal pedagogical functioning of this building were quantified, from 2017 to 2020. The associated annual average CO_2e emissions were estimated, as well as the CO_2 sequestration by the plant biomass of the respective surrounding area, in order to calculate the carbon balance. Aligning with the New European Bauhaus, which aims to create solutions that integrate sustainability, inclusion and beauty, an artistic painting activity was developed to connect young people to the threat that plastic waste poses to natural ecosystems.

2 Methodology

2.1 The Living Lab

This study was carried out during the 2020/21 school year, when Secondary School João da Rosa (Fig. 1) had 488 students aged between 12 and 18 years old, 66 teachers and 20 operational assistants. Located in the Algarve city of Olhão, this school is surrounded by disadvantaged neighborhoods with difficulties in social integration, and where family disinterest in the school career of young people is a frequent reality.

Fig. 1. Secondary School João da Rosa, Olhão (adapt. Google Earth).

The consumption of resources (water, energy, and propane gas) for normal school functioning was considered. In the data analysis, the average monthly values were estimated, analyzing the year 2020 separately, to assess any differences in resource consumption, related to the atypical functioning of the school due to the Covid-19 pandemic.

2.2 Young People's Sensitization to Eco-Efficiency

Efficient Use of Water. To introduce the topic, World Water Day was used with an audiovisual presentation, (https://www.youtube.com/watch?v=cABhd_nrsSI&t=2s) at the end of which 271 students from the 7th, 8th and 9th years calculated their WF using the WF Calculator made available online by the Portuguese Association of Water Resources (https://ech2o.aprh.pt/peghidrica/pt/), in order to become aware of their individual daily consumption of water, both direct and indirect.

Tap Water Consumption and Reduction of Plastic Packaging. The consumption of treated tap water was promoted, in collaboration with the various stakeholders, presenting audiovisual materials adjusted to this age group, with evidence of the advances in the urban water cycle in Portugal in recent decades, and the safety of tap water. Canteens were distributed to all young people to be filled, instead of using bottled commercialized water (Fig. 2).

Fig. 2. Distribution of canteens to promote the consumption of tap water.

Several videos were presented on the problem of plastic waste, particularly water bottles, with a detailed explanation of the consequences of their disposal mismanagement for the natural ecosystems, its persistence in the environment due to the formation of micro- *and nano*-plastics, and from their bioaccumulation along food chains. (e.g., https://www.youtube.com/watch?v=_6xlNyWPpB8&t=42s).

Art and the Protection of Nature. Given the proximity of this school to the Natural Park of Ria Formosa and the sea, and in order to continue exploring the contents related to the urban water cycle, an artistic activity was organized to paint the gutters of the school grounds. Taking a multidisciplinary approach and once again integrating the collaboration of stakeholders, the importance of proper management of plastic waste and its negative impacts when it enters the gutters and reaches coastal ecosystems was explored (Fig. 3).

Fig. 3. Painting of the gutters at Secondary School João da Rosa, Olhão.

2.3 Diagnosis of Resources Consumed for School Operation

The different locations inside the school building where there is water consumption (WC, kitchen, bar, and classrooms) were considered, and the flows of the respective taps and showers were measured by students and teachers. The intervention in the cisterns was not possible, as they are built into the walls. An inventory of the water consumption at the school between 2017 and 2020 was conducted, based on the values referred in the respective monthly bills. The only water used in this school is treated water for human consumption, which is captured, treated, and distributed to the high-water tanks by a high by Águas do Algarve – Águas de Portugal Group. From these tanks, the municipal company AmbiOlhão distributes water through the downtown area. The monthly energy consumption of the school building and of propane gas used in the kitchen to prepare meals, were also accessed based on the data referred in the monthly bills.

2.4 Carbon Balance

CO_2e Emissions. Carbon emissions related to energy, propane gas and water consumption were considered. Reported energy consumption from 2017 to 2020 was converted to CO_2e emissions using the recommended emission factors [23]: 298 g CO_2/ kWh in 2017, 247 g CO_2/ kWh in 2018, 249 g CO_2/ kWh in 2019, and 200 g CO_2/ kWh in 2020. To estimate the CO_2e emissions associated with the consumption of propane gas in the same time period, an Emission Factor (EF) of 63.1 kg CO_2e/ GJ and a Lower Heating Value (LHV) of 46.3 MJ/ Kg were considered [23], in order to calculate,

$$\text{Efficiency(GJ)} = \text{Mass of consumed gas x PCI}$$
$$\text{Emissions(Kg } CO_2e) = \text{Efficiency x FE}$$

To estimate carbon emissions related to water consumption, it was considered that, for each m^3 of treated water that reaches the tap of Portuguese consumers, 0.88 kWh are spent, and that 0.81 kWh are consumed for each m^3 of drained and treated effluent in the wastewater treatment plant [24]. These energy consumptions related to water usage in the school building during the period under analysis were then converted into carbon emissions using the General Directorate of Energy and Geology (DGEG) Emission Factors.

Potential for CO_2 Sequestration. To estimate the carbon sequestration of the plant biomass in the school, the outdoor spaces were characterized by quantifying permeable and impermeable areas. In the vegetated areas, trees and shrubs were counted and identified, and was measured the area occupied by spontaneous herbaceous vegetation. To make a rough estimation of each specimen sequestration potential, were used carbon sequestration factors validated in previous studies, with the same plant species and considering individuals of equivalent age in the urban environment of Barcelona [25].

3 Results and Discussion

3.1 Efficient Use of Water

Water Footprint (WF): The results obtained for the 262 valid surveys considered, point to a direct WF between 50 and 116 L, on average 279 ± 182 L, therefore higher than the per capita consumption of drinking water in the Algarve in 2019, estimated at 189 L [26]. This average WF is slightly higher than that previously obtained for a multi-age community in Lisbon, corresponding to 255 ± 126 L [15]. For young people from Secondary School João da Rosa, showers represent 82% of the total water consumption (Fig. 4), about twice as much as previously reported by Muller *et al.* [27], followed by toilet flushes (6%) about four times lower than reported in the same study. When the young people were confronted with the excessive consumption of water in the showers, they repeatedly mentioned the need to keep the water always running so as not to feel cold, which reflects the difficulties these families face in heating their homes. On the other hand, these young people exhibit lower water consumption when washing dishes, clothes, and teeth, but consume twice as much when washing hands compared to previous studies. These results suggest that the high consumption of water in hand washing may be associated with the Covid-19 pandemic scenario experienced in 2020–2021.

As expected, the indirect water footprint was about 10 times higher (2678 ± 1093 L) than the direct one, reinforcing previously obtained results for young secondary school students in Faro [14]. It was found that food was responsible for most of the indirect WP (69%), followed by clothing (30%), and teaching materials only represented 1%. It should be noted, however, that Muller *et a*l. [27], in the multi-age community study carried in Lisbon, obtained higher consumption rates for food (95%) and significantly lower rates for clothing (4%), reinforcing the importance of age and social context in water consumption patterns, in this case for indirect water footprint.

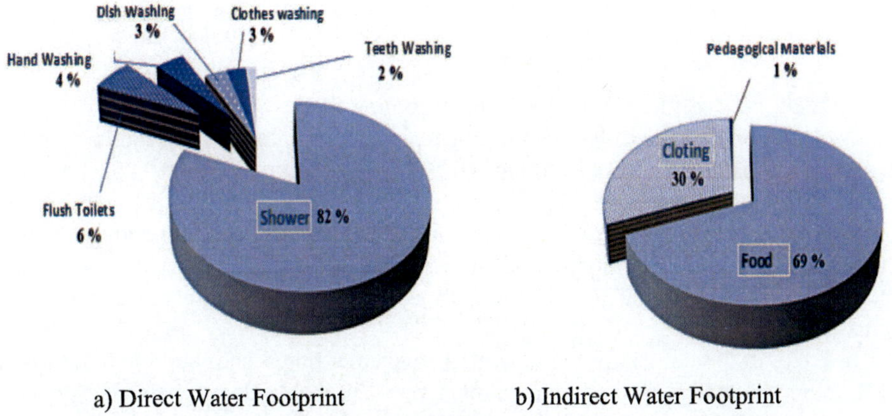

Fig. 4. Contributions to direct WF (a) and indirect WF (b) of young students.

3.2 Water Consumption in the School Building

The average monthly consumption of water at Secondary School João da Rosa between 2017 and 2020, is presented in Table 1 and indicates the existence of some water losses. It is particularly evident in the months of April and May 2017, September 2018 and July 2019, which significantly exceed the respective monthly average consumptions of those three years and are associated with high standard deviations (SD). If these months in which there were water losses are excluded, considering the monthly averages from 2017 to 2019, April corresponds to the month with the lowest consumption (at 144 m^3, excluding April 2017 when there was a water leak), and January is the month with the highest consumption (291 m^3), followed by November (221 m^3). These higher consumptions in January and November are probably associated with longer showers, which students take in these colder months.

In July 2019, water consumption was 60% higher compared to the average of the months of July 2017, 2018 and 2020 (Fig. 5). Since there is no justification in the functioning of the building for this anomalous consumption, it means that it is a loss of around 181 m^3, which corresponds to what 957 people living in the Algarve spent per day in 2019 [26].

Table 1. Monthly water consumption (m3) between 2017 and 2020 at Secondary School João da Rosa

Year/Month	Jan	Feb	Mar	Apr	May	Jun	Jul	Ago	Sep	Oct	Nov	Dec	Total Annual
2017	338	176	160	305	357	174	190	150	155	216	247	216	2684
2018	200	165	150	123	158	181	161	155	331	172	185	205	2186
2019	335	189	162	165	182	207	454	157	138	191	230	200	2610
Mean	291	177	157	198	232	187	268	154	208	193	221	207	2493
SD	79	12	6	95	109	17	161	4	107	22	32	8	269
2020	238	214	204	212	212	310	192	264	202	314	369	325	3056

In 2020, between March 13th and the beginning of the Summer holidays (end of June), classes operated in a distance learning regime, but, contrary to expectations, water consumption inside the school building increased. Also, in July and August 2020, despite the school holidays, water consumption exceeded the respective monthly averages for the previous three years (2017 to 2019). The months of October, November and December 2020 were those with the highest consumption levels, corresponding respectively to increases of 39%, 40% and 36%, respectively, compared to the monthly averages of the three previous years.

In 2020, in the context of the Covid-19 pandemic, this school building consumed 18% more water than in the previous three years, probably due to personal hygiene practices and cleaning of the building and its equipment. This fact highlights the social importance that water has for public health, particularly at critical times, such as the Covid-19 pandemic.

By intervening in the monitored equipment referred to in Table 2, by replacing damaged taps and installing flow reducers, a global saving of 270 L/min is achieved. Of these, 203 L/min will be in the WC, 3 L/min in the bar, 40 L/min in the cafeteria and 24 L/min in the classrooms. In showers, 16 L/min can be saved. Therefore, with the same level of comfort for students and staff at this school, around 219 L/min can be saved just in the WCs.

This value is highly illustrative of the existing opportunity to address wastefulness and improve water efficiency in school buildings such as this one. It should be noted that, in the Algarve regio and in the country as a whole, schools of various levels of education are operating in very different states of conservation. This fact should be reflected in very different efficiencies in the use of the necessary resources for their operation, namely in terms of water consumption. Therefore, it is highly relevant to periodically collect this type of data and critically analyze it, to ensure monitoring that allows for the reduction of water waste in the school environment.

3.3 Energy Consumption in the School Building

The monthly energy consumption of 2017 significantly exceeded the average monthly consumption of the three-year period (2017, 2018 and 2019), with a sharp decrease in

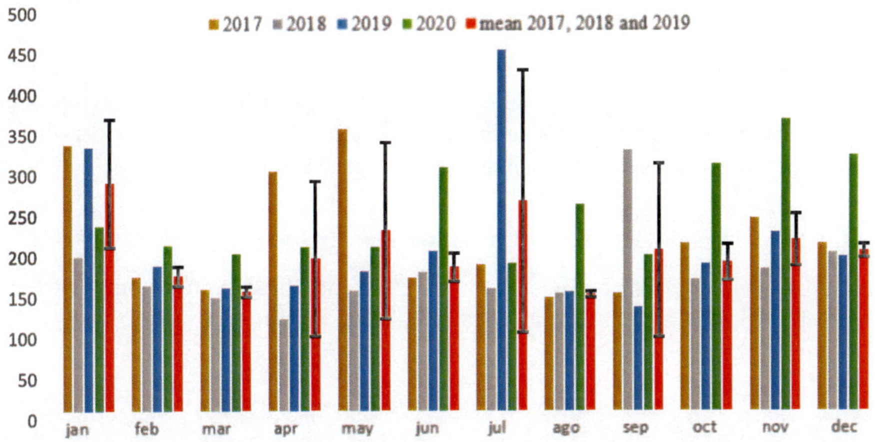

Fig. 5. Monthly water consumption (m^3) at Secondary School João da Rosa between 2017 and 2020, and average ± standard deviation of the three years without the influence of the pandemic situation.

Table 2. Flow rates monitored in the various school equipment.

	Total No. Taps	No. Taps to replace	No. of taps to place reducer	Average current flow rate (L/min)	Average intended flow rate (L/min)
WC	29	18	11	11 ± 4	4
Bar	3	1	2	10 ± 1	9
Kitchen	10	9	1	13 ± 3	9
Classrooms	2	8	9 ± 2	8	
Total	66	30	22	10 ± 3	8 ± 2
WC	**Total No. Taps**	**No. Taps to replace**	**No. of taps to place reducer**	**Average current flow rate (L/min)**	**Average intended flow rate (L/min)**
	16	16	0	9 ± 4	8

consumption between 2017 and 2018 (Fig. 6). This was probably related to the replacement of lighting systems, and the breakdown reported by operational staff, of some equipment/machines in the kitchen that have not yet been replaced. Energy consumption decreased in 2020 in the context of the Covid-19 pandemic, with confinement measures starting in mid-March and face-to-face activities resuming from mid-September. This decrease in energy consumption is clearly visible in the months of March through to August 2020.

3.4 Propane Gas Consumption in the School Building

Propane gas is used in this school exclusively for meal preparation in the kitchen, with a consumption of 1523 kg in 2017, 2420 kg in 2018, 2886 kg in 2019. However, propane gas consumption decreased to 1208 kg in 2020, a year in which the Covid-19 pandemic led to a significant decrease in the number of meals prepared in the school.

3.5 Carbon Emissions Related to Consumed Resources

Considering the resources consumed between 2017 and 2019, the functioning of this school was responsible for an average annual emission of 31,046 t CO_2e.

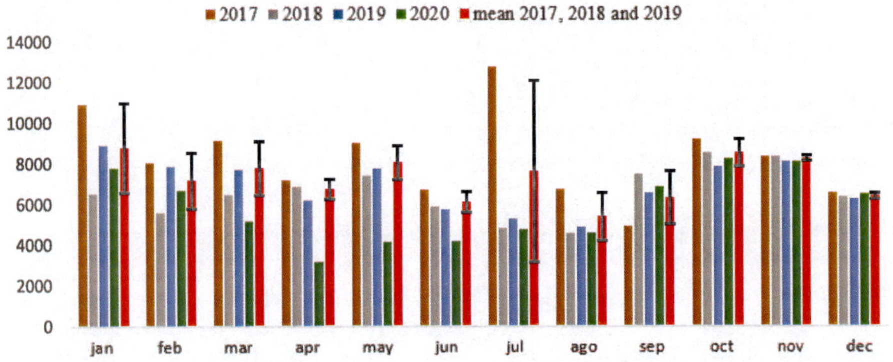

Fig. 6. Monthly energy consumption (kWh) at Secondary School João da Rosa between 2017 and 2020, and average ± standard deviation of the three years without the influence of the pandemic situation.

Energy consumption represented between 68 and 84% of emissions, propane gas consumption represented between 13 and 28% and water consumption represented the remaining 4% (Table 3). Like reported for other sectors of society [28, 29], the Covid-19 pandemic will have caused a decrease in carbon emissions related to this school functioning in 2020, quantified at about 60% of the average of the three previous years.

Table 3. Carbon emissions (kg CO_{2e}) related to consumed resources at Secondary School João da Rosa between 2017 and 2020.

Year	Energy	Propane Gas	Water	Total
2017	29 646	4450	1352	35 448
2018	19 477	7067	913	27 457
2019	20 703	8432	1098	30 233
2020	14 050	3529	1032	18 611

3.6 Potential for CO_2 Sequestration

The external area of this school is 12,852 m^2 wide, of which 10,426 m^2 are impermeable and 2,426 m^2 are permeable. Within the permeable area, 1,665 m^2 does not have vegetation, 419 m^2 correspond to spaces covered by trees, and 342 m^2 covered by shrubs. The herbaceous vegetation that exists is spontaneous and without expression in most months of the year. Therefore, for the calculation of the carbon sequestration potential, only tree and shrub vegetation were considered, which was characterized as shown in Table 4.

Table 4. Characterization of existing vegetation in the Secondary School João da Rosa and annual sequestration factor [25]

	No. of specimens	Kg CO_2/ ind
Trees		
Prunus dulcis	2	8.59
Morus nigra	1	35.30
Olea europaea	5	17.58
Jacaranda mimosifolia	1	25.32
Grevillea robusta	4	--
Ficus elástica	1	11.16
Populus nigra	7	−7.707
Pinus pinea	1	9.06
Cercis siliquastrum	2	12.70
Acer negundo	1	19.01
Casuarina cunnighamiana	54	40.63
Ficus carica	1	9.87
Melia azedarach	22	25.73
Rosa gálica	1	--
Nerium oleander	80	10.94
Myoporum laetum	2	--
Lantana camara	1	--
Pistacia lentiscus	1	9.58

The data presented allow for the estimation of an annual carbon sequestration by tree vegetation of 2,946 kg CO_2/year and by shrubs of 885 kg CO_2/year, according to previous studies previously carried out for these species, with individuals of equivalent age, in the urban environment of Barcelona [25]. Therefore, in this rough estimate, the existing vegetation in the outdoor spaces of Secondary School João da Rosa appears to have the capacity to sequester about 3,831 kg CO_2/year. The species that most contributed to this sequestering was the Australian pine (*Casuarina cunnighamiana*), which, although is not autochthonous, has long been distributed and implemented in regions far from its native land, Australia. It exhibits great resistance to strong winds and salinity, being very common in coastal areas where it adapts well to different edaphoclimatic conditions and presents rapid growth [30], which translates into its high capacity for carbon sequestration.

3.7 Carbon Balance of Secondary School João da Rosa

According to this study, on average this school emits 31,046 t CO_2e annually for its functioning (excluding the year 2020 from the annual average). The vegetation of its external spaces has a carbon sequestration potential of approximately 3,873 kg CO_2, which is about 12.5%. Thus, by calculating the carbon balance, we can conclude that this school contributes to the emission of approximately 27 t CO_2/year into the atmosphere. Therefore, a set of measures related to efficient resource use (especially energy) must be implemented, as well as considering increasing the density of native tree species (e.g. *Morus nigra, Olea europaea, and Ceratonia síliqua*) and shrubs (e.g. *Arbutus unedo*). These native species do not require irrigation, avoiding an increase in water needs, and provide shade and milder temperatures to the external school spaces, whilst sequestering carbon.

4 Final Considerations

In this study, over 500 individuals were involved in a socially deprived school environment, predominantly young students between 12 and 18 years old, in an effort to improve water use efficiency. The school acted as a living laboratory, where individual water consumption (WF) and the school community consumption were measured, revealing various situations of water loss and waste, despite the current scarcity scenario in the Algarve region. Behavioral changes and concrete measures were discussed and implemented to improve the school's water efficiency, while simultaneously correcting individual and collective daily actions. Young people talked to technicians and researchers, and adopted tap water consumption practices using canteens, instead of buying bottled water, reducing the usage of plastic packaging. Following the New European Bauhaus principles, environmental protection was addressed through art, and a painting activity was used to promote an emotional connection to the biodiversity of the Ria Formosa Natural Park. The role of Nature in the school environment and in the city of Olhão was explored by young people and teachers, who identified the plant species in their school grounds and recognized their important functions for quality of life in the school and in the city of Olhão, namely for carbon sequestration. Disseminating this type of study to other schools across the country could represent a fundamental contribution to the environmental performance of our cities and to the quality of life of present and future generations.

References

1. Galli, A., et al.: Assessing the ecological footprint and biocapacity of portuguese cities: critical results for environmental awareness and local management. Cities **96**, 102442 (2020). https://doi.org/10.1016/j.cities.2019.102442

2. Sterner, T., Barbier, E.B., Bateman, I., et al.: Policy design for the Anthropocene. Nature Sustainability **2**, 14–21 (2019). https://doi.org/10.1038/s41893-018-0194-x
3. United Nations People Division (UNPD): World Urbanization Prospects. The 2018 Revision (2018). https://population.un.org/Wup/Publications/Files/WUP2018-KeyFacts.pdf. accessed at 15 January 2022
4. Hardiman, B.S., Wang, J.A., Hutyra, L.R., Gately, C.K., Getson, J.M., Friedl, M.A.: Accounting for urban biogenic fluxes in regional carbon budgets. Sci. Total Environ. **592**, 366–372 (2017). https://doi.org/10.1016/j.scitotenv.2017.03.028
5. Tang, Y., Chen, A., Zhao, S.: Carbon storage and sequestration of urban street trees in Beijing. China. Front. Ecol. Evol. (2016). https://doi.org/10.3389/fevo.2016.00053
6. Leal Filho, W., et al.: The role of higher education institutions in sustainability initiatives at the local level. J. Clean. Prod. **233**, 1004–1015 (2019). https://doi.org/10.1016/j.jclepro.2019.06.059
7. García-González, E., Jiménez-Fontana, R., Goded, P.A.: Approaches to teaching and learning for sustainability: Characterizing students' perceptions. J. Clean. Prod. **274**, 122928 (2020). https://doi.org/10.1016/j.jclepro.2020.122928
8. Tasdemir, C., Gazo, R.: Integrating sustainability into higher education curriculum through a transdisciplinary perspective. J. Clean. Prod. **265**, 121759 (2020). https://doi.org/10.1016/j.jclepro.2020.121759
9. United Nations Educational Scientific and Cultural Organization (UNESCO): Correio da UNESCO. Muitas Vozes, um Mundo.A Educação Transforma Vidas (2021). https://pt.unesco.org/courier/january-march-2018/educacao-transforma-vidas. Acedido em 14 de janeiro de 2022
10. Galafassi, S., Nizzetto, L., Volta, P.: Science of the total environment plastic sources : a survey across scienti fi c and grey literature for their inventory and relative contribution to microplastics pollution in natural environments, with an emphasis on surface water. Sci. Total Environ. **693**, 133499 (2019). https://doi.org/10.1016/j.scitotenv.2019.07.305
11. Intergovernmental Panel on Climate Change (IPCC): Aquecimento Global de 1,5 °C. Sumário para Formuladores de Políticas. IPCC, p. 28 (2019). https://www.ipcc.ch/site/assets/uploads/2019/07/SPM-Portuguese-version.pdf
12. Freire-González, J.: Does Water Efficiency Reduce Water Consumption? The Economy-Wide Water Rebound Effect. Water Resources Management. **33**(6), 2191–2202 (2019). https://doi.org/10.1007/s11269-019-02249-0
13. Hoekstra, A.Y., Mekonnen, M.M.: The water footprint of humanity. Proc. Natl. Acad. Sci. U.S.A. **109**(9), 3232–3237 (2012). https://doi.org/10.1073/pnas.1109936109
14. Venckute, M., Silva, M., Figueiredo, M.: Education as a tool to reduce the water footprint of young people. Millenium - Journal of Education, Technologies, and Health **2**(04), 101–111 (2017). https://doi.org/10.29352/mill0204.09.00144
15. Neto, S., Moreira da Silva, M., Muller, L., Weller, K.: Social and Technological Innovation for Water Conservation – The Project ECH2O-ÁGUA. In: Increase 2019 – Book of Proceedings of the 2nd International Congress on Engineering and Sustainability in the XXI Century, pp. 786–798. Springer International Publishing, Cham, Switzerland (2020). https://doi.org/10.1007/978-3-030-30938-1_62
16. World Health Organization (WHO): Diretrizes globais de qualidade do ar da OMS: material particulado (PM2.5 e PM10), ozônio, dióxido de nitrogênio, dióxido de enxofre e monóxido de carbono: resumo executivo (2021). Disponível em: https://apps.who.int/iris/handle/10665/345334. accessed at 3 January 2022
17. Peixoto, D., Pinheiro, C., Amorim, J., Oliva-Teles, L., Guilhermino, L., Vieira, M.N.: Microplastic pollution in commercial salt for human consumption: A review. Estuarine. Coastal and Shelf Science **219**(January 2018), 161–168 (2019). https://doi.org/10.1016/j.ecss.2019.02.018

18. Agência portuguesa do Ambiente (APA): Portal do Estado do Ambiente. Produção e Gestão de Resíduos Urbanos (2021). https://rea.apambiente.pt. accessed at 15 January 2022
19. Entidade Reguladora dos Serviços da Água e dos Resíduos (ERSAR): Relatório Anual dosServiços de Águas e Resíduos em Portugal (2021): Controlo da qualidade da água para consumo humano (Vol. 2). www.ersar.pt (2021)
20. McPherson, E.G., Xiao, Q., Aguaron, E.: A new approach to quantify and map carbon stored, sequestered and emissions avoided by urban forests. Landsc. Urban Plan. **120**, 70–84 (2013). https://doi.org/10.1016/j.landurbplan.2013.08.005
21. Kuittinen, M., Moinel, C., Adalgeirsdottir, K.: Carbon sequestration through urban ecosystem services: a case study from Finland. Sci. Total Environ. **563–564**, 623–632 (2016). https://doi.org/10.1016/j.scitotenv.2016.03.168
22. Fry, J., et al.: Assessing carbon footprints of cities under limited information. J. Clean. Prod. **176**(2018), 1254–1270 (2018). https://doi.org/10.1016/j.jclepro.2017.11.073
23. Direção Geral de Energia e Geologia (DGEG) Despacho nº17313, 2ªSérie - n.122. Diário Da Républica, 27912–27913 (2008)
24. Entidade Reguladora dos Serviços da Água e dos Resíduos (ERSAR). Estudo de satisfação dos utilizadores dos serviços de águas e resíduos. www.ersar.pt (2016)
25. Chaparro, L., Terradas, J.: Ecological Services of Urban Forest in Barcelona. Centre de Recerca Ecològica i Aplicacions Forestals Universitat Autònoma de Barcelona. Disponível em: https://citeseerx.ist.psu.edu/viewdoc/download?doi=10.1.1.364.9799&rep=rep1&type=pdf (2009)
26. Entidade Reguladora dos Serviços da Água e dos Resíduos (ERSAR): Relatório Anual dos Serviços de Águas e Resíduos em Portugal. Controlo da qualidade da água para consumo humano, vol. 1. www.ersar.pt (2019)
27. Muller, L., Moreira da Silva, M., Rodriguez, C., Afonso, A.: Uso eficiente da água em Lisboa-Cálculo da pegada Hídrica. 15ª Congresso da Água, Lisboa 22 a 26 de março. https://www.aprh.pt/congressoagua2021/docs/15ca_134.pdf (2021)
28. Silva, C.M., Soares, R., Machado, W., Arbilla, G.: A Pandemia de COVID-19: Vivendo no Antropoceno. Rev.Virtual Quim. **12**(4), 1001–1016 (2020). http://rvq.sbq.org.br
29. Lobato, M. F., Rodrigues, B.M.M., Santos, A.G. dos.: Impacto da pandemia de COVID-19 nas emissões veiculares no Brasil no período de janeiro a maio de 2020. Engenharia Sanitaria e Ambiental **26**(5), 829–836 (2021). https://doi.org/10.1590/s1413-415220200261
30. Zhong, C., et al.: Past and current casuarina research and development in China. In: Haruthaithanasan, M., Pinyopusarerk, K., Nicodemus, A., Bush, D., Thomson, L. (eds.) Casuarinas for green economy and environmental sustainability. In: Proceedings of the Sixth International Casuarina Workshop held at Krabi, Thailand, 21–25 October 2019, pp. 21–30. Kasetsart Agricultural and Agro-Industrial Product improvement Institute, Kasetsart University, Bangkok (2020)

Engaging Community on Water Circularity in Culatra Island, Algarve – Portugal

M. A. Torres[1], M. Moreira da Silva[1,2,3], C. Sequeira[1,3,3](✉), and A. Pacheco[3,4]

[1] University of Algarve, Institute of Engineering, Campus Penha, 8005-139 Faro, Portugal
{msanti,cdsequeira}@ualg.pt
[2] CeiiA, Av. D. Afonso Henriques, 1825, 4450-017 Matosinhos, Portugal
[3] CIMA-ARNET, Faro, Portugal
ampacheco@ualg.pt
[4] University of Algarve, Campus Gambelas, 8005-139 Faro, Portugal

Abstract. Culatra is a small island located in the Ria Formosa Natural Park in the Algarve region at the south of Portugal, with a Mediterranean climate and facing an increasing water scarcity. The Culatra2030 Project is being developed in the island creating local actions to enhance circularity and improve the sustainability in the natural resources use. Drinking water consumed in Culatra is provided by the mainland, where it is extracted from aquatic ecosystems, treated, and transported to the island. The effluents produced on the island are drained and transported to the mainland where they are treated in a Wastewater Treatment Plant (WWTP). The transport of drinking water and effluents represents high energy consumption and therefore carbon emissions. The first stage of the Culatra2030 was to develop a Sustainable Energy Community, and since the last year we are studying measures to improve water circularity and eco-efficiency, avoiding the drinking water waste, and creating alternative water sources for non-potable uses. Two reference buildings have been chosen for community involvement, the Social Centre and the Primary School, where several educational actions were carried out e.g., the Water Footprint was calculated with teachers and young people, and some devices were installed to flow reducing on taps, showers and toilet flushers. The nexus drinking water /energy /carbon emissions was calculated before and after the installation of the flow reducers. The engagement of the local community on the sustainable water management was assessed through a survey prepared and distributed to families in collaboration with the island residents' association.

Keywords: Sustainable Water Management · Mediterranean Islands · Alternative Water Sources · Energy and Carbon Emissions in Urban Water Cycle

J. F. L. C. Semião et al. (Eds.): INCREaSE 2023, ASST, pp. 63–78, 2023.
https://doi.org/10.1007/978-3-031-44006-9_6

1 Introduction

1.1 Culatra Island as a Case Study in Mediterranean

In 2014, when publishing its 5th assessment report the Intergovernmental Panel on Climate Change (IPCC) warned about the increase of population suffering from water scarcity during the 21st century: "The fractions of the global population that will experience water scarcity (…) are projected to increase with the level of warming in the 21st century (robust evidence, high agreement)" [1].

Water in the environment follows a natural circular model i.e., the water precipitates on the Earth's surface as rain and flows over the surface. This water evaporates and eventually returns to the Earth's surface by its precipitation, closing the cycle named the water cycle. However, when thinking of water in urban water systems, these resources are managed by human decisions following an anthropogenic water cycle i.e., water is extracted from different primary sources (e.g., surface water, groundwater, and seawater) to meet urban water demands, used, qualitatively degraded, and discharged, eventually, as wastewater into the water cycle. This strategy is no longer sustainable due to its imprudent consumptions of resources and contamination of the environment.

Around the world there are 85000 islands of those almost 13000sare inhabited, representing a total of 750 million people [1]. Because of their insulation, island territories have been facing water shortage and studied water management in advance relatively to mainland, which usually benefit from various sources of water and developed infrastructures. Last century islands have evolved quickly to meet the life comfort standards of mainland resulting in improved management of basic resources such as water and energy, which are fundamentally linked. Different solutions were adopted according to each specific situation, constraint and opportunities.

Eco-efficiency is a main issue for the sustainable management of the planet, and particularly in the context of islands. Eco-efficiency is the efficiency with which resources are used to meet human needs, showing how efficient the economic activity is regarding nature's goods and services [2]. In Mediterranean during the last decades, the water supply and wastewater treatment are being aggravated by the adverse impacts of climate change, demographic growth, and consumption patterns. To face these challenges, some islands are developing their eco-efficiency testing innovative approaches, through greener solutions and with the engagement of the communities. These solutions can be latter scaled up to work on the mainland or serve as examples for other island communities [3–6]. Also, the energy dependence of the continent is a main problem that needs to be overcome, and local production of renewable energy with low environmental impact can be recommended [7–10].

One of the driving forces that led islands to test and develop new solutions is the high price of importing essential resources such as water or energy. Energy and water are linked in different ways i.e., energy is required to obtain potable water and pump it from the water treatment plant to the reservoirs and places where it will be used [11, 12]. Since the energy used is provided by fossil fuels, the related carbon emissions contribute to the global warming, acceleration of climate change, habitats degradation, and biodiversity losses [13–16].

Culatra is a small island (Fig. 1), with 7 km in length and 1.2 km wide, located in Ria Formosa Natural Park (36.9937 N; 7.839793 W), where are living about 1000 people and the main economic activities are fisheries and aquaculture [6]. The Ria Formosa is a shallow coastal lagoon system classified as a Ramsar site due to its unique biodiversity, economy, and social values. Regarding the economy, the source of income is related to fish and shellfish farming, fisheries, salt production, and tourism. Ria Formosasrepresentssthesmostsproductivesaquacultureszonesinsportugal, representing 80% of total national [17]. During the last decades tourism becomes as significant economic income, especially in Summer, when the population triples [6, 17, 18].

The initiative "Culatra 2030 – Sustainable Energy Community" aims to create a pilot community in renewable energies on Culatra Island, Ria Formosa, Algarve. The community will produce energy exclusively from renewable sources, use electric mobility, decarbonize its fishing industry and acquire sustainable habits and living practices. It will also manage its own energy system, recycle water for self-consumption and retrieve value from its waste. The value of the Culatra 2030 initiative lies in its all-encompassing strategy covering multiple aspects of green transition, including social issues such as energy poverty. Rather than the development of new technology, the key perspective is the holistic model and demonstration character of the initiative: "A truly bottom-up initiative, inspired by the smart specialization approach, which can be replicated in other communities, having recently been selected by the European Commission as the example initiative of the smart specialization strategy in the Algarve [19]".

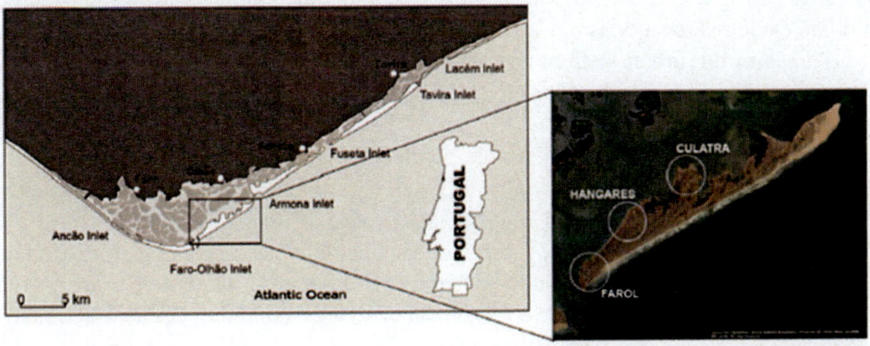

Fig. 1. Culatra Island location (36.9937°N; -7.839793°W) in the Atlantic Ocean. (adapted from Pacheco et al., 2022).

1.2 Water Cycle in Culatra Island

The responsibility for providing drinking water supply to Culatra is of AdA (Águas do Algarve– Grupo Águas de Portugal), the company that collects, treats, and transports it from the mainland to the island. Then, FAGAR, the municipal company, is responsible for distributing the water to consumers' domestic taps and for the sewage collection. According to European and National legislation (Law 152/2017) only safe water, i.e.

potable water for human consumption, can be distributed in Culatra. In 2020, AdA provided Culatra Island with 90 243 m^3 of potable water for human consumption, however, 32.7% of the water consumed in urban cycle is for external uses ERSAR, [20] that do not require potable water. Besides, the long-distance water transport from the abstraction site on the mainland (over 60 km) involves high energy consumption and thus carbon emissions. After the different uses (domestic and others), the effluents produced in Culatra are drainage and transported to be treated in a WWTP in the mainland, and the treated effluent is discharged, again in Ria Formosa lagoon.

Thus, to improve the water circularity and eco-efficiency in Culatra, the local community should evolve to a water sensitive community, where nature-based solutions are applied, i.e., ecological landscapes based on preserving or mimicking natural processes to support and overcome urban challenges such as environmental quality, climate change and socio-economic issues, among others [21, 22]. The concept of water sensitive communities represents the synergic connection between institutions and infrastructures, ensuring a sustainable water resource management focused on people needs and environmental protection. Water sensitive communities represent the climax of urban water cycle and environmental protection, bringing to the cities not only sustainability but also resilience, prosperity and ecosystem services, as improved environmental and human health, urban amenity, recreational opportunities, and decrease of the heat island effects [22, 23]. In a water sensitive community, the efficiency in water usage is a main issue as well as to diversify water sources (e.g., rainwater harvesting, water reuse and desalination) and suit its utilization for different fit-for-purpose uses e.g., external surfaces washings and agriculture irrigation [22, 23]. The integration of alternative water sources in urban cycle makes it possible to improve the irrigation of vegetation, increasing the green areas in the urban settlements. The shading provided by urban vegetation and water transpiration from plant leaves contribute to lower temperatures and reduce the effects of heat islands.

Also, education is an indispensable element of property and provides skills and knowledge to people, being a very powerful tool to promote more sustainable behaviors in the present and future [2, 11, 15, 24, 25].

The aim of this work was to promote sustainable water management on Culatra Island, improving the efficiency of the water usage and finding alternative water sources for non-potable purposes, with the engagement of all local community. The water/energy and carbon emissions nexus were also analyzed.

2 Methodology

2.1 Drinking Water Consumption and Efficiency in Use

To characterize water consumption and test water saving methods we selected two public buildings in Culatra, the Social Center and the Primary School (respectively, Fig. 2a and Fig. 2b) in working conditions during the year 2022, as shown in Table 1.

The water flows (L/min) from all the equipment in kitchen and bathrooms (taps, showers and toilet flushers) were measured and it was evaluated the need to install flow reducers devices. These faucet aerating devices reduce directly the water flow by increasing the pressure. Different flow intensity options are available for specific uses,

Fig. 2. Public buildings in Culatra Island where the water consumption was characterized: (a) Social Centre; (b) Primary School.

Table 1. Social Centre and Primary School operational conditions in 2022.

Social Centre Coordinates: 36.990800 N; –7.841100 W	Operating hours: weekdays from 9:00 am to 6:00 pm No. of teachers and support staff: 16 No. of students: 28 and once per week 11 seniors Students' age: 1–10 years-old
Primary School Coordinates: 36.992056 N; –7.840611 W	Operating hours: weekdays from 7:45 am to 5:30 pm No. of teachers and support staff: 4 No. of students: 16 Students' age: 6–10 years-old

meeting the water needs while maintaining comfort for the user [24]. In the case of taps and showers, the original filter/diffuser was replaced by the flow reducers, and in the cisterns, a vinyl bag of 2 L of volume was disposed in the reservoir, which means that the users save 2 L of water per flush.

2.2 Energy Consumption and Carbon Emissions on Culatra Water Cycle

To relate drinking water consumption with energy consumption we consider the volumes of water consumed referred in the monthly bills and the conversion factors of the national regulatory authority for water and waste services [20]. 4According to ERSAR, in Portugal it takes 1.12 kWh to abstract, treat and transport 1 m^3 of drinking water to the customer's tap; and 0.46 kWh to drain and treat in the WWTP 1 m^3 of wastewater.

To estimate the carbon emissions related to energy consumption on Culatra water cycle we used the emission factors recommended by the competent authority (DGEG, 2022) [26]. Emission Factors which were to 2017 = 298 g of CO_2e/kWh, to 2018 = 247 g of CO_2e/kWh, to 2019 = 279 g of CO_2e/kWh, and to 2020 = 200 g of CO_2e/kWh.

2.3 Alternative Water Sources in Culatra

Some alternative water sources were identified that could supply several non-potable external uses and reduce the drinking water consumption.

Groundwater: There are five aquifer systems which drain into the Ria Formosa. Before 2010 the groundwater assumed a fundamental role in supplying water to populations and

Table 2. Wells coordinates.

Well	Latitude	Longitude
A	36.993396 N	−7.841435 W
B	36.993722 N	−7.840944 W
C	36.993972 N	−7.839833 W
D	36.993889 N	−7.838556 W
E	36.992056 N	−7.840611 W
F	36.990972 N	−7.841028 W
G	36.994694 N	−7.838944 W
H	36.994722 N	−7.838667 W
I	36.994278 N	−7.837556 W
J	36.994006 N	−7.838592 W

Fig. 3. Monitored wells in Culatra Island.

tourism infrastructures. There were a series of wells scattered around the island, which supplied the population needs. However, groundwater become contaminated by urban wastewater and farming, and began to pose a threat to public health if used for human consumption.

To evaluate the feasibility of groundwater consumption in Culatra, for non-potable purposes and especially for horticulture irrigation, we selected 10 different wells (from A to J) distributed across the island (Fig. 3 Table 2) For quantification of salinity and conductivity under different tide conditions, two water samplings were performed: one at high tide and another at low tide. Salinity was quantified by refractometry (ZUZI FG 211 ATC) and conductivity by electrometry using a conductivity meter WTW Brand [28].

Rainwater Harvesting: Although mean precipitation has been decreasing in last decades in the Algarve, intense rainfall events in short periods of time are increasingly frequent. Thus, rainwater harvesting can be an alternative water source for external non-potable uses in the Social Centre and the Primary School. Culatra Island is a demo site of HYDROUSA, a European Union Horizon2020 Innovation Action Project (Grant Agreement No. 776643) under the topic *Water in the context of the circular economy*. We organized technical audits of the two public buildings to investigate the rainwater harvesting systems installed a few decades ago and later abandoned, to assess the best way to rehabilitate them. The Social Centre has a water catchment area comprising a roof of 38 m^2 in ceramic tiles, and an accessible terrace area of 95 m^2 in reinforced concrete, which was not considered for water quality reasons. The water catchment area of the Primary School is a roof of 133 m^2 in ceramic tiles.

To the sizing of storage tanks, we considered the annual volume of usable rainwater (Va) which was calculated according to ETA ANQIP 0701 [29]:

$$Va = C \times P \times A \times \eta f$$

where,

C is the runoff coefficient considering water retention, absorption and diversion on the collection surface (80% for flat or low slope waterproof roof, according to TA ANQIP 0701 and EN 16941-1:2018 [30];
P is the average annual of accumulated precipitation in Faro ($\approx$ 511.6 mm);
A is the coverage catchment area measured in horizontal projection; and.
ηf is the hydraulic filtration efficiency (in general close to 90%).

Seawater Desalination: To assess solar still efficiency (volume of produced freshwater per m^2.day) and understand how it is related with the dimensioning or functioning parameters, we tested a solar system to seawater desalination that works autonomously, without the need for chemicals or external energy supply, and monitored the water evaporation rate in a continuous manner. During July 2022, it was installed in Culatra Island a small solar desalination tank (Fig. 4), built in fiberglass reinforced by polymeric composite material, with 40 cm in height and 100 cm in width and length. The tank walls were painted black to enhance the absorption of sunlight, and was installed a glass cover with 3 cm thick and a 37° inclination angle, corresponding to the Culatra Island latitude, i.e., 37° N. The air and water temperatures were monitored using PT100 thermometers and a Mini Conductivity Probe K 1.0. Was used for measuring the water conductivity.

Fig. 4. The solar still installed on Culatra Island during July 2022.

2.4 Engagement of Local Community

Following what had already been done during the participatory diagnosis with the energy component in Culatra 2030 initiative, and to enhance the water circularity in island, priority was given to evolve the community on the process. By discussing with the community, the water needs and daily actions we can achieve the best solutions to be implemented and increase the final benefits for people and for the environment [6].

The Diagnosis Process on Water issues, conducted under Culatra2030 initiative on 2019, was able to identify several problems and needs, prioritized as follow by the stakeholder groups: (1) Responsible water consumption; (2) Water losses; (3) High cost of water; (4) Pollution of the Ria; and (5) Alternative water sources. For issue (1), Responsible Water Consumption, the following solutions were identified as a priority: make use of water from wells by identifying water quantity and quality and definition of possible uses; study different forms of circular economy of water: wells, reservoirs, rainwater collectors, return valves, etc. For issue (2), Water losses, mainly due to clogs in the water pipes, it was decided to create a training program to make the population aware of the problem of water i.e. its use, good practices, quantity and quality of water. For issue (3), High cost of water, the solution proposed was to consume less and negotiate prices with the distribution company. For issue (4), Pollution of the Ria, the stakeholder groups decided to address it in an integrated manner e.g. denounce and report situations of illegality to the competent authorities; and work to connect illegal sewage points to the wastewater management plant. For issue (5), Alternative water sources, it was considered a priority to create a group to study desalination, since producing water from desalination using environmentally friendly technology is key, as well as reuse of wells for irrigation and recover the water storage reservoirs.

With the collaboration of the resident's association (Associação de Moradores da Ilha da Culatra) we prepared and conducted a survey to families to gather information about their position and involvement in sustainable water management. The survey contained 14 questions with multiple possible answers (between 3–5), about the willingness to develop daily actions that promote a more efficient water use, and their acceptance of integrating alternative water sources for non-potable uses. These questions addressed different topics, including volume of water consumed per family, tap water uses, awareness of water scarcity, water reuse, rainwater harvesting, seawater desalination, and composting of organic wastes.

3 Results and Discussion

3.1 Water Consumption and Efficiency in Use in the Public Buildings

Drinking Water Consumption: The mean of the monthly drinking water consumption for both buildings are presented on Fig. 5. In this figure, a and b are the social center consumption prior to pandemic (2017, 2018 and 2019) and during the pandemic (2020), respectively; while c and d are the school consumption prior and during the pandemic, respectively.

In the Social Centre (Fig. 4a) during a typical year, the month with the highest water consumption is August (33.7 ± 1.5 m^3), for the remaining months, consumptions are similar, varying between 20–30 m^3, and December was the month with the lowest water consumption due to Christmas holidays. In 2020, the pattern of water consumptions was clearly different due to Covid-19 pandemic (Fig. 5 b). During April and March it was registered lower water consumptions because children were not going to the Social Center, and kindergarten teachers and other collaborators were taking turns for maintenance activities. The total water consumption in 2020 was 271 m^3 with a monthly average of 22.6 ± 7.8 m^3. These values are lower than those obtained for 2017 with total water consumption of 292 m^3 and a monthly average of 24.3 ± 4.0 m^3, 2018 with a total water consumption of 312 m^3 and a monthly average of 26.0 ± 4.1 m^3, and 2019 with a total water consumption of 303 m^3 and a monthly average of 25.3 ± 3.2 m^3.

In the Primary School (Fig. 5 c) the monthly water consumptions between 2017 and 2019 were about 50% of the of Social Centre reflecting the different number of users, respectively 20 and 44 persons (Table 1), and showed a similar pattern of variation.

August was the month with the highest water consumption (42.3 ± 2.3 m^3), even exceeding the consumption of the Social Center, since during that month various sports activities for young people take place in Primary School. In 2020 (Fig. 5 d) the monthly consumptions were less than half of those verified in the same months in the years 2017, 2018 and 2019. April, May and June of 2020 had water consumptions less than 1 m^3, because in the case of the Primary School no maintenance activities were carried out during the lockdown. In this pandemic year, August presented a total water consumption of 14 m^3.

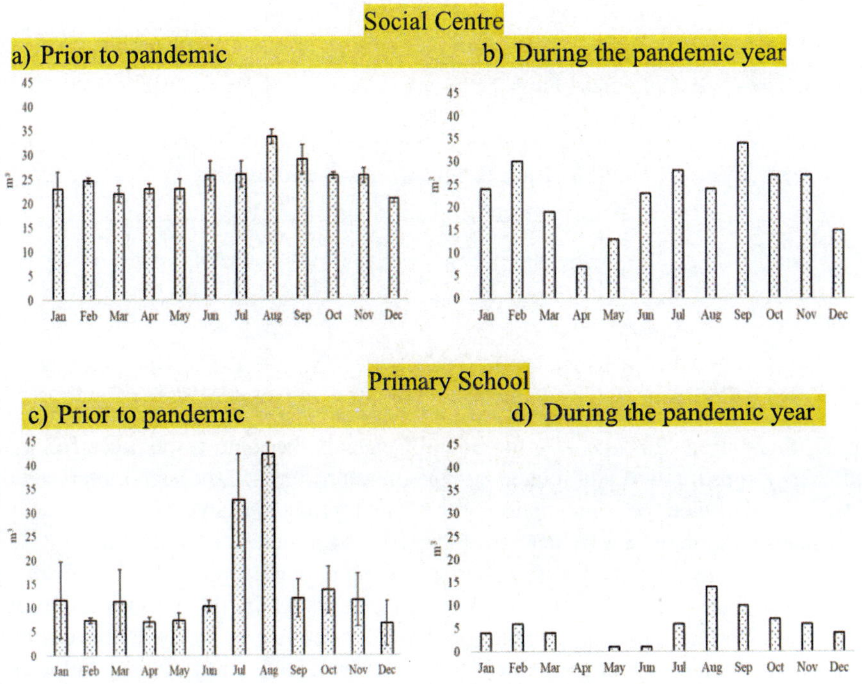

Fig. 5. Drinking water consumption: prior to pandemic (mean and standard deviation of 2017, 2018 and 2019 years) for the Social Centre and for the Primary School, respectively; and during the pandemic (year 2020), for the Social Centre and for the Primary School, respectively.

The year 2020 presented total water consumption of 63 m^3, with a monthly average of 5.3 ± 4.0 m^3. These values are lower than those obtained for 2017 of 179 m^3 and a monthly average of 14,9 ± 9,7 m^3, for 2018 of 189 m^3 and a monthly average of 15.8 ± 12.0 m^3), and for 2019 of 154 m^3 and a monthly average of 12.8 ± 13.7m^3).

Efficiency in Drinking Water Use Culatra's Public Buildings: In the Social Centre were installed flow reducers devices in the various equipment including in taps, toilet flushing devices and a shower, in 6 bathrooms and 1 kitchen (Table 3). However, in the Primary School building were just installed flow reducers devices in the taps distributed throughout the 6 existing rooms, as the toilet flushing devices are interior and inaccessible, and there are no showers.

By using these devices, we are reducing the water abstraction from aquatic ecosystems, the volume of natural water that needs to be treated for human consumption, the volume of domestic wastewater that requires treatment in a WWTP, the energy consumption for all processes and transports, and the respective carbon emissions.

Table 3. Flow reduction due to installation of the water saving devices in Culatra's public buildings.

	Equipment	Nº of installed devices	Mean Flow Reduction (%)
Social Centre	10 Toilet Flushings	7	29 ± 1
	17 Taps	11	40 ± 1
	1 Shower	1	28 ± 1
	Total	19	35 ± 1
Primary School	8 Taps	8	40 ± 1

3.2 Energy Consumption and Carbon Emissions on Culatra Water Cycle

During this period (2017–2020), the Social Center is the main responsible for water and energy consumption and related carbon emissions on Culatra Island, representing approximately twice the consumption of the Primary School (Table 4).

The installation of flow reduction devices allows for an average reduction of 35% in Social Centre' water consumption which, according to our estimates during the considered period (from 2017 to 2020), would avoid the emission of 167 kg CO_2e. In Primary School, as it was not possible to reduce the flow of toilet flushes, we do not have enough data to estimate the emissions avoided by improving the efficiency of using tap water.

These results were disseminated in the community to demonstrate how the installation of flow reducers devices can improve their efficiency in water usage at home and decrease the costs they pay for the drinking water supply, wastewater and solid waste collection and treatment taxes.

3.3 Alternative Water Sources in Culatra

Groundwater: The results obtained for the 10 monitored wells in Culatra Island (Table 5) show that there are different patterns in terms of tide influence in groundwater salinity and conductivity. During the considered period, five wells presented typical

Table 4. Nexus drinking water, energy consumption and related carbon emissions in Culatra's public buildings.

		2017	2018	2019	2020	Total
Social Centre	Water Consumption (m^3)	292	312	303	271	1178
	Energy Consumption (kWh)	461	493	479	428	1861
	Carbon Emissions (kg CO_2e)	137	122	134	86	478
Primary School	Water Consumption (m^3)	179	189	154	63	585
	Energy Consumption (kWh)	283	299	243	100	924
	Carbon Emissions (kg of CO_2e)	84	74	68	20	246

values for freshwater, even in high tide conditions (wells A, B, C, G and H). However, four wells seem to suffer slight to moderate saline intrusion (wells D, E, I and J), and one well (F) presented conductivity and salinity values that correspond to a severe saline intrusion, even in low tide. The salt contents of the well F in both tide conditions (conductivity > 3 000 μ/cm) are higher than the defined limit recommended by the Food and Agriculture Organization (FAO) of the United Nations for agriculture irrigation [28]. In terms of salt contents, the groundwater from the wells A, B, C, G and H can supply the demand for all kind of external non-potable uses in the island; while in the cases of wells D, E and I, more water samplings and analysis are needed to confirm the saline intrusion intensity, mainly during the high tide. The groundwater from wells D, E, and I in high tide, and from well F in both tide conditions, with higher salt contents, can also be used for some of non-potable purposes, as washing external pavements and waste containers.

Rainwater Harvesting: Currently the rainwater harvesting system is not functioning and needs to be rehabilitated. The results from the audits of buildings to assess the intervention to enable the use of rainwater for non-potable external uses, concluded that Primary School needs to install gutters around the roofs to collect rainwater and downpipes to transport the water for the existing reservoir, which has 25 m^3 of capacity. According to previous calculation [29, 30], this tank has an insufficient storage capacity for the annual volume of usable rainwater, which is about 48.9 m^3. The Social Centre counts on gutters to collect the water and downpipes to transport it from the roof to a tank with capacity of 18.3 m^3, which is in accordance with the previous calculations [29, 30]. We used a mean precipitation value of 511.6 mm [31], which was calculated considering the precipitation data series from 1981 to 2011 collected by the meteorological station in the International Faro Airport, 12 km from Culatra Island.

Seawater Desalination: The solar still installed in Culatra Island during July 2022 showed a low productivity, of 0.49 L/m^2.day of freshwater (salinity < 1 g/L).

To improve the productivity of the solar desalination system it was chosen to develop a thermal solar system powered by the sun with a triangular shape (Fig. 6 a), complemented by energy surplus from the photovoltaic system already installed in Social Centre allowing to heat a coil under the salt water (Fig. 6 b). It is a desalination panel function like that of a solar still with a glass cover (Fig. 6 c), but with the particularity of

Table 5. Salt contents in Culatra's groundwater in different tide conditions.

Wells	Status*	Conductivity at 20 °C µS/cm		Salinity g/L
		Min: Low Tide	Max:High Tide	
A	Freshwater	623	679	<1
B	Freshwater	374	404	<1
C	Freshwater	645	680	<1
D	Saline intrusion Slight to moderate	733	2310	1–2
E	Saline intrusion Slight to moderate	1260	1765	1–2
F	Saline intrusion Severe	**5863**	**6420**	**5–6**
G	Freshwater	528	550	<1
H	Freshwater	551	553	<1
I	Saline intrusion Slight to moderate	1039	2593	1–3
J	Saline intrusion Slight to moderate	1351	1680	1–2

* Ayers & Westcot,1985

the water flowing in a thin film over the glass. The thinning of the water film makes it possible to increase the temperature of the water and steam, improving evaporation and condensation. The existence of a higher internal temperature increases the difference with the temperature of the glass and therefore improves water condensation. So far, preliminary results of the thermal solar system showed a productivity increase of around 25%, compared to the previous system.

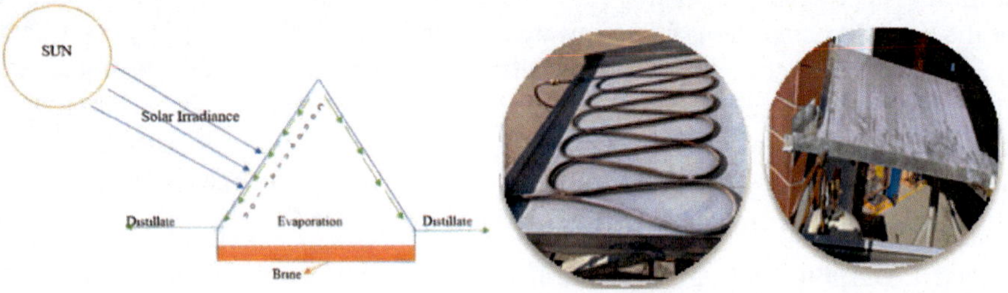

a) b) c)

Fig. 6. Scheme of thermal solar system tested in Culatra Island (a), details of electrical resistance (b) and of glass cover (c).

As no chemical products will be used in all the process, the recovered salt can also be used as a new Culatra's commercial product, or in local activities, e.g. for preserving fish, and cooking at home or in restaurants.

3.4 Engagement of Local Community

The surveys were distributed among families residing on the island (around 400) and, of these, 52 families (13%) responded and returned them complete. The analysis of these answers allows us to describe approximately how the residents of Culatra Island relate to water. Each person consumes less than 5 m^3 of drinking water per month, i.e., less than 167 L per day, and the majority do not drink tap water but use it for cooking, preferring to drink bottled mineral water. About 40% of families do not know the natural source of the drinking water.

Some families (33%) that responded to the survey consider that they already save water, especially during the shower, as they collect the cold-water while is heating.

Most families (44%) are concerned with water scarcity and consider that is important to promote alternative water sources for non-potable uses, including rainwater harvesting, seawater desalination and water reuse from washing vegetables in kitchens.

The seawater desalination is considered an important water source for most families (37%) if it is done using renewable energy. The main advantages that families see in desalination as an alternative water source is the reduction of dependence on the mainland, the creation of more jobs in Culatra and the salt production. The main constraints that families point to desalination are the environmental impact caused by CO_2 emissions and the toxicity of brine, and the high associated costs.

4 Final Considerations

Fresh water is becoming rarer on Earth but mostly it is unevenly shared around the planet resulting in water scarcity growth in many regions. The Algarve is expected to experience an increase in drought periods, with impacts on the quantity and quality of water resources. Knowing that water is a scarce and indispensable resource, societies need to value water and use it more efficiently. Water efficiency will generate new opportunities, such as new value-added services, companies, jobs, and qualified professionals.

Culatra Island has been supplied with sanitation services and safe water since 2010, pumped from the mainland, meaning high energy consumption and carbon emissions.

Water scarcity anticipation, energy consumption and greenhouse gases emission concerns are key aspects to drive Culatra into a smarter management of water, which implies the attempt to close the water loop on the island. Desalination powered by solar energy appears to be a good solution for islands which, like Culatra, are surrounded by seawater and receive an important sunlight irradiation. But it is of primary importance to include desalination in a broader water circularity plan. Indeed, desalination cannot provide enough water on its own. Its main advantage is rather relying on the fact that its production peak is happening in the summer season during which rainwater is scarce and there are more water needs for irrigation and tourism.

Desalination capacity is intrinsically related to energy availability. Non-intrusive techniques such as solar stills are a good option since both their construction and maintenance are inexpensive. It is considered a sustainable technology to provide freshwater for small-scale applications, particularly in remote arid areas. Nevertheless, they are characterized by the slowness and low productivity of their processes, since replicate the natural hydrological cycle, i.e., the seawater evaporation, vapor rising and then condensation. The main challenge is therefore to increase the low productivity of the modified solar still by transforming it from a passive system to a semi-passive system using surplus renewable energy sources (RES) to increase yields of freshwater production; and, complementary, serve as a flexibility instrument to avoid extra storage of RES surplus energy.

Thanks to the dynamism of its associations, Culatra is today a fishing community awake to environmental issues and their direct and indirect implications for the sustainability of resources and the quality of life of its people. While the implementation of the full range of solutions for Culatra 2030 is still underway, several remarkable achievements are already visible. The islanders have reached 20% of its energy autonomy, with capacity to store their own produced energy, giving the first steps to share energy profits under the concept of a renewable energy community. Investing in education is crucial to achieve sustainable development, poverty eradication, equity, and inclusion. In the sense of carbon neutrality, Culatra aims for a transition in terms of water efficiency, as the perception of the scarcity of the drinking water resource is a reality that requires in-depth work.

Acknowledgments. This study had the support of national funds through Fundação para a Ciência e Tecnologia (FCT), under the project LA/P/0069/2020 granted to the Associate Laboratory ARNET and UID/00350/2020 CIMA.

References

1. United Nations Environment Program: Chemicals & Pollution Action Island nations get new lifeline to beat pollution. https://www.unep.org/news-and-stories/press-release/island-nations-get-new-lifeline-beat-pollution. 07 Jun 2022 Press release
2. Liu, L., Bergen, J.: Green infrastructure for sustainable urban water management: practices of five forerunner cities. Cities **74**, 126–133 (2018). https://doi.org/10.1016/j.cities.2017.11.013
3. European Economic and Social Committee (EESC): TEN section: Report on the "Smart islands" project. Publications Office. https://data.europa.eu/doi/https://doi.org/10.2864/909553 (2017)
4. Kaldellis, J.K.: Clean energy in European Islands. Renew. Energy **155**, 100–101 (2020). https://doi.org/10.1016/j.renene.2020.03.160
5. Skjølsvold, T.M., Ryghaug, M., Throndsen, W.: European island imaginaries: examining the actors, innovations, and renewable energy transitions of 8 islands. Energy Res. Soc. Sci. **65**, 101491 (2020). https://doi.org/10.1016/j.erss.2020.101491
6. Pacheco, A., Monteiro, J., Santos, J., Sequeira, C., Nunes, J.: Energy transition process and community engagement on geographic islands: the case of Culatra Island (Ria Formosa, Portugal). Renew. Energy **184**, 700–711 (2022). https://doi.org/10.1016/j.renene.2021.11.115

7. Santamarta, J.C., et al.: The clean energy transition of heating and cooling in touristic infrastructures using shallow geothermal energy in the Canary Islands. Renew. Energy **171**, 505–515 (2021). https://doi.org/10.1016/j.renene.2021.02.105
8. Triantafyllou, P., et al.: Optimum green energy solution to address the remote islands' water-energy nexus: the case study of Nisyros island. Heliyon **7**, e07838 (2021)
9. Ayou, D., Ega, H., Coronas, A.: A feasibility study of a small-scale photovoltaic-powered reverse osmosis desalination plant for potable water and salt production in Madura Island: a techno-economic evaluation. Therm. Sci. Eng. Prog. **35**, 101–450 (2022). https://doi.org/10.1016/j.heliyon.2021.e07838
10. Icaza-Alvarez, D., Jurado, F., Tostado-Véliz, M., Arevalo, P.: Decarbonization of the Galapagos Islands. Proposal to transform the energy system into 100% renewable by 2050. Renew. Energy **189**, 199–220 (2022). https://doi.org/10.1016/j.renene.2022.03.008
11. Arora, M., Yeow, L.W., Cheah, L., Derrible, S.: Assessing water circularity in cities: methodological framework with a case study. Resour. Conserv. Recycl. **178**, 106042 (2022). https://doi.org/10.1016/j.resconrec.2021.106042
12. Kumar, R., et al.: Phase change materials integrated solar desalination system: an innovative approach for sustainable and clean water production and storage. Renew. Sustain. Energy Rev. **165**, 112611 (2022). https://doi.org/10.1016/j.rser.2022.112611
13. Gude, G., Nirmalakhandan, N.: Sustainable desalination using solar energy. Energy Convers. Manage. **51**(11), 2245–2251 (2010). https://doi.org/10.1016/j.enconman.2010.03.019
14. Gude, G., Nirmalakhandan, N., Deng, S.: Desalination using solar energy: towards sustainability. Energy **36**(1), 78–85 (2011). https://doi.org/10.1016/j.energy.2010.11.008
15. Gude, G.: Desalination and sustainability – An appraisal and current perspective. Water Res. **89**, 87–106 (2016). https://doi.org/10.1016/j.watres.2015.11.012
16. Wang, Z., Lin, X., Tong, N., Li, Z., Sun, S., Liu, C.: Optimal planning of a 100% renewable energy island supply system based on the integration of a concentrating solar power plant and desalination units. Int. J. Electr. Power Energy Syst. **117**, 105707 (2020). https://doi.org/10.1016/j.ijepes.2019.105707. CE4EU: Culatra Energy Transition Agenda, Clean Energy for EU Islands – European Commission, 57 p. (2019)
17. Rodrigues, M., Rosa, A., Cravo, A., Jacob, J., Fortunato, A.: Effects of climate change and anthropogenic pressures in the water quality of a coastal lagoon (Ria Formosa, Portugal). Sci. Total Environ. **780**, 146311 (2021). https://doi.org/10.1016/j.scitotenv.2021.146311
18. Cravo, A., et al.: Unravelling the effects of treated wastewater discharges on the water quality in a coastal lagoon system (Ria Formosa, South Portugal): relevance of hydrodynamic conditions. Mar. Pollut. Bull. **174**, 113296 (2022). https://doi.org/10.1016/j.marpolbul.2021.113296
19. JCR: European Commission – Smart Specialization Platform. A living lab for the sustainable energy transition ambition of the Algarve Region's S3 (2020). https://s3platform.jrc.ec.europa.eu/en/web/guest/w/a-living-lab-for-the-sustainable-energy-transition-ambition-of-the-algarve-region-s-s3
20. ERSAR: Water and Waste Services Regulation Authority. Annual Report of Water and Wast Services in Portugal. ISBN: 978-989-8360-41-0 (2021)
21. Jeuken, Y.R.H.: Nature4Cities-Nature Based Solutions Project Implementation Handbook [Review of Nature4Cities-Nature Based Solutions Project Implementation Handbook, by Breukers, S., Heijnen, V.M.D., Decker, S., Elie, M., Rugani, B., Tieman, M., Egusquiza Ortega, A., Arana Bollar, M. 57 (2020)
22. Wong, T., Rogers, B., Brown, R.: Transforming cities through water-sensitive principles and practices. One Earth **3**(4), 436–447 (2020). https://doi.org/10.1016/j.oneear.2020.09.012
23. Serrao-Neumann, S., Renouf, M., Morgan, E., Kenway, S., Low Choy, D.: Urban water metabolism information for planning water sensitive city-regions. Land Use Policy **88**, 104144 (2019). https://doi.org/10.1016/j.landusepol.2019.104144

24. Neto, S., Moreira, M., da Silva, L., Muller, K Weller: Social and Technological innovation for water conservation – the project ECH2O-ÁGUA. In: Monteiro, J., et al. (eds.) INCREaSE 2019: Proceedings of the 2nd International Congress on Engineering and Sustainability in the XXI Century, pp. 799–811. Springer International Publishing, Cham (2020). https://doi.org/10.1007/978-3-030-30938-1_62
25. Venckute, M., Silva, M., Figueiredo, M.: Education as a tool to reduce the water footprint of young people. Millenium – J. Educ. Technol. Health **2**, 101–111 (2017). https://doi.org/10.29352/mill0204.09.00144
26. DGEG, General Directorate of Energy and Geology: https://www.dgeg.gov.pt/pt/areas-setoriais/energia/energia-eletrica/ (2022)
27. Baird, B., Rice, C., Eaton, A.: Standard Methods for the Examination of Water and Wastewater. In: 23rd. Water Environment Federation, American Public Health Association, American Water Works Association (2017)
28. Ayers, R., Westcot, D.: Water quality for agriculture. Food and Agriculture Organization of the United Nations, Rome (1985). ISBN 92-5-102263-1
29. ANQIP: Technical Specification ETA 0701—Rainwater Harvesting Systems in Buildings (Version 11). ANQIP, Aveiro, Portugal (2022). (In Portuguese)
30. European Committee for Standardization: European Standard EN 16941-1—On-Site Non-Potable Water Systems—Part 1: Systems for the Use of Rainwater. CEN, Brussels, Belgium (2018)
31. Portal do clima. Climate Change Portal. http://portaldoclima.pt/en/. Accessed 4 July 2022

Climate Change, Circular Economy, and Risk Mitigation

ERA5-Land Reanalysis Temperature Data Addressing Heatwaves in Portugal

Luis Angel Espinosa[1](✉), Maria Manuela Portela[2], and José Pedro Matos[2]

[1] The Association of Instituto Superior Técnico for Research and Development (IST-ID), Civil Engineering Research and Innovation for Sustainability (CERIS), Lisbon University (UL), Lisbon, Portugal
luis.espinosa@tecnico.ulisboa.pt

[2] Instituto Superior Técnico (IST), Civil Engineering Research and Innovation for Sustainability (CERIS), Lisbon University (UL), Lisbon, Portugal
{maria.manuela.portela,jose.matos}@tecnico.ulisboa.pt

Abstract. In this research, heatwaves in Portugal were analysed using high-resolution daily minimum and maximum temperature data (Tmin and Tmax) from the European Reanalysis of Global Climate Observations (ERA5-Land) for the period from 1 October 1980 to 30 September 2021 (41 hydrological years) at four differentiated climatological locations in the country. The ERA5-Land temperature data were validated at the monthly level against ground-based observations from the Portuguese Institute for Sea and Atmosphere (IPMA), finding good agreement between the two data sets. Heatwaves were defined using the heatwave magnitude index (HWMI), which identifies a heatwave as a period of three or more consecutive days where temperatures exceed the daily threshold defined for the reference period. Additionally to an increasing trend in Tmin magnitude, more than 650 heatwave days of Tmin were identified at each of the four ERA5-Land locations, with the grid-point centred in the capital urban area, i.e., Lisbon, having the highest number of heatwave days. Regarding the heatwave days of Tmax, the locations with the highest occurrences, each with more than 830 d in the 41-year period, were in the north and interior. Both Tmin and Tmax heatwave occurrences were coupled with a kernel rate estimation technique for their annual frequency analysis. Overall, the results showed a clear increase in the frequency of heatwave days in Portugal, particularly for Tmax in the last two decades. This also evidenced geographical variations in the phenomenon's occurrence, with the southern location experiencing a higher increase in heatwave days of Tmax than the northern and interior grid-points.

Keywords: Reanalysis Data · Temperature · Heatwaves · Portugal

1 Introduction

Heatwaves are extreme weather events that can have significant impacts on human health, ecosystems, and infrastructure [10]. In recent years, the frequency

J. F. L. C. Semião et al. (Eds.): INCREaSE 2023, ASST, pp. 81–94, 2023.
https://doi.org/10.1007/978-3-031-44006-9_7

and severity of heatwaves have been increasing in many parts of the world due to climate change [20]. Portugal is no exception [16], as it has experienced several severe heatwaves in the past decade, including the historic heatwave of August 2018, which was one of the most intense and prolonged heatwaves on record in the country. The impacts of heatwaves in Portugal have been diverse and wide-ranging. In addition to causing human health problems such as heat exhaustion, dehydration, and respiratory issues [1,3], heatwaves have also led to increased risk of forest fires, reduced crop yields, and disruptions to energy and water supplies. The vulnerability of certain population groups, such as the elderly and those living in urban areas with poor air quality, further exacerbates the impacts of heatwaves.

Given the growing concerns over the impacts of heatwaves in Portugal, there is an urgent need to better understand the underlying causes and dynamics of those events. In addition, climate reanalysis datasets (CRDs) are increasingly used as surrogates for historical meteorology — the latter often presenting temporal gaps and horizontal spatial gaps [8] — but their use in temperature models or heatwave analysis is still relatively new and untested. Evaluation of temperature models using CRDs, as in the case of the European Reanalysis of Global Climate Observations (ERA5-Land) [9], in regions with heterogeneous climate and topographic conditions, such as mainland Portugal, is necessary to validate the applicability of these datasets.

Motivated by the aforementioned, this research aims to contribute to the knowledge of climate variability in Portugal by providing a preliminary analysis of recent heatwaves via a novel application of ERA5-Land daily minimum and maximum temperature data, Tmin and Tmax. The validation of temperature data from the selected ERA5-Land grid-points against temperature records from the nearest ground-based meteorological stations is a crucial part of this research.

2 Study Area

Mainland Portugal, with a total area of approximately ninety thousand square kilometres, was selected as the study case in this research. Portugal is a south-western European country located in the Iberian Peninsula, bordering Spain to the east and north and the Atlantic Ocean to the west and south (Fig. 1). The country has a diverse topography that includes mountains, plateaus, and coastal plains. The climate of Portugal is characterised by mild winters and hot summers, with significant regional variations due to the influence of oceanic, continental, and Mediterranean weather patterns [18].

The climate of Portugal is influenced by several factors, including its location on the Atlantic coast, its rugged topography, and its proximity to North Africa. The country's climate is generally classified as Mediterranean, with hot, dry summers and mild, wet winters. However, the climate in the northern region is more temperate, with cooler temperatures and more than in the southern region. Alentejo and Algarve, in the south, have a semi-arid climate, with low rainfall and high temperatures in the summer.

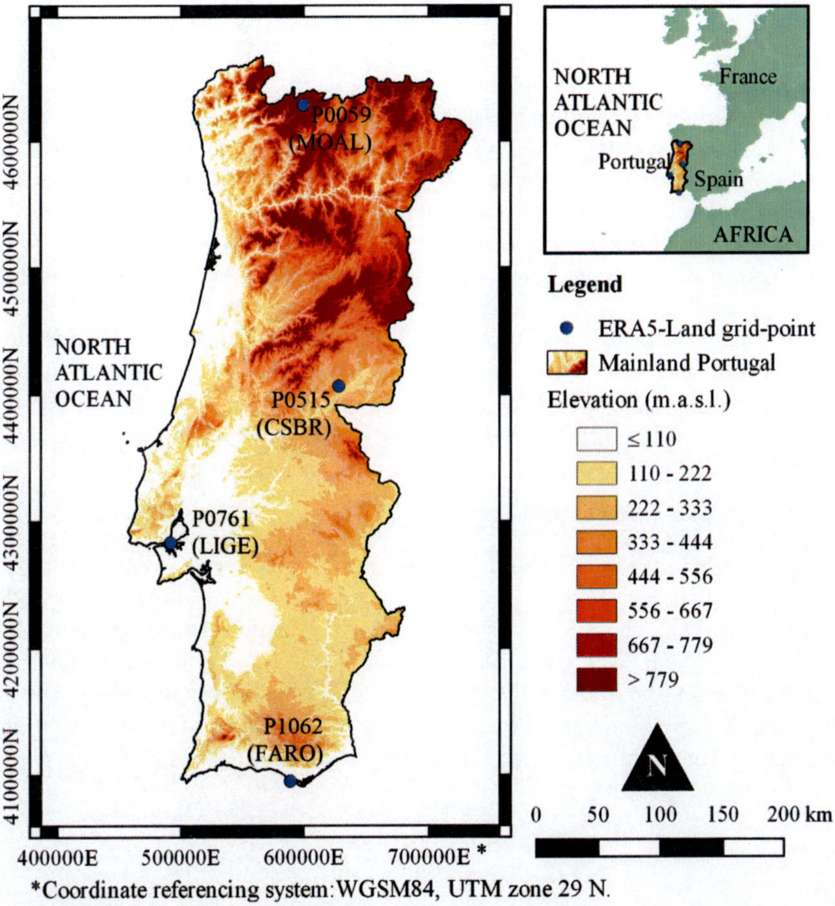

Fig. 1. Mainland Portugal elevation map in metres above sea level (m.a.s.l). Location of the four ERA5-Land grid-points used for the heatwave analysis. For each grid-point, the nearest ground meteorological station is identified between brackets (distance ranging from 2.3 to 4.6 km).

3 Material and Methods

In this research, heatwaves in Portugal were analysed using high-resolution daily temperature data — aggregated from the ERA5-Land hourly data. The heatwave analysis covered a 41-year period from 1 October 1980 to 30 September 2021. The new ERA5-Land reanalysis was created by forcing the land component of the ERA5 [4]. It is provided by the Copernicus Climate Change Service with the same temporal resolution as ERA5 (hourly resolution) but with a higher spatial resolution of 0.1 × 0.1 degrees, allowing estimates of various meteorological variables, such as temperature and rainfall [9]. Daily minimum and maximum temperature data (Tmin and Tmax) from ERA5-Land were fetched at four representative climatological locations in Portugal, namely, the grid-points P0059, P0515, P0761, and P1062 (Fig. 1).

The accuracy of the ERA5-Land temperature data at the previous grid-points was verified by comparing that dataset to the long-standing records from the Portuguese Institute for Sea and Atmosphere (IPMA), which is the national meteorological institute in Portugal. The IPMA records were retrieved from the closest stations to the four grid-points (distance to the corresponding grid-point between brackets): Montalegre, MOAL (2.3 km), Castelo Branco, CSBR (4.6 km), Lisboa Geofísico, LIGE (2.6 km), and Faro, FARO (3.4 km) — shown in Fig. 1. However, the available IPMA temperature records only go up to 2018, and just one station has daily records. Thus, the accuracy of the ERA5-Land data was addressed at monthly level via a comparison of the two datasets based on the Pearson's linear correlation coefficient, r, box plot graphs, and centroids of long-term monthly average line plots for a 39-year common period.

3.1 Heatwave Magnitude Index

Heatwaves in the daily ERA5-Land Tmin and Tmax series from 1980 to 2021 were identified using the heatwave magnitude index (HWMI) as described by Russo *et al.* [13]. The HWMI is defined as the maximum magnitude of the heatwaves in a year, where a heatwave is the period ≥ 3 consecutive days with Tmin (or Tmax) above the daily threshold. The threshold is defined as the 90th percentile of daily Tmin (or Tmax), centred on a 31-day window. Hence, for a given day d, the threshold series A_d of the dataset is given by:

$$A_d = \bigcup_{y=1980}^{2021} \bigcup_{i=d-15}^{d+15} T_{y,i}, \qquad (1)$$

where $\bigcup$ denotes the union of sets and $T_{y,i}$ is daily temperature of the day i in the year y. To obtain A_d, the daily temperature ERA5-Land data in the 15-day period before the 1 October 1980 and after the 30 September 2021 were used.

3.2 Frequency Analysis of Heatwave Days

The analysis of the changes in the annual frequency of heatwaves was addressed by applying a kernel rate estimation technique according to the stepwise approach applied by Silva *et al* [14] and Silva [15]. This included the generation of pseudodata outside of the observation interval, by the straightforward method of reflection for an amplitude of three times the bandwidth. The time data considered in the kernel technique were the dates of occurrence of the heatwave days in each of the Tmin and Tmax series.

4 Results

The ERA5-Land reanalysis provides land variables at an hourly scale, enabling the evaluation of daily climatological anomalies — as carried out in this research regarding temperature. Before addressing heatwaves in Portugal, different statistical metrics were applied at the monthly level to the ERA5-Land reanalysis data and IPMA records to evaluate the accuracy of the former in the common longest period of 39 years from January 1980 to December 2018.

4.1 Comparison Between ERA5-Land and Meteorological Stations

The adjustment between monthly temperatures, given by the ERA5 Land at the considered four grid-point and measured by IPMA at the nearest meteorological stations, was firstly assessed via the Pearson's correlation coefficient (r). In all cases, the r for Tmin and Tmax was higher than 0.95. Overall, these strong linear relationships indicate a good agreement between the reanalysis data and the IPMA temperature records over the period of 39 years. However, it is worth mentioning some specific aspects related to Tmax. The IPMA dataset recorded higher maximum temperatures in all cases, such as MOAL with 29.0°C against 26.6°C in P0059 and FARO with 31.2°C in contrast to 28.4°C in P1062. The differences between the IPMA records and ERA5 Land data are due to their temperature retrieval methods. The IPMA records refer to instantaneous maximum and minimum temperatures for each day, while ERA5-Land data to the highest and lowest temperatures over a one-hour period, resulting in smoothed values. This difference is more significant for Tmax than for Tmin, as the variation of Tmin within one-hour periods is less pronounced than that of Tmax, making Tmin closer to the minimum instantaneous temperature [5]. From Fig. 2, some of these remarks can be identified by visual inspection.

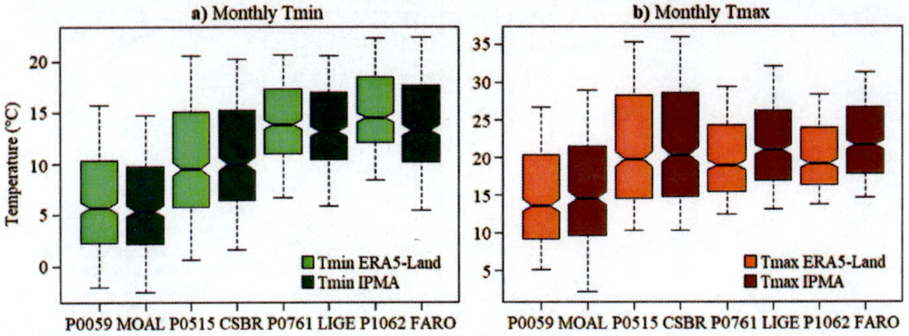

Fig. 2. Box plot graphs for monthly **a**) Tmin and **b**) Tmax from 1980 to 2018 at the coupled ERA5-Land grid-points and IPMA meteorological stations: P0059–MOAL (Montalegre), P0515–CSBR (Castelo Branco), P0761–LIGE (Lisboa Geofísico), and P1062–FARO (Faro). The box represents the interquartile range (IQR), the notch the median, and the whiskers are the maximum and minimal non-remote values.

According to the Tmin box plot graphs depicted in Fig. 2**a**, the lowest temperatures were typically recorded in the north and became progressively higher as one moved towards the south. As a whole, ERA5-Land Tmin was slightly higher, with medians higher than those from the IPMA records, except for the P0515 data compared to CSBR. In contrast, for the Tmax IPMA records, the interquartile range (IQR) is always larger and higher than the ERA5-Land data (Fig. 2**b**). These box plot graphs show the underestimate of the reanalysed Tmax which is more pronounced in grid-points located next to the sea, as in the case of P0761 and P1062 close to Lisboa Geofísico and Faro, respectively. Additionally, the graphs constructed with the Tmax from IPMA show greater variability with longer whiskers.

The long-term average line plots in Figs. 3 and 4 provide additional evidence to support the previous box plots findings. To obtain the line plots, the 39 monthly values in each of the time series were averaged. Furthermore, a centroid was estimated in both, vertical and horizontal directions, to summarise the temporal and temperature value alignments of the two datasets. It is worth noting that although the heatwave analysis adopted hydrological years, Figs. 3 and 4 were constructed using calendar years. The lines in Fig. 3 indicate that

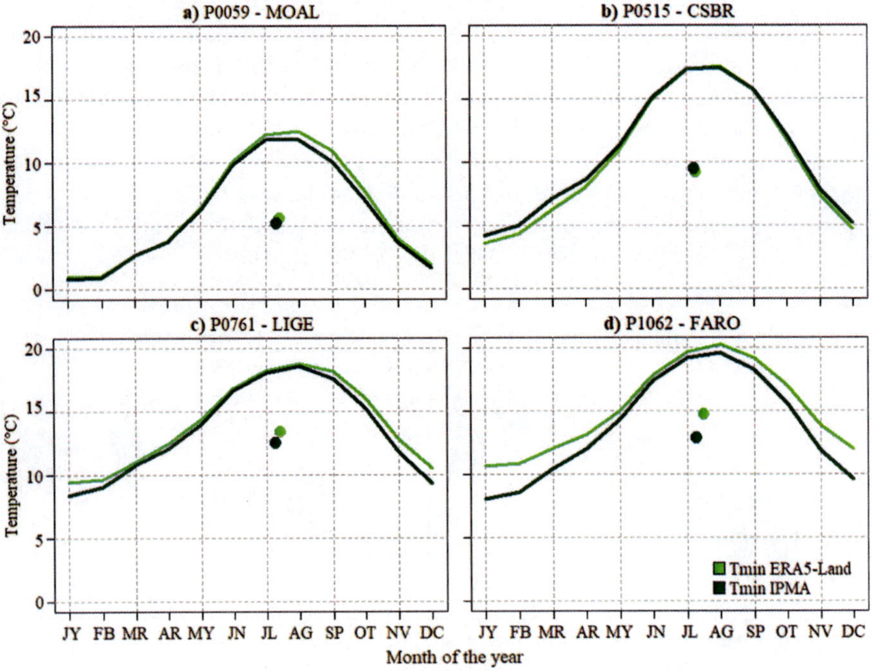

Fig. 3. For each coupled ERA5-Land grid-point and nearest IPMA meteorological station, line plots of the long-term mean monthly Tmin (continuous lines) and their centroids (circles) in the period from January 1980 to December 2018 for: **a**) P0059–MOAL, **b**) P0515–CSBR, **c**) P0761–LIGE, and **d**) P1062–FARO.

the long-term pattern of the mean monthly Tmin is maintained regardless of the dataset, with the lowest Tmin occurring in January and the highest in August. However, in Fig. 3**d**, the mean monthly Tmin at P1062 (ERA5-Land) is approximately 2.0°C higher than FARO (IPMA) during the winter (from October of one year to March of the next). The centroids for Tmin are located between July and August, with some time lag in P0761 and P1062. Regarding the mean monthly Tmax (Fig. 4), some discrepancies among the baseline data exist, except for the paired P0515—CSBR line plots with almost identical development (4**b**). The IPMA records displayed higher temperature than the ERA5-Land dataset, particularly in LIGE and FARO, as illustrated in Fig. 4**c** and **d**, respectively — being the IPMA records in average ca. 2.4°C higher than their corresponding grid-point values during the dry period. Overall, the centroids for the coupled time series of Tmin and Tmax indicated that, despite temperature discrepancies (vertical axis), the monthly ERA5-Land data and IPMA records showed good temporal pattern agreement (horizontal axis) which can be translated into a close temporal association.

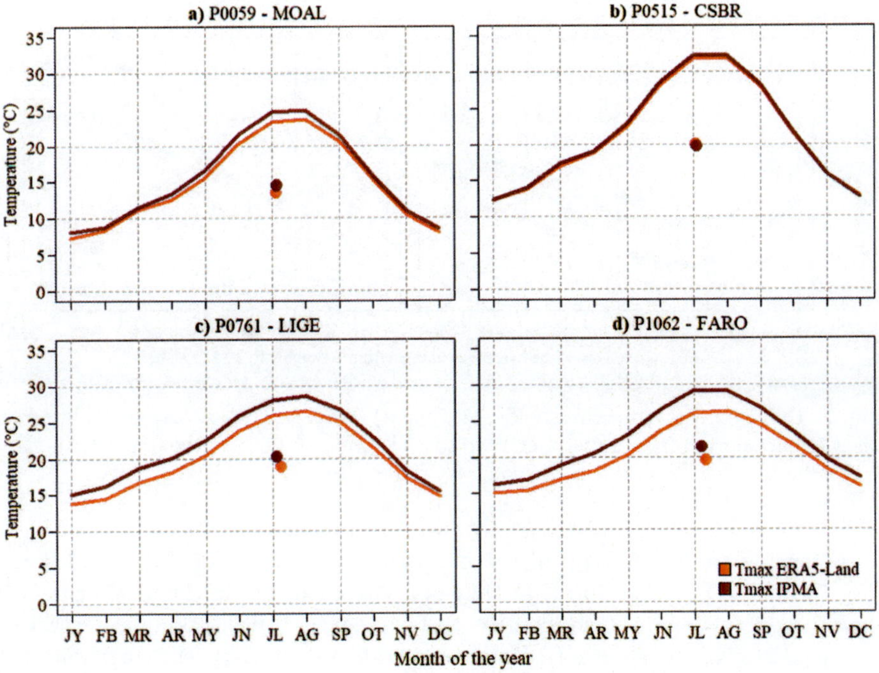

Fig. 4. For each coupled ERA5-Land grid-point and nearest IPMA meteorological station, line plots of the long-term mean monthly Tmax (continuous lines) and their centroids (circles) in the period from January 1980 to December 2018 for: **a**) P0059–MOAL, **b**) P0515–CSBR, **c**) P0761–LIGE, and **d**) P1062–FARO.

4.2 Threshold for Heatwave Identification in Tmin and Tmax

After the validation of the ERA5-Land dataset, the study addressed heatwaves via the heatwave magnitude index (HWMI) applied to that daily temperature dataset at each of the four grid-point locations (P0059, P0515, P0761, and P1062), Eq. 1 was used to determine the daily temperature thresholds for the reference period from 1 October 1980 to 30 September 2021. Figures 5**a** and **b** display the daily temperature thresholds, A_d, for Tmin and Tmax, respectively. The results indicate that P0059, located in the interior and northernmost grid-point, had the lowest threshold for both Tmin and Tmax. P0515 had a higher A_d, particularly for Tmax. During the dry period (approximately from day 180 to day 365), P0515 had the strictest threshold for Tmax out of all the grid-points. In Fig. 5**a**, the thresholds for Tmin at P0761 and P1062, located near the Lisboa Geofísico and Faro meteorological stations, respectively, exhibited remarkable similarity. This pattern was replicated for Tmax (Fig. 5**b**), with the exception that, in the case of P0761, A_d was slightly higher in the second half of the hydrological year.

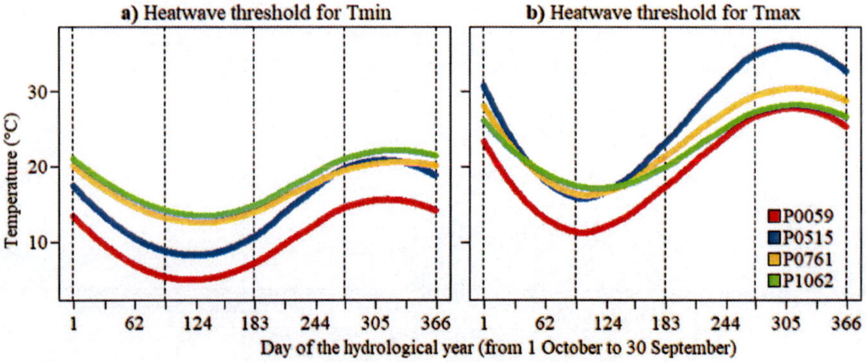

Fig. 5. Daily temperature thresholds, A_d, for **a**) Tmin, and **b**) Tmax, calculated for the reference period from 1 October 1980 to 30 September 2021 at the four ERA5-Land grid-points.

Following the calculation of the daily thresholds (A_d) for Tmin and Tmax (Fig. 5) and keeping in mind that they apply to each year of the 41-year period, the subsequent step involved identifying days that met the criteria for heatwaves. This consisted of two stages: (i) Daily temperatures exceeding the threshold were classified as hot days, and (ii) periods with three or more consecutive hot days were labelled as heatwaves, with dates of occurrence from the third day on. Out of a total of 14,965 d (excluding leap years) in the 41-year period, around 1,350 hot days (ca. 9%) of both Tmin and Tmax were identified at the four grid-point series. However, the number of heatwave days was relatively lower. Specifically, 695 occurrences of Tmin heatwave days were identified at P0059, 669 at P0515, 716 at P0761, and 642 at P1062. In the Tmax series, the total number of occurrences was higher, with 837, 838, 763, and 644 heatwave days at P0059, P0515, P0761, and P1062, respectively. After identifying the heatwave days that occurred, the subsequent task involved estimating the changes in their frequency over time.

4.3 Frequency Analysis of the Tmin and Tmax Heatwave Days

The results from the kernel rate estimation technique depicted in Figs. 6 and 7 demonstrate that the frequency of heatwave occurrences has undergone temporal changes over the 41-year period analysed. In the figures, the vertical black ticks above the horizontal axis indicate the dates of occurrence of three or more consecutive days with Tmin or Tmax exceeding the threshold A_d.

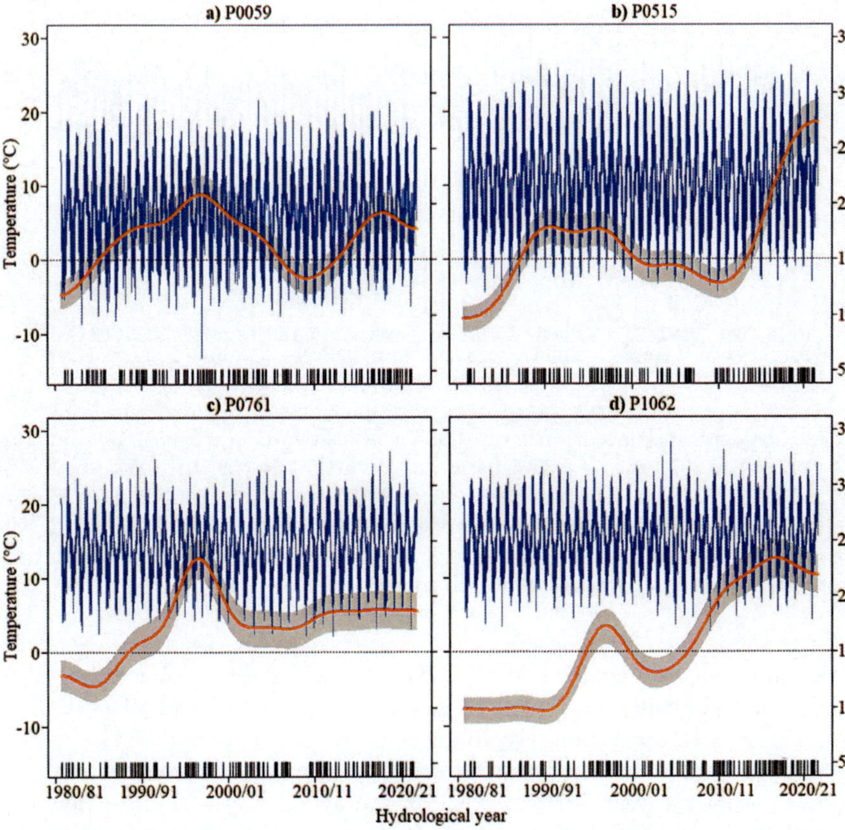

Fig. 6. Time-dependent occurrence rates of Tmin heatwave days from 1 October 1980 to 30 September 2021 at the ERA5-Land grid-points, **a**) P0059, **b**) P0515, **c**) P0761, and **d**) P1062. The left y-axis displays the ERA5-Land daily Tmin values in blue, while the right y-axis shows the number of heatwave days per year, $\lambda(t)$, in orange, with the confidence band represented by the grey area surrounding the latter. The dates of occurrence of heatwave days are indicated by vertical ticks.

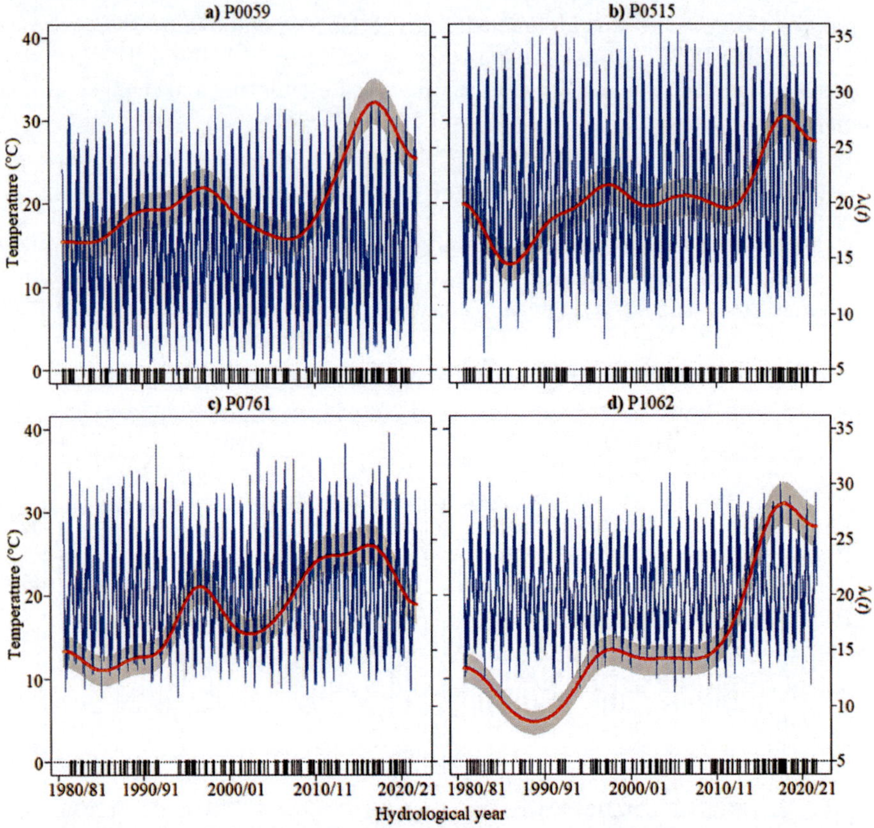

Fig. 7. Time-dependent occurrence rates of Tmax heatwave days from 1 October 1980 to 30 September 2021 at the ERA5-Land grid-points, **a**) P0059, **b**) P0515, **c**) P0761, and **d**) P1062. The left y-axis displays the ERA5-Land daily Tmax values in blue, while the right y-axis shows the number of heatwave days per year, $\lambda(t)$, in red, with the confidence band represented by the grey area surrounding the latter. The dates of occurrence of heatwave days are indicated by vertical ticks.

According to the values of $\lambda(t)$ for Tmin, the highest number of instances occurred at the grid-point P0059 (Fig. 6**a**) in the hydrological years 1998/99 and 2018/19, with approximately 20 occurrences per year. At P0515 (Fig. 6**b**), the occurrence rates remained stable from the beginning of the reference period until the end of the year 2010/11 with an average of $\lambda(t) = 10$ heatwave days per year. After this initial 30-year period, there was a sudden increase in $\lambda(t)$, which attained approximately 28 heatwave days per year in 2020/21. At the grid-point centred on Lisbon (P0761, Fig. 6**c**), a positive peak in $\lambda(t)$ was reached in 1998/99 with a value of $\lambda(t) = 24$ heatwave days per year. Subsequently, $\lambda(t)$ decreased without any evident perturbations in the following years. At P1062, the value of $\lambda(t)$ was relatively low during the first few years, with less than

9 heatwave days per year. However, in the most recent decade, this value has increased drastically, more than doubling in frequency.

In what concerns Tmax, the analysis of ERA5-Land temperature data via the coupled HWMI and kernel rate estimator revealed a significant increase in the frequency of heatwave days. In general, the results shown in Fig. 7 indicate a steady increase in the number of heatwave days over time, with an apparent positive trend observed in the most recent decade — with the steepest $\lambda(t)$ increment at the southern grid-point P1062. Yet, the four $\lambda(t)$ series in the figure exhibit distinct patterns of development, providing evidence of spatiotemporal variability in the occurrence of Tmax heatwaves across Portugal. The maximum $\lambda(t)$ was 29, 28, 24, and 28, at P0059, P0515, P0761, and P1062, respectively. Notably, at all four grid-points, the peak of $\lambda(t)$ occurred between 2017/2018 and 2018/2019, coinciding with the historic heatwave that struck Portugal in August 2018 [16]. This finding highlights the significant impact of this extreme weather event on the frequency of heatwaves in the ERA5-Land locations.

The comparison of the Tmin and Tmax heatwave day plots (Fig. 6 and Fig. 7, respectively) revealed the temporal variation of heatwaves at different levels, which was detected in both variables using ERA5-Land temperature data. Nonetheless, it is worth noting that the ERA5-Land Tmax values were underestimated compared to most of the meteorological station records, indicating that the actual values of $\lambda(t)$ for Tmax may be higher than the obtained estimates.

5 Discussion

The results presented in this study show good agreement between the ERA5-Land reanalysis data and IPMA meteorological station records for monthly Tmin and Tmax over the 39-year period from January 1980 to December 2018. The strong linear relationships found between these datasets indicate that ERA5-Land is a reliable data source for climate studies, namely for temperature analyses. This good agreement between reanalysed data and records has recently been addressed by different authors in other regions with different climatological conditions, such as in the Southern Antarctic Peninsula [17], in the Peruvian Andes glaciers [2], and in the Turkish part of the Anatolian Peninsula, and partly in the Balkan Peninsula [19]. However, some specific aspects related to Tmax should be considered. The differences observed between the IPMA records and ERA5-Land data for Tmax can be attributed to their temperature retrieval methods. While IPMA records instantaneous maximum and minimum temperatures for each day, ERA5-Land provides the highest and lowest temperatures over a one-hour period, resulting in smoothed values. In the present application, it has been observed that ERA5-Land exhibits a tendency to underestimate Tmax in comparison to IPMA records. This discrepancy is particularly noticeable at grid-points situated in close proximity to the sea, where the impact of spatiotemporal smoothing is more prominent.

The box plot graphs and long-term average line plots of the monthly ERA5-Land data and IPMA records provide additional evidence to support these findings. The box plot graphs show that the interquartile range for Tmax in IPMA records is always larger and higher than the ERA5-Land data. The analysis of the long-term average line plots suggests that the patterns of the mean monthly values of Tmin remain consistent regardless of the dataset. Furthermore, there is strong concurrence between the centroids of the long-term monthly Tmin and Tmax values obtained from both datasets.

Research on heatwaves is a newly emerging research topic within the field of climate change research with high relevance for the whole of society [7] including the Portuguese territory. Thus, the study investigated the occurrence and characteristics of heatwaves in mainland Portugal using ERA5-Land reanalysis data to overcome the lack of publicly available daily temperature data. Heatwaves were defined as periods of at least three consecutive days with daily temperatures (Tmin and Tmax) exceeding the 90th percentile of the 1980–2018 climatology. The study of heatwaves was conducted using daily data from ERA5-Land from 1 October 1980 to 30 September 2021. The dataset used in the study enabled the analysis of the frequency, intensity, and duration of heatwaves over a period of 41 years, providing valuable insights into how heatwaves have changed over time. Specifically, the annual number of heatwave days, considering both Tmin and Tmax, increased at the four grid-points from an average of 5 d in the 1980s to 12 d in the 2010s.

The study of the spatiotemporal characteristics of heatwaves in Portugal revealed that such extreme events occur with greater frequency and duration in the southern and interior regions of the country. Furthermore, certain coastal areas have also witnessed a sudden increase in the occurrence of heatwaves in recent times. Overall, these results suggest that the ERA5-Land reanalysis data can be used with confidence for temperature analyses in Portugal, despite some underestimates may occur in Tmax compared to IPMA records. However, researchers should be cautious when using the data for extreme temperature events, such as heatwaves, as the underestimation of Tmax by ERA5-Land may lead to an underrated severity of these events. Such impairment for reanalysis Tmax data has also been addressed in other applications, such as in a drought analysis in an Ethiopian river basin [11]. However, in a study for drought and water budget analysis in India, Tmax from reanalysis products outperformed, and Tmin was the underestimated variable [6].

Future studies could (i) investigate the impact of the dataset differences on the analysis of heatwaves in Portugal, (ii) explore alternative high-resolution climatological data other than ERA5-Land, and (iii) look into the physical mechanisms causing heatwaves in the country — such as done by Rousi *et al.* [12] concluding that the upward trend in the persistence of double jet events and atmospheric dynamical changes explains almost all of the accelerated recent heatwave trend in western Europe.

6 Concluding Remarks

In conclusion, the study provides valuable insights into the characteristics and impacts of heatwaves in Portugal, which can inform the development of effective adaptation and mitigation strategies to reduce the negative impacts of such extreme events on human health and the environment. The findings suggest that heatwaves are becoming more frequent in Portugal and that certain regions are more vulnerable to their impacts. The study also highlights the need for further research to better understand the drivers and mechanisms of heatwaves in Portugal, as well as their interactions with other climate variables and systems.

Acknowledgements. This research was supported by the Foundation for Science and Technology (FCT) through funding UIDB/04625/2020 from the research unit CERIS and by the European Union's Horizon 2020 research and innovation programme SCORE under grant agreement No 101003534.

References

1. Arbuthnott, K.G., Hajat, S.: The health effects of hotter summers and heat waves in the population of the United Kingdom: a review of the evidence. Environ. Health **16**, 1–13 (2017)
2. Bonshoms, M., Ubeda, J., Liguori, G., Körner, P., Navarro, Á., Cruz, R.: Validation of ERA5-land temperature and relative humidity on four Peruvian glaciers using on-glacier observations. J. Mt. Sci. **19**(7), 1849–1873 (2022)
3. Gasparrini, A., Armstrong, B.: The impact of heat waves on mortality. Epidemiol. (Cambridge, Mass.) **22**(1), 68 (2011)
4. Hersbach, H., et al.: The ERA5 global reanalysis. Q. J. R. Meteorol. Soc. **146**(730), 1999–2049 (2020)
5. Huband, N., King, S., Huxley, M., Butler, D.: The performance of a thermometer screen on an automatic weather station. Agric. For. Meteorol. **33**(2–3), 249–258 (1984)
6. Mahto, S.S., Mishra, V.: Does ERA-5 outperform other reanalysis products for hydrologic applications in India? J. Geophys. Res. Atmos. **124**(16), 9423–9441 (2019)
7. Marx, W., Haunschild, R., Bornmann, L.: Heat waves: a hot topic in climate change research. Theor. Appl. Climatol. **146**(1–2), 781–800 (2021)
8. Melton, F.S., et al.: OpenET: filling a critical data gap in water management for the western United States. JAWRA J. Am. Water Resour. Assoc. **58**(6), 971–994 (2022)
9. Muñoz-Sabater, J., et al.: ERA5-land: a state-of-the-art global reanalysis dataset for land applications. Earth Syst. Sci. Data **13**(9), 4349–4383 (2021)
10. Perkins, S.E.: A review on the scientific understanding of heatwaves-Their measurement, driving mechanisms, and changes at the global scale. Atmos. Res. **164**, 242–267 (2015)
11. Reda, K.W., Liu, X., Tang, Q., Gebremicael, T.G.: Evaluation of global gridded precipitation and temperature datasets against gauged observations over the Upper Tekeze River Basin, Ethiopia. J. Meteorol. Res. **35**, 673–689 (2021)
12. Rousi, E., Kornhuber, K., Beobide-Arsuaga, G., Luo, F., Coumou, D.: Accelerated western European heatwave trends linked to more-persistent double jets over Eurasia. Nat. Commun. **13**(1), 3851 (2022)
13. Russo, S., et al.: Magnitude of extreme heat waves in present climate and their projection in a warming world. J. Geophys. Res. Atmos. **119**(22), 12–500 (2014)

14. Silva, A.T., Portela, M., Naghettini, M.: Nonstationarities in the occurrence rates of flood events in Portuguese watersheds. Hydrol. Earth Syst. Sci. **16**(1), 241–254 (2012)
15. Silva, A.: Nonstationarity and Uncertainty of Extreme Hydrological Events. IST/UTL, Lisbon (PhD dissertation) (2017)
16. Sousa, P.M., Barriopedro, D., Ramos, A.M., García-Herrera, R., Espírito-Santo, F., Trigo, R.M.: Saharan air intrusions as a relevant mechanism for Iberian heatwaves: the record breaking events of August 2018 and June 2019. Weather Clim. Extremes **26**, 100224 (2019)
17. Tetzner, D., Thomas, E., Allen, C.: A validation of ERA5 reanalysis data in the Southern Antarctic Peninsula-Ellsworth land region, and its implications for ice core studies. Geosciences **9**(7), 289 (2019)
18. Tullot, I.F.: Climatología de España y Portugal, vol. 76. Universidad de Salamanca (2000)
19. Yilmaz, M.: Accuracy assessment of temperature trends from ERA5 and ERA5-land. Sci. Total Environ. **856**, 159182 (2023)
20. Zuo, J., Pullen, S., Palmer, J., Bennetts, H., Chileshe, N., Ma, T.: Impacts of heat waves and corresponding measures: a review. J. Clean. Prod. **92**, 1–12 (2015)

Climate Change Impact on a Green Building

Laura Almeida(✉), Keivan Bamdad, and Mohammad Reza Razavi

School of Engineering, Design and Built Environment, Western Sydney University, Penrith, Australia

{l.almeida,k.bamdad,R.Razavi}@westernsydney.edu.au

Abstract. The concept of a green building (GB) implies a building that has implemented several sustainable strategies to increase its resilience to climate change. These types of buildings are designed to be aligned with the climate and location where they stand. Therefore, the strategies used are highly dependent on the outdoor climate and may require changes in the future due to global warming. Evaluating the present energy performance in these buildings and predicting any changes in future climate scenarios is crucial to understand if the strategies adopted initially will maintain efficient. This research aims to predict the resilience of a GB to global warming. It predicts the impact of climate change and determines the thermal comfort and energy performance of a GB by comparing three future climate scenarios with present climate conditions. Hence, a university GB was simulated and calibrated with measured data. Then, to predict climate conditions in the 2050 s and 2090 s, several future weather files based on Representative Concentration Pathways (RCPs) were used. The results show a reduction in heating energy use and an increase in cooling, leading to a combined impact increase in greenhouse gas (GHG) emissions of 17.8% (RCP 8.5). The overall energy use tends to remain almost constant through the different scenarios with a maximum predicted increase of 4% in GHG emissions. Similarly, the electricity production from the 625 m^2 photovoltaic system in the GB slightly increases through the scenarios with an overall average of 2.17%. Finally, it is predicted that the case study GB maintains the required levels of thermal comfort over time given global warming.

Keywords: Climate change · Green buildings · Energy performance · Thermal comfort

1 Introduction

According to the Intergovernmental Panel on Climate Change (IPCC), global temperatures have risen by almost 1.1 °C since 1850 due to greenhouse gas (GHG) emissions caused by anthropogenic activities. This has led to climate change or global warming problem, which predicts that if no actions are taken, it is expected that temperature will exceed 1.5 °C within the next thirty years [1]. It is, therefore, expected in the future to see an increase in the frequency and intensity of extreme weather events such as heatwaves [1]. The impacts of climate change can be mitigated globally with the implementation of low-emissions solutions appropriate for energy-intensive sectors, being the building sector one of these examples.

J. F. L. C. Semião et al. (Eds.): INCREaSE 2023, ASST, pp. 95–108, 2023.
https://doi.org/10.1007/978-3-031-44006-9_8

Over 30% of black carbon emissions and 32% of the global final energy use come from buildings [2]. Additionally, taking into consideration the rapid urbanization, growth in population, and human increased access to housing and electricity, in the future, it is expected an increase in the energy demand in the building sector. However, this sector has the potential to help mitigate climate change and reduce GHG emissions by implementing cost-effective and efficient retrofitting and construction measures [2]. Therefore, green buildings may be one of the potential candidates to play a significant role in climate change mitigation, in the following year.

Green Buildings Background: Aiming to reduce greenhouse gas (GHG) emissions and resource consumption, the GB concept appeared for the first in 1990 in the United Kingdom under the Building Research Establishment's Environmental Assessment Method (BREEAM) tool [3]. A GB takes a holistic approach towards sustainable design and is a living system that translates into reality the sustainability principles defined in 2015 under the 17 Sustainable Development Goals (SDG) [4, 5]. There is a focus on planning, design, construction, operations, and buildings' end of life. These buildings aim to create environmental, social, and economic communities. GB goals comprise, among others, the minimization of environmental impacts by reduction of any type of pollution and mitigating global warming by reducing GHG emissions through the implementation of energy conservation measures. Therefore, these buildings will be more resilient to climate change [6, 7].

Green Buildings Performance: According to the International Energy Agency (IEA) and GBCA, is expected an increase in energy demand due to an extra 1.1 million homes and 400 million non-residential square meters being built by 2050. GBs consume 66% less energy and emit 50 to 55% less GHG emissions when compared to conventional buildings [8, 9]. Improving and promoting thermally comfortable and healthier buildings for occupants are other characteristics of GBs. Aligned with the main guidelines from the standard 55 ASHRAE [10], GBs incorporate under the IEQ criteria thermal comfort requirements for airflow, humidity, temperature, and type of activities performed, among other physical factors [7, 11]. These are carefully assessed during the design stage and throughout the whole lifecycle of the building. As a result, occupants' levels of satisfaction tend to be higher in GBs when compared to conventional ones [5, 12]. Therefore, when reviewing studies concerning energy use, IEQ, and occupant satisfaction in GBs, Geng, Ji [13] concluded that, on average, GBs perform better than conventional buildings in the same region. However, their actual energy consumption varies significantly. One of the reasons GBs have better performance and satisfaction levels has to do with an integrated process approach that aims to promote communication and collaboration among professionals and stakeholders throughout the whole life of a building [14]. These buildings are carefully designed, integrating an approach of 'job-sharing' where an in-depth understanding of construction practices and green strategies are shared within interdisciplinary teams to create an optimal version of a building [15, 16]. Throughout the whole process, there is an intensive interdisciplinary collaboration from the earliest stages of the building, which becomes a crucial part of the decision-making process of a GB. Therefore, besides reducing operational energy and costs, GBs are more resilient and reduce the impact of climate change [8].

Green Buildings and the impact of climate change: Building energy use and thermal comfort conditions have been shown to be significantly affected by global warming in previous studies [17–20]. For example, Haji Ismail et al. [18], based on a simulation of a prefabricated building in the UK, estimated that heating energy use in 2030 and 2080 will be 12% and 34% lower than in 2017. According to Bamdad et al. [19], cooling energy demand in office buildings will increase by 24% in Darwin and 56% in Melbourne by 2080. The effects of global warming on the energy performance of non-green buildings have been extensively studied in different climate zones [20, 21]. The full impacts on GBs are yet to be explored. GBs comprise a wide range of active and passive energy-saving strategies that intend to meet comfort levels and energy requirements [22]. The requirements for these strategies may change over time because several are dependent on outdoor climate conditions. Hence, under present climate conditions, an energy effective measure may not retain its efficiency in the future due to climate change [23], which may impact significantly the overall energy performance of a GB.

Case study: Evaluating the energy performance of GBs under present and future climate conditions is crucial to estimate if the measures implemented during the design stage will be able to face the impacts due to climate change. This research aims to evaluate the thermal comfort and energy performance of a green-rated university building (certified by the Australian Green Star system) and predict its energy performance under future scenarios representing the impacts of climate change to evaluate if in the future a GB remains green. The study will be divided by the methodology in the next section, followed by the presentation of the results and, finally, the conclusion.

2 Methodology

This section comprises the present case study, the calibration process using energy-measured data and the weather files used in the simulation model to predict future climate conditions.

2.1 Simulation Model

The building used for this case study is a university Green Building from 2016, certified according to the Australian Green Star rating system with six stars [24]. The building is intended to pertain to a representation of typical Australian construction. The building has installed several 'green' measures, which were enablers to obtain the green classification. There is a hydronic heating and cooling system that supplies the common areas, a photovoltaic system to produce electricity and several other efficient measures for energy and water, waste management and transportation measures. For example, during operation, the building aims for 98% recycling of the total waste produced and has cyclist facilities to promote alternative means of transportation.

Building simulations were used to determine the energy performance indicator for the building, with an annual temporal resolution. The main variables considered as input in the model were the rates for occupancy, type of fabric, rates for energy intensity, type of activity, technical systems and schedules.

Assumptions of the Simulation

The building is a 3-storey type of building, with a total area of 5696 m^2 and 20,563 m^3 of volume. The conditioned area corresponds to 5,181 m^2 (18,868 m^3), and offices, computer and classrooms, laboratories and corridors are the main activities performed in the building. Motion sensors, LED lights, efficient air conditioning equipment, permanent external shading devices and a photovoltaic system with 374 modules (99.1 kWp) that produces 134 MWh per annum are some of the measures installed to obtain the green classification.

Figure 1 shows the building and simulated model. The building fabric comprised U values ranging from 0.16 to 0.25 W/m^2-K, 0.35 to 1.78 W/m^2-K, and 3.50 to 5.40 W/m^2-K for the roofing, walling, and glazing systems, respectively. The windows/skylights were generically double glazed with a solar heat gain coefficient (SHGC) of 0.43. Most glazed areas are permanently shaded externally and provided with dark-colored drapes internally in some rooms. The air conditioning system consisted of a natural gas condensation boiler (90% efficiency) for the heating function and a chiller (EER = 5.5) for cooling.

Fig. 1. The actual building and simulated models.

2.2 Simulation

The building operated from 6:30 am to 8 pm with an indoor climate control strategy of fully air-conditioned. The air conditioning system set-point temperatures were 22 °C for heating and 23 °C for cooling. Except when occupants could locally control their set points, these would vary from 18 °C to 24 °C. The management system centrally managed the air conditioning system and operated from 8 am to 6 pm. Several air handling units (AHU) control indoor humidity levels and provide fresh air. To ensure compliance with the Green Star indoor air quality requirements, the AHUs are equipped with volatile organic compounds, carbon dioxide (CO_2) and motion sensors [8]. This ensures that they only operate when the building is occupied. An electric instant hot water system was installed to produce water for human use and consumption, supplying water at 55 °C. The infiltration rate was estimated to be 0.6, with an assumed minimum outdoor air per occupant of 10 L/s [25].

The appliances/equipment in the building relate to monitors, computers, printers, laboratory equipment, data centers, and other small appliances commonly found in kitchens. The average installed intensities for equipment, light and occupancy were 16.04 W/m^2, 6.07 W/m^2 and 0.16 occ/m^2, respectively.

EnergyPlus copulated with DesignBuilder as a 3D interface was used for the simulations. This software complies with the National Construction Code in Australia (NCC) and the ANSI/ASHRAE Standard 140-2017, defining the building energy simulation requirements [26, 27]. Then, based on the orientation, significance of internal loads, and activity type, the model was divided into 32 thermal zones. To simulate the impact of climate change on the photovoltaic system, it was the tool PVsyst. Figure 2 provides an example of the thermal zones used in the model.

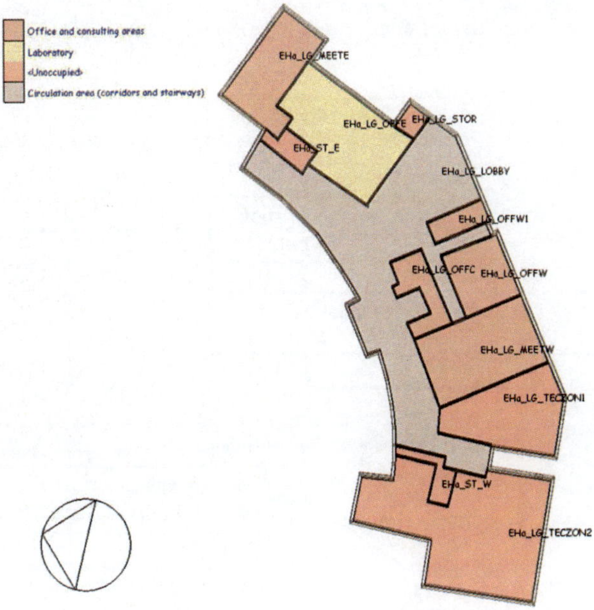

Fig. 2. Example of thermal zones.

2.3 Calibration Process

To ensure the reliability of the results, the simulation model was calibrated against actual energy monitored data, following the procedure from ASHRAE Guideline 14 [28]. Using an iterative process, described in Fig. 3, calibration aims to find deviations when comparing simulations' outputs with energy monitored data. The process stops when it is obtained an acceptable deviation between the simulated model and energy monitored data (similar to this study [29]). This deviation is translated by an error-based analysis (Eqs. 1 and 2) according to the proposed benchmark from the ASHRAE Guideline 14 and other relevant literature. Therefore, for the present case study, simulation results for electricity were compared to monitored electricity data from the university building management system, and a deviation within a range of ±10% was the criteria for acceptance [29, 30].

$$MBE(\%) = \frac{\sum_{i=1}^{n}\left(E_{s,i} - E_{mn,i}\right)}{n_{me} \times E_{av}} \times 100 \quad (1)$$

$$CV(\%) = \frac{\sqrt{\sum_{i=1}^{n}\left(E_{s,i} - E_{mn,i}\right)^2}}{n_{me} \times E_{av}} \times 100 \quad (2)$$

MBE and CV represent the errors; mean bias error and root-mean-square error coefficient of variation, respectively. $E_{s,i}$ corresponds to the energy simulated from a period i, $Em_{n,i}$ is the monitored energy use for a period i, n_{me} refers to the number of measurements, E_{av} is the average actual use of energy in a period i.

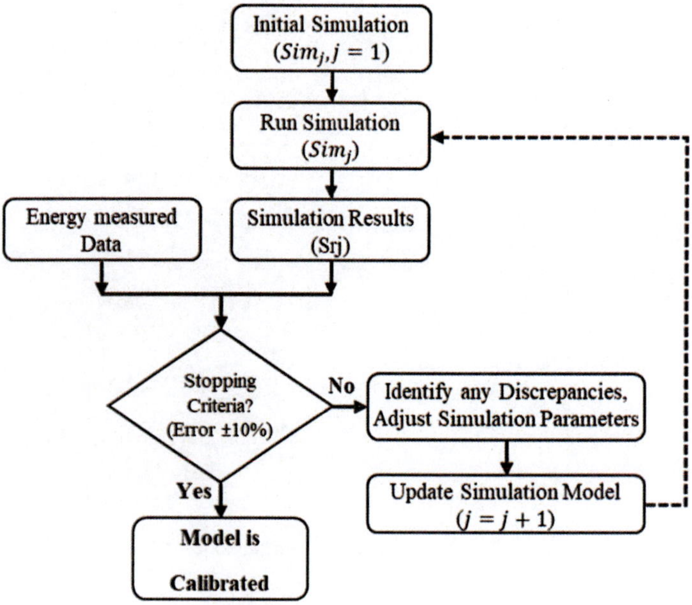

Fig. 3. Iterative calibration process.

where Sim_j corresponds to the simulation model at iteration j and S_{rj} represents the results from the simulation at iteration j.

The model was calibrated after 20 iterations, with error deviations of less than 1% and −8%, targeting the overall and HVAC electricity uses, respectively.

2.4 Present and Future Climates

The Representative Concentration Pathways (RCPs) are weather files that translate possible future climate scenarios. These RCPs were developed by the Commonwealth Scientific and Industrial Research Organization (CSIRO) [31] and are based on scenarios introduced in the IPCC fifth assessment report [32]. They are based on different GHG concentrations levels for a series of proposed future climates, resulting in different radiative forcing levels, reported to the pre-industrial era, 1750. RCP 2.6, RCP 4.5, RCP 6, and RCP 8.5 were the pathways introduced by CSIRO (Fig. 4).

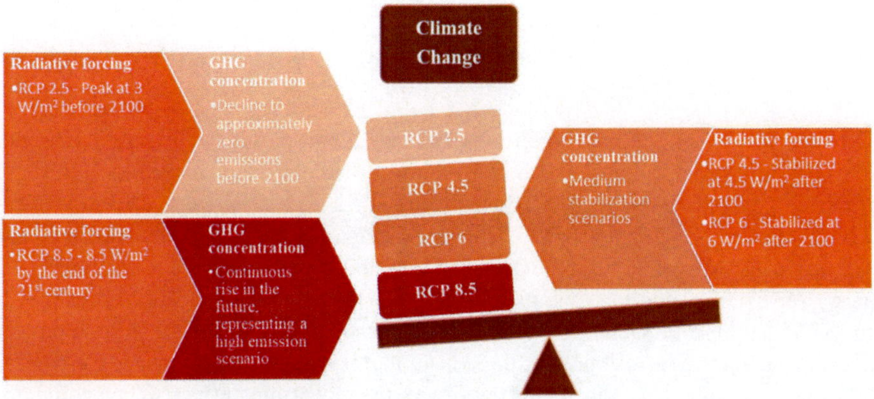

Fig. 4. Representative Concentration Pathways [31].

To assess the impact of climate change on the energy performance of the green building, lower, medium and high impact scenarios were chosen. Belcher et al. [33] morphing method and different global climate models [31] were used to generate future weather files. For the morphing method, typical meteorological year (TMY) data sets were used as baseline weather files containing historical weather data from 1990 to 2015 were used as the baseline weather files in the morphing method [31].

3 Results and Discussion

The building is located in Parramatta city NSW, a mild temperate climate with hot to very hot summers with moderate humidity and mild to cool winters with low humidity. Parramatta has approximately 17 °C as annual mean temperature, 22.2 °C as maximum monthly mean temperature (January) and 11 °C as minimum monthly mean temperature (July) under present climate conditions. For this case study, three future climate scenarios were selected: RCP 2.6, RCP 4.5 and RCP 8.5, for 2050 and 2090.

3.1 Simulation Results: Energy Performance and Peak Loads

Figure 5 shows the results from the simulations for the three scenarios considered previously. As expected, cooling represents the largest energy end use, followed by fans, heating and pumps.

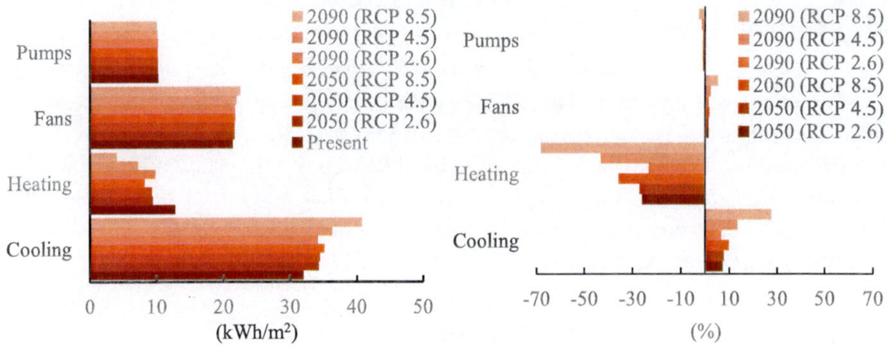

Fig. 5. Simulation results expressed in (left) energy intensity (kWh/m^2) and (right) percentage.

Under the present climate conditions, the use of energy for the heating function is 12.8 kWh/m^2, which is approximately 60% lower than cooling (32 kWh/m^2). In Fig. 5 (left) it is possible to see also that fans consume 52% more energy (21.36 kWh/m^2) when compared to pumps (10.25 kWh/m^2). Comparing the RCP 2.6, RCP 4.5, and RCP 8.5 pathways in 2050 and 2090, it is possible to see that the requirements for energy for heating, cooling, pumps and fans will be impacted by climate change. In Fig. 5 (right) these impacts were expressed in percentage of change using the present condition as baseline. As expected, there is an increase in the energy usage for cooling and a decrease in heating because of the future global warming effect. When compared to the present situation, cooling will see an increase of 9.8% and 7.2% for the high and low emission scenarios in 2050, and 27.3% and 6.7% in 2090, respectively. Opposing cooling, heating energy will decrease 35.8% and 26.1% for the high and low emission scenarios in 2050, and 68.1% and 23.5% in 2090, respectively. However, cooling demand has an expression 2.5 times more significant when compared to heating. This means that overall, the increasing impact on the total energy usage in the building will be mostly dominated by the raise in cooling demand. The decrease in heating will not have the capacity to completely offset the high use for cooling, as will be possible to see in Sect. 4.6. Fans and pumps will experience fewer impacts. There is also a slight increase of around 5% on the energy requirements for fans. This contrasts with the pumps, where there are no significant changes through the different pathways. It is predicted that the higher energy use of fans and pumps, resulting from higher cooling use, will be partly offset by a decrease in heating use in these buildings in the future.

Figure 6 presents the simulation results for the heating and cooling peak loads in the present and future climate conditions. The peak heating load is predicted to be reduced by approximately 7.8% and 40% in the low and high emissions scenarios, respectively. However, the cooling load is 0.2% under the low emissions scenario and 1.9% under the high emissions scenario, indicating that the peak cooling load is not too sensitive to climate change when compared to the heating load.

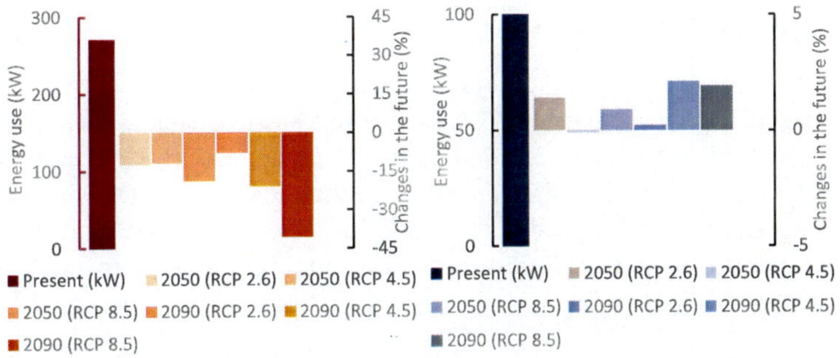

Fig. 6. Peak loads results for heating (left) and cooling (right).

3.2 Photovoltaic Production

The building has a photovoltaic system (PV) with 374 modules, totalizing 625 m^2. The modules are tilted at 3° and divided into three different arrays. The three arrays have different orientations towards 2° azimuth, 37° azimuth and 65° azimuth. The system has a 100 kWac inverter and a 99.1 kWp nominal power. Table 1 represents the predicted photovoltaic production under the present and future climate conditions.

Table 1. Predicted photovoltaic production.

		Photovoltaic system:	
		Production (MWh/yr)	Changes (%)
	Present	137.2	--
2050	RCP 2.6	140.1	2.06
	RCP 4.5	138.5	0.90
	RCP 8.5	138.5	0.94
2090	RCP 2.6	141.7	3.23
	RCP 4.5	141.1	2.83
	RCP 8.5	141.4	3.06

Under the present climate is registered the lowest solar production. However, in the future climate conditions, it is not possible to see a significant improvement change in electricity production. The highest solar production is predicted in the 2090 pathways. Compared to the present climate conditions, the production of electricity increases between 4.5 MWh to 4.2 MWh annually under pathways RCP 2.6 and RCP 8.5, respectively. It can be noticed that in both 2050 and 2090 the electricity production decreases slightly from pathway RCP 2.6 to RCP 4.5, increasing again in the pathway RCP 8.5 in 2090. Moreover, scenario RCP 2.6 represents the highest solar production in 2050 and 2090. This indicates that the predicted increase in solar radiation amongst the different scenarios is not significant. The results show that the rise in temperature penalizes solar production, which is verified by the highest production corresponding to the lower emissions scenarios.

3.3 Thermal Comfort Performance

To analyze the impact on occupancy and the overall thermal comfort performance of the building, the ASHRAE standard 90.1-2004 was used as a reference. According to the ASHRAE standard, the number of hours in a that are not able to meet in one or more conditioned zones, the temperature control range thermostat set points with a throttling range is designated as unmet load hours. Therefore, the total sum of unmet hours in a year should not exceed 300 h. If unmet hours for multiple zones coincide, one unmet hour is counted for the building.

Table 2. Unmet hours for cooling and heating

		Time setpoint not met during:	
		Occupied cooling (hrs)	Occupied heating (hrs)
	Present	24	55.5
2050	RCP 2.6	69	35.5
	RCP 4.5	66	36.5
	RCP 8.5	80.5	31.5
2090	RCP 2.6	71.5	37.5
	RCP 4.5	107	29
	RCP 8.5	246.5	13.5

Table 2 shows the unmet hours for the cooling and heating periods under present and future climate conditions. It was considered a range is 0.5 °C for thermostat throttling. As it is possible to see, in both situations, the unmet load hours are below the threshold of 300 h. This indicates that in this building, the temperature in almost all zones is kept at the desired levels, which is an indication that the required levels for thermal comfort are maintained in all future climate scenarios.

The results also show that for the heating function, there is a reduction in the number of hours that will not meet the temperature control range in the future for all scenarios. But the cooling set-point unmet hours in the future are increased in all scenarios. In the RCP 8.5 in 2090, the cooling set-point unmet hours increases substantially from 24 h to 247 h. When analyzing the total unmet hours for heating and cooling, it is possible to see that they still are below the ASHRAE threshold (260 h). This highlights that in terms of providing thermal comfort to its occupants, this green building is resilient to climate change.

3.4 Total Energy and Greenhouse Gas Emissions

To understand the environmental impact of climate change in this study, Table 3 shows the predicted total energy and greenhouse gas emissions for the current green building. As shown, the projected future climate pathways RCP 2.6, RCP 4.5 and RCP 8.5 contribute to a slight increase in greenhouse gas emissions, varying from 0.6% to around 4%, respectively. However, the total energy use has a slight decrease in almost all 6 scenarios, with the exception of RCP 8.5 in 2090, where a slight increase of 0.2% is verified.

Table 3. Total energy use and greenhouse gas emissions

		Total Energy (kWh/m^2yr)	GHG Emissions[1] ($kgCO_{2\text{-eq}}/m^2yr$)
	Present	165.5	133.5
2050	RCP 2.6	164.2	134.3
	RCP 4.5	164.6	134.7
	RCP 8.5	164.2	135.0
2090	RCP 2.6	164.0	133.9
	RCP 4.5	164.0	135.3
	RCP 8.5	165.8	138.6

[1] The GHG emissions were calculated for the natural gas and electricity end uses based on the report by the 34.Commonwealth_of_Australia, *National greenhouse accounts factors 2021, Australian National Greenhouse Accounts*, in *National greenhouse accounts factors.* 2021, Commonwealth of Australia: Australian Government Department of Industry, Science, Energy and Resources.

In the high emission scenario, RCP 8.5 in 2090, the building uses more 0.3 kWh/m^2-yr of energy and emits more 5.1 $kgCO_2/m^2$-yr GHG emissions. Additionally, it is possible to ascertain that the present case scenario represents the lowest GHG emissions, but one of the highest energy uses: 133.5 $kgCO_2/m^2$-yr and 165.5 kWh/m^2-yr, respectively. As expected, there is a slightly increasing trend for GHG emissions throughout the different pathways. Contrastingly, the overall demand for energy decreases mostly due to the balance between the heating and cooling effects. Table 4 illustrates the combined demand for heating and cooling.

Table 4. Combined demand for heating and cooling GHG emissions

		GHG Emissions ($kgCO_{2\text{-eq}}/m^2yr$)	Changes (%)
	Present	27.5	--
2050	RCP 2.6	28.5	3.6
	RCP 4.5	28.6	4.0
	RCP 8.5	28.9	5.1
2090	RCP 2.6	28.5	3.6
	RCP 4.5	29.6	7.6
	RCP 8.5	32.4	17.8

As it is possible to see, in the highest emissions scenario, the building emits almost less 18% (5 $kgCO_2/m^2$) GHG emissions when compared to the present scenario. It is possible to see a constant average increasing ratio, of around 64%, between the lowest and high emission scenarios for both 2050 and 2090. In the RCP 2.6 pathway, there are no changes in the emissions reduction in both 2050 and 2090. However, for the other pathways, it is possible to verify an increase of 3.4% and 10.8% for the RCP 4.5 and RCP 8.5, respectively, when comparing the 2050 situation with 2090.

The main limitations of this study had to do with the availability of additional monitoring energy points. Only the overall energy was being monitored, and there was not an accurate distribution of energy per end use. Moreover, the recorded, monitored data was only available for an eight-month period. With more measuring points, detailed data and a longer monitoring period (within the last 3 years), it is possible to improve the calibrated model. Moreover, this case study represents an isolated analysis. Future work may include the comparison between the current green building with other conventional and rated buildings to evaluate the significance of green strategies to help buildings resist the impact of climate change. Finally, for this study it was not accounted neither the impact of moving towards 100% electrification systems nor the future changes in the Australian electric grid with the increase implementation of renewable energies.

4 Conclusion

This paper analyzed the climate change impact on a green building energy performance in Australia. The building was calibrated against actual energy monitored data after 20 iterations, with error deviations of less than 1% and −8%. To predict future climate conditions for 2050 and 2090, three pathway scenarios were used: RCP 2.6, RCP 4.5 and RCP 8.5. As expected, in the future, there will be a decrease in the energy used for heating and an increase in the energy used for cooling due to global warming. Though, this study shows that the overall impact of climate change on energy performance of the case study GB is small. Cooling is the most energy-intensive end-use, followed by fans, heating and pumps. The energy use for cooling and heating in 2050 is 9.8% higher and 35.8% lower when compared to the high emissions scenario to the present situation, respectively. However, an increase of 27.3% in cooling energy was predicted by 2090. In contrast, heating will decrease by 68.1% under the high emissions scenario in 2090. Despite verifying an increase in the unmet hours in future scenarios, these will remain at an acceptable level. Therefore, this study shows that the GB, studied here, is somewhat not too sensitive to global warming, so that the increase in cooling energy use and its peak load is offset by the decrease in heating energy. However, when comparing the overall GHG emissions for the future climate scenarios with the present conditions, the results show an increase of 4% when comparing the high emissions and the present scenarios. Taking into consideration only the combined impact of cooling and heating, for the two high emissions scenarios in 2050 and 2090, it was verified an increase of 5.1% and 17.8%, respectively. It was also found in this case study that the GB would maintain the required levels for thermal comfort in all the future studied climate scenarios.

References

1. IPCC (eds.) Summary for Policymakers. In: Global Warming of 1.5 °C: IPCC Special Report on Impacts of Global Warming of 1.5 °C above Pre-industrial Levels in Context of Strengthening Response to Climate Change, Sustainable Development, and Efforts to Eradicate Poverty. Cambridge University Press, Cambridge, pp. 1–24 (2022)
2. Ürge-Vorsatz, D., et al.: Climate Change 2014: Mitigation. Chapter 9: Buildings. Report by the Intergovernmental Panel on Climate Change (2014)
3. He, Y., et al.: How green building rating systems affect designing green. Build. Environ. **133**, 19–31 (2018)
4. USGBC: Green Building and LEED Core Concepts, in Green Building and LEED Core Concepts Guide (n.d.)
5. Almeida, L.M.M.C.E., Tam, V.W.Y., Le, K.N.: Quantification of the energy use due to occupant behaviour collected in surveys: a case study of a green and non-green building. J. Build. Perform. Simul. **13**(6), 777–803 (2020)
6. Schmidt-Traub, G., et al.: National baselines for the sustainable development goals assessed in the sdg index and dashboards. Nat. Geosci. **10**(8), 547–555 (2017)
7. Ching, F.D.K., Shapiro, I.M.: Green Building Illustrated, ed. S. edn. John Wiley & Sons, Hoboken, New Jersey US (2021)
8. GBCA, Introducing green star | Together, we can build a sustainable future, in Building a sustainable future. Green Building Council of Australia (2020)
9. Mozingo, L., Arens, E.: Quantifying the Comprehensive Greenhouse Gas Co-Benefits of Green Buildings, in Sustainability, Whole Building Energy and Other Topics. UC Berkeley (2014)
10. ASHRAE: Thermal Environmental Conditions for Human Occupancy. In: ASHRAE Standard 55. ANSI/ASHRAE: www.ashrae.org (2013)
11. Chegari, B., et al.: An optimal surrogate-model-based approach to support comfortable and nearly zero energy buildings design. Energy **248**, 123584 (2022). https://doi.org/10.1016/j.energy.2022.123584
12. Nicol, J.: Characterising occupant behaviour in buildings: towards a stochastic model of occupant use of windows, lights, blinds, heaters and fans. In: Proceedings of the Seventh International IBPSA Conference, Rio, vol. 2, pp. 1073–1078 (2001)
13. Geng, Y., et al.: A review of operating performance in green buildings: Energy use, indoor environmental quality and occupant satisfaction. Energy Buildings **183**, 500–514 (2019)
14. Ikudayisi, A.E., et al.: Integrated design process of green building projects: a review towards assessment metrics and conceptual framework. J. Build. Eng. **50**, 104180 (2022)
15. Hwang, B.-G., Tan, J.S.: Green building project management: obstacles and solutions for sustainable development. Sustain. Dev. **20**(5), 335–349 (2012)
16. Lützkendorf, T., Lorenz, D.P.: Using an integrated performance approach in building assessment tools. Build. Res. Inform. **34**(4), 334–356 (2006)
17. Nguyen, A.T., et al.: Performance assessment of contemporary energy-optimized office buildings under the impact of climate change. J. Build. Eng. **35**, 102089 (2021)
18. Ismail, F.H., et al.: Climate change and the energy performance of buildings in the future – A case study for prefabricated buildings in the UK. J. Build. Eng. **39**, 102285 (2021)
19. Bamdad, K., et al.: Impact of climate change on energy saving potentials of natural ventilation and ceiling fans in mixed-mode buildings. Build. Environ. **209**, 108662 (2022). https://doi.org/10.1016/j.buildenv.2021.108662
20. Zhou, Y.: Climate change adaptation with energy resilience in energy districts—A state-of-the-art review. Energy Build. **279**, 112649 (2023)

21. Guo, S., et al.: Global comparison of building energy use data within the context of climate change. Energy Build. **226**, 110362 (2020)
22. Zhang, C., et al.: Passive design for green buildings by using green glass space and earth air tunnel. Energy Build. **273**, 112367 (2022)
23. Bamdad, K., et al.: Future energy-optimised buildings — Addressing the impact of climate change on buildings. Energy Build. **231**, 110610 (2021)
24. GBCA, Green Star in focus: The business case. Green Building Council of Australia (2020)
25. ASHRAE, ASHRAE Standard 62.1-2013: Ventilation for Acceptable Indoor Air Quality. American Society of Heating, Refrigerating and Air-Conditioning Engineers: Atlanta, GA (2013)
26. ANSI/ASHRAE, ANSI/ASHRAE Standard 140-2017: Standard Method Of Test For The Evaluation Of Building Energy Analysis Computer Programs, R.a.A.-C.E. American Society of Heating, Inc., Editor (2017)
27. ABCA, Handbook Energy Efficiency: NCC Volume One, ed. C.o.A.a.S.a.T.o.A. 2019. Canberra ACT: Australian Building Codes Board (2019)
28. ASHRAE, ASHRAE Guideline 14-2002 – Measurement of Energy and Demand Savings. American Society of Heating, Refrigerating and Air-Conditioning Engineers, Inc., Atlanta, GA (2002)
29. Gucyeter, B., Calibration of a building energy performance simulation model via monitoring data. In: 2018 Building Performance Analysis Conference and SimBuild co-organized by ASHRAE and IBPSA-USA, pp. 542–549 (2018)
30. Yoon, J.-H., Lee, E.-J.: Calibration Procedure of Energy Performance Simulation Model for a Commercial Building. Division of New and Renewable Energy Research. Korea Institute of Energy Research, pp. 305–343 (1999)
31. Ren, Z., Tang, Z., James, M.: Projected Weather Files for Building Energy Modelling User Guide. CSIRO, Australia (2021)
32. IPCC, Climate Change 2014: Synthesis report. In: Core Writing Team, Pachauri, R.K., Meyer, L.A. (eds.) Contribution of Working Groups I, II and III to the Fifth Assessment Report of the Intergovernmental Panel on Climate Change, p. 151. IPCC, Geneva, Switzerland (2014)
33. Belcher, S.E., Hacker, J.N., Powell, D.S.: Constructing design weather data for future climates. Build. Serv. Eng. Res. Technol. **26**(1), 49–61 (2005)
34. Commonwealth_of_Australia, National greenhouse accounts factors 2021, Australian National Greenhouse Accounts, in National greenhouse accounts factors. 2021, Commonwealth of Australia: Australian Government Department of Industry, Science, Energy and Resources

A Generic Operational Tool for Early Warning Oil Spills – Application to Cartagena Bay and the Algarve Coast

Diogo Moreira[1](✉), João Janeiro[2], Marko Tosic[3], and Flávio Martins[4]

[1] CIMA – University of Algarve, Estr. da Penha, 8005-139 Faro, Portugal
diogo.moreira1998@hotmail.com
[2] S2AQUAcoLAB, Av. Parque Natural da Ria Formosa S/N, 8700-194 Olhão, Portugal
[3] School of Sciences, Department of Earth Sciences, Universidad EAFIT, Carrera 49 #7S-50, A.A.3300 Medellín, Colombia
[4] CIMA – ISE – University of Algarve, Estr. da Penha, 8005-139 Faro, Portugal

Abstract. Over the past 10 years, the record of oil spills per year has decreased. However, the size and quantity of oil tanker ships have been growing, which will increase the hazard of a spill. Thus, spill predictions are a crucial component of monitoring because they enable a more effective response to the accident by assessing the impact beforehand and better developing a strategy to mitigate the environmental impact. These predictions are provided by Lagrangian models. The model used in this work will be able to forecast the geographical and temporal propagation of the oil spill slick with the support of the hydrodynamic and atmospheric forcing from external numerical models, which are required to compute these processes. The objective of this work is to construct a Python-based Early Warning System that will serve as the link between the user and the oil spill model. The MOHID Lagrangian module will be adapted to the region of Cartagena Bay, Colombia, as well as for the Algarve Coast, to assess oil spills events and test the adaptability of the tool. Furthermore, the operational models BASIC and SOMA will supply the hydrodynamic and atmospheric forcing for the simulations. The results in this paper showed the high level of adaptability of the tool, as well as its simplicity being adapted to users without deep knowledge in ocean modelling.

Keywords: Early Warning system · Lagrangian Simulation · Oil Spill · Cartagena Bay · Algarve Coast

1 Introduction

1.1 Oil Spills Overview

Oil spills are becoming less common in the last years, with the average number of large (greater than 7 tones) oil spills per year falling to less than 5 in the last decade (ITOPF, 2023). The most common accidents are of medium size (between 7 and 700 tons), with a low frequency of them being of large-scale (>700 tons). The majority of the oil spilled

J. F. L. C. Semião et al. (Eds.): INCREaSE 2023, ASST, pp. 109–120, 2023.
https://doi.org/10.1007/978-3-031-44006-9_9

is due to natural causes, such as oil and gas migrating to the surface through fractures or porous rock formations, contaminating surface and ground water, called natural seeps (Abrams, 2020). To minimize the impact of such human and natural accidents, technological advances have made possible to track oil spills in an effective manner, namely with the help of oil spill models, capable of predicting the spatial and temporal variation of a spill (NOAA, 2023). An oil spill incident may be caused by a ship, an accident in an oil refinery, or even in an offshore oil platform. Models of oil spills are extremely important instruments for reducing the effects of these accidents. By using these models, the authorities may plan and implement efficient mitigation efforts before an event even takes place. Oil spill models can be used to estimate the environmental effects of a spill, identify vulnerable areas that may be at danger, and evaluate the efficiency of different response strategies.

This work aims to provide a generic operational tool capable of being implemented to track oil spill pollution in any region of the world. The MOHID Modelling system is adapted to simulate a generic oil spill accident, and a generic early warning system (EWS) tool, based on Python, aiming to automate the simulations. In this work the EWS tool is demonstrated in two different systems: Cartagena Bay, Colombia, and the Algarve Coast, Portugal.

2 Methods

2.1 Oil Spill Modelling

Ocean processes are interconnected and require better understanding of their impact on the environment and human activities. Operational oceanography is a branch of science responsible for monitoring and measuring these processes in near real time. Water modelling systems, such as MOHID, can predict the hydrodynamic conditions of a certain region. These systems can compute and predict ocean drift, which is influenced by the surface currents, the stokes drift, and the wind drag effect. For that, they need to compute a variety of complex ocean processes, making it difficult to realistically simulate their behavior. The main processes responsible for the behaviour of the oil are hydrodynamic and atmospheric conditions, as well as the oil-specific characteristics. These processes include evaporation, dissolution, emulsification, dispersion, microbial degradation, and the interaction between oil and sediments. Realistically simulating these processes requires a high level of complexity, requiring operational modeling systems to be coupled with forcing models and validated with in situ data.

To compute the simulations, the models use the concept of tracers, which can be a water mass, a sediment particle, a molecule, or an oil particle. The movement of these tracers is determined by the hydrodynamic module's velocity field, the wind from the surface module and the spreading velocity from the oil dispersion module. The major factor responsible for particle movement is generally the mean velocity. The spatial coordinates are given by the definition of velocity:

$$\frac{dx_i}{dt} = u_{i(x_i,t)} \tag{1}$$

where u represents the mean velocity and x represents the particle position (Suh, 2006). Tracers are "born" at origins and have the same set of properties as tracers from the same origin. The pattern of tracer emission varies depending on the origin, which can be classified as "Point Origin" (emits at a specific point), "Box Origin" (emits over an area), or "Accident Origin" (emits in a circular pattern around a point). For this study, the third option was chosen to better represent the unpredictability of an oil spill accident (Braunschweig et al., 2005).

In order to compute the transport of the tracers, the models need the hydrodynamic and atmospheric forcing, which are provided by external operational modelling systems. These consist in computer-based models used to simulate and forecast the behaviour of intricate natural systems in real-time. They are used in a variety of sectors, such as energy generation, water management, transportation, and weather forecasting. For this work, the operational models BASIC and SOMA will be used to force the Lagrangian models in their domains (Cartagena Bay and Algarve Coast respectively). These models have been validated and calibrated over several years, and are currently running in the virtual machine infrastructure at the University of Algarve's servers. These systems are managed by a tool called SMS-Coastal, a python system capable of managing simulations of coastal operational models developed by Mendonça (2020).

2.2 Geographical Context

Cartagena Bay is located on the north coast of Colombia and is a major commercial region. It is bordered by the Magdalena River basin, part of which flows through the Dique Canal, exchanging water with the Caribbean Sea through two bars. It has a highly stratified upper water column with low salinity and high turbidity, and a less dynamic lower layer of saline water with maximum depts of about 30 m. Puerto Bahía is Colombia's most modern maritime multipurpose terminal, located at the southeast region of Cartagena Bay. However, it's primarily function is as a marina and residential development, not a commercial port for oil export, being just a connection point to the principal port. The other port, Puerto de Cartagena, is a major commercial port on Colombia's Caribbean coast as well as Colombia's largest port in terms of total cargo volume, with five terminals, four of which are for general cargo and one for containers. It is strategically located and well-equipped, providing access to shipping lanes connecting North America, South America, Europe, and Asia, and a key entry point for goods such as oil, coal, and other agricultural products. It also has a specialized bulk oil and liquid terminal. This makes its a strategically located and well-equipped commercial port that plays a vital role in the economy of Colombia and the Caribbean region. Due to this, pollution issues have been observed in the sediments, biota, and waters due to industrialization pressure, and this has led to studies of the water's circulation and contaminant transport.

Algarve is a region in the southern region of Portugal known for its beautiful beaches, warm weather, and coastline. It is a popular touristic destination, with millions of visitors each year. It is affected by a variety of oceanic processes, such as waves, tides, and currents, all of which are influenced by the region's proximity to the Atlantic Ocean. The coastline of the region is distinguished by a variety of coastal features, including long sandy beaches, rocky cliffs, and estuaries, which create complex hydrodynamic conditions. These conditions have significant implications for the local environment, influencing coastal erosion rates, sediment transport, and marine organism distribution. The coastal region of Algarve is also influenced by the region's climate, which is defined by a Mediterranean climate with mild, wet winters and hot, dry summers. The climate of the region can cause varying sea conditions, with calmer seas in the summer and more intense waves and currents in the winter. The Coast of Algarve is crossed by an important shipping corridor, connecting the Mediterranean Sea to the Atlantic Coasts of Europe. Fortunately, there is no records of highly impactful oil spill accident affecting the coast, however, Algarve is not immune to the impacts of pollution, and over the past years the probability of occurring one is significantly high. The Portuguese government announced plans to explore for oil and gas in the region in 2018, raising concerns about potential threats to the region's coastal ecosystems and the possibility of future oil spills.

3 Methodology

3.1 Oil Spill Model – MOHID

The MOHID Oil Spill model is used to simulate the evolution of possible oil spills providing forecasts of spill trajectories, computing the weathering processes and identifying possible sources of pollution from a detected spill. The MOHID Water Modelling System is an integrated modeling tool that computes the transport of oil particles and their fate, by simulating the physical and biochemical processes of the ocean. It was written in ANSI FORTRAN 95, integrating various numerical models. It is organized in a hierarchical modular structure and currently consists of over 60 modules, each of which is responsible for managing a specific type of information while performing a specific function or process (Braunschweig et al., 2005). It uses a multi-grid approach that divides the study area into multiple smaller grids with different resolutions, allowing for a more detailed representation of the coastal environment, with higher resolution in areas where the oil spill is expected to have a greater impact. This approach also enables the model to simulate the oil's transport and fate more accurately, as well as the potential impacts on marine resources and ecosystems.

The model must extract forcing data from hydrodynamical and atmospheric prediction models to provide spatial and temporal variation of wind, currents, water temperature, and waves. The model attempts to integrate all the possible processes to make the simulation close to reality as possible, by computing particle transport and oil weathering equations using the hydrodynamic and atmospheric conditions. These hydrodynamical and atmospheric conditions are provided by the operational model BASIC for Cartagena Bay and by SOMA for the Algarve region. The parameters are written in two separate DAT files instructing MOHID to execute the simulations. The first file is called Lagrangian.dat, and includes all the physical parameters from the operational models needed to compute the ocean drift, as well as the characteristics and processes of the oil. The second file, Model.dat, determines the start and end of the simulation, which for this case was set to a maximum of 3 days, which is the maximum range of the forecast models.

In this work, an automation program was developed to run the Oil Spill model in an operational way, automatizing the download and setup of all the necessary forcing sources, coastline, and major oil types being transported in the region. The user inputs are provided to generate run files which are then automatically generated to successfully run the model with minimal user intervention and independent of the user's modelling knowledge. The user just needs to fill an online entry form with all the necessary information and then receives a report from the simulation that was requested.

3.2 EWS Tool

The Early Warning system Tool (or EWS Tool) consists in a program, written in python, developed to automatize the Lagrangian simulation for the oil spills, providing easy access to nontechnical users, and producing spill reports. An illustrative diagram of the functionalities of the program can be seen in Fig. 1 with letters referring to each process. As it starts running, it will keep checking a specific email (A) designed to receive a notification every time the online form is filled (B) and submitted, working as the trigger for the rest of the program. Once it checks that a new notification was received, it reads the CSV file associated to the online form (C) with the most recent written line and extracts all the user inputs into separate variables. The required variables are the user's name, the user email (to send the results), the institution\company of the user, the hour of the submission, the simulation start date and time, the coordinates of the spill origin, the volume of the spill and the type of oil. At the moment, 3 oil types are inplemented: Cano Limon, Cusiana and Vasconia. MOHID is very strict with the format of certain parameters, so the program will follow the conversion of the volume from US Gallons to Cubic Meters and separate the data with blank spaces, to allow MOHID to understand them. Following this, the hydrodynamic forcing from the operational system BASIC or from SOMA is extracted (D), described in the section above, and the atmospheric forcing (E). For the forcing data it compiles the daily files (F) into a 3-day HDF5 single file. After all this is done, the program creates the MOHID files Lagrangian.dat and Model.dat with the instructions to run the simulation (G). Finally, the program executes the MOHIDWater.exe (H) which is executed in a specific virtual machine hosted in the UAlg servers. After the simulation is done, the program checks the model logs for possible errors (I). If an error is detected, the user receives an error message (J). In case the simulation ran without errors, the program will create a report of the results (K), with maps of the oil spill behaviour along the three days and send it to the user email (L).

4 Results

4.1 Cartagena Bay Simulations Reports

The tool described in the precedent sections was developed and tested in Cartagena Bay, Colombia, as well as in the Algarve Coast, using the operational models BASIC and SOMA as the providers of the forcing fields, respectively. The final result from the tool after a successful simulation in Cartagena Bay is a report with the information about the requested simulation by the user. Figure 2 shows three examples of reports of 3 different simulation requests, with different volumes, oil types, start days and coordinates inside the bay. The reports are presented in Spanish, because they are designed for the Colombian Authorities and regional oil companies.

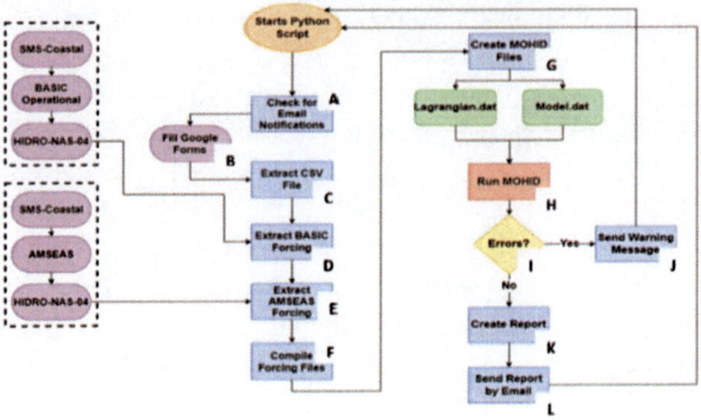

Fig. 1. Diagram of the EWS Tool.

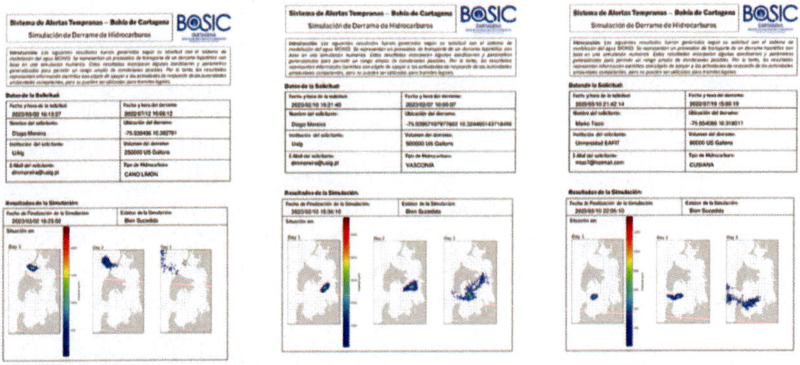

Fig. 2. Reports of three different simulation requests for Cartagena Bay.

The simulations are being performed in a virtual machine host in the University of Algarve with four Intel Xeon cores, 16 GB RAM and 200 GB high performance SSD disk. The average time of the whole process, from filling the form until receiving the report, is about 13 min.

4.2 Algarve Simulation Reports

The tool presented in this article is generic, thus, the code can be adapted to any other region of the world. To further test the tool's adaptability, it was tested for the Algarve region, using the SOMA Operational System as the provider of the forcing fields to perform the oil spill simulations. The resulting reports, shown in Fig. 3, followed the same methodology as the Cartagena Bay reports. The reports will be presented in English in this case because they are destined for an international project called ASTRiiS, the main goal of which is to develop technical and scientific knowledge in order to conceptualize and implement a set of seamlessly integrated and customizable information services and products, applied and exploited in key sectors of the blue economy with high development and value creation potential.

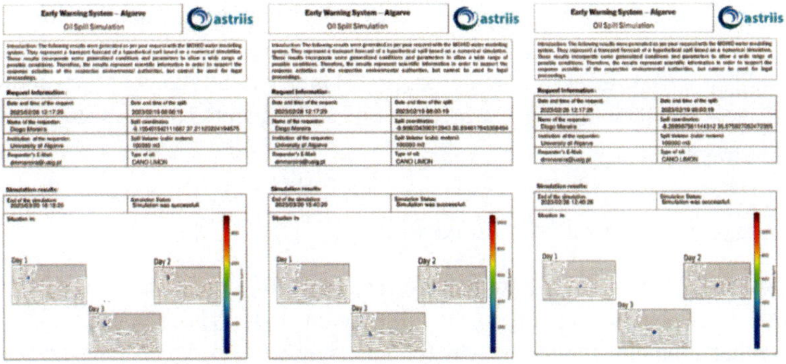

Fig. 3. Reports of three different simulation requests for Algarve.

Just as for the previous simulations, these are made in a similar virtual machine, hosted in the University of Algarve. The average time of the whole process, from filling the form until receiving the report, is about 20 min, which differs from the previous results. The explanation for this is related with the SOMA data files as explained in Sect. 5.

5 Discussion

5.1 Python Script

The final version of the EWS Tool consists of a script that connects four different functions: one that triggers the main functions, one to get the information for the simulation and run it, one to extract and compress the BASIC forcing files, and one to create simulation images and write the reports. The first function is referred to as Email Trigger, and it responsible for checking a specific email account, set up to receive new submitted forms for a simulation. When it detects a new email, it triggers the rest of the functions. The second function is called Simulation Tool, and it is the most important. This function is responsible for running the simulation, creating reports (as shown in Fig. 2) and sending them by reading the user information from the submitted entry form. This function is also responsible for the extraction of the forcing data required for the simulation, as well as for the creation of the maps using the function, Simulation Maps.

The reports in Fig. 3 followed the same code as previously described, with a few exceptions. Firstly, the program will be running in a different virtual machine, and as such, the directories in the code must be changed to match the new ones. The entry form must also be different, and thus a new link to assess it must be provided to the code so that it can read it and extract the correct information from the user. Furthermore, the email that will initiate the simulation is different, so the credentials to log in into the email account must be changed. The SOMA Operational Model, which is a hydrodynamical model similar to BASIC, capable of forecasting a variety of parameters, now provides the forcing data. The output data files differ significantly from those obtained from BASIC in that they are obtained every half hour, resulting in twice the number of files per day. This increase in the number of files also increases the total time it takes for the report to arrive, because it takes twice as long to compile all the files. The files are stored in Network Attached Storage (NAS) units, which have a directory and credentials to assess them, so this must be correctly configured, as this is an essential part of the simulations. So, with a few small changes in the program, it can be adapted and capable of perform without issues to any domain.

5.2 Cartagena Bay Simulation Reports

The simulations shown in Fig. 2 were performed in various locations throughout the bay to provide a better understanding of the oil behavior within Cartagena Bay. The first is in the northern region of the Bay, which is also home to Puerto de Cartagena, one of the Caribbean's most important ports. Despite the relatively high volume of oil spilled, minor concentrations of oil enter the port, with the majority circulating outside the bay. The second map was in the Bay's middle region, with a very high volume of oil spilled of 500000 US Gallons, approximately 670 tons. Since there remains a high level of oil inside the bay after three days, it is possible to say that the impact on the environment is more severe in this scenario due to the oil's presence in the bay for more than three days. The third and last report, displays a scenario in the Bay's southernmost region, with approximately 110 tons of oil spilled. This scenario is very similar to the first one in that the majority of the oil spilled leaves the Bay by the third day, although by the south bar (Boca Chica).

5.3 Algarve Simulation Reports

The reports in Fig. 3 differ from those in Cartagena Bay in their structure. The first point to mention is how the maps are organized, which was done in a pyramid-like structure to maximize space, given the aspect ratio of the images. The simulations in the reports followed a similar methodology to the Cartagena Bay simulations, with three different simulation requests in different locations of the model domain, for the same day and with the same volume of oil spilled to better understand the behaviour of the oil slick.

The Algarve region is heavily influenced by the Mediterranean, with a dominant current flowing from East to West and some vortices formation, and strong upwelling currents, which are quite demanding for the model. The first discharge is located on the west coast of Algarve, where there is a strong northward current, preventing the oil slick from reaching the coast. The oil slick's origin point was near the coast, but by the third day, it was visible moving north away from the coast. The second discharge was requested in front of Sagres, and due to the high volume, some of the oil reaches the coast by the second day, even if the concentration is very low. The third day of the simulation shows the oil slick approaching the coast of Sagres, with the thickness of the oil increasing, which can cause severe problems for marine life and humans. The third discharge is located in front of Portimão, where the oil slick is heavily influenced by the Mediterranean's strong currents, which push it away from the coast. The interaction of the currents with the oil greatly dissipates it, causing the oil concentration to decrease over time.

6 Conclusion

The tool facilitates local authorities' response to an oil spill incident by allowing them to act before the spill occurs. By evaluating the effectiveness of some of those strategies with hypothetical simulations for any location within the bay on any day prior to the request, this tool can test several strategies and approaches, resulting in the appropriate response for the accident at hand. Oil spill companies in the region can use the tool to assess the impact of an accident in their operations, and thus minimize it by implementing security measures. The simulation for the Algarve region demonstrates the tool's high adaptability; with minor changes in the code, the tool can perform nearly identically to the one created for Cartagena Bay. This allows the simulation to be performed not only for any region of the world, but also for any other pollutant or object.

The primary goal of this research was to create a generic tool for operating Lagrangian simulations with MOHID, for hydrocarbons or other pollutants in any location in the world. It was demonstrated that by combining a variety of Python libraries in a program, it is possible to create a useful tool capable of automating oil spill reports, from data analysis to creating images of oil spill simulations, to creating the report itself and sending it to the person who requested it. So, the study's goal was achieved, and with a few modifications, this tool could be very useful as an early warning system to help reduce the impact of pollutants in the environment, or even to aid in Search and Rescue operations. Some improvements to the tool can be suggested given some of its limitations, in order to continue this work. One of its limitations is that parallel simulation cannot be processed by the tool. This necessitates additional programming work as well as an increase in the computational capabilities of the server that hosts the system. Excessive delays in receiving simulation reports must be considered if the number of requests is very high. Other limitation is the fact that the modelling system may need additional calibration and validation, in order to improve the system's accuracy, with dedicated measurement campaigns, such as using low-cost georeferenced drift buoys. Additionally, the program would gain with the inclusion of more types of hydrocarbons or other hazardous substances, which can be easily included, for example from the ADIOS database.

It is thus possible to say that this tool can be used as a helpful alie in dealing with the ocean pollution, by assessing the impact of certain pollutants in the water before they even take place, resulting in a great Early Warning System Tool.

Acknowledgements. This work had the support of national funds through FCT, under the project LA/P/0069/2020 granted to the Associate Laboratory ARNET and UID/00350/2020 CIMA. It had received funding from the European Union's Horizon 2020 research and innovation programme under grant agreement No. 101000825 (NAUTILOS); from the INTERREG/POCTEP program CIU3A under grant agreement No GA 0754_CIU3A_5_E; from the AEROS project (reference POCI-01–0247-FEDER-045911), co-financed by ERDF through COMPETE 2020, PORTUGAL 2020, LISBOA 2020 and by FCT under the International Partnership programme MIT Portugal and from the ASTRiiS project (No. 046092) co-financed by ERDF through COMPETE 2020, PORTUGAL 2020, LISBOA 2020 and ALGARVE 2020.

References

Ari, N., Ustazhanov, M.: Matplotlib in python. In: Proceedings of the 11th International Conference on Electronics, Computer and Computation, ICECCO 2014 (2014)

Abrams, M.A.: Marine seepage variability and its impact on evaluating the surface migrated hydrocarbon seep signal. Mar. Pet. Geol. **121**, 104600 (2020). https://doi.org/10.1016/j.marpetgeo.2020.104600

Braunschweig, F., Leitão, P.C., Fernandes, L., Pina, P., Neves, R.J.J.: The object-oriented design of the integrated water modelling system MOHID. Dev. Water Sci. **55**(2), 1079–1090 (2005)

Brekke, C., Solberg, A.H.S.: Oil spill detection by satellite remote sensing. Remote Sens. Environ. **95**(1), 1–13 (2005)

Berry, A., Dabrowski, T., Lyons, K.: The oil spill model OILTRANS and its application to the Celtic Sea. Mar. Pollut. Bull. **64**(11), 2489–2501 (2012)

Bellomo, L., et al.: Toward an integrated HF radar network in the Mediterranean Sea to improve search and rescue and oil spill response: the TOSCA project experience. J. Oper. Oceanog. **8**(2), 95–107 (2015)

Cordes, E.E., et al.: Environmental impacts of the deep-water oil and gas industry: a review to guide management strategies. Front. Environ. Sci. **4**, 58 (2016). https://doi.org/10.3389/fenvs.2016.00058

Caballero-Gallardo, K., Alcala-Orozco, M., Barraza-Quiroz, D., De la Rosa, J., Olivero-Verbel, J.: Environmental risks associated with trace elements in sediments from Cartagena Bay, an industrialized site at the Caribbean. Chemosphere **242**, 125173 (2020)

Chen, S., et al.: A new transient model for motion and variation of gas phase in pipelines applying collocated multi-grid method with the Eulerian-Lagrangian approach. J. Petrol. Sci. Eng. **208**, 109761 (2022)

Delvigne, G.A.L., Sweeney, C.E.: Natural dispersion of oil. Oil Chem. Pollut. **4**(4), 281–310 (1988). https://doi.org/10.1016/S0269-8579(88)80003-0

Daling, P.S., Brandvik, J., Mackay, D., Johansen, O.: Characterization of crude oils for environmental purposes. Oil Chem. Pollut. **7**(3), 199–224 (1990)

Esau, I., Miles, V., Varentsov, M., Konstantinov, P., Melnikov, V.: Spatial structure and temporal variability of a surface urban heat island in cold continental climate. Theoret. Appl. Climatol. **137**(3–4), 2513–2528 (2019)

Fingas, M.F.: A literature review of the physics and predictive modelling of oil spill evaporation. J. Hazard. Mater. **42**(2), 157–175 (1995)

Flores, J.G., Sarica, C., Chen, T.X., Brill, J.P.: Investigation of holdup and pressure drop behavior for oil-water flow in vertical and deviated wells. J. Energy Resour. Technol. **120**(1), 8–14 (1998)

Fingas, M.F.: Modeling evaporation using models that are not boundary-layer regulated. J. Hazard. Mater. **107**(1–2), 27–36 (2004)

Giannaros, T.M., Melas, D., Daglis, I.A., Keramitsoglou, I.: Development of an operational modeling system for urban heat islands: an application to Athens, Greece. Nat. Hazard. **14**(2), 347–358 (2014)

Harsem, Ø., Eide, A., Heen, K.: Factors influencing future oil and gas prospects in the Arctic. Energy Policy **39**(12), 8037–8045 (2011)

Jordi, A., et al.: Scientific management of Mediterranean coastal zone: a hybrid ocean forecasting system for oil spill and search and rescue operations. Mar. Pollut. Bull. **53**(5–7), 361–368 (2006)

Jernelv, A.: The threats from oil spills: now, then, and in the future. Ambio **39**(6), 353–366 (2010)

Kingston, P.F.: Long-term environmental impact of oil spills. Spill Sci. Technol. Bull. **7**(1–2), 53–61 (2002)

Keramea, P., Spanoudaki, K., Zodiatis, G., Gikas, G., Sylaios, G.: Oil spill modeling: a critical review on current trends, perspectives, and challenges. J. Mar. Sci. Eng. **9**(2), 181 (2021). https://doi.org/10.3390/jmse9020181

Lehr, W., Jones, R., Evans, M., Simecek-Beatty, D., Overstreet, R.: Revisions of the ADIOS oil spill model. Env. Modell. Softw. **17**(2), 189–197 (2002)

Leitão, J.M.: Development of an operational tool for oil spill forecast: Application to oil exposed regions (2014)

Liu, Z., Liu, J., Gardner, W.S., Shank, G.C., Ostrom, N.E.: The impact of Deepwater Horizon oil spill on petroleum hydrocarbons in surface waters of the northern Gulf of Mexico. Deep-Sea Res. Part II: Topical Stud. Oceanogr. **129**, 292–300 (2016)

Mackay, D., Mcauliffe, C.D.: Fate of hydrocarbons discharged at sea. Oil Chem. Pollut. **5**(1), 1–20 (1988)

Mendonça, F.: A python based system to manage simulations of coastal operational models. Universidade do Algarve (2020)

Mooney, M.: The viscosity of a concentrated suspension of spherical particles. J. Colloid Sci. **6**(2), 162–170 (1951)

Nam, Y.-W., Kim, Y.-H.: Prediction of drifter trajectory using evolutionary computation. Discrete Dyn. Nat. Soc. **2018**, 1–15 (2018)

Parga-Lozano, C.H., Marrugo-Gonz, A.J., Fern, A., Andez-Maestre, R.: Baseline hydrocarbon contamination in cartagena bay. Colombia **44**(1), 4–71 (2002)

Ragasudha, I., Saravanan, M.: Secure automatic question paper with reconfigurable constraints. In: 2021 International Conference on Wireless Communications, Signal Processing and Networking, WiSPNET 2021, pp. 16–20 (2021)

Sebastiao, P., Soares, C.G.: Modeling the fate of oil spills at sea. Spill Sci. Technologr Bdelin **2**(2–3), 121–131 (1995)

Spaulding, M.L.: State of the art review and future directions in oil spill modeling. Mar. Pollut. Bull. **115**(1–2), 7–19 (2017)

She, J., et al.: Baltic sea operational oceanography—a stimulant for regional earth system research. Front. Earth Sci. **8**, 7 (2020). https://doi.org/10.3389/feart.2020.00007

Sánchez Catalán, F., Mayo Mancebo, G., Rivero Hernandez, J., Mercado Páez, J.P., Johnson Restrepo, B.G.: Caracterización fisicoquímica, determinación de mercurio total e hidrocarburos disueltos y dispersos en aguas y sedimentos de la bahía de Cartagena, Colombia. Bol. Cient. CIOH **39**(2), 41–50 (2021). https://doi.org/10.26640/22159045.2020.505

Seyedrezazaribaf, N., En, G.: Agustin Sanchez-Arcilla Conejo Computing plastic Lagrangian trajectories from real time Eulerian hydrodynamic data. Application to the Barcelona coastline (2021)

Sotillo, M.G.: Ocean modelling in support of operational ocean and coastal services. J. Mar. Sci. Eng. **10**(10), 1482 (2022)

Tosic, M., Restrepo, J.D., Izquierdo, A., Lonin, S., Martins, F., Escobar, R.: An integrated approach for the assessment of land-based pollution loads in the coastal zone. Estuar. Coast. Shelf Sci. **211**, 217–226 (2018)

Tosic, M., Martins, F., Lonin, S., Izquierdo, A., Restrepo, J.D.: Hydrodynamic modeling of a polluted tropical bay: assessment of anthropogenic impacts on freshwater runoff and estuarine water renewal. J. Environ. Manage. **236**, 695–714 (2019)

Tosic, M., Restrepo, J.D., Lonin, S., Izquierdo, A., Martins, F.: Water and sediment quality in Cartagena Bay, Colombia: seasonal variability and potential impacts of pollution. Estuar. Coast. Shelf Sci. **216**, 187–203 (2019)

Viveros, D.: Development of a new oil biodegradation algorithm for NOAA's oil spill modelling suite (GNOME/ADIOS) Droplet and bubble formation from blowout View project Factors Limiting Biodegradation of Exxon Valdez Oil Oil Spill View project Dalina Viveros National Oceanic and Atmospheric Administration (2015) van Sebille, E., et al.: The physical oceanography of the transport of floating marine debris. Env. Res. Lett. **15**(2), 023003 (2020)

Wang, Z., Fingas, M., Page, D.S.: Oil spill identification. J. Chromatogr. A **843**(1–2), 369–411 (1999)

Zhang, N.: Research on data acquisition of novel list based on Python. In: Proceedings – 2021 2nd International Conference on Artificial Intelligence and Education, ICAIE 2021, pp. 675–679 (2021)

Manufacture of an Acoustic Absorption Veil by Using Recycled Materials of Agro-industrial Sector

José Antonio López-Marín[1], Daniel Espinosa-Corbellini[2], and Miguel Suffo[1](✉)

[1] Department of Mechanical Engineering and Industrial Design, High Engineering School, Universidad de Cádiz, Campus Río San Pedro s/n, 11510 Puerto Real, Cádiz, Spain
miguel.suffo@uca.es

[2] Department of Automatic Engineering, Electronics, Architecture, and Computer Networks, High Engineering School, Universidad de Cádiz, Campus Río San Pedro s/n, 11510 Puerto Real, Cádiz, Spain

Abstract. The use of wastes from agro-industrial activities is an environmental and economic diversification challenge within the circular economy action plan of the European Union. This paper shows a process of design and manufacture some insolation acoustic screens, whose composites came from by-products of the agricultural industry, such as cork/dust powder (agrocomposites). A bibliographic search had been carried out in which the different eco-friendly materials and processes actually used in bioconstruction and acoustic sectors were compiled. Sound absorption had been calculated according to the norm UNE-EN ISO 10534-2:2002, this property was compared with insulating materials that are currently used in buildings, with the aim of finding eco-friendly solutions for bioconstruction (e.g. Polyisocyanurate, PIR). For sound absorption test, the specimens were cylindrical with a diameter of 40 mm and different thicknesses over the acoustic impedance tube (Kundt Tube), furthermore, they were coated with an acoustic veil made by Carbocal® provided by AB Azucarera using an adhesive provided by major producers. The use of Carbocal appears to achieve improvements in the normal incidence absorption coefficient (α), using the core formed by the agrocomposite based on cork powder (CoP), a high viscosity resin, and a lower thickness than the synthetic PIR material.

Keywords: Acoustic screens · agrocomposites · bioconstruction · acoustic insulation · Carbocal®

1 Acronyms

- Aquence CG 979 (ACG).
- Aquence ENV 1658–22 (ENV).
- Basalt fibre (BF).
- Binder and hardener (B&H).
- Carbocal ($CaCO_3$).

J. F. L. C. Semião et al. (Eds.): INCREaSE 2023, ASST, pp. 121–137, 2023.
https://doi.org/10.1007/978-3-031-44006-9_10

- Carboxymethylcellulose (CMC).
- Chicken feathers (CF).
- Core of cork powder coated with hardener Macroplast UK 8101 (CoP:M).
- Core of cork powder coated with resin SikaBiserin (CoP:S).
- Cork dust (CoD).
- Cork powder (CoP).
- Expanded polystyrene (EPS).
- Glass wool (GW).
- Impulse response (IR).
- Polyethylene (PE).
- Polyisocyanurate (PIR).
- Polyurethane (PUR).
- Polyurethane foam (PUF).
- Reflection coefficient (r).
- Rock wool (RW).
- Rock wool fibre (RWF).
- Sheep wool (SW).
- Sheet Moulding Compound (SMC).
- Sound absorption coefficient (α).
- Specimen of CoP:M core coated with Carbocal to be used in the impedance tube (M2).
- Specimen of CoP:M core without Carbocal to be used in the impedance tube (M5).
- Specimen of CoP:S core coated with Carbocal to be used in the impedance tube (M3).
- Specimen of CoP:S core without Carbocal to be used in the impedance tube (M6).
- Specimen of PIR core coated with Carbocal to be used in the impedance tube (M1).
- Specimen of PIR core without Carbocal to be used in the impedance tube (M4).
- Transfer function (H13).

2 Introduction

The construction's sector continues being the largest consumer of energy and also produce high greenhouse gas emissions. In the same way, this same sector is a major consumer of synthetic materials (non-biodegradable) and, therefore, is one of the biggest generators of wastes that, mostly, goes to landfill or incineration, processes that are also energy consumers and not eco-efficient [1], this triggered the entry into force of Directive 2010/31/EU of the European Parliament and of the Council on to the energy performance of buildings [2]. At the same time, inhabitants demand a level of comfort and safety in their homes that is usually solved by insulating materials such as expanded polystyrene (EPS), rock wool (RW) or polyurethane (PUR) [3, 4], which are not easily recyclable at the end of their useful life, as stated in the circular economy action plan, which stands out the need to use recycled materials due to the high pollution and the lack of the materials currently used [5]. In view of this situation, arose the need to develop and implement new "eco-friendly" construction materials that, thanks to their insulating properties and high recyclability, could help to reduce the carbon footprint and the greenhouse gas emissions of buildings [6–9]. These materials, are part of a new concept called "bioconstruction" [10–13].

The bibliography offers numerous studies applied to the design and manufacture of sandwich panels that provide properties to meet the needs of insulation or conditioned environments [14–17]. However, Ramon Miralbes et al. [18], tried to mix cork and EPS to reduce the negative impact of EPS on the environment, and applied it to the manufacture of protective helmets for motorcycles, bicycles, snow sports, horseback riding, etc. They claimed that cork is a material with a renewable origin which could be a suitable substitute for EPS, in applications where it would be necessary to absorb energy from a point impact load [18, 19]. For the manufacture of cork panels, it is recommended to use cork dust (CoD) or cork powder (CoP) because a smaller particle size improves the homogeneity of the panel, preventing the appearance of internal defects [20–23].

The bibliography is equally extensive on vegetable fibres such as flax [24], hemp [25], jute [26] and sisal [26–28] and fibres such as hazelnut [29], walnut, almond [30] or peanut [31]. Especially, Ankit et al. [32] compiled the physical properties of polymers reinforced with natural fibres, compared with the physical properties of synthetic fibres and Uppal et al. [38] published some specifications of physical and chemical properties of natural fibres. The different combinations between natural and synthetic fibres are published in Prabhu et al. [39].

All the previous studies share the lignocellulosic materials as the main components used in the manufacture of hybrid composites due to its good results, they are biodegradable and eco-friendly and they increase Young's modulus, toughness, tensile, flexural, compressive and shear strength of hybrid composites. A. Samadam et al. [33] applied lignocellulosic compounds and polyethylene (PE) for the manufacture of composites using the "hand lay-up" technique, which consists in applying layers of resin and fibre in an interspersed way until the desired thickness is reached. Reducing the consumption of PE by replacing it with cellulose is interesting in view of PE's low biodegradability and the high environmental impact. J. Guerra-Garcés et al. [34] mixed cellulosic fibres obtained from recycled tetra packs with calcium carbonate ($CaCO_3$) to make hybrid composites. $CaCO_3$ is cheap, non-toxic and biodegradable, and also good corrosion, wettability, hydrophobicity and particle adhesion properties have been observed in it. They obtain $CaCO_3$ by precipitation using carboxymethylcellulose (CMC) as a surfactant. The problem is that the process of separating and obtaining $CaCO_3$ is too expensive. However, [35] demonstrated the utility of Carbocal as a substitute by-product of $CaCO_3$, said by-product was composed of over 80% of this mineral.

From the process of obtaining sugar beet pulp, dried beet and Carbocal are obtained as wastes, which can be used as a by-product, dried beet has been used for the manufacture of bricks [36], but is currently being investigated to find new by-products applications [35, 37, 38], in turn, the sugarcane bagasse waste fibres are used as a thermal and acoustic insulating material in sustainable buildings [39].

Observing the advantages and disadvantages of the different materials mentioned on this document, and taking into account the information obtained from the bibliography, in this paper, it has been decided to use cork powder as a natural fibre to form the core of the acoustic screen. An adhesive will be applied to the core and a layer of Carbocal will be adhered to the adhesive like an acoustic veil/fabric [40]. Regarding the acoustic veil, the veil can be made of any type of material, although rock wool or laminated fiberglass is commonly used. The more porous the acoustic veil material, the better the acoustic insulation is. The acoustic insulation depends on the geometry of the material, if it has micro-pores, it facilitates the dispersion of noise [40–43]. Although the core can be manufactured by "compounding [33], "felting" [44] or "hand lay-up" [33, 45] in the present paper a compression process will be carried out to compact the core of the panel. This paper proceeds to develop an acoustic screen to protect from unwanted external noises which will be used in buildings. To comply with Directive 2010/31/UE [2], the acoustic screen to be manufactured will be composed of residual materials from companies that belong to the province of Cádiz.

3 Materials, Process and Methods

3.1 Materials

The materials used to manufacture the core were commercial polyisocyanurate (PIR) and cork powder (CoP), obtained as a waste of the packaging industry. The material used as an outer layer of the acoustic screen was Carbocal provided by AB Azucarera [35], which is a by-product of sugar beet processing, with a particle size smaller than 1 mm or between 1 and 2 mm. Furthermore, two types of water-based adhesives, Aquence CG 979 (ACG) [46] and Aquence ENV 1658-22 (ENV) [47], provided by major producers, were used to bond the acoustic veil to the core. The properties of these adhesives are shown in (see Table 1). Besides, a binder and a hardener were used to unify and harden the core of the acoustic screen according to Table 2.

Table 1. Properties of the adhesives used in this study.

Adhesive	Supplier	Technology	Drying time [min]	Application grammage [g/m^2]	Viscosity [MPa·s]	pH value	Application temperature [°C]
Aquence CG 979	Henkel	Synthetic resin	85	-	2700–3200	4.0–6.0	18–25
Aquence ENV 1658–22	Henkel	Acrylic ester	55	20–25	3600–5800	4.0–6.0	18–25

Note: The drying times had been measured for both adhesives, ENV and ACG, according to the left column of Fig. 2

3.2 Agrocomposites Manufacturing Process

Two types of agrocomposite cores were used to manufacture the insulation screens, these cores can be seen in Fig. 1 b) y c) and the Table 2 specifies which one is which. The core in Fig. 1 a) consists of commercial PIR material, which will be compared with the previous two.

Observing the Fig. 2, on the one hand, the middle column shows the result of first applying the adhesive to the PIR specimen and then applying the Carbocal by compacting it onto the adhesive. On the other hand, the adhesive was mixed with Carbocal and the

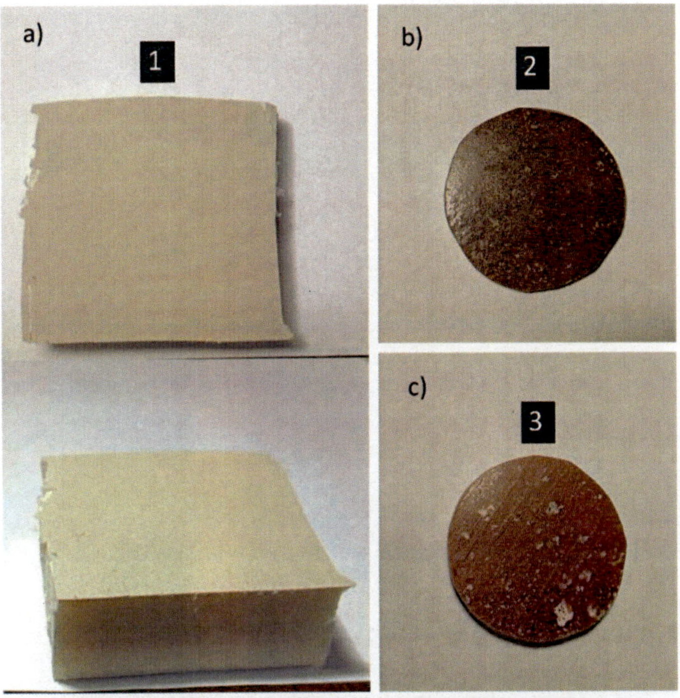

Fig. 1. Specimens' agrocomposite cores a) PIR b) CoP:S c) CoP:M.

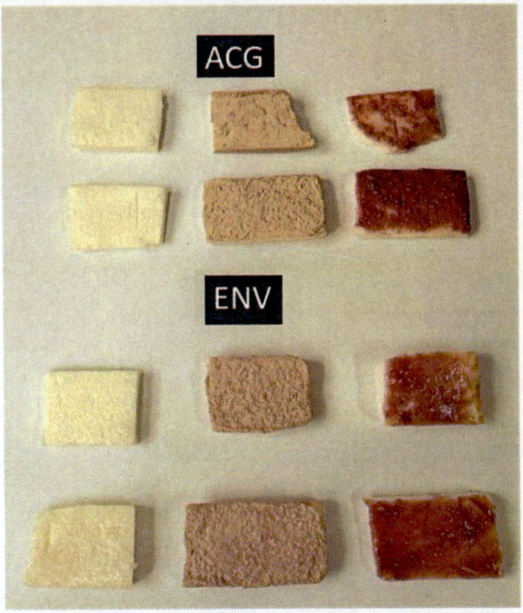

Fig. 2. Specimens of PIR with different types of coating.

Fig. 3. Specimen of PIR core coated with different particle sizes of Carbocal.

Fig. 4. Specimens of CoP:S cores coated with different particle sizes of Carbocal.

mixture was applied to the PIR specimen as shown in the right column. Specimens were compared and the most favorable result was found in the middle column because a homogeneous outer layer of Carbocal was obtained.

For this reason, eight types of specimens have been manufactured according to the middle column process, four of them with PIR cores (see Fig. 3) and the rest of them with CoP:S cores (see Fig. 4).

Table 2. Manufactured panels, core materials, adhesives and outer layers.

Numbers	Core's material	Binder - Hardener (B&H)	Viscosity of (B&H) [MPa·s]	% (B&H)	% CoP	Core density [kg/m^3]	Adhesive	Carbocal size [mm]
1	PIR	-	-	-	-	54.894 ± 0.017	-	-
2	CoP:S	Sika Biserin CR75 [48]	1800	85	15	1054.129 ± 0.051	-	-
3	CoP:M	Macroplast UK 8101 [49] – LOCTITE UK 5400 [50]	8000 ± 2000 - 250 ± 100	85	15	1241.315 ± 0.017	-	-
4	PIR	-	-	-	-	54.894 ± 0.017	ACG	0–1
5	PIR	-	-	-	-	54.894 ± 0.017	ENV	0–1
6	PIR	-	-	-	-	54.894 ± 0.017	ACG	1–2
7	PIR	-	-	-	-	54.894 ± 0.017	ENV	1–2
8	CoP:S	Sika Biserin CR75 [48]	1800	85	15	1054.129 ± 0.051	ACG	0–1
9	CoP:S	Sika Biserin CR75 [48]	1800	85	15	1054.129 ± 0.051	ENV	0–1
10	CoP:S	Sika Biserin CR75 [48]	1800	85	15	1054.129 ± 0.051	ACG	1–2
11	CoP:S	Sika Biserin CR75 [48]	1800	85	15	1054.129 ± 0.051	ENV	1–2

3.3 Methods

It was possible to verify the good adhesion of Carbocal to CoP cores, so for the acoustic absorption test, specimens of CoP:M had been also prepared. If we observe the Fig. 5, in the top row, from left to right, we can see a PIR specimen (M1), a CoP:M specimen (M2) and a CoP:S specimen (M3), all of them coated with Carbocal, while in the bottom row, the same materials can be observed in the same order but in this case without

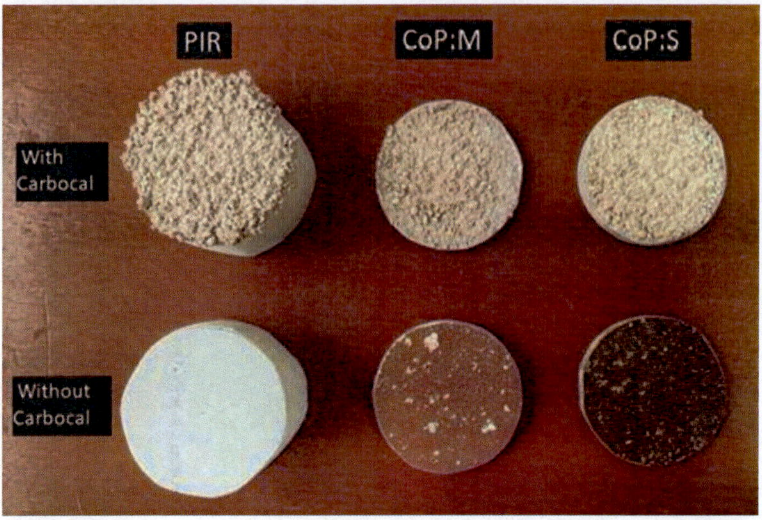

Fig. 5. Specimens of PIR, CoP:M and CoP:S prepared for the impedance tube.

Carbocal coating (M4, M5 and M6 respectively). To be used in the impedance tube, all the specimens had to have a cylindrical shape and a diameter of 40 mm. It was decided to use the adhesive ENV due to its shorter drying time and its greater adhesion once dried compared to the adhesive ACG and a Carbocal particle size between 1 and 2 mm because it can provide a better distribution and packing of the particles on the surface and it's thought that a better acoustic insulation can be achieved with this particle size.

The panels' cores were manufactured by compression, the mould compresses by means of a SMC (Sheet Moulding Compound) pneumatic cylinder, model CP96SDB80-600C, 80 mm diameter that exerts a pressure of 50 MPa. The compression is carried out with the help of a thermal source that favors the polymerization of the materials and binders to a point where it remains united by compression [51].

Panels' properties were measured according to the standard established for sound absorption in particle boards [52]. This property was measured on cylindrical specimens, with different thicknesses and a diameter of 40 mm.

The method followed to evaluate the acoustic behavior of acoustic screens was experimental, conducting measurement tests at the Physical Technologies Centre: acoustics, materials and astrophysics, at the Higher Polytechnic School of Gandia, Universitat Politècnica de València, Spain, it consists in determining the sound absorption coefficient of a material under normal incidence and it's based on the acoustic impedance tube (Kundt Tube) according to UNE EN ISO 10534-2:2002 "Determination of sound absorption coefficient and impedance in impedances tubes; Part 2: Transfer-function method (ISO 10534-2:2002)" [52].

This method uses an impedance tube (see Fig. 6), two microphone positions in impedance tube (1 and 3) and a digital signal processing system for the determination of the sound absorption coefficient (α) of sound absorbers for normal incidence. For the digital signal processing system, Dual Channel analyser hardware, and FFT analysis software of Brüel & Kjaer (PULSE), were used to determine the Impulse Response (IR) of signals from two flat response microphones.

For this, an impedance measurement tubes Kundt set was used with two flat response microphones, for measurements from 50 Hz to 6.4 Hz. Which consists of:

- A large tube with a diameter of 100 mm.
- A small tube with a diameter of 29 mm.
- Sample holder with diameters of 29 and 100 mm.
- Extension tubes with diameters of 29 and 100 mm.

They can be assembled in two configurations different:

- Large tube for frequencies from 50 Hz to 1.6 Hz.
- Small tubes for frequencies from 500 Hz to 6.4 Hz.

The large tube has a speaker at one end and a frequency weighting unit with three positions: High Pass, Linear and Low Pass. The measurements are made with microphones 1/2″ condenser type with flat response, especially designed to reduce errors due to leakage of pressure at high frequencies.

The analysis of signals was performed with the Fourier transform in order to obtain the transfer function of the system (H13) and the pressure reflection coefficient (r) from which the sound absorption coefficient (α) can be obtained. This technique requires a testing procedure to minimize differences in amplitude and phase characteristics between the two microphones positions. The same microphone was used alternatively in positions 1 and 3.

PULSE Material Testing application for acoustic characterization of pipe materials standing waves were used. It allows to measure properties of acoustic noise control materials according to ISO and ASTM standards. It works with B&K 4206 and 4206-A impedance tubes assemblies, with the transmission loss tube type 4206-T and with any other custom metering tube.

It was also made of a compact and high-performance measurement power amplifier of two-channel, designed for electroacoustic and general purposes applications. It has two independent 35 W output channels with high quality Speakon output terminals. In addition, it has two gain settings user-selectable 0 dB and 20 dB, which makes it ideal for most audio test apps. It was also made of a compact and high-performance measurement power amplifier of two-channel, designed for electroacoustic and general purposes applications. It has two independent 35 W output channels with high quality Speakon output terminals. In addition, it has two gain settings user-selectable 0 dB and 20 dB, which makes it ideal for most audio test apps.

It has an estimated frequency range of 100 to 3150 Hz and an estimated air flow resistivity between 8 and 10 kPa·s/m^2, in Normal Incidence Sound Absorption Coefficient.

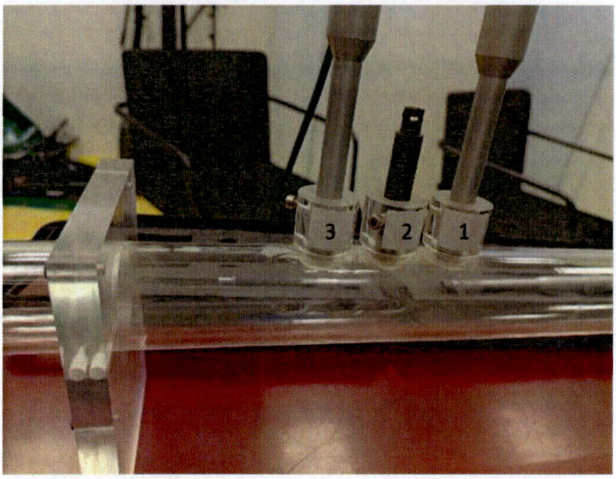

Fig. 6. Impedance tube-based instrumentation bench.

4 Results and Discussions

The tests were carried out during the month of April 2023 according to mentioned ISO standard, and the results for Normal Incidence Sound Absorption Coefficient (α) from the M1 to the M6 specimens can be seen represented (in thirds of an octave) in Fig. 7.

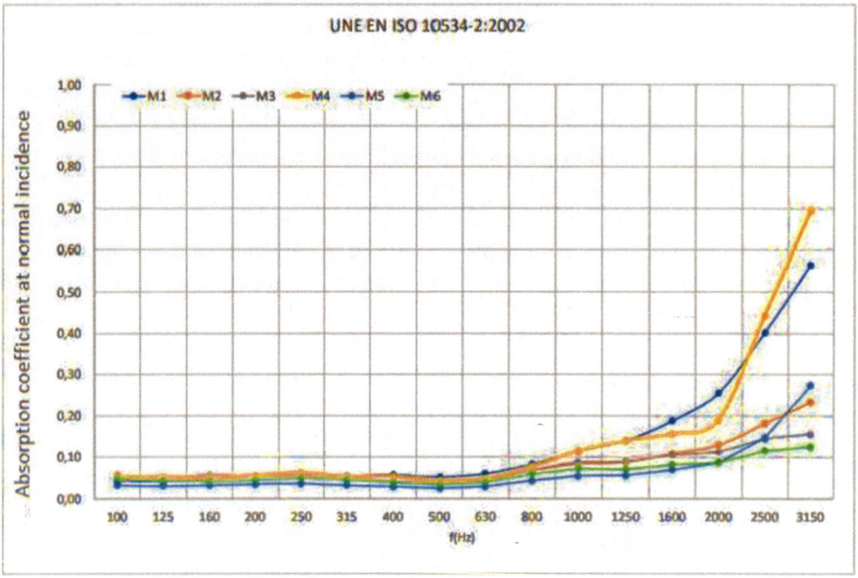

Fig. 7. Coefficient of Absorption at Normal Incidence for specimens M1 to M6.

Table 3. Classification of the sound absorption coefficient (α).

Type	(α)
A	0.90 – 0.95 – 1.00
B	0.80 – 0.85
C	0.60 – 0.65 – 0.70 – 0.75
D	0.30 – 0.35 – 0.40 – 0.45 – 0.50 – 0.55
E	0.25 – 0.20 – 0.15
Without classification	0.10 – 0.05 – 0.00

As shown in Fig. 7, the values of the absorption coefficient at normal incidence for the various materials tested from 100 Hz to 500 Hz do not even exceed the (α) value of 0.05 regardless of the thicknesses used in the specimens, which indicates a poor performance as low-frequency absorption material. However, above 500 Hz their behavior is not directly comparable due to the different thicknesses of the specimens by pairs of thicknesses M1-M4, M2-M5 and M3-M6. This leads to a detailed comparative study by pairs of equivalent specimens with and without the Carbocal coating as an acoustic veil that allows us to see its effects in a comparative way in the different PIR samples (M1-M3), CoP:M (M2-M5) and CoP:S (M3-M6).

If we look globally the Fig. 7, we can see that in general the behavior depends on the frequency and the thickness of the material as expected. However, it can be concluded that the type of recycled and sustainable material used in the specimens it´s not suitable as an acoustic absorber at low frequencies, because despite its thickness, its (α) coefficients are almost testimonial. However, it is also observed that from 800 Hz the absorption begins to be sensitive to thickness and frequency with values that exceed what gives hope of use as an absorber.

It's important to highlight that, the current regulations [53] stablished that the sound absorption coefficient values can be classified according to the Table 3. The more near to 1 is the coefficient, the more sound is able to absorb.

Observing the results of specimens M1 and M4, it can be deduced that the growth of the absorption is greater with the Carbocal treatment (M1) above a frequency of 1250 Hz compared to not using the Carbocal treatment. However, it does so at the cost of lowering the performance above 2300 Hz. The inflection point of the M4 material can be observed with an irregular behavior that the Carbocal linearizes.

When CoP:M is used, the lines M2 and M5 show the best behavior with the Carbocal (M2) treatment throughout the entire band of the spectrum studied with a more linear and predictable behaviour than without it.

Something similar can be deduced from lines M3 and M6 that also improves the absorption coefficient (α) from the average frequencies without counterparts for the thickness studied. This indicates that CoP:S is a possible base material susceptible to an uncompensated improvement in the absorption coefficient (α) for the addition of the Carbocal (M3) treatment.

In everything commented before, it must take into account that the measurements of the different specimens do not have a standardized size or an equal thickness for comparison purposes and their adjustment to the impedance tube may have consequences on the measurements made [54].

If we observe the Figs. 8, 9 and 10 obtained from the bibliography, we can appreciate the results of the acoustic absorption coefficient at low frequency tests of different insulation materials.

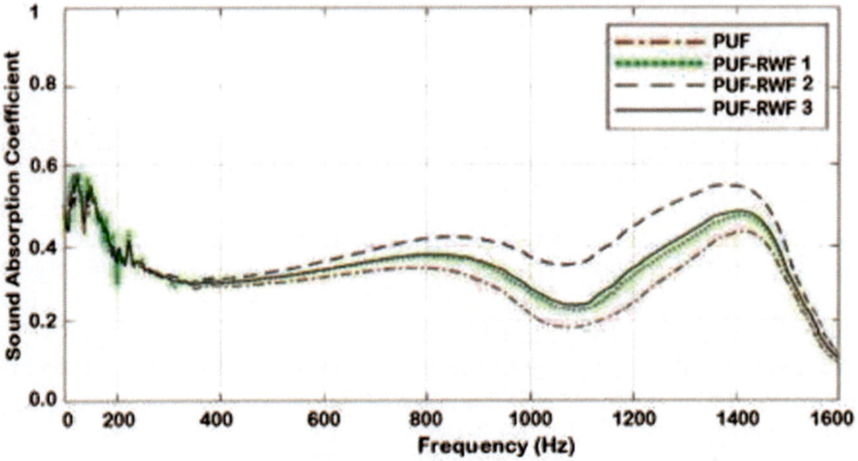

Fig. 8. Sound absorption coefficient of polyurethane foam (PUF) reinforced with rock wool fibres (RWF) contents in the frequency range of 50 to 1600 Hz [55].

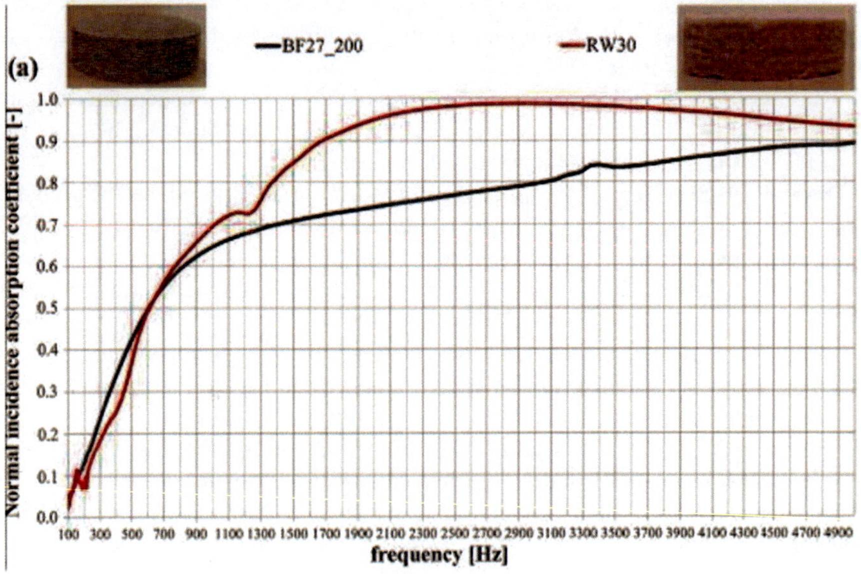

Fig. 9. Sound absorption coefficient of basalt fiber (BF) and rock wool (RW) in the frequency range of 100 to 4900 Hz [56].

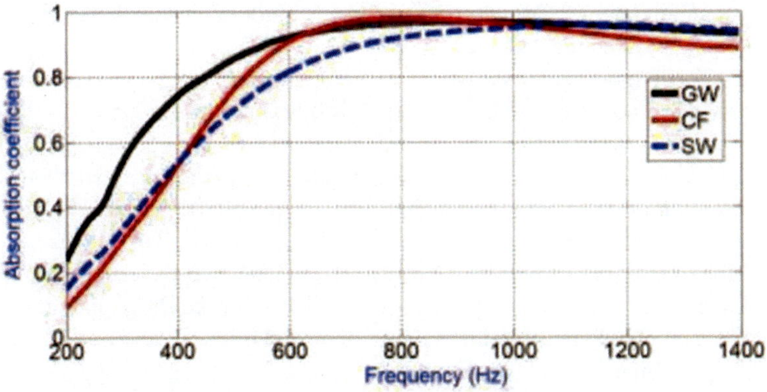

Fig. 10. Sound absorption coefficient of glass wool (GW) compared with chicken feathers (CF) and sheep wool (SW) [57].

In the Fig. 8, it can be observed that the values of acoustic absorption coefficient for specimens of PUF reinforced with RWF fluctuate between 0.2 and 0.6. These fluctuations can be linearized through the use of Carbocal coating on this material, as has been done with the PIR specimens in this study.

Figures 9 and 10 depict 5 types of materials, BF, RW, GW, CF and SW, all of them present a very good quality acoustic absorption coefficient. A Carbocal coating can be applied to these materials to further enhance these results, as demonstrated in this study with CoP:S and CoP:M specimens.

5 Conclusions

On the one hand, several interesting things can be concluded from the study. The first is that the core materials studied, of a sustainable nature and driver of a circular economy are still far from improving the absorption results at normal incidence (α) of those core materials that are already used and that are known as high efficiency materials at low frequency (at least with those tested thicknesses).

On the other hand, the highest values of the Fig. 7 show that the PIR core material used for the measurements has a behavior with an inflection point that could allow its use as a sustainable absorber only at high frequencies from 2300 Hz. The use of new measurements with a Kundt tube that could also measure the upper octave would make it possible to confirm this behavior. The use of Carbocal shifts the absorption coefficient towards low frequencies, which is an improvement in absorption performance from an energetic point of view.

The use of Carbocal seems to achieve improvements in the absorption coefficient at normal incidence (α) with the appropriate cores. With CoP:S and CoP:M cores, they show very encouraging results, since Carbocal with these cores improves throughout the entire spectrum in a very uniform and linear way. A more detailed study of these combinations of materials with a greater thickness could illuminate a better acoustic solution useful for the acoustic treatment in building acoustics and other industrial applications considering the ecological and sustainable factor.

The Carbocal developed in this study can be used as a coating on conventional cores that present good acoustic insulation results, linearizing and further enhancing these results.

It's a primary stage of the investigation and better results are expected to be obtained if a deeper and more detailed investigation than the one on the present study is carried out, regarding insulation materials.

Nevertheless, we can conclude that by using by-products from agricultural industry, similar properties to the commonly used insulation products could be obtained. In addition, a reduction in carbon footprint and environmental impact could be achieved while maintaining a similar level of comfort.

Acknowledgements. The authors of this work would like to thank "Azucarera Iberia S.L.U.", the group "Nueva Comercial Azucarera S.A." and "Henkel Ibérica" for the supply of products. This work is supported by the AGROCOM project funded by the Call for Operational Groups 2020 of the Junta de Andalucía (GOPC-CA-20-0004).

References

1. Revuelta-Aramburu, M., Verdú-Vázquez, A., Gil-López, T., Morales-Polo, C.: Environmental analysis of the use of plant fiber blocks in building construction. Sci. Total Env. **725**, 138495 (2020). https://doi.org/10.1016/j.scitotenv.2020.138495
2. Diario Oficial de la Unión Europea núm. 153. Directiva 2010/31/UE del Parlamento Europeo y del Consejo, de 19 de mayo de 2010, relativa a la eficiencia energética de los edificios. https://www.boe.es/doue/2010/153/L00013-00035.pdf (2010). Accessed 02 Dec 2022
3. Yanes-Armas, S., de Castro, J., Keller, T.: Long-term design of FRP-PUR web-core sandwich structures in building construction. Compos. Struct. **181**, 214–228 (2017). https://doi.org/10.1016/j.compstruct.2017.08.089
4. Proença, M., Garrido, M., Correia, J.R., Gomes, M.G.: Fire resistance behaviour of GFRP-polyurethane composite sandwich panels for building floors. Compos. Part B: Eng. **224**, 109171 (2021). https://doi.org/10.1016/j.compositesb.2021.109171
5. Comunicación de la comisión al parlamento Europeo al consejo. Plan de acción para la economía circular COM-2020-98. Brusels 11-3-2020, Bruselas. https://eur-lex.europa.eu/legal-content/EN/TXT/?uri=COM%3A2020%3A98%3AFIN (2020). Accessed 08 Jan 2023
6. Sala, B., et al.: Creep behaviour of eco-friendly sandwich composite materials under hygrothermal conditions. Compos. Part B: Eng. **247**, 110291 (2022). https://doi.org/10.1016/j.compositesb.2022.110291
7. Salunkhe, S., Patil, C., Thakar, C.M.: Exploring the potential of natural materials as eco-friendly sound absorbers. Mater Today Proc. (2023). https://doi.org/10.1016/j.matpr.2023.03.098
8. Silva, G., Kim, S., Aguilar, R., Nakamatsu, J.: Natural fibers as reinforcement additives for geopolymers – A review of potential eco-friendly applications to the construction industry. Sustain. Mater. Technol. **23**, e00132 (2020). https://doi.org/10.1016/j.susmat.2019.e00132
9. Oliveira, P.R., May, M., Panzera, T.H., Hiermaier, S.: Bio-based/green sandwich structures: a review. Thin-Walled Struct. **177**, 109426 (2022). https://doi.org/10.1016/j.tws.2022.109426

10. Yadav, M., Agarwal, M.: Biobased building materials for sustainable future: an overview. Mater. Today: Proc. **43**, 2895–2902 (2021). https://doi.org/10.1016/j.matpr.2021.01.165
11. Weiss, M., et al.: A Review of the environmental impacts of biobased materials. J. Ind. Ecol. **16**(SUPPL. 1), S169–S181 (2012). https://doi.org/10.1111/j.1530-9290.2012.00468.x
12. Ouakarrouch, M., Bousshine, S., Bybi, A., Laaroussi, N., Garoum, M.: Acoustic and thermal performances assessment of sustainable insulation panels made from cardboard waste and natural fibers. Appl. Acoust. **199**, 109007 (2022). https://doi.org/10.1016/j.apacoust.2022.109007
13. Fischer, H., Frühwald, F., Korjenic, A.: Ecological comparison of hygrothermally safe floor constructions based on renewable raw materials for multi-storey buildings. J. Build. Eng. **57**, 104899 (2022). https://doi.org/10.1016/j.jobe.2022.104899
14. Davies, J.M.: Sandwich panels. Thin-Walled Struct. **16**(1–4), 179–198 (1993). https://doi.org/10.1016/0263-8231(93)90044-B
15. Boccarusso, L., et al.: Design, manufacturing, and characterization of hybrid carbon/hemp sandwich panels. J. Mater. Eng. Perform. **31**(1), 769–785 (2022). https://doi.org/10.1007/s11665-021-06186-1
16. Stefan, A., Pelin, G., Pelin, C.-E., Petre, A.-R., Marin, M.: Manufacturing process, mechanical behavior and modeling of composites structures sandwich panel. INCAS Bull. **13**(1), 183–191 (2021). https://doi.org/10.13111/2066-8201.2021.13.1.19
17. Studziński, R.: Optimal design of sandwich panels with hybrid core. J. Sandwich Struct. Mater. **21**(7), 2181–2193 (2019). https://doi.org/10.1177/1099636217742574
18. Buil, R.M., Angulo, D.R., Ivens, J., Blasco, J.O.A.: Experimental study of natural cork and cork agglomerates as a substitute for expanded polystyrene foams under compressive loads. Wood Sci. Technol. **55**(2), 419–443 (2021). https://doi.org/10.1007/s00226-020-01254-6
19. Gomes, M.G., Flores-Colen, I., Melo, H., Soares, A.: Physical performance of industrial and EPS and cork experimental thermal insulation renders. Constr. Build. Mater. **198**, 786–795 (2019). https://doi.org/10.1016/j.conbuildmat.2018.11.151
20. Pintor, A.M.A., et al.: Use of cork powder and granules for the adsorption of pollutants: a review. Water Res. **46**(10), 3152–3166 (2012). https://doi.org/10.1016/j.watres.2012.03.048
21. Reis, P.N.B., Silva, M.P., Santos, P., Parente, J.M., Bezazi, Abderrezak: Viscoelastic behaviour of composites with epoxy matrix filled by cork powder. Compos. Struct. **234**, 111669 (2020). https://doi.org/10.1016/j.compstruct.2019.111669
22. Magalhães da Silva, S.P., Oliveira, J.M.: Cork powders wettability by the Washburn capillary rise method. Powder Technol. **387**, 16–21 (2021). https://doi.org/10.1016/j.powtec.2021.04.005
23. Gil, L.: "PII: s0961–9534(97)ooo33–0 Cork Powder Waste: An Overview (1997)
24. Arbelaiz, A., Fernández, B., Valea, A., Mondragon, I.: Mechanical properties of short flax fibre bundle/poly(ε-caprolactone) composites: influence of matrix modification and fibre content. Carbohydr. Polym. **64**(2), 224–232 (2006). https://doi.org/10.1016/j.carbpol.2005.11.030
25. Bhowmik, R., Das, S., Mallick, D., Gautam, S.S.: Predicting the elastic properties of hemp fiber – a comparative study on different polymer composite. Mater. Today Proc. **50**, 2510–2514 (2022). https://doi.org/10.1016/j.matpr.2021.09.562
26. Sai Shravan Kumar, P., Viswanath Allamraju, K.: Experimental study and characterization of glass, jute & sisal fiber reinforced polymer matrix composites. Mater. Today Proc. **44**, 941–948 (2021). https://doi.org/10.1016/j.matpr.2020.11.003
27. Kumar, N., Grewal, J.S., Kumar, S., Kumar, N., Kashyap, K.: Mechanical and thermal properties of NaOH treated sisal natural fiber reinforced polymer composites: Barium sulphate used as filler. Mater. Today Proc. **45**, 5575–5578 (2021). https://doi.org/10.1016/j.matpr.2021.02.310

28. Badyankal, P.V., Manjunatha, T.S., Vaggar, Gurushanth B., Praveen, K.C.: Compression and water absorption behaviour of banana and sisal hybrid fiber polymer composites. Mater. Today: Proc. **35**, 383–386 (2021). https://doi.org/10.1016/j.matpr.2020.02.695
29. Balart, J.F., Fombuena, V., Fenollar, O., Boronat, T., Sánchez-Nacher, L.: Processing and characterization of high environmental efficiency composites based on PLA and hazelnut shell flour (HSF) with biobased plasticizers derived from epoxidized linseed oil (ELO). Compos. B Eng. **86**, 168–177 (2016). https://doi.org/10.1016/j.compositesb.2015.09.063
30. Pirayesh, H., Khanjanzadeh, H., Salari, A.: Effect of using walnut/almond shells on the physical, mechanical properties and formaldehyde emission of particleboard. Compos. B Eng. **45**(1), 858–863 (2013). https://doi.org/10.1016/j.compositesb.2012.05.008
31. Güler, C., Büyüksarı, Ü.: Effect of production parameters on the physical and mechanical properties of particleboards made from peanut (Arachis hypogaea L.) HULL. BioResources **6**(4), 5027–5036 (2011)
32. Ankit, Rinawa, M., Chauhan, P., Suresh, D., Kumar, S., Santhosh Kumar, R.: A review on mechanical properties of natural fiber reinforced polymer (NFRP) composites. Mater. Today Proc. (2021). https://doi.org/10.1016/j.matpr.2021.07.275
33. Samadam, A., Vallepalli, R., Neeraj Kumar, K., Sreekanth, M., Raman Goud, R.: Mechanical properties evaluation and behaviour of cellulose-HDPE composite. Mater. Today: Proc. **62**, 3405–3410 (2022). https://doi.org/10.1016/j.matpr.2022.04.271
34. Guerra-Garcés, J., et al.: Morphologically diverse CaCO3 microparticles and their incorporation into recycled cellulose for circular economy. Mater. Today Sustain. **19**, 100166 (2022). https://doi.org/10.1016/j.mtsust.2022.100166
35. Suffo, M., de la Mata, M., Molina, S.I.: A sugar-beet waste based thermoplastic agro-composite as substitute for raw materials. J. Cleaner Prod. **257**, 120382 (2020). https://doi.org/10.1016/j.jclepro.2020.120382
36. Harb, E., et al.: Thermal performance of starch/beet-pulp composite bricks for building insulation at a wall scale. Case Stud. Constr. Mater. **18**, e01851 (2023). https://doi.org/10.1016/j.cscm.2023.e01851
37. Usmani, Z., et al.: Valorization of sugar beet pulp to value-added products: a review. Bioresour. Technol. **346**, 126580 (2022). https://doi.org/10.1016/j.biortech.2021.126580
38. Garcia Gonzalez, M.N., Björnsson, L.: Life cycle assessment of the production of beet sugar and its by-products. J. Cleaner Product. **346**, 131211 (2022). https://doi.org/10.1016/j.jclepro.2022.131211
39. Mehrzad, S., Taban, E., Soltani, P., Samaei, S.E., Khavanin, A.: Sugarcane bagasse waste fibers as novel thermal insulation and sound-absorbing materials for application in sustainable buildings. Build. Environ. **211**, 108753 (2022). https://doi.org/10.1016/j.buildenv.2022.108753
40. Sealy, C.: Acoustic fabric detects sound in a heartbeat. Nano Today **44**, 101495 (2022). https://doi.org/10.1016/j.nantod.2022.101495
41. Darwish, E.A., Midani, Mohamad: The potential of date palm midribs-based fabric acoustic panels for sustainable interior design. Ain Shams Eng. J. **14**(6), 102100 (2023). https://doi.org/10.1016/j.asej.2022.102100
42. Karimi, F., Soltani, P., Zarrebini, M., Hassanpour, A.: Acoustic and thermal performance of polypropylene nonwoven fabrics for insulation in buildings. J. Build. Eng. **50**, 104125 (2022). https://doi.org/10.1016/j.jobe.2022.104125
43. Tang, X., Kong, D., Yan, X.: Facile dip-coating method to prepare micro-perforated fabric acoustic absorber. Appl. Acoust. **130**, 133–139 (2018). https://doi.org/10.1016/j.apacoust.2017.09.021
44. Dénes, T.-O., et al.: Analysis of Sheep wool-based composites for building insulation. Polymers (Basel) **14**(10), 2109 (2022). https://doi.org/10.3390/polym14102109

45. Uppal, N., Pappu, A., Gowri, V.K.S., Thakur, V.K.: Cellulosic fibres-based epoxy composites: from bioresources to a circular economy. Ind. Crops Prod. **182**, 114895 (2022). https://doi.org/10.1016/j.indcrop.2022.114895
46. AQUENCE CG 979. www.henkel.com (2018)
47. FT AQUENCE ENV 1658-22. Adhesin J 1658-22. www.henkel.com. Accessed: 12 Apr 2023
48. SikaBiresin ® CR75 SikaBiresin ® CR75 (RSF 816 G). Accessed: 25 Apr 2023. www.sika.de
49. Macroplast UK 8101. www.henkel.com. Accessed: 25 Apr 2023
50. LOCTITE UK 8103/LOCTITE UK 5400. www.henkel.com
51. Suffo, M., Molina-Pérez, J.F., Lloret, F.: Recycled hybrid material for use as shielding in operations with ionizing radiation. Cleaner Mater. **7**, 100175 (2023). https://doi.org/10.1016/j.clema.2023.100175
52. UNE-EN ISO 10534-2:2002 Acústica. Determinación del coeficiente de absorción acústica y de la impedancia acústica en tubos de impedancia. Parte 2: Método de la función de transferencia.
53. UNE-EN ISO 11654:1998. Acústica. Absorbentes acústicos para su utilización en edificios. Evaluación de la absorción acústica (1998)
54. Lou, H., Li, M., Fu, Y., Min, H.: Measurements of the normal incidence sound absorption coefficient for Non-standard sized absorbers in impedance tubes. Measurement **203**, 111989 (2022). https://doi.org/10.1016/j.measurement.2022.111989
55. Mohammadi, B., Safaiyan, A., Habibi, P., Moradi, G.: Evaluation of the acoustic performance of polyurethane foams embedded with rock wool fibers at low-frequency range; design and construction. Appl. Acoust. **182**, 108223 (2021). https://doi.org/10.1016/j.apacoust.2021.108223
56. Moretti, E., Belloni, E., Agosti, F.: Innovative mineral fiber insulation panels for buildings: thermal and acoustic characterization. Appl. Energy **169**, 421–432 (2016). https://doi.org/10.1016/j.apenergy.2016.02.048
57. Bousshine, S., Ouakarrouch, M., Bybi, A., Laaroussi, N., Garoum, M., Tilioua, A.: Acoustical and thermal characterization of sustainable materials derived from vegetable, agricultural, and animal fibers. Appl. Acoust. **187**, 108520 (2022). https://doi.org/10.1016/j.apacoust.2021.108520

Sustainable Energy Generation and Management. Information Technology and Artificial Intelligence Applied to Sustainability

The Influence of the Fan-Controlled Chimney Draft on the Pollutants Emission to the Environment

Krzysztof Drozdzol[1](✉), Robert Junga[2], Damian Beben[1], Thor Kielland[3], and Pawel Jarzynski[4]

[1] Faculty Civil Engineering and Architecture, Opole University of Technology, Katowicka 48, 45-061 Opole, Poland
{k.drozdzol,d.beben}@po.edu.pl

[2] Faculty of Mechanical Engineering, Opole University of Technology, Mikolajczyka 5, 45-271 Opole, Poland
r.junga@po.edu.pl

[3] Pipe Eksperten As, B R A veien 55, 1782 Halden, Norway
Thork@halden.net

[4] Jawar sp. z o.o., Sonska 89, 06-400 Ciechanow, Poland
pawel.jarzynski@jawar.com.pl

Abstract. Chimney draft affects the proper operation of heating devices. The burning rate of fuel, the efficiency of heating devices, and the number of pollutants in the exhaust depend on this parameter. This determines the amount of emission of harmful substances into the atmosphere, including e.g. particle matter. Appropriate chimney draft has an impact on reducing the concentration of harmful substances emitted in exhaust gases into the atmosphere during the operation of heating devices. This article presents experimental studies showing differences in the concentration of pollutants emitted to the atmosphere during the operation of a wood stove. The analyses were performed for a stove connected to a chimney with a natural draft and to a fan – which changed the draft depending on the stage of fuel combustion. The main attention in the performed analyses was focused on the concentration of particle matter emitted with exhaust gases. Tests have shown that equipping the chimney with a fan generating a variable chimney draft can reduce the particle matter emission of pollutants into the atmosphere. As part of the tests, in addition to particle matter, the concentrations of CO, NO_x were measured. These studies may be the beginning of a new approach to reducing pollutants emitted to the atmosphere during solid fuel combustion, and as result, better protection of the environment.

Keywords: draft generator · stove pollution · fan · environmental protection

J. F. L. C. Semião et al. (Eds.): INCREaSE 2023, ASST, pp. 141–156, 2023.
https://doi.org/10.1007/978-3-031-44006-9_11

1 Introduction

1.1 Description of the Problem

During the operation of heating devices, including in particular those fired by solid fuels, pollutants are emitted into the atmosphere. This is a global problem, especially in countries, where wood-burning stoves have a large share in the emission of such pollutants [1–3]. If fuel (ex. wood) is burned in an inappropriate way, concentrations of exhausted pollutants such as particle matter, CO, NO_x, etc. will occur. This poses a danger to human health [4–6]. During operation of heating devices, such as stoves or fireplaces, the emission can increase when:

- the fuel is of inappropriate quality (ex. the moisture concentration in wood is too high),
- the heating device is damaged or improperly used or maintained,
- the heating device is improperly ventilated.

1.2 Impact of Air Supply to Stove Combustion Chamber on Pollutant Emissions

Ventilation of the stove combustion chamber, apart from its design, depends on the amount of air supplied to the stove, the amount of burned fuel, fuel bed size and combustion time. In view of the above, it can be concluded that the chimney draft has an impact on the number of pollutants emitted. As part of the research carried out so far at the Auburn University, measurements of pollutant emissions in flue gases and unburned residues in the bottom ashes from three types of fireplace devices and cast-iron stoves burned with seasoned and "green" wood were carried out. The results showed that the air excess ratio has a positive effect on the combustion with little loss of fuels. In addition, it was shown that cast iron stoves had a more negative impact on the emission of harmful substances into the atmosphere [7]. In addition to the type of fuel and the construction of the stove, as mentioned above, the amount of air supplied to the stove has a significant impact on the emission of pollutants [8]. The amount of air supplied to the stove is closely related to the method of operating the stove and adjusting the amount of air supplied by closing or opening the damper. It turns out that in the case of combustion with insufficient air (oxygen) supply (the so-called cold or closed-damper combustion); particles that consist mainly of organic carbon with negligible amounts of trace elements are generated. In contrast, hot (damper—open) combustion produces aerosols with relatively lower total carbon content.

1.3 Influence of Chimney Structure and Chimney Draft

As part of the work [9, 10], the impact of the chimney system and its diameter on the concentration of particle matter in the flue gas was examined [11]. In addition, these studies have shown that the supply of pre-heated air from the flue gas through a chimney with an air space does not significantly affect the concentration of soot in the flue gas. Soot concentration was determined as soot number (smoke number). Soot number was specified by extraction of the exhaust gas via a filter paper, on the surface of which a so-called soot stain is formed. Then with the aid of the comparison nine point scale

(Bacharach scale), the soot spot is evaluated, and the so-called soot number is determined as a result.

This work also concerns the subject of the efficiency of heat recovery from the chimney with air space, which in effect acts as a counter-current exchanger and allows the supply of pre-heated air to the stove. The chimney draft usually exceeds 12 Pa for stoves, which are in accordance with the test procedure EN 13240:2001 + AC:2003 [12]. A change in the chimney draft may have a positive or negative effect on various types of combustion chambers, in each case, a greater chimney draft decreased the device's efficiency. In the work [13], it was shown that the mechanism controlling the regulation of the inlet air to the combustion chamber based on temperature and O_2 measurements could significantly reduce CO emissions to the atmosphere.

2 Description of the Tests

The main objective of the research was to determine the concentrations of pollutants emitted during the combustion of wood in a stove with a fan with chimney draft control and in the same stove, but with a natural draft.

As part of the research carried out in the first stage, a number of measurements were taken to determine the concentration of pollutants in the flue gases, which are discharged through the flue pipes. The study presents the results of measurements of concentrations of gaseous substances (CO_2, CO, and NO_x) and particulate matter (on-line waveforms) in a wood-burning stove. The series of tests lasted approximately 15 min each. The testing involved 3 measurement sessions each for the stove with and without a fan. Based on the measurement results obtained, the outlet loss and the incomplete combustion loss were calculated using simplified methods (e.g. SEGIERT's formula) and assuming the predominance of carbon monoxide (CO) in the flue gases [14]. In the simplified method the combustion efficiency is calculated using formula:

$$\eta = 100 - qA - qB, \tag{1}$$

where:

qA – flue gas loses, %,
qB – incomplete loses, %.

Flue gas loses are estimated by the "Siegert" formula:

$$qA = (T_s - T_o)\left(\frac{A}{CO_2} + B\right), \tag{2}$$

where:

T_s – flue gas temperature, K,
T_o – ambient temperature, K,
CO_2 – carbon dioxide concentration in the flue gas, %,
A, B – Siegert's specific factors for a given fuel, -.

The incomplete losses are calculated according to the equation:

$$qB = \frac{\alpha \cdot CO}{CO + CO_2}, \tag{3}$$

where:

CO – carbon monoxide concentration, %,
α – fuel factor, % (69 for solid fuel).

Other non-combusted hydrocarbons were not taken into account in these calculations. On this basis, the efficiency of the heating device was determined. The study was performed for two cases:

1. For the natural draught. Three separate fifteen-minute measurement sessions were taken to analyse the resulting measurements from the time the wood was replenished. The flue gas flow pressure was 4–5 Pa.
2. For a forced draught with a fan controlled by the controller. This phase of the study involved a 45-minute measurement consisting of three 15-minute sessions, where fuel was replenished after 15 min. The studies were carried out in one cycle to get a better illustration of the process changes in relation to the pressure increase caused by faster fan operation during the refilling of wood.

The parameters of the measurement equipment used during the study are summarized in Table 1.

3 Description of the Test Stand

3.1 Fuel, Stove and Test Condition

The main objective of the research was to determine the concentrations of pollutants emitted during the combustion of wood in a stove with a fan with chimney draft control and in the same stove, but with a natural draft.

During the tests, wood (birch) with moisture in the range of 12–20% was burned on the stove. The wood used was properly seasoned for drying beforehand. This fuel complies with the requirements of the manufacturer of the heating device.

All measurements were carried out in real conditions, but efforts were made to perform the tests in similar weather conditions. The test stand was performed according to the PN-EN 15259: 2011 standard [15]. Location of the measurement points is similar to those used for testing heating devices in accordance with EN 13240 [16], NS 3058-59 [17], NS 3059 [18] and described in the report by Sevault et al. [19] and in the publications of Seljeskog et al. [20] and Drozdzol [9, 10]. Moreover, the research was conducted in a manner analogous to those described in [9].

The measurements were carried out for the chimney with natural and forced by a fan the draft, regulated by an adjustable prototype controller. The results were determined for the heating device (stove) with a power of 4 kW and an efficiency of approx. 83%. The device met the requirements of European standards and was equipped with an afterburning function. This kind of stove has a secondary air inlet. Secondary burn

Table 1. Specifications of the equipment used during the tests.

Parameter	Method	Range	Unit	Accuracy	Resolution
Flue gas analyser TESTO 330					
O_2	electrochemical	0–21	vol. %	±0.2	0.1
CO, H_2-comp	electrochemical	0–8000	ppm	±10 ppm or ±10%* (0…200) ±20 ppm or ±5%* (201…2000) ±10% (2001…8000) 8000-30000 ppm (automatic dilution)	1
NO	electrochemical	0–300	ppm	± 0.2 ppm (0-39.9) ± 5% (>39.9)	0.1
Temperature	k-type thermocouple	0…1000	°C	±0.5 °C (0.0 … 100) ±0.5% (>100)	0.1
Particle matter meter TESTO 380					
PM	gravimetric	0–300	mg/m^3	**	0.1

Note: * - higher value is valid, ** - the extended uncertainty for the total system is reported after each measurement series in the data acquisition software TESTO.

works by burning off the initial smoke produced from the fire that otherwise would have gone up the chimney. The aim of the research was to determine the changes in the concentration of the selected gases in the exhaust flue after the fan installation, and not to assess the emission of the heating device, therefore the paper intentionally omits the name of the manufacturer and model of the heating device.

3.2 Chimney Description

A test installation was designed to carry out the research, consisting of a ceramic-concrete chimney (Fig. 1). The installation used consisted of (i) expanded clay concrete block, (ii) mineral wool insulation and (iii) a pipe made of isostatic ceramics. The internal diameter of the chimney was 160 mm. During the tests, the chimney system was equipped with an exhaust fan and a draft regulator. The air to the hearth was supplied from the outside of the building.

Thermal resistance presented in Table 2 characterized the individual elements of the chimney system.

3.3 Description of Extraction Fan

During the tests, an Exodraft RS type exhaust fan was used, regulated by a prototype control (Fig. 2). The set's task was to change or maintain the target pressure according to the following algorithm:

a) b)

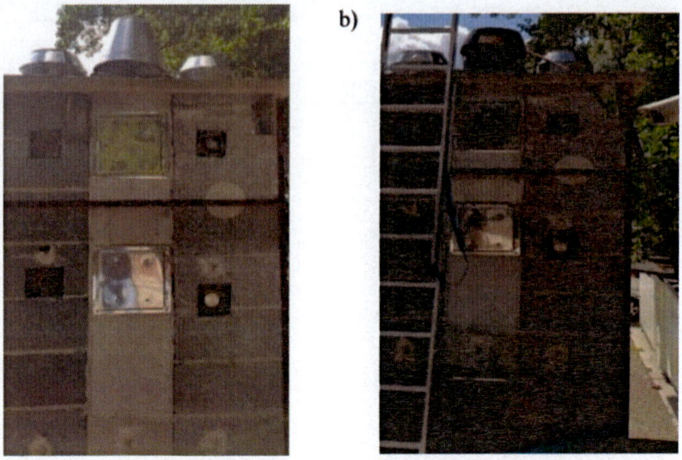

Fig. 1. Photos of the upper part of the chimney during tests: a) without a fan, b) with fan.

Table 2. The thermal resistance of materials used to build chimney systems.

Thermal resistance, m^2 K/W			
Ceramic chimney pipe, 15 mm thick	Mineral wool insulation	Airspace	Hollow blocks
0.1	0.98	~0.1	~0,8

1. Start 30–40 Pa; gradual slowing down of the fan operation within 10 min to that proposed by the manufacturer (in our case 13 Pa); maintaining a constant draft.
2. Refuelling: Open door (30–40 Pa); gradual slowing down of the fan operation within ~5 min to that proposed by the manufacturer (in our case 13 Pa).

4 The Course of Testing

4.1 Test Conditions

The tests were carried out on days when the weather conditions were similar. The wind speed ranged from 0.22–0.69 m/s, air humidity 33.8–44.5%, outside air temperature 19–25 °C, and atmospheric pressure 990.5–1000.3 hPa.

4.2 Kindling and Refuelling

Before kindling, two larger logs of wood were placed on the grate. A pile of smaller wood was placed on top of the logs, and a firelighter was lit at the top. The flame lit the wood from top to bottom (top – downstream). During kindling, the air supply was fully open and the door was left ajar (open approx. 1 cm). After ignition (about 3–5 min),

Fig. 2. View of the prototype control used.

the door was closed and the air inlet damper was changed to the working position, as recommended by the manufacturer.

By refuelling, pieces of wood in the amount of a maximum of 0.7 kg were placed in the vicinity of the embers on the grate. After the wood had ignited for even combustion, the secondary air was re-adjusted.

4.3 Conducted of Measurements

Before the measurements were taken, in order to stabilize the system, the fire in the stove was kept on for about 3 h. After that time, tests were started and measurements were made. Attempts were made to perform the measurements each time in the second phase of the wood combustion degassing phase). The exhaust gas velocity during the tests was 1.01–1.61 m/s (value resulted from the pressure control practiced). The use of a fan did not significantly change the flow. This measurement was carried out with a Pitot tube connected to the TSI 9565 m (Fig. 3). The exhaust gas temperature at the inlet of the chimney was constantly monitored, and was in the range of 180–190 °C. The optimal flue gas temperature, defined by the manufacturer of the heating device is 243 °C, which was difficult to achieve and only occurred temporarily. The position of the measuring points on the chimney duct is shown in Fig. 4.

The research was performed in two scenarios:

1. After installing the calibrated fan control. At this stage, measurements were performed for the chimney draft (pressure) changed according to the algorithm presented in Sect. 3.3. This action has no effect on the change of the flue gas parameters as compared to the changes made automatically in the device. At this stage, tests were additionally carried out with the installed draft regulating flap (mechanical regulator). In practice, the draught was stable, so the flap did not open during the tests. The draft deviation noted during the tests was ~1.8 Pa. The difference in draft between the set on the controller and the actual draft recorded in the chimney was a maximum of 3 Pa (Table 3).

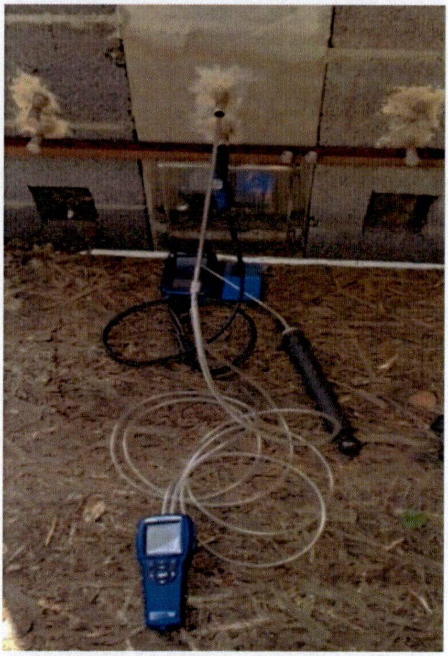

Fig. 3. Measuring equipment used.

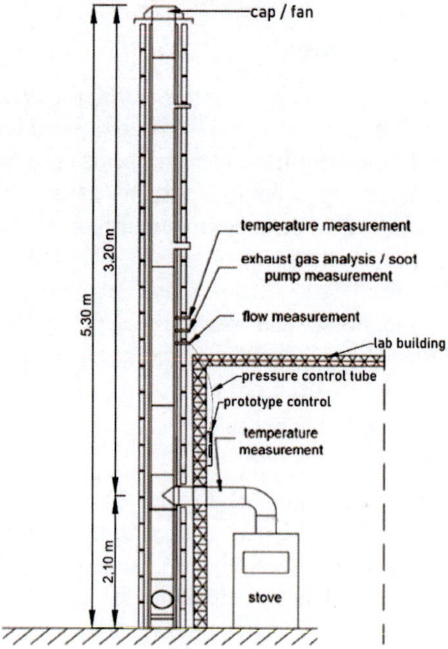

Fig. 4. Design of the research installation with sensors used.

Table 3. The difference between the measured draft and the draft set on the controller for scenario no 1

Draft measured in the chimney, Pa	Draft set on the controller, Pa	Difference Pa
16.1	14.0	2.1
17.0	14.0	3.0
15.4	13.0	2.4
15.9	13.0	2.9
15.0	12.0	3.0
14.8	12.0	2.8
13.9	11.0	2.9
13.9	11.0	2.9

2. Without using the fan, with about 3–4 Pa of natural draft. The required 12 Pa was not obtained. As weather conditions have a significant influence in this case, the tests were carried out on two different dates with 3 measurement sessions each.

5 Tests results

The graphs below (Figs. 5, 6, 7, 8 and 9) shows the results of the exhaust gas analysis and soot number performed for the Scenario no. 1 with the fan equipped with control set and for the Scenario no. 2 in a natural draft were compared. Parameters are measured at some point in the test after the system has stabilized. The measurement was always made in the second phase of the combustion process. The following sections show the observations from the values obtained during the tests:

a) The oxygen (O_2) content in the exhaust gas increased by an average of 1.6%; the maximum oxygen content noted during the tests was 18.8% for Scenario no. 1 with a controlled fan and 16.9% for Scenario no. 2 without a fan (Fig. 5).

b) The carbon dioxide (CO_2) content presented in Fig. 6 decreased by an average of 1.55% after the use of the fan (Scenario no. 1). The maximum CO_2 value measured for the installation with a fan was 3.6%, and for the installation without a fan, 5%, which means that the maximum CO_2 value decreased by 1.4 percentage points after using the fan.

c) The carbon monoxide (CO) content (Fig. 7). The average content of carbon monoxide in the exhaust gas increased by 1202 [mg/m^3 for 13% O_2] after the use of the fan. The maximum CO value for the system with a fan was 3441 [mg/m$_3$ for 13% O_2], and for the system without a fan, it was 1439 [mg/m^3 for 13% O_2] (series 2). The maximum CO value increased by 2002 [mg/m^3 for 13% O_2] after the use of the fan.

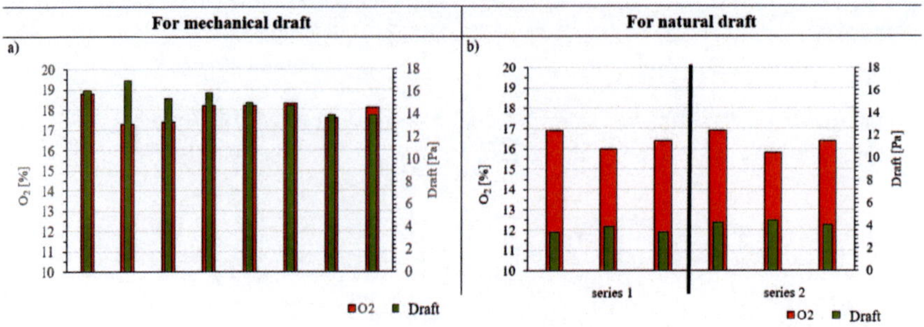

Fig. 5. Oxygen content of the flue gases for: a) mechanical draft, b) natural draft.

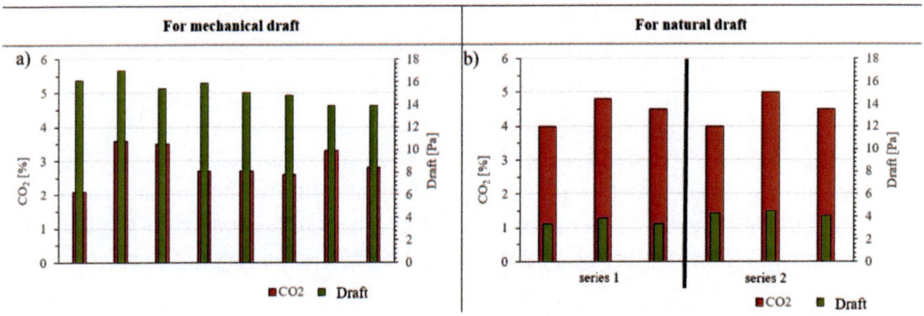

Fig. 6. CO_2 content in flue gases for: a) mechanical draft, b) natural draft.

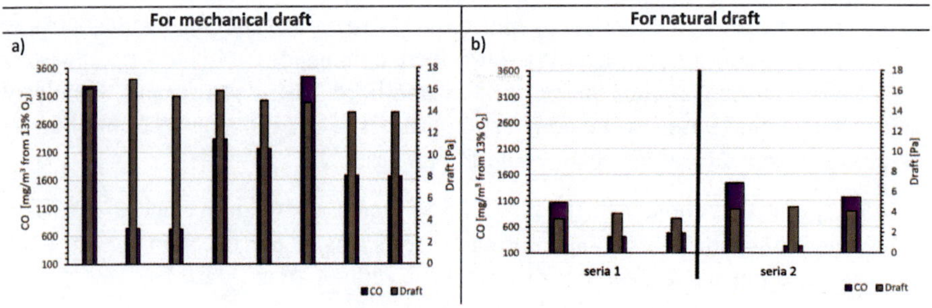

Fig. 7. CO content in flue gases for: a) mechanical draft, b) natural draft.

d) The nitrogen oxides (NO_x) content is presented in Fig. 8. After using the fan, the average amount of nitrogen oxides was increased by 6.25 [mg/m^3 for 13% O_2]. The maximum value that was recorded using the fan is 139 [mg/m^3 for 13% O_2], and 143 (series 2) [mg/m^3 for 13% O_2] for natural draft. The maximum NO_x value was reduced by 3.6 [mg/m^3 for 13% O_2] with the use of a fan.

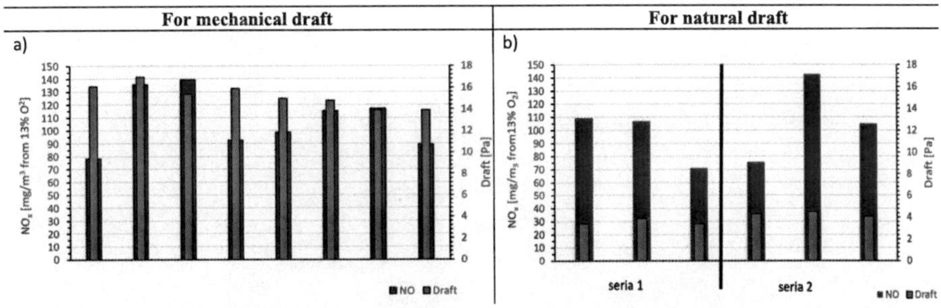

Fig. 8. NO_X content in flue gases for: a) mechanical draft, b) natural draft.

e) Soot number is shown in Fig. 9. The average soot number after the application of the fan decreased by 0.11 on the Bacharach scale. The highest recorded value of the soot number for installations with a fan was 7.8 according to the Bacharach scale and 7.5 without a fan.

6 Emission Tests with On-line Registration of Results

To carry out the tests was used a particle matter meter testo 330-2 + testo 380 analyser and gas analyser (Fig. 10) Wohler A450 L. The following diagrams (Figs. 11 and 12) summarise the results of the measurements carried out: PM in g/m^3; carbon monoxide CO in mg/Nm^3 for 13% O_2; nitrogen oxides NO_X in mg/Nm for 13% O_2; carbon dioxide CO_2 in %; efficiency ɲ in %; flue gas temperature T_S in °C.

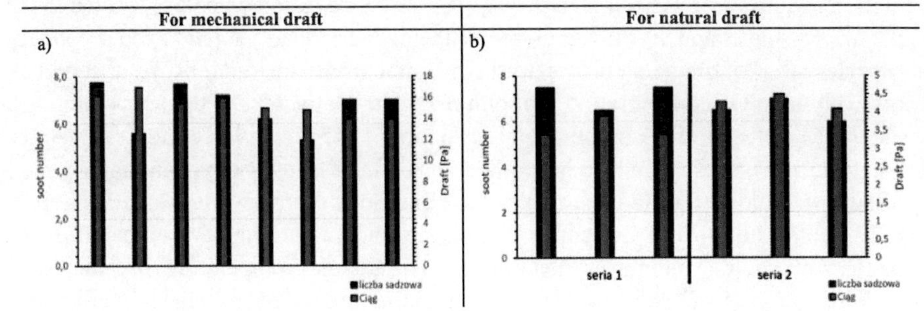

Fig. 9. Soot number for: a) mechanical draft, b) natural draft.

Figures 11a–c show the gaseous and particle bound flue gas concentrations, the flue gas temperature and the efficiency measured during combustion tests on a woodstove connected to the chimney with natural draught (Scenario no. 1). An average O_2 concentrations of approximately 16.6%, 16.0% and 15.0% were measured during 1, 2, and 3 series, respectively. The very high O_2 concentration in the flue gases resulted in relatively low CO_2 content being in the range 4.2–5.7%. CO emissions were relatively high at approximately 2789 mg/m^3 in series 1, 1294 mg/m^3 in series 2, and 1735 mg/m^3 in series 3 as the excess air dosed to the process cooled the furnace. Both, low CO_2 and

Fig. 10. Test stand during particle matter measurements.

high CO emissions led to inefficient combustion, below over 80% in all trials. It should be noted that the over-air effect is due to the design of the stove and the guidelines for its operation, where secondary air is supplied over the flame through nozzles located in the upper area of the chamber. Furthermore, all the primary air is routed to the secondary air nozzles after kindling, according to the manufacturer's instructions, resulting in an excessively high O_2 concentration in the flue gas. In the context of the above findings, the high PM concentrations in the flue gases of about 127, 74 and 261 mg/m^3 for series 1, 2, and 3, respectively, can also be explained as effects of over-oxygenation.

The installation of a fan even makes the conditions more severe as can be seen in Fig. 12. Both, the O_2 and CO concentrations in flue gas are higher compared to the scenario with natural draught in the chimney. The flue gas temperature also increased slightly by about 10°C. All these variations together reduce the combustion efficiency by a few percent. Nevertheless, the compensation of the chimney draught due to the fan operation clearly influences the PM level despite the unfavourable ventilation conditions of the system. It can be seen that the concentration of PM is clearly above 100 mg/m^3 during the initial phase of combustion, to remain constantly below this level after about 500 seconds until the next batch of fuel is added. The average PM values in the tests with the fan were $PM_a = 105$ mg/m^3, $PM_b = 78$ mg/m^3 and $PM_c = 94$ mg/m^3.

In summary, a proper and stable air/fuel ratio should be determined for the tested system, by appropriate air distribution below and over the grate. The total air stream fed to the combustion chamber and regulated by the chimney fan should better stabilize the draft, especially during the first phase of the fuel combustion.

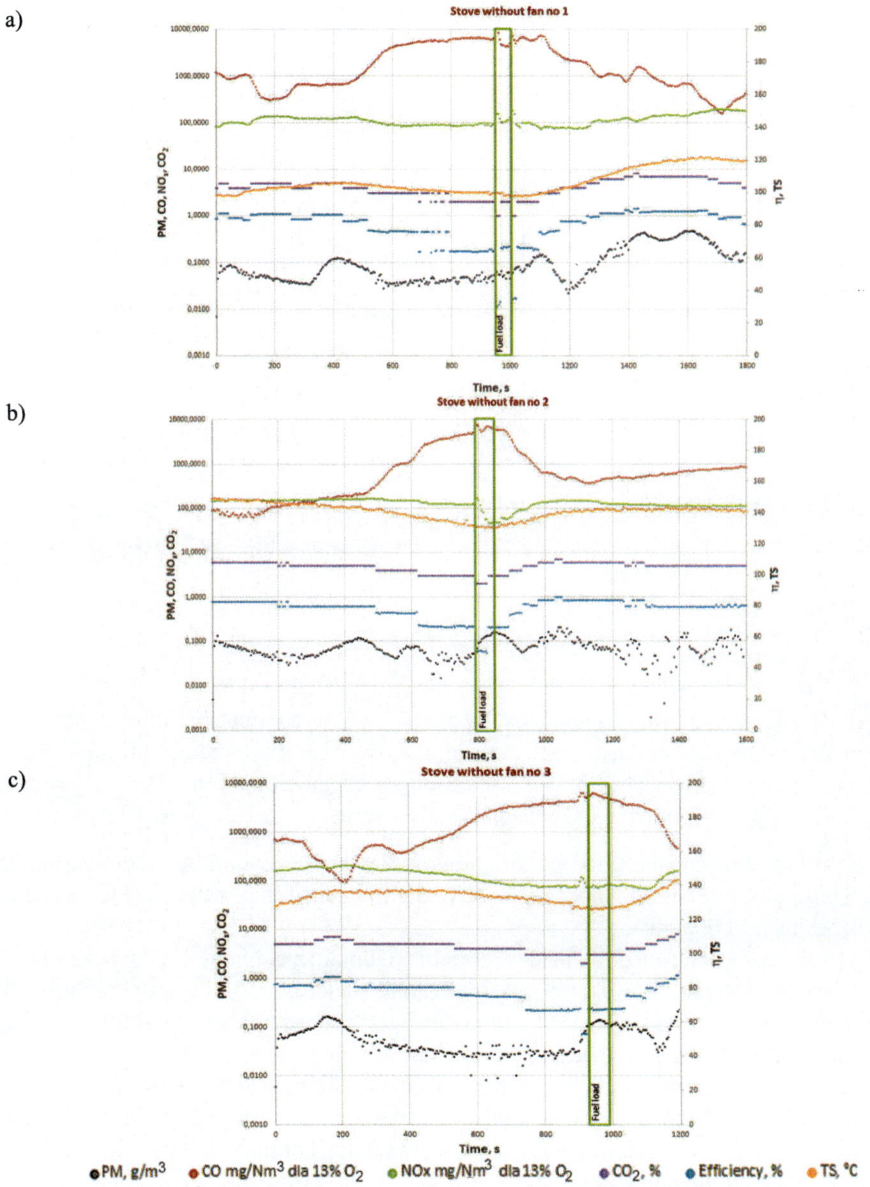

Fig. 11. The results of flue gas analysis and particle matter measurement for installations without a fan in 15 min series no.: a) 1; b) 2; and c) 3.

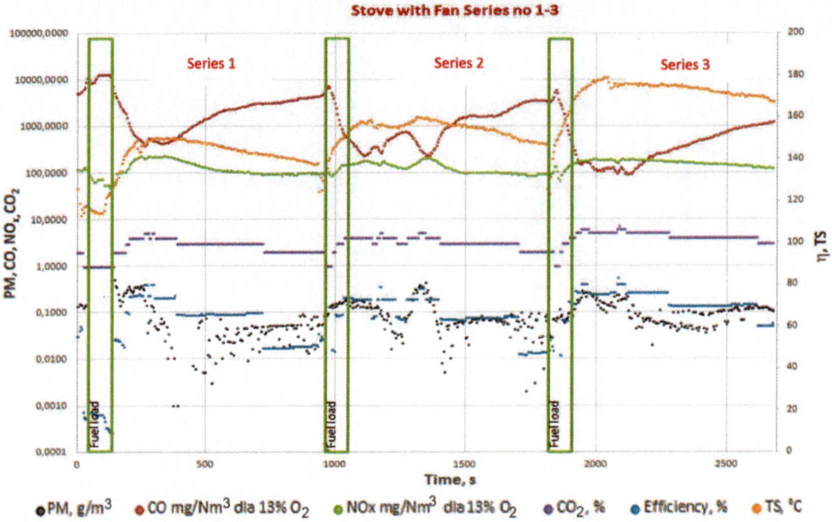

Fig. 12. The results of flue gas analysis and particle matter measurement for installations with a fan in three 15 minutes series (no. 1, 2, 3).

7 Conclusions

Taking into account the obtained results of research on the concentration of pollutants emitted during the combustion of wood in a stove with the forced draft, the possibility of their reduction was noticed, which improves the condition of the atmosphere. Additionally, the following conclusions can be drawn:

1. The use of a fan in the wood-burning stove indicates the potential to reduce particulate emissions, due to PM values below 0.01 mg/m^3 appearing, which was not observed with natural draught.
2. The use of a controlled fan made it possible to obtain a chimney draft greater than 12 Pa required by the manufacturer for the heating device used for the tests. Additionally, after using the exhaust fan, the ignition time was reduced by more than 60%. However, the use of a fan reduces the stove's efficiency.
3. Temporary increase of pressure up to 40 Pa in the chimney with forced draught during refuelling of the stove negatively influences the particulate emissions, increasing their content in the flue gas. Reducing the pressure in the chimney when adding fuel, may solve the problem in the future.
4. During the tests, a mechanical draft regulator (air supply flap) was used. The draught set by the prototype control connected to the fan was so stable that the mechanical regulator flap did not open during the tests. In addition, the differences in draft between the value set on the controller and the pressure in the chimney amounted to a maximum of 3 Pa. The draft during the tests was very stable and its difference did not exceed 1.8 Pa.

5. With a temporary (3 min) increase in the pressure in the chimney during refuelling to 40 Pa, the carbon monoxide content increases. On the other hand, the amount of nitrogen oxides is at a similar level for measurements with a fan and for a natural draught.
6. The emissions presented by the manufacturer's specifications, that correspond to the situation of mechanical draught (as it is in accordance with the test standard) are different from the emissions verified with natural draft that correspond to the conventional chimney assembly. Can also be concluded that the mechanical draft in this type of stove should be required.

With reference to the above conclusions, it is advisable to repeat the measurements at a reduced chimney draught when adding fuel (e.g. up to 20 Pa) and to carry out additional measurements on another model of the stove. Therefore, the execution of gravimetric measurements is not justified, because it does not indicate a continuous recording of the particle matter concentration in the flue gas over time (on-line).

Acknowledgments. The authors would like to thank JAWAR and PipeEksperten for their help in carrying out this research. The authors also would like to thank Ms Edyta Komendzińska for technical support. The research was co-financed under the Innovation Norway grant.

References

1. Fuller, G.W., Tremper, A.H., Baker, T.D., Yttri, K.E., Butterfield, D.: Contribution of wood burning to PM10 in London. Atmos. Environ. **87**, 87–94 (2014)
2. Sarigiannis, D.A., et al.: Total exposure to airborne particulate matter in cities: the effect of biomass combustion. Sci. Total Env. **493**, 795–805 (2014)
3. Caseiro, A., Bauer, H., Schmidl, C., Pio, C.A., Puxbaum, H.: Wood burning impact on PM10 in three Austrian regions. Atmos. Environ. **43**, 2186–2195 (2009)
4. Kappos, A.D., et al.: Health effects of particles in ambient air. Int. J. Hyg. Environ. Health **207**, 399–407 (2004)
5. Chapman, R.S., Watkinson, W.P., Dreher, K.L., Costa, D.L.: Ambient particulate matter and respiratory and cardiovascular illness in adults: particle-borne transition metals and the heart–lung axis. Environ. Toxicol. Pharmacol. **4**, 331–338 (1997)
6. Peden, D.B.: The epidemiology and genetics of asthma risk associated with air pollution. J. Allergy Clin. Immunol. **115**, 213–219 (2005)
7. Hall, R.E., DeAngelis, D.G.: Epa's research program for controlling residential wood combustion emissions. J. Air Pollut. Control Assoc. **30**, 862–867 (1980)
8. Rau, J.A.: Composition and size distribution of residential wood smoke particles. Aerosol Sci. Technol. **10**, 181–192 (1989)
9. Drozdzol, K., Jarzyński, P.: Impact of chimney on low emission. In: Beben, D., Rak, A., Perkowski, Z. (eds.) 3rd Scientific Conference Environmental Challenges in Civil Engineering, vol. 174, MATEC Web of Conferences (2018)
10. Drozdzol, K.: Development of modern chimney systems designed for solid fuel devices. In: Monteiro, J., João Silva, A., Mortal, A., Aníbal, J., Moreira da Silva, M., Oliveira, M., Sousa, N. (eds.) INCREaSE 2019, pp. 487–501. Springer, Cham (2020). https://doi.org/10.1007/978-3-030-30938-1_39
11. Reichert, G., et al.: Effect of draught conditions and ignition technique on combustion performance of firewood roomheaters. Renew. Energy **105**, 547–560 (2017)

12. EN 13240:2001 + AC:2003: Room heaters fired by solid fuel. Requirements and test methods. European Committee for Standardization. Brussels (2003)
13. Illerup, J.B., et al.: Performance of an automatically controlled wood stove: thermal efficiency and carbon monoxide emissions. Renew. Energy **151**, 640–647 (2020)
14. Verma, V.K., et al.: Agro-pellets for domestic heating boilers: standard laboratory and real life performance. Appl. Energy **90**(1), 17–23 (2012)
15. PN-EN 15259:2011: Air quality. Measurements of emissions from stationary sources. Requirements for measurement sections and measurement sites, measurement purpose and plan, and measurement report. Polish Committee for Standardization, Warsaw (2011)
16. PN-EN 13240:2008: Solid fuel space heaters. Requirements and tests. Polish Committee for Standardization, Warsaw (2008)
17. NS 3058: Enclosed wood heaters. Smoke emission, Part 1: Test facility and heating pattern, Part 2: Determination of particulate emission. Norwegian Standards, Oslo (1994)
18. NS 3059: Enclosed wood heaters. Smoke emission. Requirements. Norwegian Standards, Oslo (1994)
19. Sevault, A., Seljeskog, M., Rydså, L., Østnor, A., Rønning, B., Rishaug, M.: Report Effect of maintenance on particulate emissions from residential woodstoves, SINTEF, Trondheim (2016)
20. Seljeskog, M., Sevault, A., Østnor, A., Skreiberg, Ø.: Variables affecting emission measurements from domestic wood combustion. Energy Procedia **105**, 596–603 (2017)

Analytical and Numerical Simulation of a Commercial Thermoelectric Module

André van der Kellen and Pedro J. Coelho(✉)

IDMEC, Mechanical Engineering Department, Instituto Superior Técnico, Universidade de Lisboa, Av. Rovisco Pais, 1049-001 Lisboa, Portugal
pedro.coelho@tecnico.ulisboa.pt

Abstract. This work reports a comparison of three different mathematical models used to evaluate the performance of a thermoelectric module: a linear model that relies on the assumption of temperature-independent thermoelectric properties and two nonlinear models that account for the temperature dependence of the properties. The linear model solves analytically the equations for the conservation of energy, electric potential and electric charge. The nonlinear models solve the conservation equations using either the homotopy perturbation method or the finite volume method. The models are applied to a commercial thermoelectric module and compared with the performance data provided by the manufacturer. The results demonstrate that the temperature dependence of the material properties, as well as the electrical contact resistances, must be taken into account to obtain accurate predictions. The assumption of constant properties is reasonable at low temperatures of operation, where the variation of the thermoelectric properties with temperature is negligible, but at high temperatures, this assumption overestimates the performance. Both non-linear models yield a good agreement with each other and with the data obtained from the manufacturer.

Keywords: Thermoelectric Generator · Seebeck Effect · Waste Heat Recovery

1 Introduction

Thermoelectrics is defined as the science and technology associated with thermoelectric generation and refrigeration [1]. Thermoelectric devices are based on the Seebeck effect and can either act as generators, where electrical power is produced, or as thermoelectric coolers for refrigeration. They allow for the direct conversion of thermal energy into electrical energy or vice versa. Thermoelectric devices have no moving parts and require no maintenance thus making them suitable for a broad range of applications, such as waste heat recovery from power plants and automotive vehicles but also refrigeration and temperature control in electronic packages and medical instruments. They are usually made up of a stacked assembly of several thermoelectric modules.

Thermoelectric technology has received considerable attention for waste heat recovery in energy conversion devices. To predict and optimize the performance of thermoelectric systems, a mathematical model may be used for the analysis, accompanied

J. F. L. C. Semião et al. (Eds.): INCREaSE 2023, ASST, pp. 157–172, 2023.
https://doi.org/10.1007/978-3-031-44006-9_12

by a deep understanding of the heat and electrical current transfer phenomena and the selection of the right materials according to the temperature range of operation.

A typical modern thermoelectric module consists of several *n*-type and *p*-type semiconductors, forming thermoelectric units, also commonly referred to as thermocouples, connected electrically in series between two electrical conductors. The conductors are protected by an electrical insulator, usually ceramic plates. In *p*-type semiconductors, the absence of electrons creates a positive charge and in *n*-type semiconductors the excess of electrons creates a negative charge, thus leading to a flow of electrons across the junctions of the thermocouple. The device operates as a thermoelectric generator (TEG) and supplies electrical power to an external load provided that a temperature difference is maintained across the module, while if electric current passes through the module, heat is absorbed at one junction and rejected at the other, acting as a thermoelectric cooler [2].

Several models have been employed in the literature to estimate the performance of thermoelectric devices. Fraisse *et al.* [3] compared four different one-dimensional steady-state models to analyse the coefficient of performance (COP) and efficiency, as well as the voltage and thermal/electrical power of thermoelectric couples made of bismuth telluride (Bi_2Te_3). They concluded that there were no significant differences between the simplified and improved models in the thermoelectric generator and cooler operation modes. However, the maximum efficiency was overestimated in the simplified linear model when operating in TEG mode. Slightly improved accuracy was observed when considering the Thomson effect in the calculations if the Seebeck coefficient is very dependent on the temperature. It was also shown that the results of the electrical analogy model agree very well with those of simulations performed using the finite element method.

Zhang [4] studied the effect of the material temperature dependence when evaluating the performance of a Hi-Z thermoelectric module and solved the non-linear heat conduction equation based on the homotopy perturbation method, as described in Zhang [5, 6]. The non-linear analytical solution provided consistent results with the electrical analogy method and good estimations of the temperature variation along the thermoelement leg, absorbed power at the cold end when operating in the cooling mode, and power output at the hot end when operating in the TEG mode.

Niu et al. [7] studied the effect of the leg geometry on the temperature gradient using two three-dimensional mathematical models, based on different formulations and boundary conditions, to analyse heat and electricity transfer. It was shown that one model was more precise for power output predictions and suitable for simulations with defined output voltage or current and the influence of the leg cross-sectional area could also lead to significant improvement in power output.

As shown in the reviewed literature, a great variety of assumptions can be made when applying analytical or numerical models to evaluate the performance of a thermoelectric module. To correctly predict these parameters, non-linearities must be taken into account since the thermoelectric properties of materials have a strong temperature dependence. This work compares three models aimed at calculating the performance of a commercially available thermoelectric module. Data obtained from the manufacturer [8] were used to assess the accuracy of the predictions.

2 Theory

In this section, the fundamental concepts regarding thermoelectricity are presented in order to understand the models applied to evaluate the performance of the thermoelectric module. Three thermoelectric effects occur in the presence of a temperature difference between the junctions of a thermoelectric unit (thermocouple): Seebeck, Peltier and Thomson effects. These three effects are related among them.

In a thermoelectric material, when a temperature difference is applied across a conductor, the hot region produces more free electrons and natural diffusion of these electrons occurs from the hot region to the cold one. The resultant electromotive force generates an electric current flowing against the temperature gradient. This is known as the Seebeck effect [9]. If a temperature gradient is maintained at the junctions of two wires, A and B, joined at both ends, an electromotive force is produced. The Seebeck electric potential V_{Sbk} [V] in each wire can be written as:

$$\nabla V_{Sbk} = -\alpha \nabla T \tag{1}$$

where α [V·K^{-1}] is the Seebeck coefficient and T the temperature. Depending on the sign of α, the Seebeck voltage, also referred to as open-circuit voltage, will be negative or positive.

As an electric current flows across a junction between two wires of dissimilar materials, heat must be continuously added or subtracted at the junction to keep its temperature constant. This is known as the Peltier effect. The heat rate absorbed or dissipated at a junction is given by

$$\dot{Q}_{Peltier} = \pi_{AB} I \tag{2}$$

where π_{AB} [V] is the Peltier coefficient and I [A] represents the electric current intensity across the junctions.

The third thermoelectric effect is the Thomson effect. Heat is absorbed or dissipated across a wire depending on the material and the direction of the current. The Thomson heat rate is defined as

$$\dot{Q}_{Thompson} = \tau \nabla T \cdot \vec{J} \tag{3}$$

where τ [V·K^{-1}] and $\vec{J}$ [A·m^{-2}] are the Thomson coefficient and the current density vector on a wire, respectively. The sign of τ is positive if heat is absorbed in the wire and negative if heat is dissipated.

When an electric current flows in a wire, heat is also dissipated due to the irreversible Joule effect, which is not a thermoelectric effect. The Joule heat rate is given by

$$\dot{Q}_{Joule} = \rho \|\vec{J}\|^2 \tag{4}$$

where ρ [Ω·m] is the electrical resistivity of the material.

The three thermoelectric effects are related by the Kelvin (or Thomson) relationships [10], which may be written as

$$\pi_{AB} = \alpha_{A,B} T = (\alpha_A - \alpha_B) T \tag{5a}$$

$$\tau = T\frac{d\alpha}{dT} \tag{5b}$$

Applying the Kelvin relationships, the Peltier and Thomson heat rates can now be defined, respectively, as

$$\dot{Q}_{Peltier} = \alpha_{A,B}TI \tag{6}$$

$$\dot{Q}_{Thompson} = T\frac{d\alpha}{dT}\nabla T.\vec{J} \tag{7}$$

3 Thermoelectric Module

The thermoelectric module HZ-14 manufactured by Hi-Z Technology [11] is considered in the present work. It is composed of several thermocouples connected electrically in series and thermally in parallel to increase the output voltage of the module.

3.1 Thermocouple

The thermocouples forming a thermoelectric module consist of *p*-type and *n*-type semiconductor materials. A basic representation of a thermocouple is shown in Fig. 1, where L is the length of the thermocouple legs, A the area of the cross-section of the legs, ρ the electrical conductivity, k thermal conductivity, T_h the hot side temperature and T_c the cold side temperature. Subscripts p and n refer to the *p*-type and *n*-type materials, respectively.

The thermocouple can act as a thermoelectrical generator if it is connected to an external load and delivering electrical power. The *n*-type and *p*-type materials are connected by a conductor strip, usually made of copper. Figure 2 schematizes a basic electrical circuit of a TEG.

The heat flux at the hot junction is due to the heat absorbed due to the Peltier effect and the heat transferred by conduction through the legs. Using Fourier's law to evaluate the heat conducted across the *n*-type and *p*-type legs, the heat flux is given by:

$$\vec{q}_p = \alpha_p T\vec{J} - k_p\nabla T \tag{8}$$

$$\vec{q}_n = -\alpha_n T\vec{J} - k_n\nabla T \tag{9}$$

The voltage drop in the thermocouple due to the electrical resistivity can be quantified using Ohm's law, which may be written as

$$\nabla V_{Ohm} = -\rho\vec{J} \tag{10}$$

The total ohmic electric potential drop is given by $V_{Ohm} = V_{Ohm,p} + V_{Ohm,n}$. The voltage drop across the load, V, in an electrical circuit is a function of the open-circuit voltage, $V_{oc} = V_{Sbk}$, and the ohmic voltage, V_{Ohm}, in the thermocouple due to the internal electrical resistance, R_L (see Fig. 2). The voltage drop across the load is given by

$$V = V_{sbk} - V_{Ohm} \tag{11}$$

The Seebeck (open-circuit) voltage represents the highest voltage possible. The electrical power supplied to the load, $\dot{W}$, can be written as a function of V or R_L:

$$\dot{W} = VI = R_L I^2 \tag{12}$$

The conversion efficiency of the TEG is given by

$$\eta = \dot{W}/Q_h \tag{13}$$

where Q_h is the total heat rate supplied to the thermocouple by an external source.

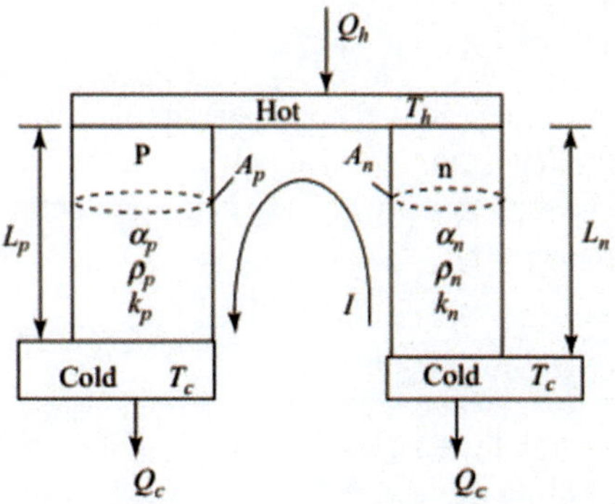

Fig. 1. A basic *p*-type and *n*-type thermocouple [9].

Fig. 2.Thermoelectric generator with a single thermocouple [9].

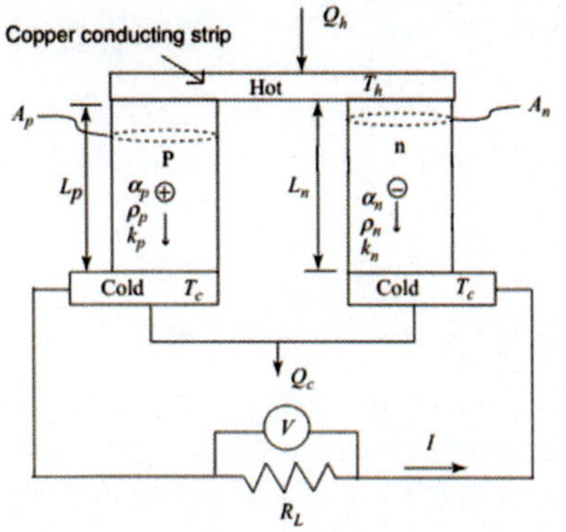

Fig. 2. Thermoelectric generator with a single thermocouple [9].

3.2 Thermoelectric Module

A thermoelectric module is composed of several thermocouples connected electrically in series through a copper conducting strip, and thermally in parallel, as illustrated in Fig. 3. The module, however, must be electrically insulated from the heat source and the heat sink while also allowing high thermal conductivity to minimize the temperature difference between the heat source/sink and the thermocouple surface. Usually, alumina ceramic plates are used for this purpose.

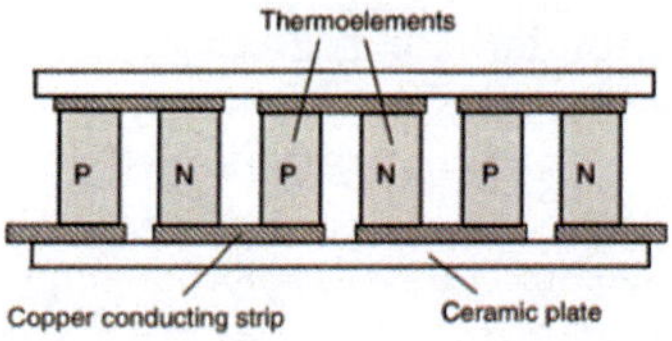

Fig. 3. Basic configuration of a thermoelectric module.

In a module with *N* thermocouples, the voltage, the electrical power, and the heat rate are equal to *N* times the voltage, the electrical power, and the heat rate in a thermocouple, respectively, while the current intensity and the conversion efficiency of the module are equal to those in a thermocouple.

3.3 Hi-Z Thermoelectric Modules

The Hi-Z thermoelectric modules manufactured by Hi-Z Technology, Inc. Are optimized for waste heat recovery. They use bismuth telluride (Bi_2Te_3) based alloys as thermoelectric materials. The thermocouples are electrically and thermally insulated by a special frame referred to as "egg crate" that fills the air gaps between thermocouples, which is manufactured through injection moulding to make the modules less expensive to fabricate [11]. The evaluation of the performance of these modules reported below does not consider the presence of heat exchangers and ceramic plates.

The Bi_2Te_3-based alloys used in Hi-Z modules strongly depend on temperature. The Seebeck coefficient, thermal conductivity and electrical resistivity were evaluated using polynomials as a function of temperature. The coefficients of each polynomial are available in reference [8].

3.4 Module Performance Calculator

The module performance calculator [8] is a free software developed by Hi-Z Technology, Inc., that allows the user to determine the performance of the thermoelectric modules. The inputs required by the user are the hot-side and cold-side temperatures and the module type, which can be selected among those commercially available by Hi-Z Technology, Inc. or a module with custom-defined parameters.

4 Analytical and Numerical Models

This section describes the models used to evaluate the thermoelectric module performance. Three different models were used. The first model solves the conservation equations assuming constant properties of the thermoelectric material. The second model solves the heat conduction equation using the homotopy perturbation method (HPM), which is an approximate analytical method. The third model solves the governing equations numerically using the finite volume method (FVM).

4.1 Conservation Equations

The governing energy conservation equation may be written as follows:

$$\nabla.(k_i \nabla T_i) + T_i \frac{d\alpha_i}{dT} \nabla T_i.\vec{J} + \rho_i \|\vec{J}^2\| = 0 \tag{14}$$

where subscript i stands for either p or n, depending on the leg of the thermocouple under consideration. The three terms on the left side of the above equation represent heat conduction, the Thomson effect and the Joule effect, respectively.

The conservation equation of the electric potential is obtained by taking the divergence of both sides of Eq. (1), yielding

$$\nabla.\left(\nabla V_{Sbk,i}\right) + \nabla.(\alpha_i \nabla T_i) = 0 \tag{15}$$

The conservation equation for the electric charge or the current intensity is obtained by applying the divergence operator to both sides of Eq. (10) and noting that the divergence of the current density is equal to zero for steady state, which yields

$$\nabla.\left(\sigma_i \nabla V_{Ohm,i}\right) = 0 \tag{16}$$

where $\sigma = 1/\rho$ is the electrical conductivity of the material.

4.2 Simplified Linear Model

The simplified linear model (SLM) presented in this section solves analytically the conservation equations by assuming constant thermoelectric material properties. Since the temperature gradient in a cross-section of a thermocouple leg is negligible, the analysis can be reduced to a one-dimensional problem. Hence, the conservation equations may be simplified as follows:

$$\overline{k}_i \frac{d^2 T_i}{dx^2} + \frac{\overline{\rho}_i I^2}{A^2} = 0 \tag{17}$$

$$\frac{d^2 V_{Sbk,i}}{dx^2} + \overline{\alpha} \frac{d^2 T_i}{dx^2} = 0 \tag{18}$$

$$\frac{d^2 V_{Ohm,i}}{dx^2} = 0 \tag{19}$$

where $\overline{k}_i$, $\overline{\rho}_i$ and $\overline{\alpha}_i$ are the thermal conductivity, electrical resistivity and Seebeck coefficient, respectively, evaluated at the average temperature in the i (p or n) thermocouple leg, and x denotes the coordinate along the leg. The temperatures at the hot ($x = 0$) and cold ($x = L$) junctions are set to T_h and T_c, respectively. A reference potential of 0 V is prescribed at the hot junction for the Seebeck and Ohmic electric potentials, while the electric field and the current density are prescribed at the cold junction. The analytical solution of Eqs. (17), (18) and (19) is the following:

$$T_i(x) = -\frac{\dot{q}_i}{2\overline{k}_i}x^2 + \left(\frac{T_c - T_h}{L} + \frac{\dot{q}_i L}{2\overline{k}_i}\right)x + T_h \tag{20}$$

$$V_{Sbk,i}(x) = \overline{\alpha}_i \frac{\dot{q}_i}{2\overline{k}_i}x^2 - \overline{\alpha}_i\left(\frac{T_c - T_h}{L} + \frac{\dot{q}_i L}{2\overline{k}_i}\right)x \tag{21}$$

$$V_{Ohm,i}(x) = -\overline{\rho}_i J x \tag{22}$$

where

$$\dot{q}_i = \frac{\overline{\rho}_i I^2}{A^2} \tag{23}$$

4.3 Homotopy Perturbation Method

The homotopy perturbation method (HPM) was used to solve the non-linear heat conduction equation. Equation (14) is rewritten as follows for one-dimensional heat conduction:

$$kA\frac{d^2T}{dx^2} + A\frac{dk}{dT}\left(\frac{dT}{dx}\right)^2 - IT\frac{d\alpha}{dT}\frac{dT}{dx} + \frac{\rho I^2}{A} = 0 \tag{24}$$

where the subscript i was omitted for conciseness. Third-degree polynomials were used to express α, k and ρ as a function of the temperature. A normalized temperature and a normalized coordinate are defined as

$$\theta = (T - T_c)/(T_h - T_c) \tag{25}$$

$$\xi = x/L \tag{26}$$

Inserting the third-degree polynomials and the non-dimensional parameters in Eq. (24) yields

$$f(\theta)\left(\frac{d\theta}{d\xi}\right)^2 + g(\theta)\frac{d^2\theta}{d\xi^2} - h(\theta)\frac{d\theta}{d\xi} + q(\theta) = 0 \tag{27}$$

where the functions $f(\theta)$, $g(\theta)$, $h(\theta)$ and $q(\theta)$ are polynomials whose coefficients depend on the thermocouple geometry, hot junction and cold junction temperatures, material properties, and electric current intensity. Equation (27) may be expressed as

$$L(\theta) + N(\theta) - C(\xi) = 0 \tag{28}$$

where L and N include the linear and the non-linear terms of Eq. (27), respectively, and C is a known function. Using the homotopy technique, a homotopy function can be constructed as follows [12, 13]

$$H(v, \beta) = (1 - \beta)(L(v) - L(u_o)) + \beta[L(v) + N(v) - C(\xi)] \tag{29}$$

where u_o is the initial approximation for the solution v, and $\beta \in [0,1]$ is an embedding small parameter. The solution θ can be written as a power series in β [12]:

$$\theta = v = v_o + \beta v_1 + \beta^2 v_2 + \ldots \tag{30}$$

Introducing Eq. (30) into Eq. (29) and collecting terms at the zeroth and first order of p, the resulting equations are

$$L(v_o) - L(u_o) = 0 \tag{31}$$

$$L(v_1) + L(u_o) + N(v_o) - C(\xi) = 0 \tag{32}$$

Equation (31) yields $v_o = u_o = \theta_o$, where θ_o is the initial approximation of θ, which satisfies the boundary conditions. Zhang [5, 6] define the initial approximation θ_o as

$$\theta_o = \xi^2 + a\xi(1 - \xi) \tag{33}$$

where a is obtained from the weak form solution of Eq. (28) for θ_o:

$$\int_0^1 \xi(L(\theta_o) + N(\theta_o) - C(\xi))d\xi = 0 \tag{34}$$

and by satisfying the prescribed temperature boundary conditions. Equation (32) is a linear homogeneous differential equation that can be solved by standard methods, leading to

$$v_1 = v_{1g} + v_{1p} \tag{35}$$

where v_{1g} and v_{1p} are the general and particular solutions, respectively, of Eq. (32). The first-order approximate analytical solution of Eq. (27) is given by

$$\theta(\xi) = \theta_o + v_{1g} + v_{1p} \tag{36}$$

Reverting the non-dimensional temperature $\theta(\xi)$ profile back to dimensional form $T(x)$, the Seebeck and ohmic electric potentials can be evaluated, respectively, from Eqs. (1) and (10):

$$V_{Sbk}(x) = -\int_{T_h}^{T(x)} \alpha dT \tag{37}$$

$$V_{Ohm(x)} = -\int_0^x \frac{\rho I}{A} dx \tag{38}$$

4.4 Finite Volume Method

In the finite volume method (FVM), the computational domain is divided into a finite number of control volumes and the governing equations are integrated for each control volume. The integration over a control volume of the energy conservation equation leads to

$$\int_{x_w}^{x_e}\left(\nabla.(k\nabla T)+T\nabla\alpha.\vec{J}+\rho\|\vec{J}\|^2\right)dx=0 \tag{39}$$

where subscripts w and e denote the west and east faces of the control volume. Applying the Gauss' divergence theorem to the first term on the left hand side of Eq. (39), the discretised equation for nodal point P is written as

$$k_eA_e\left(\frac{dT}{dx}\right)_e-k_wA_w\left(\frac{dT}{dx}\right)_w+\frac{T_PI}{A}\left(\frac{d\alpha}{dx}\right)_P\Delta x+\frac{\rho_PI^2}{A^2}\Delta x=0 \tag{40}$$

where $\Delta x = x_e - x_w$ and subscript P denotes the grid node at the centre of the control volume. The discretised form of this equation is obtained by evaluating the fluxes and the properties at the cell faces w and e using the central differencing scheme [14]. In the case of control volumes adjacent to the west and east boundaries, the heat fluxes at the boundaries are calculated using forward and backward finite differences, respectively. Applying the same method to the conservation equations of the electric potential and electric charge yields

$$A_e\left(\frac{dV_{Sbk}}{dx}\right)_e-A_w\left(\frac{dV_{Sbk}}{dx}\right)_w+\alpha_eA_e\left(\frac{dT}{dx}\right)_e-\alpha_wA_w\left(\frac{dT}{dx}\right)_w=0 \tag{41}$$

$$\sigma_eA_e\left(\frac{dV_{Ohm}}{dx}\right)_e-\sigma_wA_w\left(\frac{dV_{Ohm}}{dx}\right)_w=0 \tag{42}$$

5 Results

The predicted performance of the HZ-14 thermoelectric module [8] is presented in this section. Figures 4, 5 and 6 show the electrical power, output voltage and conversion efficiency, respectively, as a function of the electric current, and for prescribed temperatures of the heat source and heat sink. This analysis was based on an ideal scenario where there was no electrical contact resistivity between the copper conducting strips and the top and bottom surfaces of both the n-type and p-type units, and therefore no drop in potential at the hot and cold junctions. Heat conduction through the egg crate material was also neglected.

Figures 4, 5 and 6 show that the HPM and FVM yield similar results and predict lower voltage, power and efficiency than the SLM. However, there is a large difference between the results obtained from all the models and the data obtained from the module performance calculator [8]. Hence, it is clear that neglecting the contact effects and the egg crate material leads to a large overestimation of the module performance.

The copper conducting strip, the contact effects between this strip and the thermoelectric p-type and n-type materials, and the conduction across the egg crate need to be considered to accurately predict the thermoelectric module performance. The internal resistance of the copper strip and the contact electrical resistivity are both obtained from the data provided by the manufacturer of the HZ-14 module. The voltage drop due to these resistances may be evaluated using Ohm's law, noting that there are two copper strips, one on each side of the thermocouples, and four contact resistances per thermocouple.

The heat transfer rate across the TEM is the sum of the heat transfer across the legs of the thermocouples and the heat transfer across the egg crate, even though the latter is much smaller than the former. The heat transfer across the egg crate is evaluated using Fourier's law. The thermal conductivity of the egg crate material is obtained from the available data [8].

Figures 7, 8 and 9 show that the results agree very well with those determined from the module performance calculator [8] when the above effects, particularly the contact effects, are considered. The overestimation of the performance by the SLM is expected since the material properties exhibit a strong variation with the temperature for the temperatures considered in the analysis. The difference between the linear and the non-linear models becomes more evident at matched load conditions (maximum power).

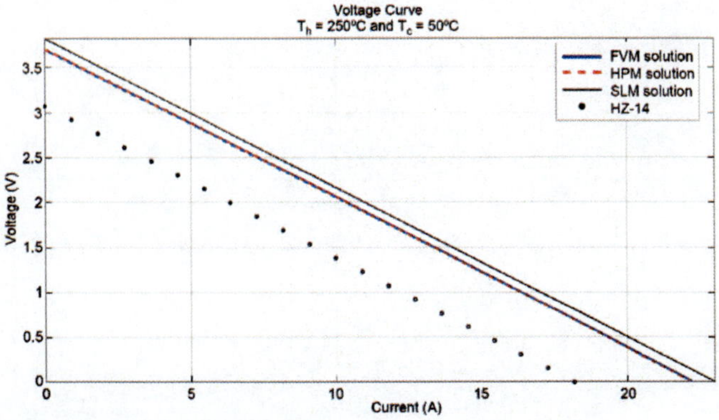

Fig. 4. Relationship between the voltage and the electric current for $T_h = 250$ °C and $T_c = 50$ °C.

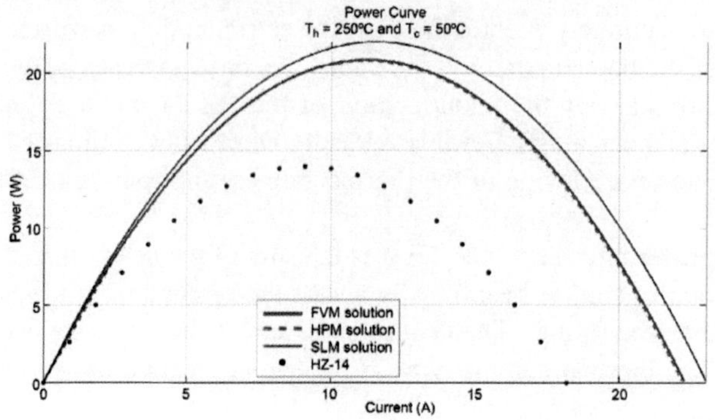

Fig. 5. Relationship between the power and the electric current for $T_h = 250$ °C and $T_c = 50$ °C.

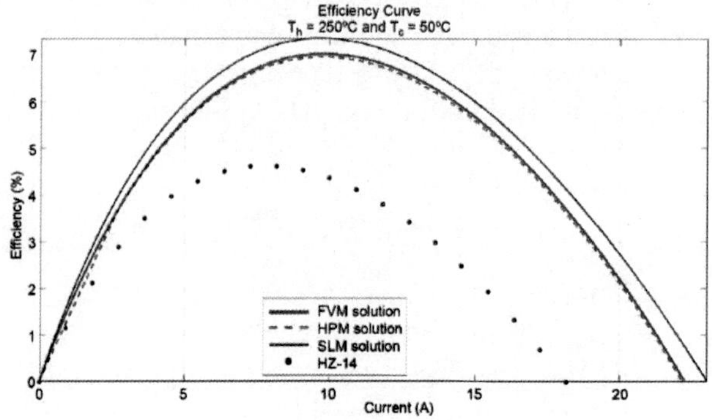

Fig. 6. Relationship between the efficiency and the electric current for $T_h = 250$ °C and $T_c = 50$ °C.

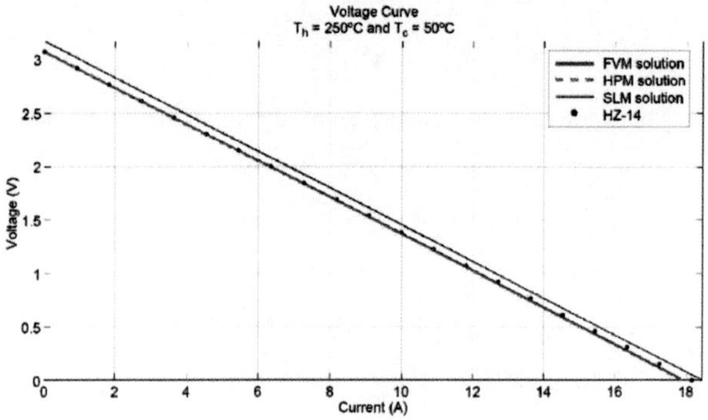

Fig. 7. Relationship between the voltage and the electric current for $T_h = 250$ °C and $T_c = 50°$ (with contact losses).

Under matched load conditions, the output power, voltage and efficiency were evaluated using the SLM and the HPM and compared with the module performance calculator,

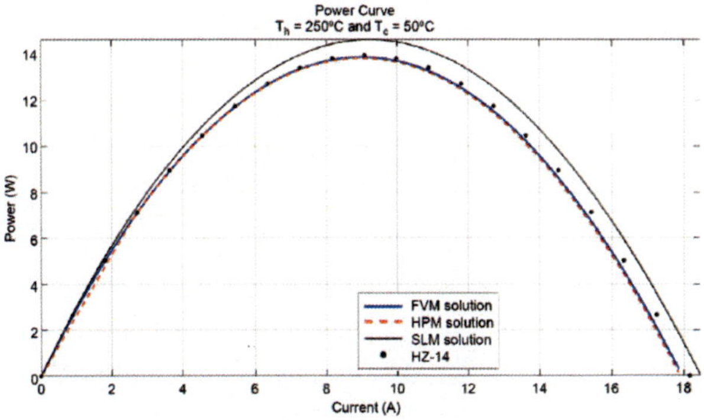

Fig. 8. Relationship between the power and the electric current for $T_h = 250$ °C and $T_c = 50$ °C (with contact losses).

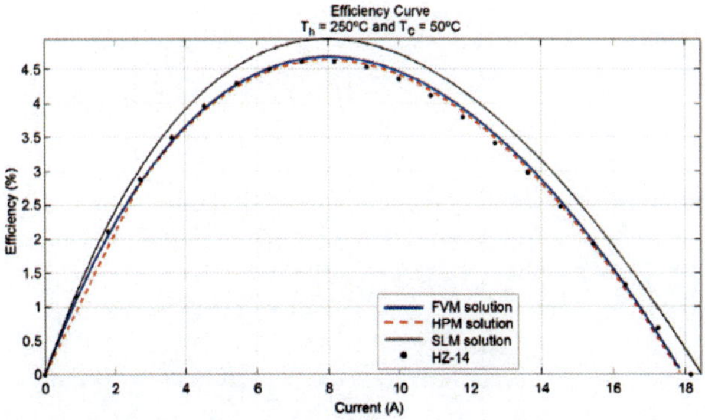

Fig. 9. Relationship between the efficiency and the electric current for $T_h = 250$ °C and $T_c =$ 50 °C (with contact losses).

as a function of the temperature difference between the heat source and heat sink, $\Delta T = T_h - T_c$, and the results are shown in Figs. 10, 11 and 12.

The matched load curves reveal that both the SLM and the HPM agree very well with the data from reference [8] until $\Delta T \cong 150$ °C for all the values of T_c considered, for both power and voltage predictions. In the case of higher ΔT, the SLM overestimates the performance parameters since the non-linearity of the properties of the materials becomes stronger and so the SLM results exhibit differences when compared to the HPM and the manufacturer results. However, for matched load conditions, the HPM predictions agree well with the data from the manufacturer with a few marginal deviations when 250 °C $\leq \Delta T \leq$ 300 °C. The differences in power and voltage observed in the latter range of ΔT and in the efficiency for almost all values of ΔT can be explained by the calculation method used in the module performance calculator [8], which uses a discretised form of Eq. (1) to evaluate the Seebeck potential with 20 discrete intervals regardless of the ΔT selected by the user. As ΔT increases, so does the non-linearity of the properties of the materials and so the number of intervals used to discretise Eq. (1) should be larger to increase the accuracy. This explains why the predictions of the non-linear models used

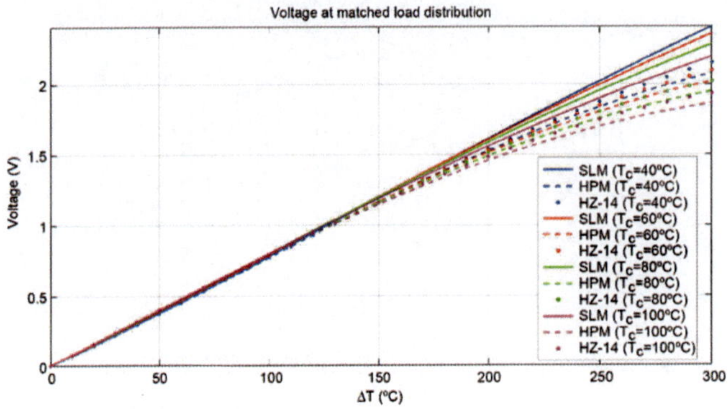

Fig. 10. Voltage as a function of the temperature difference between the hot and cold sides at matched load.

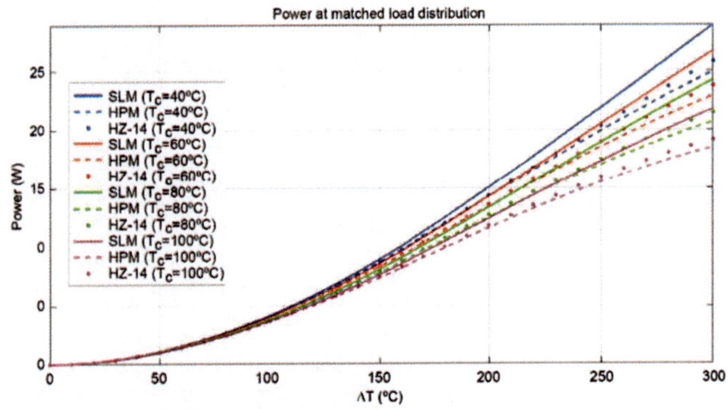

Fig. 11. Power as a function of the temperature difference between the hot and cold sides at matched load.

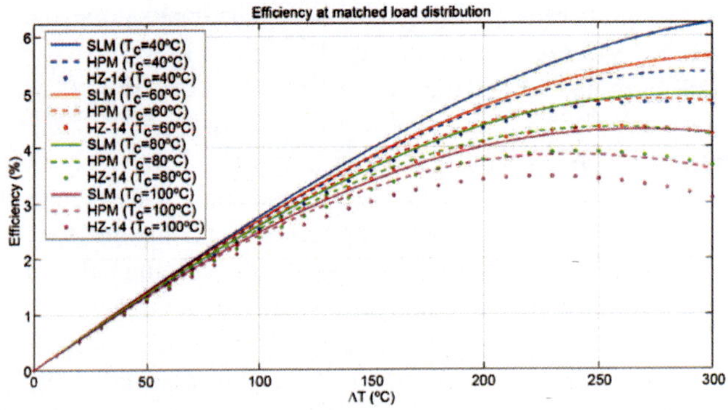

Fig. 12. Efficiency as a function of the temperature difference between the hot and cold sides at matched load.

in the present work are in better agreement with those of the manufacturer for lower values of ΔT.

The three models presented in this work are useful to predict the performance of thermoelectric modules depending on the application. The SLM offers a fast and efficient way to estimate the performance at lower temperature differences between the hot source and the cold sink, relying on simplified equations that are easy to implement. The non-linear approximate HPM evaluates the performance by accounting for temperature-dependent thermoelectric material properties with more complex equations but is computationally faster than the FVM.

6 Conclusions

Three different mathematical models were used to evaluate the performance of a commercial thermoelectric module. The simplified linear model assumes constant thermoelectric properties of the materials. The second model is a non-linear approximate analytical solution based on the homotopy perturbation method, which takes into account the variation of the properties of the material with the temperature. The third model is based on the numerical solution of the energy conservation equation using the finite volume method. The models proposed were validated by comparing the predictions with data obtained from a free software provided by the manufacturer.

The results calculated without taking the electrical contact effects into account significantly overestimate the voltage, power and efficiency of the thermoelectric module, no matter the model employed. However, when these effects are considered, as well as the resistance of the copper conducting strip and the heat conduction across the egg crate, the predictions are greatly improved and reproduce the data of the manufacturer.

It was shown that, up to a temperature difference between the hot source and the cold sink of 150 °C, the assumption of constant thermoelectric properties can provide a reasonable estimate of the thermoelectric module performance since the variation of the properties with the temperature is reasonably small for that range of temperatures. In the case of larger temperature differences, the thermoelectric properties of the *n*-type and *p*-type semiconductors exhibit stronger non-linear behaviour, which means that the assumption of constant properties no longer holds and the simplified linear model does not predict the module performance correctly. In contrast, the non-linear models yield more accurate results for the prediction of the performance. The non-linear approximate analytical solution based on the homotopy perturbation method and the non-linear numerical solution based on the finite volume method agree very well with each other. Moreover, the results obtained by these two methods show good agreement with those obtained from the module performance calculator provided by the manufacturer, particularly for small temperature differences. The slightly higher discrepancies observed at higher temperatures can be explained by the crude discretization employed in the module performance calculator.

Acknowledgements. The support of the Portuguese Science Foundation (FCT) through IDMEC, under LAETA, project UID/50022/2020 is acknowledged.

References

1. Rowe, D.M.: Thermoelectrics Handbook: Macro to Nano. CRC Press, Boca Raton (2006)
2. Rowe, D.M.: CRC Handbook of Thermoelectrics. CRC Press, Boca Raton (1995)
3. Fraisse, G., Ramousse, J., Sgorlon, D., Goupil, C.: Comparison of different modeling approaches for thermoelectric elements. Energy Convers. Manage. **65**, 351–356 (2013)
4. Zhang, T.: New thinking on modeling of thermoelectric devices. Appl. Energy **168**, 65–74 (2016)
5. Zhang, T.: Effects of temperature-dependent material properties on temperature variation in a thermoelement. J. Electron. Mater. **44**(10), 3612–3620 (2015)
6. Zhang, T.: Analytical solution of non-linear thermoelectric heat transport equation using homotopy perturbation method. J. Comput. Intell. Electron. Syst. **4**, 59–68 (2015)
7. Niu, Z., et al.: Elucidating modeling aspects of thermoelectric generator. Int. J. Heat Mass Transf. **85**, 12–32 (2015)
8. Module performance calculator, https://hi-z.com/performance-calculator/, last accessed 2023/03/12
9. Lee, H.S.: Thermal Design: Heat Sinks, Thermoelectrics, Heat Pipes, Compact Heat Exchangers, and Solar Cells. John Wiley & Sons Inc, New Jersey (2010)
10. Apertet, Y., Goupil, C.: On the fundamental aspect of the first Kelvin's relation in thermoelectricity. Int. J. Therm. Sci. **104**, 225–227 (2016)
11. Kushch, A.S., Bass, J.C., Ghamaty, S., Elsner, N.B.: Thermoelectric development at Hi-Z Technology. In: Proceedings ICT2001. 20 International Conference on Thermoelectrics (Cat. No.01TH8589), pp. 422–430. Beijing, China (2001)
12. He, J.-H.: Homotopy perturbation technique. Comput. Methods Appl. Mech. Eng. **178**, 257–262 (1999)
13. Liao, S.-J.: An approximate solution technique not depending on small parameters: a special example. Int. J. Non-Linear Mech. **30**(3), 371–380 (1995)
14. Malalasekera, W., Versteeg, H.K.: An Introduction to Computational Fluid Dynamics: The Finite, vol. Method. Pearson-Prentice Hall, Glasgow (2007)

Implementation Process of a Local Energy Community in Portugal – The Case of Culatra Island

Jóni Santos[1](✉), André Pacheco[2], and Jânio Monteiro[3]

[1] Universidade Do Algarve, Faro, Portugal
jnsantos@ualg.pt
[2] CIMA, Universidade Do Algarve, Faro, Portugal
[3] Universidade Do Algarve and INESC-ID Lisbon, Faro, Portugal

Abstract. This work describes the challenges to create a local energy community at Culatra Island, a small coastal community located in the south of Portugal. Although the creation of local energy communities is recently possible in Portugal after the transposition of the EU directive to the national law, there are no real case studies to follow. The study starts by describing the Portuguese legal framework for these communities, including the defined energy sharing methodologies. Subsequently, it researches into the specific context of Culatra Island, exploring both consumption patterns and renewable energy production. The aim is to propose an energy sharing scheme that prioritizes energy costs reduction for the community while increasing the energy sustainability of island, ensuring fairness and proportionality among the community members. A new economic model is proposed to maximize consumption of locally produced energy, demonstrating how members can simultaneously benefit from discounts on energy bills and invest on increasing future production and energy self-sufficiency. The model will guarantee greater savings for the members during a temporary horizon of decades and can be further replicated on other local energy communities with different social, economic and environmental conditions.

Keywords: Local Energy Communities · Pilot Project · Smart Grids · Demand Response · Participatory Economic Model

List of Acronyms

CETA	Clean Energy Transition Agenda
CP	Consumption Point
CSC	Collective Self-Consumption
CSCME	Collective Self-Consumption Management Entity
CUC	Clube União Culatrense
DPC	Delivery Point Code
DSO	Distribution System Operator
EMS	Energy Management System
ERSE	Regulatory Entity of Energy Services

J. F. L. C. Semião et al. (Eds.): INCREaSE 2023, ASST, pp. 173–191, 2023.
https://doi.org/10.1007/978-3-031-44006-9_13

IPCC	Intergovernmental Panel on Climate Change
ISC	Individual Self-Consumption
LEC	Local Energy Community
PEG	Public Electrical Grid
PQ	Production Quota
PV	Photovoltaic
REPU	Renewable Energy Production Unit
RES	Renewable Energy Sources
SC	Self-Consumption

1 Introduction

Based on the simple idea of a smart usage of resources, sustainable development principles aim at improving living conditions of our generation without hindering those of the next ones. The concept started to emerge only three decades ago at the Earth Summit of Rio in 1992. Although sustainable development is an ideal concept on paper, it can rapidly find limitations. The first one that comes to mind is our responsibility for climate change as reported by the Intergovernmental Panel on Climate Change (IPCC) technical assessment [1], recognized as an existential threat to many coastal cities and communities as they grapple with rapidly increasing flooding, more frequent and intense storms, saltwater intrusion, etc. Thankfully, and as highlighted in the latest assessment of the IPCC report Working Group III [2], it is still not too late to act and there are possible pathways to limit our impact on the environment such as replacing carbon fossil energy generation by Renewable Energy Sources (RES), which seem to be one of the most promising solutions to building a sustainable development [3–5].

However, independently of our present actions, past emissions will cause major sea level rise which is expected to expose world coasts to unprecedented levels of coastal flood risk and erosion. These coastal areas are home to a large and growing proportion of the world's population where important economic activities take place and where the most ecological resources are in danger from global climate change. Today, with vast investment and opportunities flowing into large cities, coastal communities are often neglected. The limited resources and infrastructures put these local coastal communities at risk due to their natural, social and economic vulnerabilities. Creating new roadmaps that enhance circular business models and accelerate energy transition, are crucial to build resilience in coastal communities, that highly depend on singular and highly vulnerable ocean activities (e.g., fishing), by providing them novel sources of income.

The objective of the present paper is to describe the challenges of creating a Local Energy Community (LEC) and explain how the creation of a decentralized energy market can empower coastal communities on becoming economically self-sufficient, resilient and adaptable to future climate change scenarios. The test case is a coastal community located on a fragile ecosystem, the Culatra Island in the heart of Ria Formosa, the most important wetland of Portugal. The reason to choose an island is twofold: (1) the Paris Agreement [2] recognizes that islands are particularly vulnerable to climate change and are highly dependent on fossil fuels and energy imports. Due to their intrinsic nature, islands face barriers that restrain their self-sufficiency; (2) these territories are understood as micro versions of larger society. The laboratory dimension of an island enables addressing experimentally the entire system. As living laboratories, they are essential sites for validating and spreading new concepts, processes and technology beyond their territories [6], and can serve as showcases at an international level, i.e., be replicable and scalable, to address the future challenges that coastal communities will face.

After the introduction, the paper has the following structure: in Sect. 2, is described LEC legislation in Portugal, including the structure, entities and sharing mechanisms; in Sect. 3 is addressed the context of the Culatra Island, including the consumption characterization of the island and the installed Renewable Energy Production Units (REPU); in Sect. 4 is described the objectives of the Culatra LEC; in Sect. 5 are discussed the implementation details of the LEC; and finally, in Sect. 6, is concluded the paper with the final considerations and ongoing work.

2 Legislative Context of Local Energy Communities in Portugal

Energy systems are undergoing a transformation from centralized to decentralized structures, and many countries are changing their legislation to enable RES implementation through what is known as LEC. In Europe, a strong legislative framework, the "Clean Energy for All Europeans" Package [4], places the final consumer at the centre of the energy system, where the concept of Self-Consumption (SC) is being expanded from homes and small businesses to local communities and municipalities. The European Commission is also working towards improving the existing legislative framework – the Green Deal – needed to facilitate the energy transition, required to achieve the goals set for 2050, i.e., become the World's first climate-neutral continent and assume a central role in the transition to a greener, more inclusive, democratic, transparent and participatory energy system [5].

The legal Portuguese regime for the consumption of electricity produced by one or more REPU, and performed by one or more of these so-called prosumers, or self-consumers, was firstly approved by Decree-Law no. 162/2019 [7], which also approved the SC in a collective way and therefore the constitution of a LEC, partially transposing into national law the Directive 2018/2001 of the European Parliament and of the Council directive of 11th December 2018 about the promotion of the use of energy from RES [8]. Subsequently, this regime was repealed and integrated into the legal regime of the National Electric System by the new Decree-Law no. 15/2022 [9].

2.1 Architecture

In Portugal, the activity of SC can be performed individually, or collectively which includes the LEC [9, 10]. When performed individually, Individual Self-Consumption (ISC), the REPU is associated with only one Consumption Point (CP) [9, 10], i.e., associated only with a Delivery Point Code (DPC) as exemplified in Fig. 1. The DPC is the code that identify the meter installed in a certain facility.

Unlike ISC, in Collective Self-Consumption (CSC) one or more REPU are associated with two or more CP, i.e., when one or more production units are associated with more than one DPC [9, 10]. Figure 2 shows two CSC examples schemes. The first one, on left, the energy is shared with the participants through the building internal electrical grid. In the second example, on right, is used the Public Electrical Grid (PEG) to share the energy to the participants, which results in the payment of grid use tariffs. In CSC

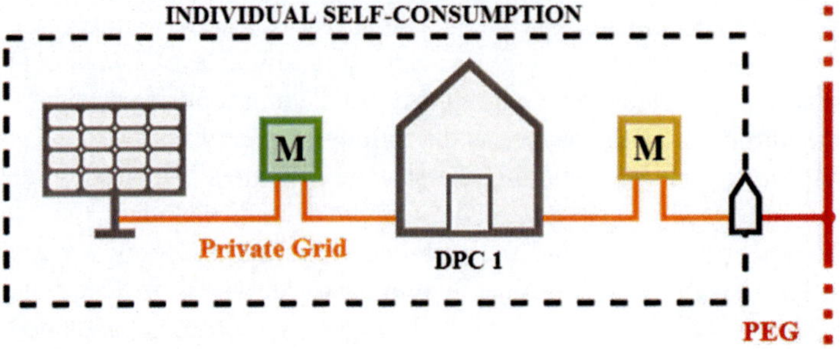

Fig. 1. Individual Self-Consumption example scheme. The (M) symbolizes the meters, the green one is associated to the REPU and the orange one is associated to the CP. In ISC the REPU is usually connected to the CP by an internal grid, i.e., private grid.

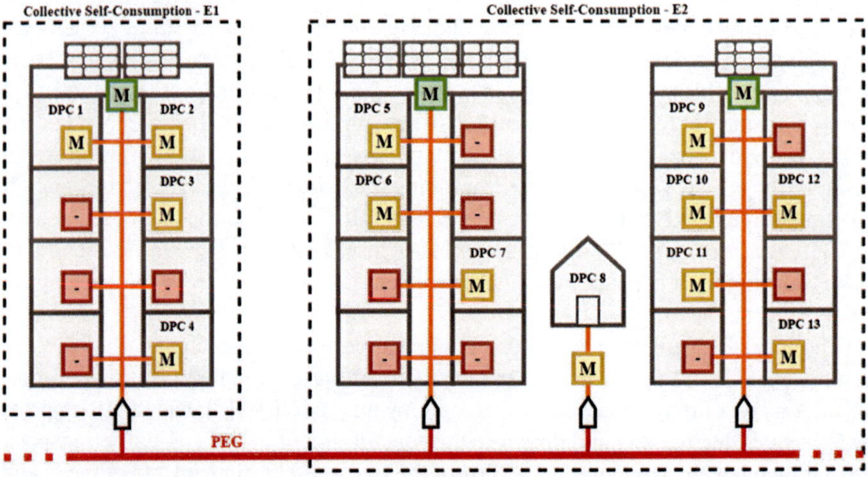

Fig. 2. Collective Self-Consumption Examples Schemes. In E1 (left) the energy is shared with the members through the building internal grid; In E2 (right) the PEG is used, and grid use tariffs must be paid. The red boxes in the schemes represent the meters of the CP outside of the CSC activity.

regime, when there are two or more REPU, as exemplified in the second example (E2) of Fig. 2, the total energy production of the system is the summatory of the production of each REPU accounted by its own meter [9, 10].

To perform the activity of CSC, the concept of proximity should be considered. This concept consists of the geographical distance, established through the voltage levels, between the REPU and the CP that it supplies. This means that to apply the rules of CSC, the CP cannot distance more than 2 km from the REPU in the case of low voltage, 4 km in the case of medium voltage, 10 km in high voltage and 20 km in very high voltage [9], as defined in Table 1.

Still, in CSC, it must be defined a member, or an independent entity, as the responsible representative of the activity, called the Collective Self-Consumption Management Entity (CSCME). This entity is responsible for the practice of operational management

Table 1. The four levels of proximity concept definition based on the distance between the REPU and the CP.

Voltage Level	Distance between REPU and CP
Low Voltage	2 km
Medium Voltage	4 km
High Voltage	10 km
Very High Voltage	20 km

acts of the activity such as: (1) the articulation with the SC electronic management platform; (2) the connection to the PEG; (3) the relationship to adopt for surplus energy (e.g., storage, sale or lost to the grid); (4) celebrate a supply contract with a supplier for the purchase of energy, a PEG use contract with the Distribution System Operator (DSO) for payment of grid use tariffs, and a sale of surpluses contract with a supplier or aggregator for the sale of energy; (5) share among its CSC members the costs associated with the activity, such as grid use tariffs when the PEG is used to share energy; (6) share with its CSC members the benefits derived from the sale of surplus energy; and (7) access to the consumption and production data of its CSC members. In addition to these responsibilities, CSCME is also responsible for defining the energy sharing modes, through the definition of the respective sharing coefficients.

In order to ensure that all CSC members fulfil their obligations, the CSCME must elaborate a Collective Self-Consumption Internal Regulation, which all members must comply. This is a mandatory document that must define at least: (1) the access requirements of new members; (2) the exit of existing members; (3) the energy sharing mode that will be used; (4) the payment of grid use tariffs, when applicable; (5) the destination of self-consumption surpluses; (6) and the commercial relationship policy to be adopted and, when applicable, the destination of the respective revenue [9, 10].

2.2 Energy Sharing Modes

In the Portuguese legislative framework of CSC, energy can be shared in different modes. This is a matter of great importance because the choice of these sharing methodologies is what will define the economic viability of a CSC project. The following subsection analyse this regulated sharing methodologies.

Fixed Coefficients. This energy sharing mode consists in the division of the energy generated by the REPU, on fixed percentages pre-established by CSCME, i.e., each CP participating in the CSC must have an associated fixed percentage of the energy generated by the production units during a quarter-hour period. These coefficients can be differentiated by working days, holidays or weekends, and may or may not consider the seasons [9, 10].

The calculation of the energy shared with a consumption point i by means of fixed coefficients (ES_{CP_i}) is expressed by Eq. 1, where FC_{CP_i} represents the fixed coefficient associated of a consumption point i, and E_{REPU_j} represents the energy produced by a production unit *j*.

$$ES_{CP_i} = \frac{FC_{CP_i}}{\sum_i FC_{CP_i}} \times \sum_j E_{REPU_j} \tag{1}$$

Figure 3 gives a practical implementation of the energy sharing mode based on fixed coefficients. In this example, which takes place during a quarter-hour period, two renewable energy production units with a total production of 35 kWh are considered, as well as four consumption points that sums up an energy consumption of 35 kWh. It means that the production is adjusted to the consumption. The 35 kWh produced are shared by the 4 consumption points according to their fixed coefficients for that period (i.e., CP 1: 40%, CP 2: 25%, CP 3: 25%, CP 4: 10%), which translates into an energy sharing of 14 kWh for CP 1, 8.75 kWh for CPs 2 and 3, and 3.5 kWh for the CP 4, respectively. However, as can be seen in Fig. 3, CP 3 has a consumption close to 0 during this quarter-hour period, which means that the energy shared with that point is treated as energy surplus, i.e., the energy will be lost to the grid at zero cost or sold to an energy supplier much lower than its purchase price.

According to Fig. 3, only 26.25 kWh of the 35 kWh produced are consumed, as the remaining 8.75 kWh are treated as energy surplus. To suppress the lack of those 8.75 kWh, the same amount of energy is supplied by an energy supplier in order to partially satisfy the total consumption of each CP. This energy sharing mode turns out to be the simplest, where the energy is shared proportional to what is transmitted by the CSCME. It proves to be the fairness option if the system is financed by members, since

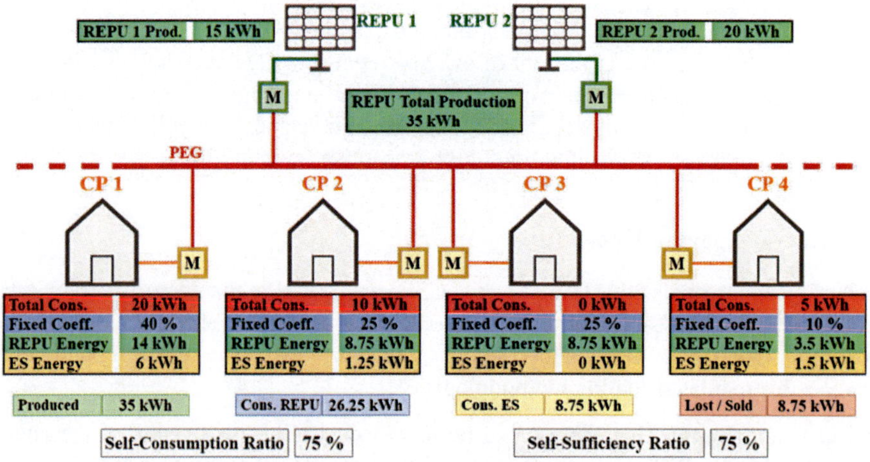

Fig. 3. Fixed Coefficient sharing mode example, involving two renewable energy production units (REPU) and four consumption point (CP), during a quarter-hour period where the production is adjusted to the consumption. Each CP is characterized by its total consumption (Total Cons.), associated fixed coefficient (Fixed Coeff.), amount of energy shared by the REPU (REPU Energy), and amount of energy consumed from an energy supplier (ES Energy). The system is characterized by the total energy produced (Produced), the total amount of energy consumed from the REPU (Cons. REPU), the total amount of energy consumed from the energy supplier (Cons. ES) and the total of energy sold or lost to the grid (Lost/Sold).

each member can keep a fixed coefficient according to the percentage of investment that each one made. However, it loses efficiency as can be observed by the Self-Consumption Ratio of 75% and the Self-Sufficiency Ratio of 75%, due to the individual energy surplus that ends up being injected into the grid (sold or lost) if the consumption in the CP is low, therefore not taking full advantage of the concept of energy sharing.

Consumption-Proportional Coefficients. The energy sharing based on consumption-proportional coefficients consists in the division of the energy produced by percentages equivalent to the consumption measured at each CP during each period of 15 min. In this case, CSCME has no control over the sharing coefficients, as these are defined based on the consumption of each CSC member. It means that the member with the highest consumption will also have a higher percentage of the energy produced by the REPU [9, 10]. The calculation of the energy shared with a consumption point i by means of consumption-proportional coefficients (ES_{CP_i}) is expressed by Eq. 2, where EC_{CP_i} represents the energy consumed by a consumption point i and E_{REPU_j} represents the energy produced by a production unit j.

$$ES_{CP_i} = \frac{EC_{CP_i}}{\sum_i EC_{CP_i}} \times \sum_j E_{REPU_j} \tag{2}$$

Figure 4 gives a practical implementation of this energy sharing mode. However, in this case, the 35 kWh produced are shared by the 4 points according to their consumption (CP 1: 57%, CP 2: 29%, CP 3: 0%, CP 4: 14%). In practice, the production is adjusted

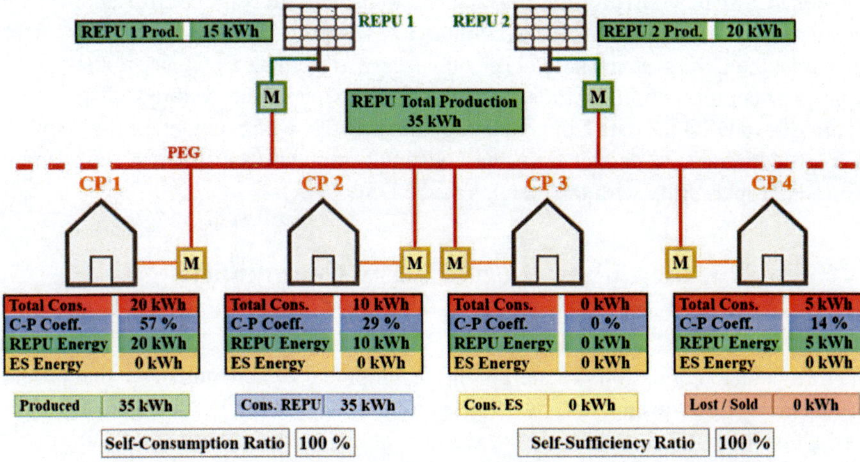

Fig. 4. Consumption-Proportional Coefficients Sharing Mode, involving two renewable energy production units (REPU) and four consumption point (CP), during a quarter-hour period where the production is adjusted to the consumption. Each CP is characterized by its total consumption (Total Cons.), associated consumption-proportional coefficient (C-P Coeff.), amount of energy shared by the REPU (REPU Energy), and amount of energy consumed from an energy supplier (ES Energy). The system is characterized by the total energy produced (Produced), the total amount of energy consumed from the REPU (Cons. REPU), the total amount of energy consumed from the energy supplier (Cons. ES) and the total of energy sold or lost to the grid (Lost/Sold).

to consumption and, without energy consumed at point 3, the system adapts, and all the energy produced is consumed by the remaining 3 CPs.

Analysing Fig. 4, the Self-Consumption Ratio is 100% because all the energy produced is consumed, and the Self-Sufficiency Ratio is also 100% because the total energy consumed in the system comes from the REPU. Overall, it can be assumed that this energy sharing mode is efficient because it maximizes the consumption of the energy produced, while minimizing the energy injected into the grid. However, there is no longer control on the sharing coefficients by CSCME, which becomes unfair in CSC systems financed by members, since most of the energy produced is delivered to those who consume more and not to those who contributed more to the investment of the REPU.

Mixed Coefficients. In addition to the two modes examined, the sharing of energy can also be based on mixed coefficients, which results in a proposal that combines the previous two modes and complies with national Regulatory Entity of Energy Services (ERSE) regulations [8], i.e., where efficiency and optimization are combined, with some control by the CSCME. In this case, the calculation of the energy shared with the consumption point i through mixed coefficients (ES_{CP_i}) would be expressed by Eq. 3, where FC_{CP_i} represents the fixed coefficient associated to a consumption point i, EC_{CP_i} represents the energy consumed by a consumption point i and E_{REPU_j} represents the energy produced by a production unit j.

$$ES_{CP_i} = \frac{FC_{CP_i} \times EC_{CP_i}}{\sum_i (FC_{CP_i} \times EC_{CP_i})} \times \sum_j E_{REPU_j} \tag{3}$$

Dynamic Coefficients. This mode is usually proposed when conducting pilot projects in order to test new sharing models that allow the monitoring, control and dynamic management of energy, in real time, with a view to optimize energy flows. To perform this mode, the CSC system shall: (1) ensure access to the data necessary to the DSO for its proper operation and operationalization, in particular to the readings of the meters; (2) provide to DSO the data from the energy shared with each member, or the respective sharing coefficient, to deduct from the consumption measured in the meters; and (3) ensure interoperability with the systems of the DSO [9].

3 Context of the Culatra Local Energy Community

3.1 Implementation Area

The European Commission is fostering the development of new initiatives that promote the adoption of best available technologies on islands, including exchange of best practices in financing, legal, and regulatory regimes, and in energy for transport. One of those initiatives is the Clean Energy for EU Islands Programme, where the local communities are being challenged to search for a stronger control over their traditional activities, overcoming external factors that can lead to high costs of energy supply and production. In 2019, and under the scope of this initiative, six pilot islands were identified to each develop their own "Clean Energy Transition Agenda" (CETA), a strategic roadmap for the transition process towards clean energy. One of those islands was Culatra (Fig. 5), a small island located in one of the five barrier islands that compose the Ria Formosa Natural Park, a multi-inlet barrier island system in southern Portugal (Algarve).

Fig. 5. Culatra Island, at Ria Formosa, Portugal with its three settlements: Farol, Hangares and Culatra, the latter standing as the case study for this paper.

The island is around 7 km long and has a maximum width of 1.2 km, comprising a total area of 4.34 km^2. This paper focuses on the case study of Culatra village, which is the largest of the three settlements on the island and has a population of 759 inhabitants, representing 80% of the island's population. Fishing is at the centre of the village of Culatra and reflects the adaptation of the population to the space and environment in which it operates, the Ria Formosa. The Ria Formosa, the most important wetland of Portugal, is unique and remarkable, and is amongst the most studied national coastal areas on a wide range of topics from geomorphology and coastal dynamics to ecology, biodiversity, economy and social values.

The initiative today known as "Culatra2030 – Sustainable Energy Community" is a pilot transition initiative, uniting support from the key stakeholders, including the Minister of the Sea, the Coordination Commission for the Algarve Region, the Faro City Municipality, energy and environmental authorities, companies and non-governmental associations. The project has the ambition to be a pivot step on behalf of sustainability issues in Portugal, particularly on clean energy and smart grid flexibility, since a 100% renewable energy future will intelligently combine variable RES, integrated into a smart grid system providing continuous energy supply. The innovative starting point of Culatra2030 was the development of the Culatra' CETA [11], a roadmap for energy transition until 2030, through a community participatory diagnosis [12], which proved essential for defining priorities and generating ideas for actions with a high degree of commitment by all local actors, i.e., island community, public authorities, academia and commercial companies.

3.2 Characterization of the Consumers

The three settlements, Culatra, Farol and Hangares, which compose Culatra Island have different realities in which regards their energy system. Culatra is electrically interconnected with the mainland of Portugal through a 15 kV submarine cable, installed in 1992. This cable provides electricity to parts of the island but has broken down several times in the past causing power outages of large parts of the island. While most of the houses of the main settlement, Culatra village – which are the current focus of the transition process due to their legal status – are connected to the interconnected electrical distribution grid, 25% of all houses of Culatra Island do not have grid-connected electricity supply. Part of the houses of the Farol settlement and all the houses of the Hangares settlement are completely autonomous.

The entire electrical grid on the island is low voltage, which is why local power losses and potential losses due to the transport of electricity from the mainland through a submarine cable could be significant. Since historically there has been problems with the submarine cable causing blackouts, a diesel generator (410 kVA, 380 V) was installed on Culatra village. In fact, the village was the focus of some projects from the Portuguese DSO – E-REDES. These projects have resulted that 316 of the houses on the island are connected to the electricity grid and have a smart meter installed.

Using the available data and taking as example the year of 2019 (pre-pandemic year), Culatra village consumed a total of 1181 MWh [12]. The energy consumption was higher in summer months and lower in winter (Fig. 6B) and do not change much in distinct days of the week (Fig. 6A). The consumption is yet very sparse over the day and thus the generation profile should take into consideration this characteristic.

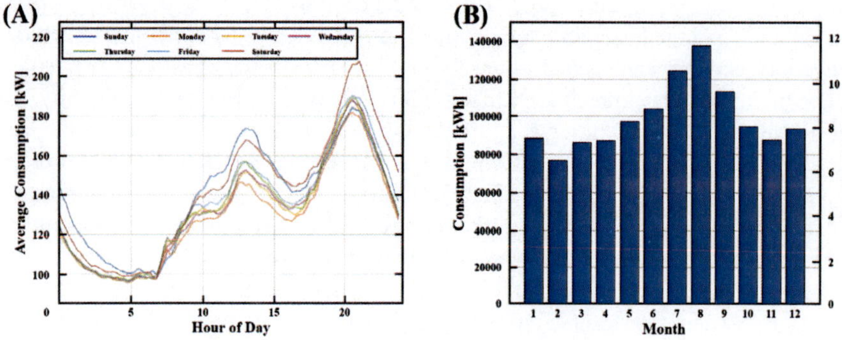

Fig. 6. Consumption pattern of Culatra Island. (A) Average Daily Consumption; and (B) Monthly Energy Consumption of the Culatra habitational nucleus (data provided by the DSO, adapted from [12]).

3.3 Renewable Energy Production Units Installed

On a precedent analysis [12] authors concluded that the Culatra energy needs during the daylight are about 484.384 MWh, and in order to produce it using solar energy, would be necessary to install around 309 kWp in photovoltaics (PV). Resulting from applications within the scope of the Culatra 2030 Initiative, Culatra village currently has five autonomous PV production units (Fig. 7), totalizing 85.3 kWp capable of producing a yearly average of 155.6 MWh (Table 2), which is more than 1/4 of its average energy needs during the daylight [12]. In addition to the installed capacity for energy generation, the initiative also acquired funding for 46 kWh of Lithium batteries storage, and the technological development of a solar boat to be used on oyster farming and a connection point to the electricity grid, allowing interaction of the boat with the remaining PV facilities.

Fig. 7. Photovoltaic units installed in the Culatra village. From top to bottom and left to right: units installed at the shading areas, unit installed at the Social Centre, unit being installed at the Elementary School, installation at the headquarters of the Clube União Culatrense (CUC), Solar boat for the transportation of bivalves produced in Ria Formosa.

Table 2. Installed capacity and estimated average yearly energy production of the five renewable energy production units installed at Culatra village.

REPU	Location	Capacity Installed	Yearly Production
1	Shadow Platform 1 – Harbour Zone	20 kWp	37.13 MWh
2	Shadow Platform 2 – Harbour Zone	40 kWp + 46 kWh	73.26 MWh
3	Social Centre	5.5 kWp	10.00 MWh
4	CUC Headquarters	6 kWp	11.80 MWh
5	Elementary School	13.8 kWp	24.21 MWh

4 Objectives of the Culatra Local Energy Community

The vision of Culatra 2030 Initiative is to create a LEC that manages its own energy and shares it, according to the rules that it establishes. The intention is to define and implement a model of participatory economics, which allows the distribution of costs and the income that results from renewable generation, supporting the involvement of the community, investment in equipment and its operation, in a temporary horizon of decades. This should be combined with a reduction of energy costs in the community, and an increase of the energy sustainability of island. To accomplish this, high levels of self-consumption and self-sufficiency are required. However, high levels of fairness and proportionality between community members also need to be considered. In the following is described each of these parameters.

Self-consumption translates the ratio between the energy that is generated by a renewable energy system, at a given period, and the part of that energy that is consumed on-site. In other words, in a self-consumption scenario, when a renewable energy unit, such as a solar photovoltaic array, produces electricity, the ratio of energy that is directly consumed by the household, or building, translates the self-consumption ratio. Any excess electricity that is not consumed immediately can be stored in batteries for later use or fed back into the grid. However, these options are normally not as cost effective as direct consumption. Thus, self-consumption is seen as a way to increase the efficiency and cost-effectiveness of renewable energy systems, with direct impact in the associated payback periods of investments.

Self-sufficiency translates the ratio between the energy that is generated by a renewable energy system, at a given period, and the total energy that is consumed on-site during that period. Higher levels of self-sufficiency contribute to the reduction of energy costs in the community members, as the expected costs that result from renewable energy generation are lower than the tariff costs of buying it to a supplier. However, overproduction, i.e., generating more than what is consumed, needs to be avoided.

In the context of a LEC, members that have an equal share in the community should get, on average, equal benefits from its participation and/or investment. This is defined as fairness. Benefits can be quantified though energy allocation, or economic return. For instance, a consumer that buys a share in the community, but that is not consuming its share of energy, can be economically compensated, because its energy is being allocated to other members. Finally, if two members buy two distinct share sizes in the LEC they should get, on average, benefits that are proportional to the investments made. The fairness case is in fact a special case of proportionality, where investments are equal. The assurance of fairness and proportionality between LEC members is critical to assure cohesion between members.

5 Project and Implementation Details

The Culatra Island LEC was submitted to the ERSE and approved in January 2023 as a pilot project. The aims of the pilot are to test the technical and economic feasibility of a LEC, allowing the inclusion of innovative technologies, with results that can be reflected in legal and regulatory development proposals. Using existing equipment and structures, the Culatra Island LEC will support the following research activities: (1) implement new forms of consumption management, which include the integration of energy management systems, and which allow controlling the consumption of equipment by LEC members, integrating demand forecast and renewable generation, considering the limitations of the existing electrical grid; (2) implement a model of participatory economy, which allows the distribution of costs and benefits derived from renewable energy production, and which supports community involvement, investment in new equipment and its operation in a temporary horizon of decades; (3) implement an intelligent system for charging electric boats, capable of controlling their charging periods and intensity according with the generated power, respecting user preferences and maximizing self-consumed energy.

The achievement of the proposed goals is considered essential to create long-term conditions for the residents to live in this fragile environment, to continue their traditional economic activities, preserving their identity and cultural values, and to seek new economic activities compatible with the rational use of natural resources. Having into consideration the global results of the participatory exercise (transcending energy issues, from water and waste to local economy and social welfare, among others), the nature of such jointly owned institution has been outlined in the form of a cooperative, i.e., an autonomous association of persons united voluntarily to meet their common economic, social, and cultural needs and aspirations.

5.1 Establishment of the C-COOP

The Culatra LEC promotor is the C-COOP, Cooperative for the Sustainability of Culatra Island, a cooperative formally created in 2022 exclusively by residents. It has a broader social purpose, that can cover all the socioeconomic activities of the island. The cooperative will work as a mechanism of community responsibility, in order to protect, conserve, promote, develop, and especially to maintain the identity of Culatra. In addition to an administrative office, a registered office, organized accounting, etc., the cooperative has an important role in the energy activity of the island as the CSCME. It means that C-COOP is the responsible for sharing the produced energy in the community. This multisectoral cooperative aims to create a decentralized energy market able to empower this coastal community on becoming economically self-sufficient, resilient and adaptable to future climate change scenarios, as well as fight energy poverty, promote social cohesion and create jobs.

5.2 Participatory Economic Model

As the promotor of the Culatra LEC, the C-COOP is designing a participatory economic model, to distribute the benefits derived from renewable energy production to its members, and drive the community to investments in the installation of more RES. The model is based on the principle of partitioning the PV capacity already installed (85.3 kWp) into small fractions called Production Quotas (PQ) that can be acquired by the LEC members. The partitioning of the installed PV capacity into small PQs, guarantees that the whole community can benefit from the renewable energy production, obtaining a reduction on their electric energy bills.

The first challenge was finding a fair PQ value, that could bring real benefits to the members. In this context, several metering's were made on Culatra buildings, during periods of several weeks, with a periodicity of one minute. These metering's allowed to conclude that most of these buildings (residential ones) show the same pattern of consumption, i.e., an uninterrupted power consumption resulting from constant loads, around 220 W, as shown in the graph of Fig. 8.

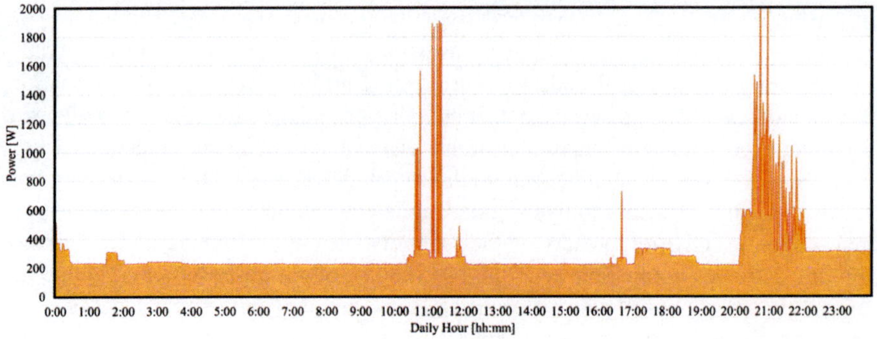

Fig. 8. Representative graph of a common daily energy consumption in a residential building in Culatra village, where it can be observed a constant load during the day around 220 W.

Based on this average energy consumption pattern, a fixed a PQ value of 200 Wp was defined and complemented with the following analysis: (1) Culatra has a PV capacity of 85 300 Wp installed, which divided by 200 equals to 426 available PQ of 200 Wp each to distribute; (2) there are 316 buildings in the community that already have smart meters installed, i.e., the C-COOP can thus guarantee that all the potential member can acquire at least one PQ, and benefit from a direct discount on the electricity bill; (3) even if one PQ is allocated to every building, remaining PQs will still exist, which can be acquired by members with higher consumptions, like local businesses as restaurants. This value can be perfectly applied to the fairness concept of proportional investments as a member can acquire more than one PQ, in case of availability, in accordance with their energy needs, i.e., 400 Wp, 600 Wp and so on.

With the value of PQ defined, the second requirement was to evaluate the annual energy production associated with it. The PVsyst software was used to compute the annual energy production of a PQ. This assessment was made using the meteorological data from Culatra, and real installation data, considering that all the PV units in Culatra have similar installation characteristics, i.e., the panel tilt of 15° and the azimuth facing south. The results are presented in Fig. 9 and show that a 200 Wp PQ can produce an estimated energy of 368 kWh per year, producing in average more energy during the summer months and less during winter months, following the behaviour of the consumption pattern of the island seen in Fig. 6B.

The model was designed to maximize the consumption of the energy produced by the PQ, i.e., reach 100% Self-Consumption Ratio, which results in a potential energy saving of 368 kWh per year. Figure 10 shows that a LEC member with a common consumption pattern, who acquires one PQ, can maximize its Self-Consumption Ratio during the wintertime and summertime.

According to OMIE, the average price of the energy during the year of 2022 was 0.16 €/kWh without taxes [13]. With taxes the value is 0.1968 €/kWh, which means that 368 kWh/year corresponds to a saving of 72.43 €/year per member. From the community perspective, savings of 30855 €/year can be targeted.

All these savings can be used to fund new installations, as all the PV production units installed in Culatra have resulted from applications made by Culatra2030 Initiative and are already paid. But to maintain those units operating and to invest in new ones, the

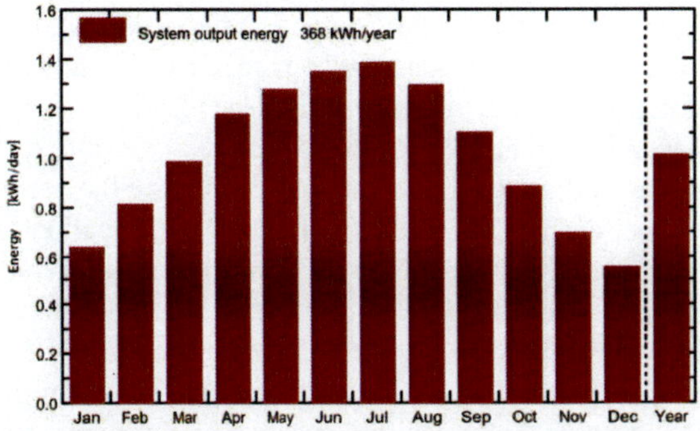

Fig. 9. Results from the annual electrical production assessment made using the PVsyst software for the 200 Wp production quota, producing in average more energy during the summer months and less during winter months.

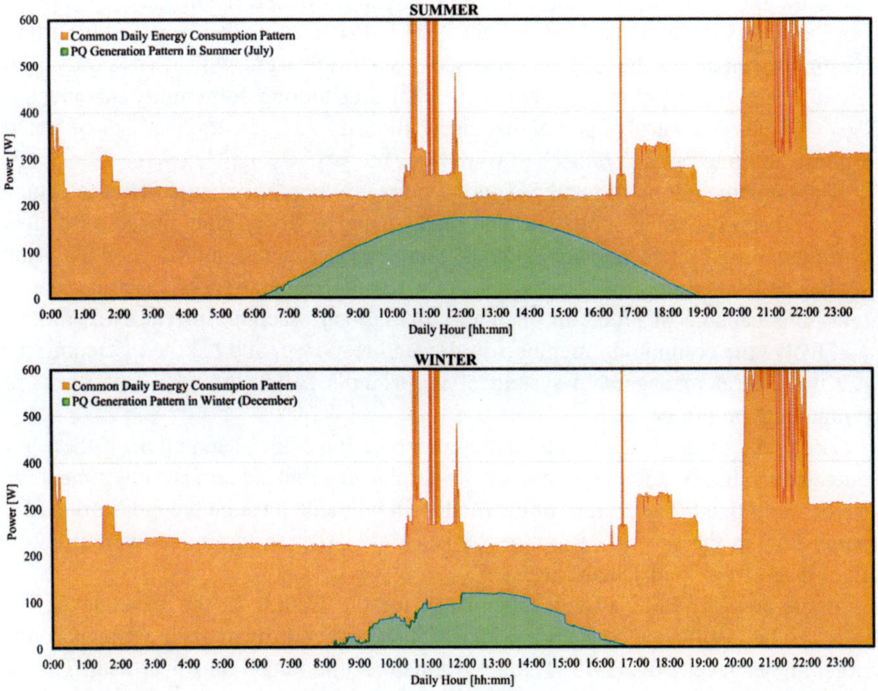

Fig. 10. 200 Wp PQ generation pattern overlaid on the common daily energy consumption pattern. The top graph shows a 200 Wp PQ generation pattern of a day in July, which is the representative month for summertime, i.e., the best month for PV energy production. The bottom graph shows a 200 Wp PQ generation pattern of a day in December, which is the representative month for wintertime, i.e., the worst month for PV energy production. Both generation patterns fit under the consumption pattern.

model must be economically sustainable. So, the last step was to assign a price to the PQ. In this context a market study was made based on the real prices that the local companies charge for the installation of photovoltaic systems. The results have shown that the prices range from 2 to 4 €/Wp for small PV installations, i.e., one to two PV panels. Taking into consideration these values, the cooperative is evaluating the option of charging 1 € per Wp, i.e., 200 € per each PQ of 200 Wp. The PQ will be allocated to the member during a period of 10 years. A single PQ will give to the members the advantages of using a system that is already installed and ready to use, with no maintenance costs. The payback period of the investment in one PQ is approximately of 3.5 years, which, in a 10 years period, corresponds to 6.5 years of profits. Those profits will be invested in more PV units to increase the self-sufficiency, which consequently will increase the energy savings of the community.

5.3 Integration of Smart Grid mechanisms

The aim of the Smart Grid mechanisms, in the scope of the proposed project, is to define and implement a community-based Energy Management System (EMS). That EMS will allow the adjustment of the consumption pattern of the whole LEC, accordingly with the energy generated by the distinct REPU, taking into consideration the different investments made by the LEC members. As such, the LEC will act not only in the definition and testing of energy sharing mechanisms among community members, but also in the energy control of part of that distribution. To do it, control methods of demand response and Internet of Things hardware must be created.

The implementation of that EMS will require: (1) household equipment to communicate and control home appliances and loads (using smart infrared or radio frequency controllers, smart plugs, among others), (2) the development of optimization algorithms for the load scheduling of these appliances, (3) the integration of machine learning algorithms capable of predicting demand, (4) human machine interfaces to motivate and involve the community members in demand response, and (5) the interconnection with the central management system of the island's LEC, through IP networks and communication protocols.

The focus of the EMS is the maximization of the self-consumption of the energy generated by the PV units. By reducing the surplus, the return period of investments will also be minimized. The control of the loads will be made through the monitoring of the energy flow in the several sectors of the island, in order to minimize overcurrent that might result from load scheduling.

A hierarchical EMS system is foreseen for the island. At the lower level, local EMS will be installed at consumer premises, that will coordinate their action according with a higher level LEC EMS. The involvement of the inhabitants is crucial. They are expected to participate in the control of demand, directly or through equipment installed in their homes. Innovative mobile interfaces should be used to allow the selection of the more profitable usage profiles for different flexible loads, respecting the desired comfort levels. Decision support systems should inform the LEC member about the ideal level of participation in the LEC, which allows them to maximise their self-sufficiency and minimise their costs, considering their actual daily, weekly and monthly consumption profiles.

6 Conclusion and Future Developments

LEC are emerging throughout Europe, in a context where many technical, legislative and economic challenges need to be tackled. At the legislative level, the Renewable Energy Directive 2018/2001/EU requires EU Member States to create enabling frameworks for the development of LEC, which was mirrored in the Portuguese Decree-Law no. 15/2022. The current legislation allows two energy sharing mechanism among members, which are still far from supporting the maximization of self-consumption and self-sufficiency at energy community level, and fairness and proportionality between community members.

In order to overcome this gap, the present study intends to use the pilot project of LEC on a living laboratory as Culatra Island to investigate and create solutions to tackle the technical and non-technical restrictions that limit the penetration of energy generated from RES. Regarding technical constraints, it is important that the solutions are developed in partnership with the Portuguese DSO – E-REDES, by testing new energy sharing models where decentralized production systems can be implemented without creating grid stability problems. Regarding non-technical constraints, more related to economic, market, social and regulatory issues, it is important to find solutions to finance the REPU. The participatory economic model proposed within the present study exemplifies how LEC participants can benefit from electricity bill discounts and invest those saving in the community to increase energy production and next year discounts i.e., allowing the model to run on a temporary horizon of decades. In addition to the investment, energy management is an objective that must be ensured by all LEC members. In monthly terms, there is a high correlation between consumption and generation, which favours the levels of self-consumption and self-sufficiency. However, energy generation intermittency requires direct consumption management measures and/or indirect measures through the involvement of the community.

All the developments made at Culatra in this context are available in an open science format to encourage other communities to create their own LEC and prove that sustainable development will have an increasing impact on society, especially in communities that live in conditions of great adversity such as the Culatra island.

Regarding future developments, it is intended to increase the PV generation capacity on the island, namely in a project of requalification and modernization of the fishing working area. This project will allow to install around 200 kWp, approaching Culatra to the energy self-sufficiency during the daylight. Completed, Culatra2030 Initiative, supported by C-COOP, will create a decentralized energy market that will allow fighting energy poverty, promote social cohesion, and create employment.

The transition team also intends to design and operate solar boat platforms and terrestrial vehicles, install a set of charging points for these electric vehicles, allowing charging them according to the needs of the owners, but also in line with the electrical management of the island. Finally, it is also envisaged the installing and operating of a water desalination station and a waste treatment plant, powered with RES, and integrated with the island's electrical management system.

Acknowledgements. This work was supported by the Portuguese Foundation for Science and Technology (FCT) through the individual research grant 2021.08721.BD and 2020.00994.CEECIND.

References

1. Riahi, K., et al.: Mitigation pathways compatible with long-term goals. In: Shukla, P.R., et al. (eds.) IPCC, 2022. Climate Change 2022: Mitigation of Climate Change. Contribution of Working Group III to the Sixth Assessment Report of the Intergovernmental Panel on Climate Change. Cambridge University Press, Cambridge, UK and New York, NY, USA (2022). https://doi.org/10.1017/9781009157926.005
2. IPCC: Climate Change 2022: Mitigation of Climate Change. In: Shukla, P.R., et al. (eds.) Contribution of Working Group III to the Sixth Assessment Report of the Intergovernmental Panel on Climate Change. Cambridge University Press, Cambridge, UK and New York, NY, USA (2022). https://doi.org/10.1017/9781009157926
3. Dincer, I.: Renewable energy and sustainable development: a crucial review. Renew. Sustain. Energy Rev. **4**(2), 157–175 (2000). https://doi.org/10.1016/S1364-0321(99)00011-8
4. EC: 'Clean Energy for all Europeans' – Communication from the EC to the European Parliament, the Council and the European Economic and Social Committee, the Committee of regions and the European Investment Bank, 13pp. (2016)
5. EC: 'A European Green Deal – Striving to be the first climate-neutral continent'– Communication from the EC to the European Parliament, the Council and the European Economic and Social Committee, the Committee of regions and the European Investment Bank, 24pp. (2019)
6. Skjølsvold, T.M., Ryghaug, M., Throndsen, W.: European island imaginaries: examining the actors, innovations, and renewable energy transitions of 8 islands. Energy Res. Soc. Sci. **65**, 101491 (2020). https://doi.org/10.1016/j.erss.2020.101491
7. Presidência do Conselho de Ministros: Decreto-Lei n.o 162/2019, de 25 de outubro," Diário da República n.o206/2019, Série I 2019-10-25. https://dre.pt/dre/detalhe/decreto-lei/162-2019-125692189 (2019)
8. European Commission: DIRECTIVE (EU) 2018/2001 OF THE EUROPEAN PARLIAMENT AND OF THE COUNCIL of 11 December 2018 on the promotion of the use of energy from renewable sources. Off. J. Eur. Union. https://eur-lex.europa.eu/legal-content/EN/TXT/?uri=CELEX%3A32018L2001 (2018)
9. Presidência do Conselho de Ministros: Decreto-Lei n.o 15/2022, de 14 de janeiro," Diário da República n.o10/2022, Série I 2022-01-14, 2022. https://dre.pt/dre/detalhe/decreto-lei/15-2022-177634016
10. Entidade Reguladora dos Serviços Energéticos. Regulamento n.o 373/2021, de 5 de maio. Diário da República n.o 87/2021, Série II 2021-05-05. https://dre.pt/dre/detalhe/regulamento/373-2021-162753427 (2021)

11. CE4EU: Culatra Energy Transition Agenda, Clean Energy for EU Islands – European Commission, 57 p. (2019)
12. Pacheco, A., Monteiro, J., Santos, J., Sequeira, C., Nunes, J.: Energy transition process and community engagement on geographic islands: the case of Culatra Island (Ria Formosa, Portugal). Renew. Energy **184**, 700–711 (2022). https://doi.org/10.1016/j.renene.2021.11.115
13. OMIGROUP: "OMIE," Resultados do Mercado. https://www.omie.es/pt. Accessed 10 Mar 2023

Comparing Noise Vessel Azimuth Tracking with a Planar Hydrophone Array and a Single Vector Sensor

Cristiano Soares[1(✉)], Friedrich Zabel[1], Paulo Santos[2], and António Silva[2]

[1] MarSensing Lda. Centro Empresarial Pav. B1 Campus de Gambelas, 8005-139 Faro, Portugal
contact@marsensing.com

[2] LARSyS — University of Algarve Campus de Gambelas, 8005-139 Faro, Portugal
{pjsantos,asilva}@ualg.pt

Abstract. Vector sensors are appealing for monitoring underwater noise due to its inherent directivity. While acoustic pressure sensors are ambiguous in all directions, vector sensors permit the aquisition of directional information through the measurement of particle velocity, which enables the possibility of azimuth tracking of underwater noise sources.

Underwater acoustic systems based on vector sensors can play an interesting role in Marine Protected Areas where integrated marine observatories are needed to assess the evolution of the environmental state. The MARREAL marine observatory is a marine observatory equipped with a number of sensors and subsystems, including an acoustic acquisition system made of four hydrophones and a vector sensor. The observatory was deployed in September 2022 in Sagres, Portugal, near the Baleeira Port which is accessed by fisher boats and recreational boats. This paper shows preliminary results on azimuth estimation of boats passing in the deployment area, obtained independently with a 4-hydrophone planar array and a vector sensor. The results indicate that a single vector sensor can provide fair results on azimuth tracking of boats passing in the area. At high signal-to-noise ratio (SNR) the vector sensor is able to yield results similar to those obtained with the planar array. When the SNR is low the planar array outperforms the vector sensor with actual processing methods used.

1 Introduction

Over the time, the Direction of Arrival (DOA) estimation have been determined from pressure sensors using linear arrays of hydrophones. Recently the literature has been presenting works using the particle velocity information, as a way to resolve ambiguous acoustic pressure information. To measure particle velocity two-dimensional or three-dimensional, a vector sensor (VS) is used, and the benefit of using this comes from measuring three-dimensional components containing information about the direction of the sound source. While a pressure

J. F. L. C. Semião et al. (Eds.): INCREaSE 2023, ASST, pp. 192–203, 2023.
https://doi.org/10.1007/978-3-031-44006-9_14

sensor is omnidirectional, a VS is inherently directional. The VS devices measure the particle velocity by means of three orthogonal axes (spatial derivative of the pressure) and the acoustic pressure, when an hydrophone is collocated. The particle velocity can also be obtained from acceleration using a tri-axial accelerometer, currently the most common device. Classically, DOA estimation has been performed with arrays of hydrophones with different shapes, including linear, planar or volumetric arrays. From an operational point-of-view linear arrays are most commonly used for its practical application, either as a towed array or in vertical moorings, while planar arrays are more difficult to handle mechanically. While horizontal linear arrays suffer from the left/right ambiguity in azimuth tracking applications, a vertical array would suffer from a 360° ambiguity in this context. Planar arrays are appealing for eliminating azimuthal ambiguities inherent to linear arrays.

Since 90 s, theoretical works involving VS appeared in the literature, first for sound propagation in the air [1] and then for underwater acoustic sound propagation [2,3]. The spatial filtering capabilities of VS for DOA estimation clearly outperforms acoustic pressure only (scalar) hydrophones. The combination of several VS in an Array configuration (VSA) can be used to estimate both azimuth and elevation angles, eliminating the well know left/right ambiguity inherent to the response pattern of linear hydrophone arrays [4]. Taking advantage of its directionality and its high performance in DOA estimation, the use of VS became a subject of investigation [5–8]. In 2022, Smith et al.[9] reported on azimuth estimation from a VS moored on the seafloor at a depth of approximately 900 m for almost 2 years, outside of Monterey Bay, California near a major shipping lane. The analysis of the vector sensor data demonstrated the ability to accurately determine bearings to merchant vessels at ranges up to 60 km.

In several occasions, it has been demonstrated that the use of VS has not only advantages for DOA estimation but also for other applications, such as, underwater communications [10] and geo-acoustic inversion [11,12].

In September 2022 the MARREAL marine observatory was deployed in the channel between Baleeira Port in Sagres, Portugal and the Martinhal Islets. The MARREAL is a cabled observatory equipped with several sensors and subsystems to acquire environmental data and an acoustic array made of hydrophones and a VS. The aim of this paper is to present preliminary results on azimuth angle estimation for boat radiated noise and comparing results obtained from a 4-hydrophone planar array with the outputs from a single vector sensor (using one pressure and two particle velocity components). The data analysed consists of noise radiated by small fisher or recreational boats during local passages in the area of deployment.

This paper is organised as follows: in Sect. 2 the data and processing methods are described; in Sect. 3 are shown results on boat azimuth estimation; and finally, in Sect. 4 are drawn conclusions from the actual preliminary results and future work is planned.

2 Data Analysis Methods

The acoustic data analysed herein was collected by an acoustic array, the digitalHyd TP-1/4/VS equipment, developed by MarSensing. This system is composed of 4 hydrophones and a single Vector Sensor (VS). Each of the hydrophone elements is mounted on an arm spaced 90° with a length of 75 cm from the center, all on the same plane, shaping a horizontal planar array. The VS is mounted on a central pilar 75 cm above the plane.

The digitalHyd TP-1 is a multichannel acoustic acquisition system with the capability to stream acquired data in real-time over Ethernet to a remote computer. In this setup both power and communications were made available through the MARREAL infrastructure [13], an underwater observatory which is shore connected through an electro-optical cable. The MARREAL infrastructure was developed in the University of the Algarve, and was deployed at a depth of 20 m, near the small islands of Martinhal outside the bay of the port of Baleeira, Sagres, Portugal. This infrastructure was deployed with a video camera, a CTD, an acoustic MODEM, and an Acoustic Back Scatter, and a junction box, which was designed to allow the connection of other user defined sensors for real-time ocean observations, equipment testing and development. Other objectives may be the long-term observation underwater environmental physical parameters or phenomena. In the future, the shore connection shall grant remote access to the observatory.

The complete observatory was deployed at the end of September 2022. The left panel of Fig. 1 shows a map of the deployment area, outside the port of Baleeira. The acoustic system was deployed away from the main platform, approximately in the middle of the channel half way from the seawall extremity and the Martinhal Islets, where the water column is approximately 20 m deep. This area is mainly cruised by fisher and leisure boats that use the Baleeira Port, or boats sailing along the coast line. The right panel depicts the acoustic array just after deployment, where one can see the planar section made of four hydrophones, and on top the VS. The structure was such that the planar array was approximately 1.3 m above the seafloor, and the VS another 75 cm above the planar array. For beamforming the working frequency of the planar array is 500 Hz. The user can include the hydrophone contained in the vector sensor, although it is raised above the plane, for azimuth estimation. In that case the working frequency would be 1000 Hz. The hydrophones' calibration curves of the actual system were obtained by means of laboratory measurements of the amplification chain and the transducers sensitivity provided by the manufacturer.

Figure 2 shows drawings of the acoustic system as it was deployed. The planar array is in parallel with the XY horizontal plane. Hydrophone H11 has the same orientation as the z-component accelerometer and hydrophone H14 is aligned with the y-component of the accelerometer. The x-component of the accelerometer points upwards.The vector sensor is located in a horizontal plane 75 cm above the planar array. The left panel of Fig. 2 shows a top view of the array.

Fig. 1. The MARREAL Experimental setup in Sagres, off the Baleeira port: observatory deployment area; the position is marked by the * (left); acoustic system during the deployment (right).

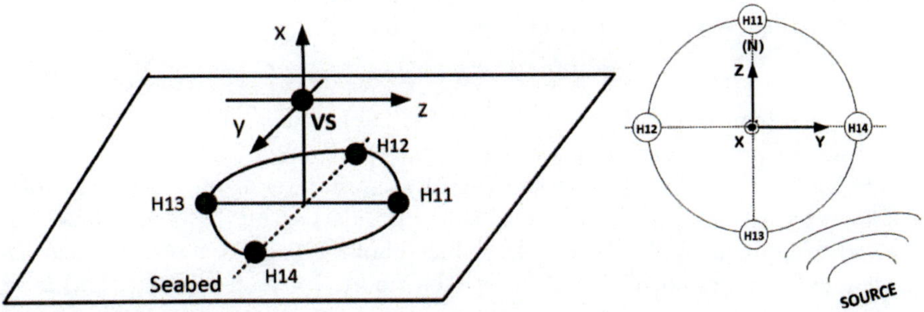

Fig. 2. Drawing of the orientation of the acoustic system both with the 4-hydrophone planar array and the vector sensorThe MARREA: front view (left); top view (right).

2.1 Data Description

This data set has a special feature due to the employment of a planar array of 4 hydrophones in tandem with a VS, also consisting of 4 channels—one omnidirection pressure channel and three orthogonally oriented acceleration channels. The 8 channels are sampled at a rate of 10547 samples per second. The acoustic data is stored in audio WAV files of 180 s, including the eight sensor channels in each file. Figure 3 shows a brief analysis of the acoustic array channel 1 of the planar section of the array for the sake of showing the dynamics of the soundscape in the observation area, for an interval of a complete week, beginning at the 24th of October 2022, which was Monday. The top panel shows a spectrogram in the 5 kHz frequency band, and the bottom panel shows an estimation of the sound pressure level over that band as a function of time. Each time bin in the plot represents a period of 180 s, which is a complete audio file. To estimate the power spectral density respective to each audio file, periodogram averaging using the Welch spectral analysis method was carried out. For periodogram averaging, the Fourier Transform was computed with Hanning observation windows of 1024

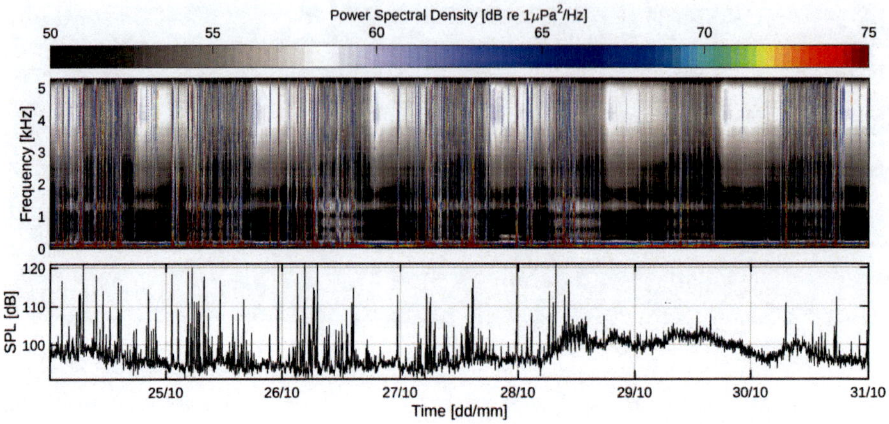

Fig. 3. Soundscape in the MARREAL deployment area: spectrogram of the data collected during the week starting at 24th of October 2022.

samples each (equivalent to a time window of 0.097 s). The SPL is an integration of the instant power spectrum over the entire band.

The spectrogram shows periods of natural noise interchanged with episodes of ship noise contaminating the whole band. The passing of boats cause the received broadband SPL to peak at values above 110 dB when boats pass in the channel to access or to leave the Baleeira port. In the frequency band from about 2.5 to 5 kHz a level variability can be consistently observed, which rises immediately after sunset and falls off with sunrise. This dynamics is typical of coastal areas and is usually of biotic origin. From these results, it is apparent that the local boat traffic occurs mainly during day light, while at night the ship traffic is comparably reduced. The 29th and the 30th of October were weekend days, which may be the reason for almost no traffic passing near the acoustic observatory during the nights from Friday to Sunday, and also during the day, where the sound pressure level peaking is significantly reduced.

2.2 Azimuth Estimation with Conventional Beamforming

The MARREAL acoustic observatory has a section of four hydrophones held by horizontal arms holding these elements on a horizontal plane. These receivers can be used for azimuth estimation by means of a beamforming technique. Beamforming is a space-time array processing technique used to estimate the direction-of-arrival (DOA) of a signal impinging onto an array of sensors [14]. By steering the array in one direction at a time, one can combine the received signals and measure the output power, whereas the steering angle with the maximum output power yields the DOA estimate. For the uniforme planar array of the actual system, the steering vector used to combine the signals for different hypothetical directions-of-arrival represented by θ. Considering the geometry of the uniform planar array, the steering vector is given as

$$\underline{a}_{\text{up}}(\theta) = [e^{-jkd\sin\theta}\ e^{-jkd\cos\theta}\ e^{jkd\sin\theta}\ e^{jkd\cos\theta}]^{\text{T}}, \tag{1}$$

where d is half of the planar array aperture, which is the distance from the origin to each hydrophone on the horizontal plane; $k = \omega/c$ is the wavenumber, ω is the angular frequency, and c is the sound speed at the array position. The steering vector is directly related to the array geometry considering the XY horizontal plane.

To measure the degree of adjustment for a hypothetical angle, herein, the conventional (or Bartlett) beamformer is used. The conventional beamformer was implemented in the frequency domain as

$$P_{\text{c}}(\theta) = \sum_{k=1}^{K} \frac{\underline{a}^H(\theta, \omega_k)\mathbf{C}_{YY}(\omega_k)\underline{a}(\theta, \omega_k)}{\underline{a}^H(\theta, \omega_k)\underline{a}(\theta, \omega_k)}, \tag{2}$$

where K is the number of discrete frequencies, ω_k is the kth frequency, and $\hat{\mathbf{C}}_{YY}(\omega_k)$ is the spectral density matrix (SDM) for frequency ω_k. In practice, a sample matrix will be obtained based on array data collected over a given observation time. Vector $\underline{a}(\theta)$ is the steering vector $\underline{a}_{\text{up}}(\theta)$ of Eq. (1) to combine the signals for different hypothetical directions-of-arrival represented by θ. The value of θ that maximizes $P_{\text{c}}(\theta)$ is the DOA of the impinging signal,

$$\hat{\Theta}_S = \arg\max_{\theta} P_{\text{c}}(\theta), \tag{3}$$

where the azimuth angle θ is in the interval from $-\pi$ to π.

2.3 Azimuth Estimation with a Vector Sensor

The azimuth angle estimation can be also obtained with the single VS of the MARREAL observatory and was evaluated using the Intensity Based Estimator described in [15]. The four channels of the vector sensor are constituted of three acceleration components and one pressure component. Each acceleration component was converted to their respective pressure equivalent particle velocity component, in a first step by using the Fourier Transform to frequency domain (ω), and then through:

$$\hat{V}_i(\omega) = \frac{\rho}{jk} A_i(\omega) = \frac{\rho c}{j\omega} A_i(\omega), \tag{4}$$

where $\hat{V}_i(\omega)$ is the pressure equivalent particle velocity component, $A_i(\omega)$ is the acceleration component being $i = x, y$ or z-axis, $k = \frac{\omega}{c}$ is the wavenumber, c is the water sound speed, ρ is the water density and ρc is the scaling factor according to the definition of acoustic impedance. After this, the outputs are converted back to the time domain and the Intensity estimator is applied [15], where the pressure $p(t)$ is cross-correlated at lag 0 with $v_y(t)$ and with the $v_z(t)$

particle velocity components. In this case, the VS was deployed with the x-axis oriented to the surface and the y and z-axis in the horizontal plane.

Then, an estimation of the azimuthal direction of the source signal, Θ_S at large signal to noise ratio (SNR) is given by:

$$\hat{\Theta}_S = \arctan \frac{\langle v_z(t)p(t)\rangle}{\langle v_y(t)p(t)\rangle}, \tag{5}$$

where $\langle\rangle$ stands for time averaging and the azimuth angle Θ is in the interval from $-\pi$ to π.

3 Experimental Results

This section shows preliminary experimental results obtain for boats passing in the deployment area in two occasions. The objective is to perform an assessment of the vector sensor directional processing in comparison with the 4-hydrophone planar array conventional beamforming. The data processed was collected during passage and was selected arbitrarly. There was no control on the trajectory of the vessels passing, and no visual observation.

The planar array beamforming was performed with the conventional beamformer according to Eq. (2). For each azimuth estimate, to compute the sample SDM, 3 s of data were considered, and divided in segments of 4096 samples (or 388.4 ms) for Fourier Transform, which gives 7 snapshots. Discrete frequencies with a 5 Hz resolution in the band 250 to 450 Hz were taken for beamforming. This was repeated every 2 s.

The directional processing with the VS was carried out according to Eq. (5), using 5 s data segments. Before converting the acoustic data from the time domain to the frequency domain, the two acceleration channels and the acoustic pressure channel were filtered using a 4th order Chebyshev Type I filter with the central frequency at 375 Hz and a bandwidth of 300 Hz, which gives a pass band from 225 Hz to 525 Hz. This was repeated for every 1 s. Figure 4 shows the data and the azimuth estimation results. The panels on the top show spectrograms of the data processed for azimuth estimation, in two occasions, in order to provide an idea on how the signal power spectral density evolves, as the boat passes in the vicinity of the acoustic system. The data on the left side is from the 28th October, comprising an interval of 6 min. It shows the broadband acoustic interference pattern due to the close passage of a boat radiating underwater noise, called Lloyd's mirror. When a ship moves relative to the receiver, the time difference between the acoustic multipaths changes, and the broadband interference pattern changes symmetrically around the point of closest approach. The lower left panel shows the results for azimuth estimation. The red curve shows the estimation with planar array conventional beamforming, and the blue curve is the estimation for VS with based on the intensities, as a function of time. The beamformer provides a very stable tracking of the azimuth over the whole interval, and is taken as the reference, herein. The VS estimation performs fairly well, showing some slight instability at the beginning of the track. Around

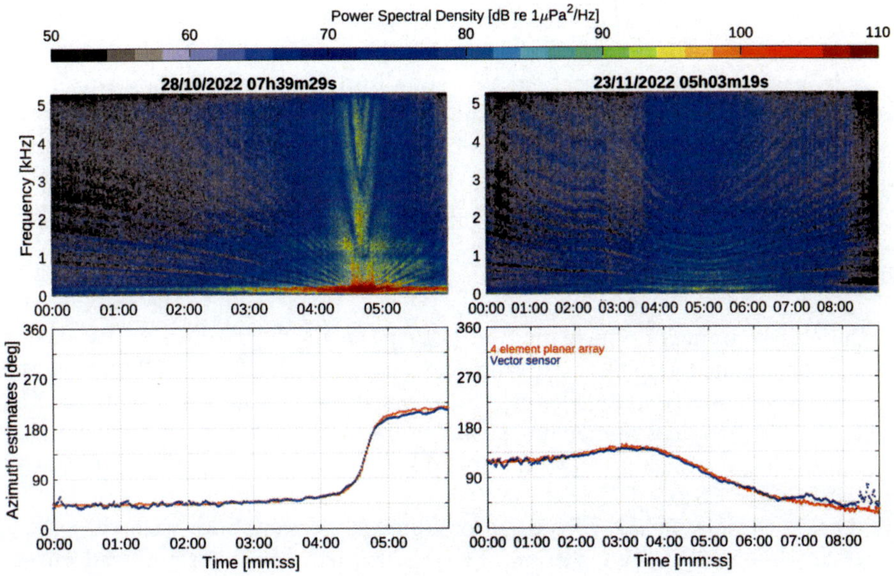

Fig. 4. Experimental results on vessel radiated noise pressure received in two occasions: spectrograms of the received acoustic pressure field (upper line); azimuth estimation with conventional beamforming and vector sensor based on received intensities (lower line).

the closest point of approximation, when the acoustic power is maximum, both processors yield very similar results.

In the second case, shown on the right side of Fig. 4, a 9-min interval of data collected on 23rd November is considered. Again a Lloyd's mirror pattern is observed. The acoustic power radiated by the boat can be distinguished from the background noise during the whole time. The conventional beamformer yields a tracking estimate of the azimuth that appears to be smooth for most of the time except during the last minute where some variability is seen due to increasing source-receiver range. The processing of the VS performs well most of the time except at the extremities, during the first minute, and during the last 2 min when it yields azimuth estimates with some offset from the beamformer estimates.

The Port of Baleeira harbours fisher ships equipped with Automatic Identification System (AIS). The analysis of AIS data reveals that there is a set of boats that leave the port heading south and some head towards southwest, and west. Figure 5 shows the results on azimuth tracking of an identified fisher boat leaving the port at UTC time 04:51 [HH:MM] of 21 of October with an acoustic time series of approximately 21 min. The boat track was available on AIS data for an interval of about 14 min. The green circles in the lower panel of Fig. 5 represent the azimuth estimated from these AIS data relative to the position of the acoustic array. It is unknown why the AIS signal is irregular and limited in time, since the boat was out at sea after 15 min, when the AIS signal vanished.

One possible reason is the poor AIS coverage in the area due to the reduced number of AIS receivers on land. At the beginning of that AIS sequence the boat is inside the port and behind the seawall protecting the port. Hence there is no direct acoustic path between the radiating boat and the receiver array. The upper panel shows the power spectral density of the acoustic series. At the beginning only background noise is detected. At data time 03:29 [mm:ss] the received power suddenly increases by more than 10 dB across the whole spectrum, which is when the boat becomes in line of sight from the array point of view. At that instant the azimuth of the boat is about 254° and the distance to the array is 530 m. Then the boat moves towards the array into the channel and at the closest point of approximation (CPA) the the boat to array distance is 85 m at data time 06:30 [mm:ss]. From the data times 03:29 to 11:52 [mm:ss] the boat keeps moving steadily whereas the azimuth varies from values approaching azimuth 270° to values aproaching 90°. At data time 11:52 [mm:ss] the boat is at 1.5 km from the acoustic array and at the final AIS position at data time 14:43 the distance is 1.8 km.

This data portion was processed for azimuth tracking both with the planar array section and the VS section. The planar array beamformer (red curve) consistently yields azimuths between 25 and 45° until data time 02:00. It is likely that ambient noise is dominated by the acoustic and bioacoustic activity occuring at the islets. Then it picks at azimuth 264° even before data time 03:29. Detailed observation of the spectrum shows a slight increase in acoustic power at data time 02:12 for some frequencies around 500 Hz by about 10 dB, which is possibly related to the boat radiated noise being propagated indirectly. After that, the beamforming tracks the azimuth of the boat in a very consistent way most of the time, until the AIS information vanishes. Around the CPA portion, there is some deviation, which is not fully explained yet. One possible cause is some error in the array position or boat position, which can cause an error on the azimuth estimate when the distance is reduced. Also in the interval from data times 11:38 to 13:12 when the boat stalled, and hence the received acoustic power reduced, the beamformer was able to keep the azimuth tracking altough with some uncertainty. After data time 16:00 the acoustic power progressively reduces. It is unknowns which path the boat took and whether it was moving or not. We assume that the boat continued moving to west with heading to 90° The beamformer kept tracking the boat's azimuth consistently.

Concerning the VS directional processing, in general, the processor is able to track the boats azimuth, provided that the acoustic signal is received with power above a given threshold. For the data interval from 03:20 to 11:38 the VS processor yields results matching the AIS estimated azimuth and the planar array beamformer. Out of that data interval the acoustic power is less, causing the VS to yield results with increase variability. When the boat stalled, in the interval from data times 11:38 to 13:12 VS processor was unable to keep tracking. After data time 16:00 the uncertainty of the VS estimated progressively increased.

In comparative terms the planar array was able pick up the tracking before the VS processor, and maintain the tracking during the whole time. These results

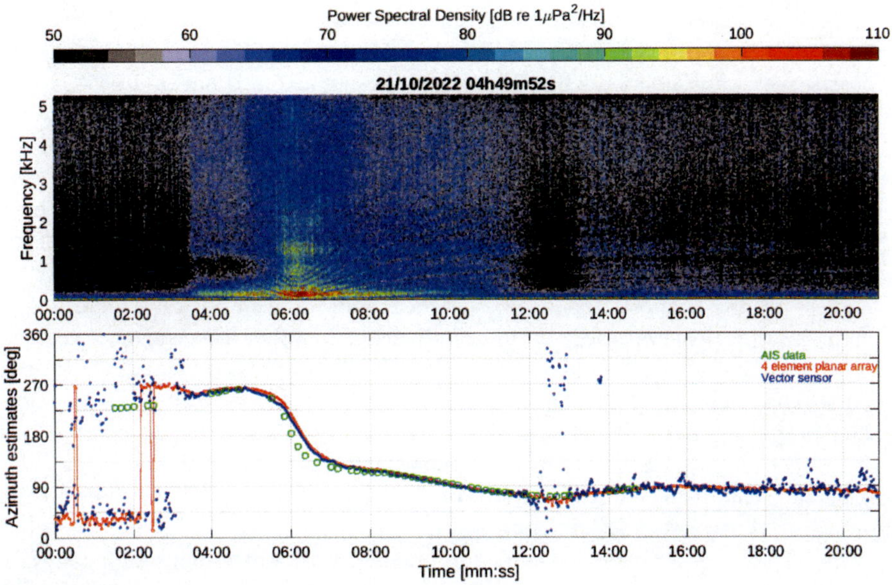

Fig. 5. Experimental results on the azimuth tracking of a fisher boat equipped with an AIS system: spectrogram of the received acoustic pressure field (upper panel); AIS estimated azimuth superposed with the azimuth estimation with the conventional beamforming and vector sensor based on received intensities (lower panel).

indicate that the actual VS intensity based directional processing can perform quite well when the SNR is above a certain threshold. In comparison to the planar array, considering that the VS is a single device, although three channels must be considered, the performance is remarkable. This result shows that a single VS can be an interesting choice for the integration in marine observatories, when acoustic arrays can not be conveniently integrated. Beamforming has always a trade-off between the maximum operating frequency and the beamformer sensitivity. To increase the working frequency, the array aperture must be reduced, which has reduces the beamformer sensitivity due to the decrease in the differences of the arrival times at the different receivers composing the array.

4 Conclusions

This paper provides preliminary experimental results on the azimuth estimation from data acquired by the MARREAL marine observatory deployed in Sagres, Portugal, in September 2022, which was operational for about 2 month. The acoustic data was collect with an acoustic receiver system made of a total of 8 channels, including a horizontal planar array of four hydrophones and a vector sensor with an acoustic pressure sensor and three orthogonal accelerometers.

Three data segments comprising the passage of small boats were selected for attempting azimuth estimation. The planar section of the array and the vector sensor were independently processed for azimuth estimation in the global frequency range of 250 to 650 Hz, within time scales of up to 9 min.

The planar array conventional beamforming yielded very stable azimuth tracking results over the entire data processed. The VS intensity based processing performed well nearly over the entire data, except for data portions where the signal-to-noise ratio was relatively reduced, where some degree of instability was observed.

These results suggest that acoustic vector sensors are appealing for integration in marine observatories, or for application in other moorings, in marine protected areas, and other areas where restrictions apply, or where there is a need to know what is going on in terms of traffic or maneuvering. These compact devices are possibly more resilient to increased sea level, in comparison to arrays with a more complex mechanical structure, and capable of causing less perturbation to the marine environment. Additionally, a VS may permit for monitoring bioacoustic activity and noise.

The analysis described here is part of ongoing work of thorough processing of the complete data set, which shall serve the purpose of a full assessment of the VS capability in tracking all boat passages in the area. The full spectrum of vessel radiated noise shall be exploited for improved detection and tracking accuracy. For the actual data set, another topic to be approached is the processing of all array elements together, including the planar array section and the vector sensor for the purpose of evaluating the gain brought by the VS to a planar array in azimuthal tracking of opportunistic acoustic sources.

References

1. Nehorai, A., Paldi, E.: Vector-sensor array processing for electromagnetic source localization. IEEE Trans. Signal Process. **42**(2), 376–398 (1994)
2. Nehorai, A., Paldi, E.: Acoustic vector-sensor array processing. IEEE Trans. Signal Process. **42**(9), 2481–2491 (1994)
3. Hawkes, M., Nehorai, A.: Acoustic vector-sensor beamforming and capon direction estimation. IEEE Trans. Signal Processing **46**(9), 2291–2304 (1998)
4. Wan, C., Kong, A., Liu, C.: A comparative study of DOA estimation using vector/gradient sensors. In: Proceedings of Oceans06, pp. 1–4, Asia, Pacific (2007)
5. Shipps, J.C., Abraham, B.M.: The use of vector sensors for underwater port and waterway security. In: Proceedings of Sensors for Industry Conference, pp. 41–44, New Orleans, Louisiana, USA (2004)
6. Krishna, K.M., Anand, G.V.: Narrowband detection of acoustic source in shallow ocean using vector sensor array. In: Proceedings of Oceans 2009 MTS/IEEE, pp. 1–8, Biloxi, USA (2009)
7. Santos, P., Felisberto, P., Jesus, S.M.: Vector sensor arrays in underwater acoustic applications. In: Camarinha-Matos, L.M., Pereira, P., Ribeiro, L. (eds.) DoCEIS 2010. IAICT, vol. 314, pp. 316–323. Springer, Heidelberg (2010). https://doi.org/10.1007/978-3-642-11628-5_34

8. Felisberto, P., Santos, P., Jesus, S.M.: Traking source azimuth using a single vector sensor. In: Proceedings of 4th International Conference on Sensor Technologies and Applications, pp. 416–421. Venice, Italy (2010)
9. Smith, K.B., et al.: Acoustic vector sensor analysis of the monterey bay region soundscape and the impact of covid-19. J. Acoust. Soc. Am. **151**(4), 2507–2520 (2022)
10. Abdi, A., Guo, H.: Signal correlation modeling in acoustic vector sensor arrays. IEEE Trans. Signal Process. **57**(3), 892–903 (2009)
11. Peng, H., Li, F.: Geoacoustic inversion based on a vector hydrophone array. Chin. Phys. Lett. **24**(7), 1997–1980 (2007)
12. Santos, P., Rodríguez, O.C., Felisberto, P., Jesus, S.M.: Seabed geoacoustic characterization with a vector sensor array. J. Acoust. Soc. Am. **128**(5), 2652–2663 (2010)
13. Marreal website (2022). https://www.marreal.pt/
14. Krim, H., Viberg, M.: Two decades of array signal processing research: the parametric approach. IEEE Signal Process. Mag. **13**(4), 67–94 (1996)
15. Felisberto, P., Rodríguez, O.C., Santos, P., Ey, E., Jesus, S.M.: Experimental results of underwater cooperative source localization using a single acoustic vector sensor. Sensors **13**(7), 8856–8878 (2013)

Development of Artificial Intelligence Applied to the Production of Biopolymers with a Focus on Sustainability

Audelis de Oliveira Marcelo Júnior[1], Arthur Brito Gomes[1], Eder Paulus Moraes Guerra[1](✉), and Enio pontes de Deus[2]

[1] State University of Vale Do Acaraú, Av. Padre Francisco Sadoc de Araújo, 850 - Alto da Brasília, Sobral, CE 62010-295, Brazil
{oliveira_marcelo,eder_paulus}@uvanet.br

[2] Federal University of Ceará, Av. Humberto Monte, s/n - Pici,, Fortaleza, CE 60440-593, Brazil

Abstract. The use of chitosan as a sustainable alternative in food packaging has been studied due to its biodegradable and antibacterial properties. To increase the resistance of chitosan, an alternative is to incorporate natural fibers such as sisal into the polymeric matrix, resulting in a more resistant composite material. However, the precise evaluation of the mechanical properties of these biomaterials is a challenge, considering their non-homogeneity and the number of variables that interfere with the model. With the evolution of microprocessors, performing complex calculations has become a less exhausting task and allows for better analysis. Nevertheless, creating an algorithm capable of verifying the veracity of a hypothesis is a complex task that requires the elaboration of new propositions by an expert. It is in this context that the work in development arises. Using techniques of artificial intelligence and computer vision, it aims to extract information from sisal fibers with a chitosan matrix to efficiently evaluate not only the macrostructure but also the microstructure of these materials, providing more concise information. It is possible to estimate the elastic modulus with the volumetric fraction of the phases, verify the critical length of the fiber and the load supported per phase. The algorithms used for this solution were k-Means, Mask RCNN and U-Net. It is important to emphasize that, being generic models, it is expected that in future stages of the project it will be possible to develop more comprehensive solutions with applications in the construction and industry sectors.

Keywords: Sustainability · Artificial intelligence · Mechanical properties · Computer vision

1 Introduction

The field of study that describes the ability of machines to learn like humans can be defined as Artificial Intelligence (AI). Since it was driven by the market, the technology and business fields have changed rapidly [1]. Therefore, this work sought to apply the techniques of this area in polymers, to contribute to the evaluation of these materials by

J. F. L. C. Semião et al. (Eds.): INCREaSE 2023, ASST, pp. 204–218, 2023.
https://doi.org/10.1007/978-3-031-44006-9_15

providing more accurate information in a short time interval. Among the consolidated techniques, K-Means algorithms and Neural Networks stand out as alternatives for the study.

K-Means is an iterative procedure that partitions N objects into K distinct clusters. It is a relevant clustering method in the literature that uses centroids to present clusters [2]. It is common to use Euclidean distance as a parameter to define the proximity between each data entry.

The artificial neural network is a computing technique designed to simulate the human brain in problem solving [3]. Deep Learning (DL) is a division of Machine Learning (ML) that is based on a set of algorithms that try to model high-level abstractions in data. These algorithms develop a hierarchical architecture in layers of learning and data representation. This hierarchical learning architecture is inspired by AI emulating the learning process of the human brain [4]. It is common to use neural networks in the area of computer vision for image segmentation. Among the most used algorithms, we can mention Mask-RCNN and U-Net.

Mask R-CNN is a convolutional neural network architecture that combines object segmentation techniques with object detection, as evidenced by [5]. Segmentation involves identifying each object in an image individually, while detection focuses on recognizing instances of these objects by drawing a mask around them.

On the other hand, U-Net is a convolutional neural network architecture that is known for its success in segmenting biomedical images, as evidenced by [6]. [7] worked with its application in body bones from computed tomography images. However, other studies have shown that the network can be used in different situations. [8] and [9] used U-Net in the area of materials, such as atom segmentation and texture segmentation. In addition, [10] discusses works that applied ML techniques in composites in which not only limits can be predicted in fatigue, buckling, compression and drop impact tests, but also contributed to the design of materials in addition to reducing computational effort during calculations.

Industry in general is increasingly looking for new materials that are less harmful to the environment, widely available in nature and with properties equivalent to or superior to existing ones. Therefore, these reasons encouraged the use of chitosan film composite polymer reinforced with sisal fiber for the characterization of materials using the mentioned techniques. It is important to emphasize that the work is in development, focusing on the generation of trained models to segment the fibers in order not only to obtain the average length and diameter, but also the volumetric fractions of each phase, since they are important factors to obtain the modulus of elasticity in polymers.

The work discusses the design of the films, the dataset used, the networks considered and the metrics for validating the models. Finally, it will be discussed briefly about the prediction of diameter and length by the algorithm.

2 Materials and Methods

2.1 Data

The dataset used consists of 15 images containing samples of a composite with a chitosan matrix reinforced with sisal fibers. Mercerization of sisal fiber was performed according to the methodology of [11]. After mercerization and drying, the fibers were crushed in a knife mill (Pulverisette 16, Fritsch, Idar-Oberstein, German) with 2, 1 and 0.5 mm meshes.

The chitosan solution was performed using chitosan powder with 75% degree of deacetylation supplied by Polymar Ciência e Nutrição S/A (Fortaleza, Brazil). The solutions and films were produced with 1.5% w/v of chitosan following the methodology of Vale et al. (2021) [12], which consists of producing a 1.5% w/v solution of chitosan dissolved in a 1% v/v acetic acid solution with subsequent addition of 0.6% v/v glycerol as a plasticizer. The solutions are poured into petri dishes and taken to a laboratory oven at 35 °C for 48 h for drying and production of the films. To produce the fiber-reinforced films, the mercerized sisal fibers were mixed with the solution before drying the molds, using a proportion of 10% of the mass of chitosan.

Some of the most important advantages of using biopolymers as a matrix in composites is that they are renewable and carbon neutral. However, the challenge in increasing the applicability of such composites is conditioned by the enormous variation in their properties as they are affected by a number of variables, such as: environmental condition of their source, type of fiber modification and processing techniques. From the computational point of view, the analysis and design process becomes complex due to the dimensions of the input data.

In a larger perspective, polymeric composites are classified into different categories based on the type of reinforcement material used, associated with shape, size, chemical composition, amount of reinforcement and the way in which the reinforcement agents are added along with the composite manufacturing process can greatly affect the general properties resulting from the composite materials [10].

To teach a network how to segment fibers, it is necessary to use supervised training. This technique consists of a supervisor defining the correct output for a given data input. In view of this, the delimited patterns are obtained automatically after training. Consequently, in the area of computer vision, it is necessary to previously define in a set of images the limits of the regions for learning the neural network.

Thus, the regions of interest for training the models were defined. Figure 1 shows the result of manual fiber segmentation using the Visual Geometry Group Image Annotator (VIA) program. This tool generates a json file with the regions delimited by the user to create the masks used in learning the algorithms.

In view of the above, the identification of patterns will be done automatically in the network through the adjustment of weights through the processes of feed forward and backpropagation. Feed forward makes the network calculate the probability that each pixel contains an object. The starting weights for this process are randomly entered. On the other hand, the backpropagation checks whether the first definition of weights

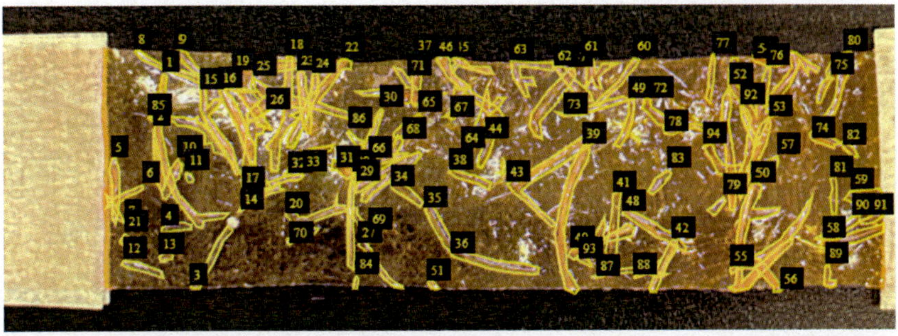

Fig. 1. Fibers segmented using VIA.

was consistent by comparing the result obtained with the one delimited by the supervisor based on the error calculation. After that, the weights are adjusted again using optimization algorithms to find the smallest error as a function of the weights.

Overfeeting and underfeeting are problems that can occur during training depending on the dataset used. Overfitting occurs when the neural network sharply adjusts the weights to the input data, which can happen during training with a very high number of epochs. On the other hand, underfitting occurs when the network has not been trained for a sufficient number of epochs to adjust the weights or when there is a deficiency of training images.

Apply a technique known as data augmentation is one of the alternatives to fill the lack of diversified images. This technique involves applying a filter or rule to each pixel in the image. For that, the Albumentations library [13] was used, with the classes HorizontalFlip, VerticalFlip, ElasticTransform, GridDistortion and OpticalDistortion. Thus, a total of 75 images were obtained, which were used not only for training but also for validation. The filters were applied to the original image and the respective mask, as can be seen in Fig. 2 .

Fig. 2. Applying the ElasticTransform class to the image and ground truth.

2.2 Architectures

K-Means

K-Means is not a deep learning algorithm like the other networks mentioned. Therefore, there is no neural network that will have weights adjusted. A calculation, specifically

Euclidean distance, will always be performed for segmenting similar information. As sisal fibers tend to have numerically similar pixel intensities, the algorithm was designed to classify fibers as similar parts of an image. Figure 3 provides an example of K-Means clustering.

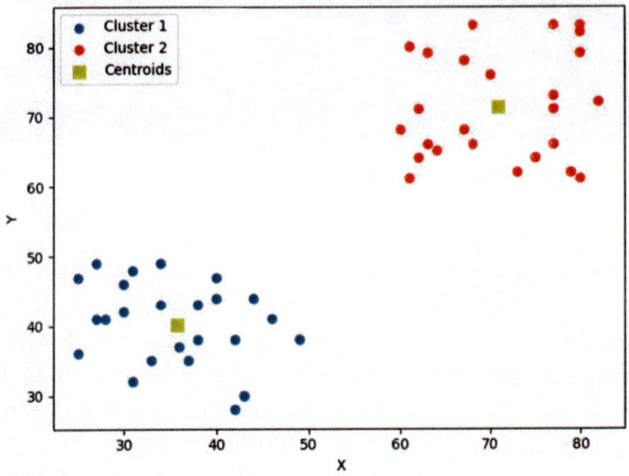

Fig. 3. K-Means example clustering.

U-Net

The U-Net architecture consists of three key components: the convolutional block, the encoder block, and the decoder block. The convolutional block performs a multiplication with a matrix (kernel) with random values that generates 64 different images. The encoder applies max pooling for a given kernel. In each iteration, only the maximum value within the matrix is stored in a new matrix. The decoder block reverses the max pooling process that reduced the image proportionally to the kernel size. The architecture can be seen in Fig. 4.

Mask-RCNN

According to [5] Mask R-CNN, extends Faster R-CNN by adding a branch for predicting segmentation masks on each Region of Interest (RoI), in parallel with the existing branch for classification and bounding box regression. The mask branch is a small FCN applied to each RoI, predicting a segmentation mask in a pixel-to pixel manner. Figure 5 provide more information about head architectures of this neural network.

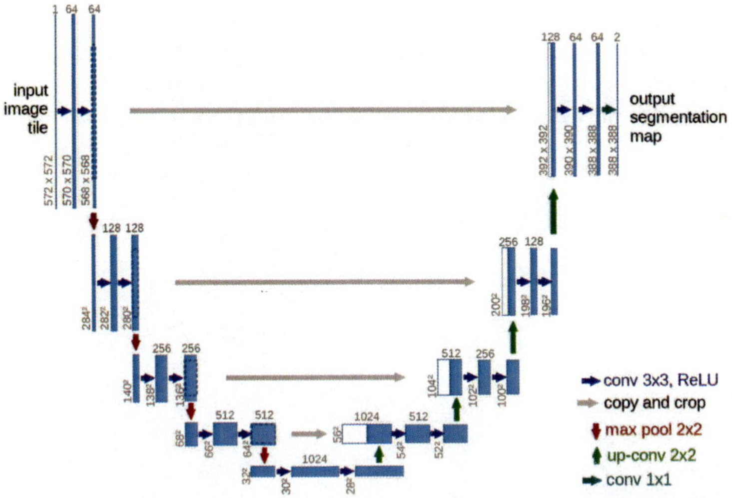

Fig. 4. Representation of the U-Net convolutional neural network architecture [6].

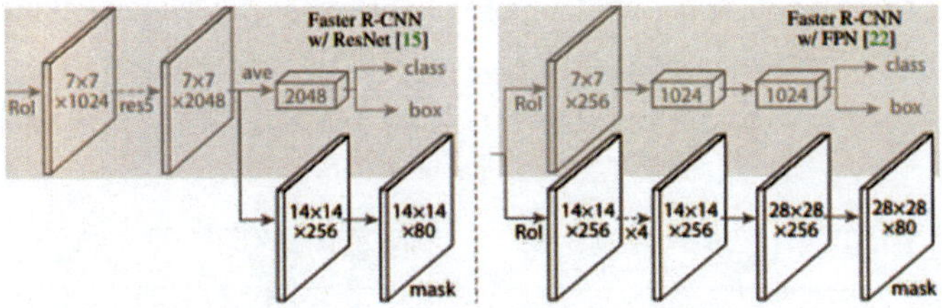

Fig. 5. Head architecture Faster RCNN and Mask-RCNN [5].

2.3 Training

U-Net and Mask-RCNN training was performed on an Nvidia Tesla T4 GPU, running Chromium OS 13.0 with kernel 5.10.147 +. The K-Means algorithm does not require training, it just performs the Euclidean distance calculation to verify the clusters. The dataset was divided into 90% images for training and 10% for validation. The image resolution was 512 × 512 and 1024 × 1024 respectively for U-Net and Mask-RCNN. Furthermore, U-Net was developed using tensorflow [14] over 150 epochs, while for Mask-RCNN detectron2 [15] was used with 270,000 iterations.

2.4 Metrics

The metrics used in the work were: Pixel Accuracy (PA) and Intersection Over Union (IOU) according to Jiang et al. (2020) [16] and Dice Coefficient (DC) Trutti et al. (2021) [17]. These metrics are widely used in the area of computer vision because they are interesting parameters to verify the capacity of predictions obtained by the model. PA measures the percentage of pixels that were correctly classified by checking each matrix position, where k is the total number of categories labelled in advance, p_{ij} is the total number of pixels in each category i, and the predicted category is j. The IOU, on the other hand, counts the area of intersection between the prediction of the network and the real object in relation to the union of these areas. The DC also quantifies the intersection between 2 areas (A and B) and the obtained value is doubled to be related to the sum of the areas.

$$PA = \frac{\sum_{i=0}^{k} p_{ii}}{\sum_{i=0}^{k}\sum_{j=0}^{k} p_{ij}} \tag{1}$$

$$IOU = \sum_{i=0}^{k} \frac{p_{ii}}{\sum_{j=0}^{k} p_{ij} + \sum_{j=0}^{k} (p_{ji} - p_{ij})} \tag{2}$$

$$DC = \frac{2|A \cap B|}{|A| + |B|} \tag{3}$$

3 Results

Each image name was defined by an image code and a filter code number. The code refers to the image in question, while the filter specifies which augmentation was used on the coded image. Table 1 specifies this.

Table 1. Image name specification.

Class	Filter Code
HorizontalFlip	1
VerticalFlip	2
OpticalDistortion	3
ElasticTransform	4
GridDistortion	5

3.1 K-Means

Starting from Fig. 6, the histogram of Fig. 7 was generated. Note that the frequency of the brightest pixels is lower than that of the darkest ones. However, their arrangement is not uniform to the point that a simple thresholding accurately segment the fibers. Thresholding is a technique that segments the image from a specific value between 0 and 255 resulting in 2 subdivisions. The K-Means exceeds the threshold using the K-factor which can define the number of subdivisions to be generated.

As K-Means will group the most similar pixels into a single cluster, the results tend to be better as the fibers are similar. To determine the number of clusters, the previous histogram was used so that the amount of noise per interval was minimized. The number of clusters was defined by the coefficient K = 10. The result obtained is shown in Fig. 8.

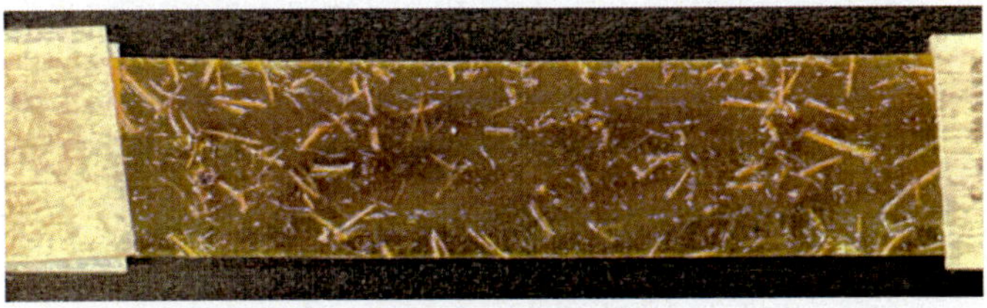

Fig. 6. Sample 7453_1.

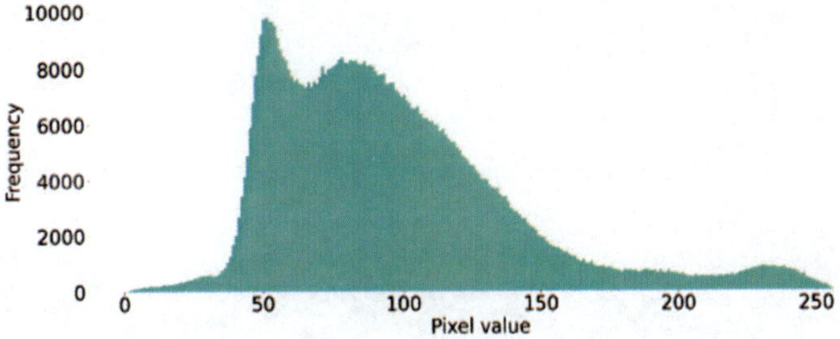

Fig. 7. Sample 7453_1 histogram.

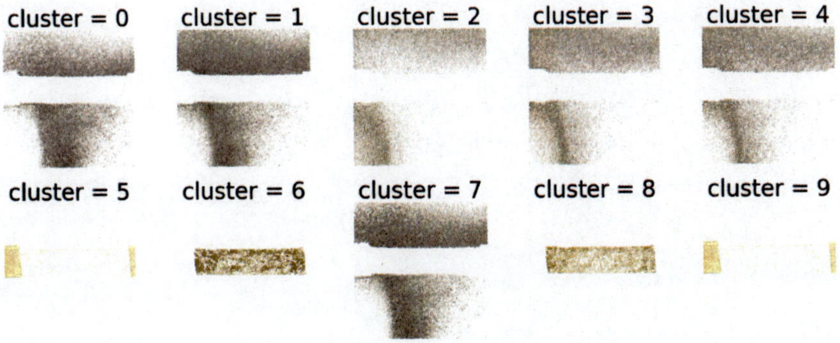

Fig. 8. Sample 7453_1 K-Means clustering.

Figures 10 and 12 provide a more detailed view of the clusters that have the highest IOU among the segmentations while Figs. 9 and 11 the correct region of interest. These clusters can be evaluated as the most accurate considering the metric used. However, it is important to note that the results were classified as unsatisfactory due to the persistently high noise level. The use of techniques such as erosion and dilation can result in loss of information. Erosion is an algorithm that reduces the predicted segmentation pixels, while dilation increases them. Both are handled by a kernel that defines a number of pixels to be considered throughout the iterations. Therefore, these techniques can worsen the noise level results when using an inadequate kernel. Another factor to consider is that the kernel and K parameters must be adjusted for each image which removes the automation from the process .

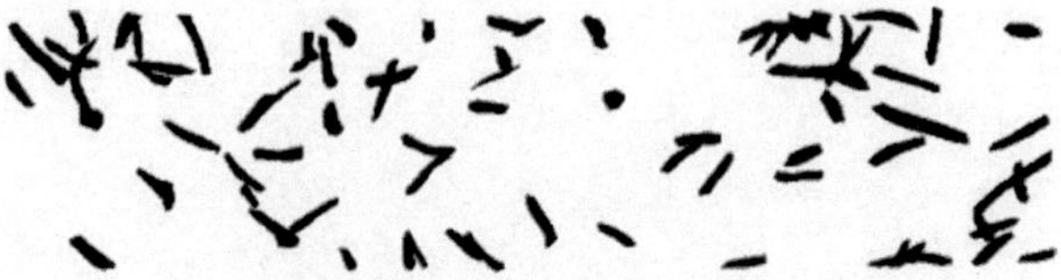

Fig. 9. Sample 7453_1 ground truth.

Fig. 10. Sample 7453_1 prediction.

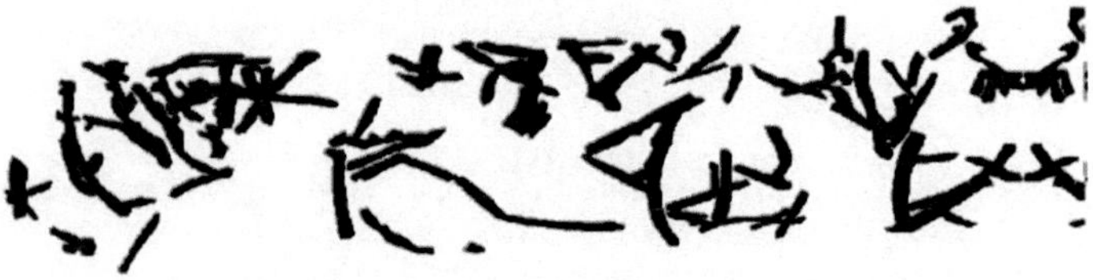

Fig. 11. Sample 7441_5 ground truth.

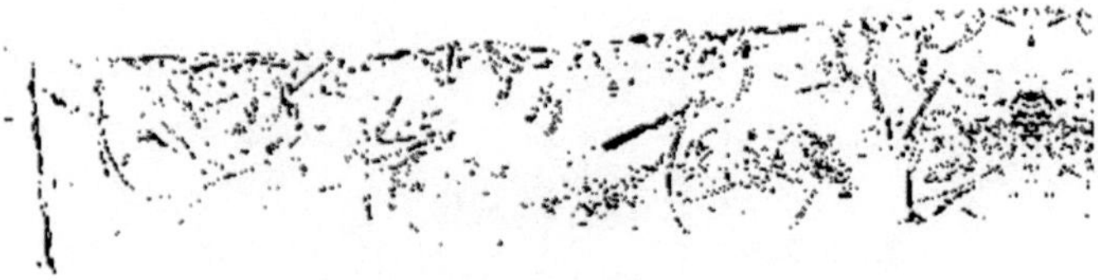

Fig. 12. Sample 7441_5 prediction.

Despite being an unsupervised algorithm, the human factor still stands out, as it is necessary to identify the best metrics to define the best cluster. Even if the cluster is defined based on the highest score per metric, it is possible to get inconsistent results.

Table 2 was calculated using the previously discussed metrics from the cluster with the best result per metric. This scenario considers the best possible result among the segmentations obtained.

As evidenced, the PA obtained high values although the model did not present satisfactory results in the segmentation. This statement can be observed through the IOU and DC metrics that reveal the fragility of the segmentation, in addition to highlighting the lack of equity in the image data.

Table 2. K-Means metrics.

Image	PA	IOU	DC
7441_5	0.94	0.56	0.31
7465_4	0.95	0.60	0.32
7443_1	0.97	0.58	0.50
7441_1	0.95	0.56	0.26
7453_1	0.98	0.55	0.33
7465_2	0.98	0.55	0.10
7451_4	0.98	0.55	0.34
7443_3	0.93	0.55	0.44
Mean	0.96	0.56	0.32

3.2 Mask-RCNN

With the use of the MASK RCNN R-101 FPN 3x network, which is a more robust network in which computational effort is high. The training carried out in this work for the network lasted about 2 days with a base learning rate equal to 0.001. It can be seen in Fig. 13 that the network weights did not adjust satisfactorily during training. Therefore, considering the results, the metrics for the model were not calculated.

Fig. 13. Mask-RCNN trained model predictions.

During the training of this network, it was observed that the learning rate did not remain constant. It is common to use coefficients that modify the learning rate over time to favor the descent of the gradient and improve the accuracy of neural networks. Gradient descent is the network's way of finding the smallest possible error. However, it was observed throughout the training that the learning rate started with values around 10–5 and ended at 10–4. Even so, it is worth mentioning that the fibers started a segmentation, even if small, and one of the factors that may have collaborated with this is that the number of pixels containing the fibers is lower than the rest of the image. As one of the metrics evaluated to train the network is the PA, the network may have masked the accuracy from this metric. However, better results are expected for a larger number of epochs. It is important to note that although the neural network did not adjust correctly for training, it is expected to adjust the model by performing another training.

3.3 U-Net

Among the analyzed algorithms, U-Net obtained the best results. Figures 15 and 17 show the prediction result of processing by the neural network. The obtained prediction can be compared with the previously defined mask through Figs. 14 and 16. Based on this, it becomes necessary to use metrics to assess the model's accuracy.

One factor that can lead to discrepancies in the results is the presence of unbalanced data in the network input. This statement can be observed by visualizing the number of pixels containing fibers in relation to other image elements. If the model fails to detect any fibers, the PA value may still be high. This characteristic can cause the PA to obfuscate the lack of precision during segmentation. Therefore, the prioritized metrics to evaluate the performance of the models were IOU and DC, as they are more rigorous in terms of image segmentation.

Fig. 14. Image 7453_1 ground truth.

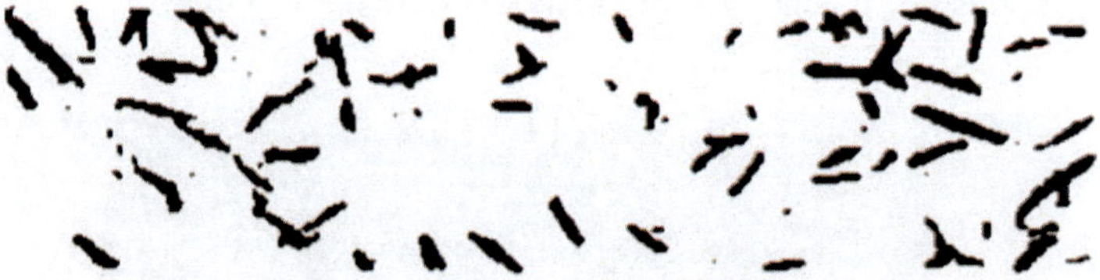

Fig. 15. Sample 7453_1 prediction.

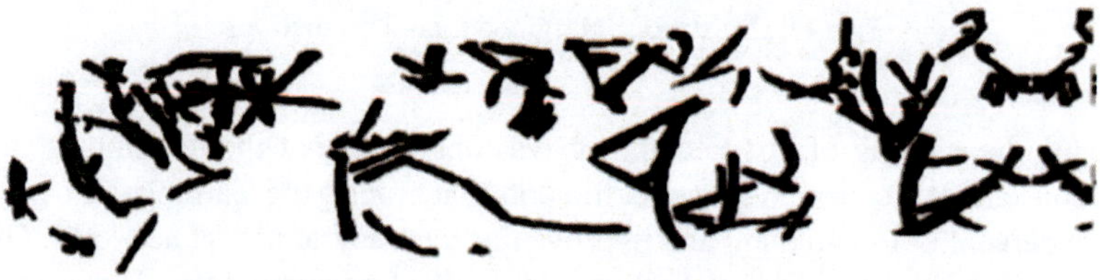

Fig. 16. Sample 7441_5 ground truth.

As the results obtained indicate in Table 3, which reveals the metrics calculated for each sample, the PA metric is insufficient to validate the model during training, as it can indicate a high level of precision even without a careful segmentation. From the IOU metric, it is possible to observe that the neural network weights better adjusted to the required segmentation. This fact was confirmed by DC, which also remained at an acceptable average.

Fig. 17. Sample 7441_5 prediction.

Table 3. U-Net metrics.

Image	PA	IOU	DC
7441_5	0.98	0.74	0.68
7465_4	0.98	0.77	0.71
7443_1	0.98	0.76	0.71
7441_1	0.96	0.71	0.63
7453_1	0.99	0.78	0.72
7465_2	0.98	0.77	0.72
7451_4	0.99	0.75	0.68
7443_3	0.95	0.74	0.71
Mean	0.98	0.75	0.69

3.4 Average Diameter and Length Forecast

Table 4 shows the sampling error comparing the K-Means and U-Net algorithms based on the average of the previously discussed metrics. The algorithm with the lowest sampling error was defined based on this analysis.

Table 4. Standard error from the mean of the metrics.

	PA	IOU	DC	Error
K-Means	0.96	0.56	0.32	0.19
U-Net	0.98	0.75	0.69	0.09

For the calculation of the parameters, only the test images were used and the enlarged images were not considered. Therefore, classes with filters 3, 4 and 5 that change the dimensionality of the image needed to be removed from the propositions.

In the literature the diameter of sisal fiber is concentrated between 0.100 and 0.300 mm [18, 19], although it can also reach different values, but in lesser frequency [20, 21]. However, as the waxes, lignin and hemicellulose from the outermost layer of the fiber has been removed during the mercerization process, it is expected that the diameters will be smaller, usually below 0.2 mm, and less dispersed [19, 21, 22]. It is also important to emphasize that sisal does not have a regular and round section, so it is also common to find in the literature refer to the average cross-section area instead of the average diameter [23]. Besides, the sisal fibers have an internal cavity, the lumen, and usually the calculated cross-section area is based in the contour line and does not take this cavity into account [23, 24]. Table 5 shows the values found for fiber diameter. The predicted mean value of the fibers using the net was 0.25 mm while the expected 0.27 mm for the manually traced segments.

Table 5. Fiber diameters.

Image	Predicted (mm)	Expected (mm)
7441_1	0.22	0.28
7443_1	0.26	0.28
7453_1	0.25	0.27
7465_2	0.27	0.26
Mean	0.25	0.27
Error (%)	1.13	0.50

The length obtained for the fibers can be seen in Table 6. Each length of the fibers depends on the sieves used during milling in the production of the samples. Among the images in question, only 7441_1 and 7443_1 correspond to the same sieve. In them, it is possible to verify that you hear an interesting approximation. 1.87 mm predicted for 1.80 mm expected in sample 7441_1 and 2.15 mm predicted for 2.72 mm expected for sample 7443_1. Even so, it is possible to observe an accentuated sampling error for the length calculation algorithm, presenting values of 42.05% for the predicted and 35.87% for the expected. However, one must consider in the analysis that the images result from different sievings, so this accentuated error is expected.

Table 6. Fiber lengths.

Image	Predicted (mm)	Expected (mm)
7441_1	1.87	1.80
7443_1	2.15	2.72
7453_1	2.14	2.11
7465_2	3.71	3.43
Mean	2.47	2.51
Error (%)	42.05	35.87

4 Conclusions

The use of AI techniques proves to be a robust alternative to optimize the polymer characterization process, and the presented techniques can contribute to the automation of this process. However, the K-Means and Mask-RCNN algorithms did not provide enough evidence throughout this study to justify their use, since the fiber phase did not show consistent results in the validation images. However, it is important to highlight that the Mask-RCNN network can still be improved, as it is observed that the network did not properly adjust the weights during the training performed, indicating a possible underfeeting. On the other hand, the U-Net algorithm proved to be a promising alternative for the characterization of materials, since not only did it present interesting PA, IOU and DC results, but also the diameter and length calculations are within a small margin of error.

As this work is still under development, more relevant information such as the polymer's modulus of elasticity, which directly depends on the volumetric fraction of the fibers and matrix, has not yet been obtained. The focus of the study at this stage concentrated efforts on fiber segmentation as it is a more challenging task. Therefore, it is expected that in future stages of the project, with the improvement of the networks for the mentioned segmentations, new analyzes with better results will be obtained.

References

1. Alsedrah, M.K.: Artificial intelligence, American university of middle east, middle east. Research(2017). https://doi.org/10.13140/RG.2.2.18789.65769
2. Oti, E.U., Olusola, M.O., Eze, F.C., Enogwe, S.U.: Comprehensive review of k-means clustering algorithms. Int. J. Adv. Sci. Res. Eng. **07**, 64–69 (2021). https://doi.org/10.31695/ijasre.2021.34050
3. Zayegh, A., Al Bassam, N.: Neural network principles and applications. In: Digital Systems. IntechOpen (2018)
4. Hordri, N.F., Yuhaniz, S.S., Shamsuddin, S.M.: Deep Learning and its Applications. Nova Science Publishers, New York (2021)
5. He, K., Gkioxari, G., Dollár, P., Girshick, R.: Mask R-CNN. IEEE Trans. Pattern Anal. Mach. Intell. **42**, 386–397 (2020). https://doi.org/10.1109/TPAMI.2018.2844175
6. Ronneberger, O., Fischer, P., Brox, T.: U-Net: convolutional networks for biomedical image segmentation. undefined. IEEE Access **9**, 16591–16603 (2015). https://doi.org/10.1109/ACCESS.2021.3053408
7. Klein, A., Warszawski, J., Hillengaß, J., Maier-Hein, K.H.: Towards whole-body CT bone segmentation. In: Maier, A., Deserno, T., Handels, H., Maier-Hein, K., Palm, C., Tolxdorff, T. (eds.) Bildverarbeitung für die Medizin 2018. Informatik aktuell, pp. 204–209. Springer, Heidelberg (2018). https://doi.org/10.1007/978-3-662-56537-7_59
8. Maksov, A., et al.: Deep learning analysis of defect and phase evolution during electron beam-induced transformations in WS2. npj Comput. Mater. **5**(1), 12 (2019). https://doi.org/10.1038/s41524-019-0152-9
9. Chen, D., Sun, D., Fu, J., Liu, S.: Semi-supervised learning framework for aluminium alloy metallographic image segmentation. IEEE Access **9**, 30858–30867 (2021). https://doi.org/10.1109/ACCESS.2021.3059505

10. Sharma, A., Mukhopadhyay, T., Rangappa, S.M., et al.: Advances in computational intelligence of polymer composite materials: machine learning assisted modeling, analysis and design. Arch. Comput. Methods Eng. **29**, 3341–3385 (2022). https://doi.org/10.1007/s11831-021-09700-9
11. Rokbi, M., Osmani, H., Imad, A., Benseddiq, N.: Effect of chemical treatment on flexure properties of natural fiber-reinforced polyester composite. Procedia Eng. **10**, 2092–2097 (2011). https://doi.org/10.1016/j.proeng.2011.04.346
12. Vale, D.A., Vieira, C.B., Vidal, M.F., et al.: Chitosan-based edible films produced from crab-Uçá (Ucides cordatus) waste: physicochemical, mechanical and antimicrobial properties. J. Polym. Environ. **29**, 694–706 (2021). https://doi.org/10.1007/s10924-020-01913-6
13. Buslaev, A., Iglovikov, V.I., Khvedchenya, E., et al.: Albumentations: fast and flexible image augmentations. Information **11**, 125 (2020). https://doi.org/10.3390/info11020125
14. Abadi, M., et al.: TensorFlow: Large-scale machine learning on heterogeneous systems (2022). www.tensorflow.org
15. Wu, Y., et al.: Detectron2 (2019). https://github.com/facebookresearch/detectron2
16. Jiang, X., Yu, H., Lv, S.: An image segmentation algorithm based on a local region conditional random field model. Int. J. Commun. Netw. Syst. Sci. **13**, 139–159 (2020). https://doi.org/10.4236/ijcns.2020.139009
17. Trutti, A.C., Fontanesi, L., Mulder, M.J., et al.: A probabilistic atlas of the human ventral tegmental area (VTA) based on 7 Tesla MRI data. Brain Struct. Funct. **226**, 1155–1167 (2021). https://doi.org/10.1007/s00429-021-02231-w
18. Okeola, A.A., Abuodha, S.O., Mwero, J.: Experimental investigation of the physical and mechanical properties of sisal fiber-reinforced concrete. Fibers **6**, 53 (2018). https://doi.org/10.3390/fib6030053
19. Jiang, A., Xi, J., Wu, H.: Effect of surface treatment on the morphology of sisal fibers in sisal/polylactic acid composites. J. Reinf. Plast. Compos. **31**, 621–630 (2012). https://doi.org/10.1177/0731684412441867
20. da Costa, L.L., Loiola, R.L., Monteiro, S.N.: Diameter dependence of tensile strength by weibull analysis: part i bamboo fiber. Rev Mater **15**, 97–102 (2010). https://doi.org/10.1590/s1517-70762010000200004
21. Fernandes, E.M., Mano, J.F., Reis, R.L.: Hybrid cork-polymer composites containing sisal fibre: Morphology, effect of the fibre treatment on the mechanical properties and tensile failure prediction. Compos. Struct. **105**, 153–162 (2013). https://doi.org/10.1016/j.compstruct.2013.05.012
22. Bekele, A.E., Lemu, H.G., Jiru, M.G.: Experimental study of physical, chemical and mechanical properties of enset and sisal fibers. Polym. Test. **106**, 107453 (2022). https://doi.org/10.1016/j.polymertesting.2021.107453
23. de Silva, F., A, Chawla N, Filho RD de T,: Tensile behavior of high performance natural (sisal) fibers. Compos. Sci. Technol. **68**, 3438–3443 (2008). https://doi.org/10.1016/j.compscitech.2008.10.001
24. Orue, A., Jauregi, A., Peña-Rodriguez, C., et al.: The effect of surface modifications on sisal fiber properties and sisal/poly (lactic acid) interface adhesion. Compos. Part B Eng. **73**, 132–138 (2015). https://doi.org/10.1016/j.compositesb.2014.12.022

Smart and Sustainable Buildings and Cities. Special Session on Technological Advances in Construction Sustainability

Santa Catarina's Tiles: An Artisanal Heritage to Preserve. Manufacturing Process and Characteristics

Miguel Oliveira[1](✉), Marta Marçal Gonçalves[2], Elisa da Silva[1], Fátima Farinha[3], and Elisabete Rocha[4]

[1] CEPAC-Ualg, Algarve University, Campus da Penha, 8005-139 Faro, Portugal
mjolivei@ualg.pt

[2] CEAACP-Ualg, Algarve University, Campus da Penha, 8005-139 Faro, Portugal

[3] CEPAC-UAlg/C-MADE, Algarve University, Campus da Penha, 8005-139 Faro, Portugal

[4] Cerâmica Alberto Rocha, 8800-156 Santa Catarina da Fonte Do Bispo, Portugal

Abstract. The heritage related to the artisanal production of ceramic building materials, was officially recognized in 1999 by ICOMOS, with the "Charter of Vernacular Built Heritage". From the beginning of the XX century, the Santa Catarina clay soil gain expression in the Algarve region, since several small and family factures, start to produce artisanal ceramic products for building construction. These products were mainly roof tiles, floor tiles and solid bricks. With the industrialization, the factories were abandoned and just a few survive, pursuing the goal of maintaining the tradition alive. Nowadays, with the growing demand for ecological and sustainable materials, with a strong aesthetic value, the return of this ceramic tiles opens new possibilities of growth and expansion in the construction sector, as well as in the region. In fact, this traditional ceramic industry is in danger of disappearing, being including in the "Red List of Crafts in the Algarve". It is urgent to preserve this heritage and the know-how of the professionals in this industry. Santa Catarina tiles requires no special maintenance, they are fireproof, resistant to shock and temperature gradients, resistant to sunlight even in places of great exposure. Ceramic tiles are also versatile and made with natural and ecological clay soils. It is hoped that, with this work, it will be possible to sensitize decision-makers and stockholders to its importance, maintaining this local culture so that the future generations can discover, enjoy, and perpetuate it.

Keywords: Santa Catarina's Tiles · Manufactured Ceramic Product · ICOMOS

1 Introduction and Motivation

Santa Catarina da Fonte do Bispo is a parish in the municipality of Tavira, located in southern mainland Portugal (Fig. 1), in the region of Algarve (South Portugal). From the beginning of the XX century, the Santa Catarina clay soil gain expression in the Algarve region, since several small and family factures, start to produce artisanal ceramic products for building construction. These products were mainly roof tiles, floor tiles and solid

J. F. L. C. Semião et al. (Eds.): INCREaSE 2023, ASST, pp. 221–233, 2023.
https://doi.org/10.1007/978-3-031-44006-9_16

bricks (in Portuguese *tijolo de burro*). The floor tiles are considered as the brand product, due to its pictorial characteristics, being the object of study of this work. These floor tiles are designated in Portuguese as *ladrilhos*.

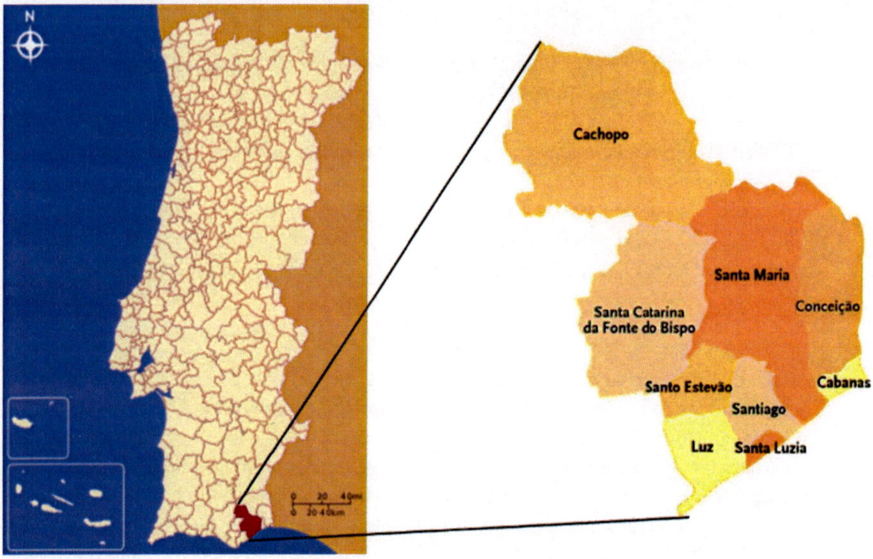

Fig. 1. Location of the municipality of Tavira in Portugal [1] and the parish of Santa Catarina da Fonte do Bispo in the municipality and Tavira [2]

The word ceramic is derived from the Greek term "keramike", which means, clay, being also used to describe products developed out of clay. Clay has the property of molding itself easily while paste, then acquiring hardness and strength after cooked in the oven.

Pottery already existed in Portugal before the arrival of the Romans, but later on, with the takeover by the Arabs, this activity develops significantly. They were the ones that introduced many of the techniques used in the manufacturing process of ceramic and which are still used in the production of many different ceramic products. The Moorish pieces were produced with clay that was extracted on site, and its fabrication essentially manual. In Algarve, as well as Alentejo, the building construction of the past was based essentially upon ceramic.

Nowadays, with the growing demand for ecological and sustainable materials the return of this ceramic tiles, with an excellent hygrothermal performance, due to its porosity, associated with an exceptional aesthetic appearance, open new possibilities of growth and expansion in the construction sector. When used on terraces, for example, they isolate without muffling. The *tijoleira* tile is less affected by the heat and can keep a colder ambience, therefore suitable for the southern regions, where the summer is very hot.

2 Heritage Value of the Floor Tiles of Santa Catarina

The heritage related to the artisanal production of ceramic building materials in the parish of Santa Catarina da Fonte do Bispo is of the vernacular type. This type of heritage was officially recognized in 1999 by ICOMOS, with the "Charter of Vernacular Built Heritage" [3]. This heritage has, in particular, a lot of value both at the immaterial level, as at the material level, either movable and immovable.

The floor tiles of Santa Catarina are manufactured in the "sheds" (*telheiros*), a term that designates "manufacturer of roof tiles" [4] but, according to the same author, it can be also designated as "porch". In the Algarve, the term *telheiro* refers to a traditional ceramic industry, where, under a shed, roof tiles, floor tiles and massive bricks are manufactured.

The use of sheds, under which ceramic elements are manufactured, is relatively recent, as they did not exist at the beginning of the activity. Initially the manufacturing was done in the open air, but there was a need to build a porch, since the direct incidence of sunlight can lead to a greater risk of tile cracking, even before being cooked [5].

This industry has also associated a large diversity of vocabulary and specific tools, which enriches the immaterial and material facet of this heritage [5, 6].

Nowadays, when builders, engineers and architects refer to this material as "floor tile of Santa Catarina", everyone in the profession knows that this designation refers to square or rectangular ceramic elements, handmade and with certain pictorial characteristics (Fig. 2).

Fig. 2. Tiles of Santa Catarina applied to a floor.

Initially these products were only designated as regional tiles, with no association to a specific and region or geography. However, Mr Alberto Rocha, owner of one of the *telheiros*, decided to start designating these products as "of Santa Catarina", and thus creating a regional brand. This classification was generalized, and from the 60's to the current days, in the specifications, this designation appears. Depending directly on the climatic conditions, the production usually takes place every working day from March to October, being, consequently, a seasonal activity although the annual sustenance is guaranteed because in the remaining months the produced pieces were also sold [5]. Nowadays the production can be performed all year round.

The lands where the sheds are located did not always belong to the people in charge of the production or to the owners of the *telheiro*. They were often rented to the owners of the land, who were paid with a share of the production [5].

As it was necessary to have some purchasing power to acquire ceramic elements, such as roof tiles, floor tiles and solid bricks, these pieces were often reused for other purposes and were even part of the inheritance, left as goods with financial value [5].

With the automation of the ceramic industry, the placement of ceramic elements on the market originated a higher offer at lower price, leading to the crashing and almost disappearance of this regional and traditional activity. This risk is so real that the *telheiros* are included in the Red List of Handicraft Activities of the Algarve, regarding the intangible heritage of this region, and the main threats to this art are the "aging of the trade" and the "workforce" [7], which compromises the future of this activity.

Along with agriculture, this industry was, and still is, one of the most important in the parish of Santa Catarina da Fonte do Bispo, employing many of its inhabitants, having marked its economic dynamics in the last decades [5].

3 Borrow Sites: The Red Clay Soils

Algarve is a complex Mesozoic sedimentary basin and its lithology is very rich. The sandstones, conglomerates and sands, associated to the sedimentary and more recent formations (Cretaceous, Miocene, Pleistocene, Pliocene and Holocene) are the dominant type of rock massifs along the South coastline and Southeast. Clay shales and graywacke prevail on the North, being the oldest formation (Paleozoic). In the middle, where the faults are significant, due to the tectonic movements, it prevails limestones, sandstones and dolomites, formed during the Upper Triassic and Jurassic [8] (Fig. 3).

Santa Catarina is located on this last geologic and lithologic zone, where clayey sandstones, limestones usually dolomitic, marls and marly limestones are the predominant formations [9] (Fig. 4).

According to the Geologist and Paleontologist, Stefan Rosendahl, the clays that serve as raw material for the Santa Catarina ceramic industry of bricks and tiles, are located in a narrow strip of deposits of the ancient Rheic Ocean [10]. This Rheic Ocean separated the two major palaeocontinents, Gondwana and Laurussia and it was one of the principal oceans of the Palaeozoic. Its closure resulted in the assembly of the supercontinent Pangaea [11].

The Paleo-Tethys Ocean was an ocean located along the northern margin of the paleocontinent Gondwana that started to open during the Middle Cambrian, grew throughout the Paleozoic, and finally closed during the Late Triassic [12]. Paleo-Tethys was a precursor to the Tethys Ocean, also called the Tethys Sea or the Neo-Tethys. This was a prehistoric ocean during much of the Mesozoic and early Cenozoic Eras, located between the ancient continents of Gondwana and Laurasia, before the opening of the Indian and Atlantic oceans during the Cretaceous Period.

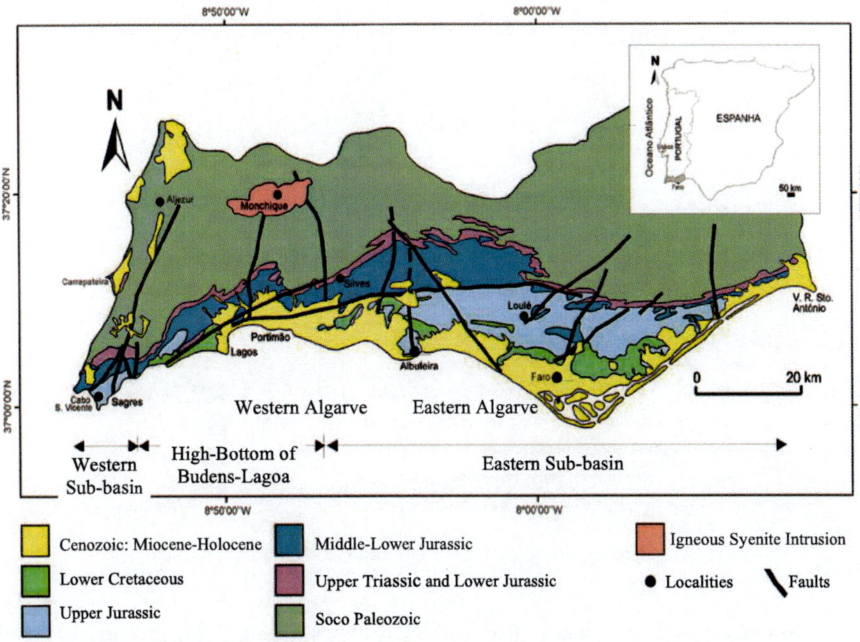

Fig. 3. Simplified geological map of Algarve (source [13], based upon [14]).

Tethys closed about 50 million years ago, during the Cenozoic. Remnants of the Tethys Sea remain today as the Mediterranean, Black, Caspian and Aral seas. Limestones are dominant sedimentary facies of Tethys, and often very rich in fossils, indicating an abundant and diverse tropical marine fauna [12].

Jorge Carvalho, in 2009, presented a spatial distribution of the Algarve region resources, including rocks, soils and minerals exploitations for Industry purposes [15] (Fig. 5). As it can be seen, the clays resources are located precisely on the Upper Jurassic, constituted by conglomerates, sandstones, limestones, dolomitic and marls.

In fact, two types of clay are used, red and white, and after combined in the right proportion, the raw material to produce tiles and other products is obtained. The white clay comes from the salt flats of Castro Marim, being added in smaller quantities. It is a fatty clay which increases workability, giving plasticity and reducing shrinkage.

The extraction process is supervised, so that clays are not contaminated with stones, dirt or any other material that may damage its quality. Both types of clay, red and white, are transported to the ceramic factory and are stored in two separate lots, being mixed according to the ceramic purpose [6, 16].

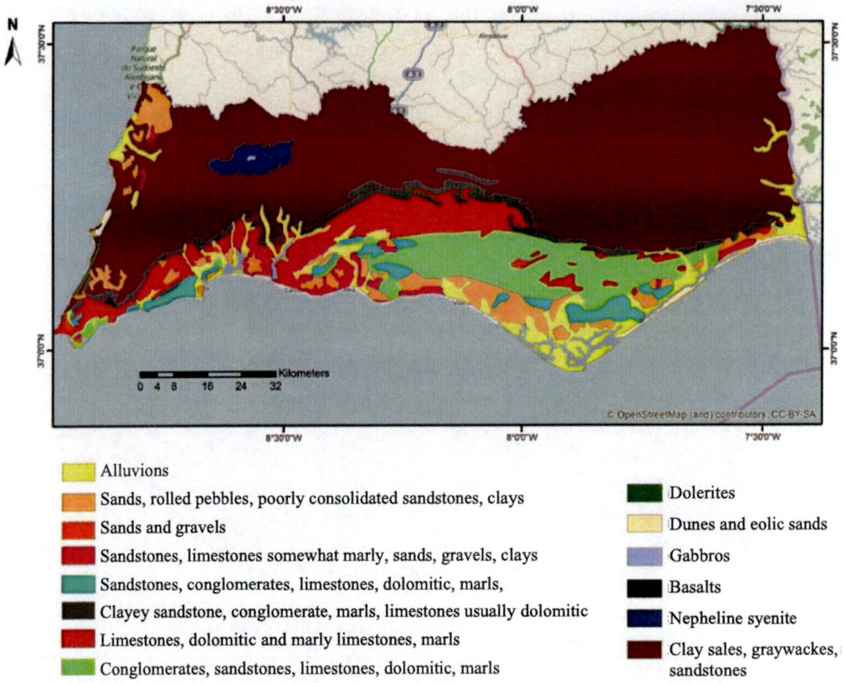

Fig. 4. Algarve lithological map (source [13], adapted from [8]).

The borrow sites (Fig. 6) were "discovered" by clay seekers, who knew how to "read" the land [5], which can only be achieved with a lot of experience and knowledge, which further enriches this heritage.

4 Manufacturing Process

The clay soil is taken from natural clay deposits located close to the artisanal factory and then is placed through a backhoe loader in a dozer, which, in turn, is connected to a crusher to break up the larger clods. Then the material goes on the conveyor belt to the hammer mill, transforming the clay into powder. Again, through the carpet, it goes to the kneader where the water is mixed and turned into clay. From here, the clay goes through wheelbarrows to the place where it will be molded manually. Through a wooden or stainless-steel mold, the tiles are molded manually (Fig. 7).

At this stage, a mixture of ash and sand is also used to facilitate molding without the clay sticking to the base and mold. When they come out of the mold, the tiles are placed on the floor where they are left to dry (Fig. 8).

During the drying process, the clay suffers shrinkage and therefore the tiles must be beaten to avoid this natural deformation. Wooden planks are used to prevent the tiles from warping, and that's the reason for the threshing floors.

Once dry, the pieces are transported to the oven (Fig. 9), where they are placed in an appropriate order (Fig. 10).

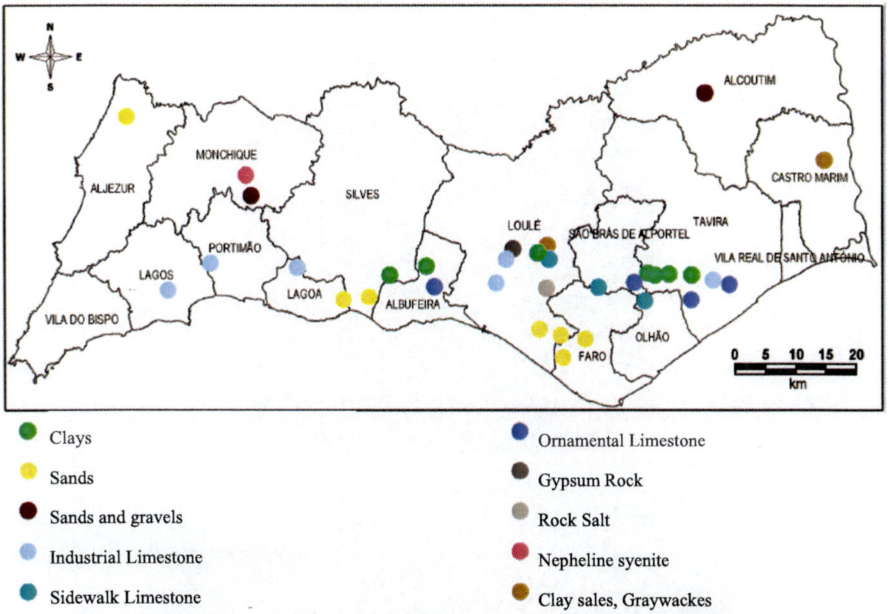

Fig. 5. Algarve major rocks, soils and minerals exploitations (adapted from [15]).

Fig. 6. Borrow site.

The fuel used in the ovens are mainly olive pit (bagasse) waste, a residue obtained from olive oil production as well as almond shells and firewood (Fig. 11).

After cooked, the pieces are left to cool inside the oven, ranging from 2 to 4 days. Lastly the tiles are packed on pallets (Fig. 12), ready to be placed in the commercial circuit.

Fig. 7. Wooden or stainless-steel mold frame for tiles.

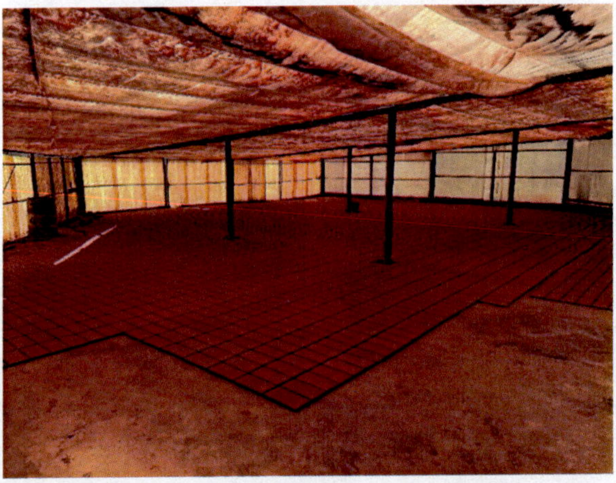

Fig. 8. Tiles drying on the floor, protected from sun and rain.

5 Tiles Performance

Ceramic coverings have a long tradition in Portugal, constituting one of the most used materials in the construction industry, particularly in their most common forms such as ceramic tiles, usually applied on floors, interior walls and facades.

The basic material of ceramic tiles are clayed soils, but some secondary elements, correctives or additives, can be added during manufacture, in order to obtain better desired characteristics.

The surface of the ceramic tile is subject to a wide range of actions depending on the use, depending on the element to be covered (floor or wall), its location (indoor or outdoor), the type of building, the type of covered space and the intensity of use

Fig. 9. Artisanal and traditional oven.

Fig. 10. Placement of tiles inside oven.

or occupation of the space. Furthermore, these actions depend on the activity of the users and the fixed or mobile furniture – bending, mechanical shock, abrasion, etc.; environmental conditions – temperature, humidity, etc.; and cleaning or maintenance operations – chemical or abrasive aggressiveness of methods or products used.

Fig. 11. Storage of olive pit waste (bagasse) and almond shells.

Fig. 12. Storage of the final product.

As building materials, ceramic floors have the regulatory framework Regulation (EU) Nº 305/2011 of the European Parliament and the Council of 9 March 2011 [17], establishing harmonized conditions for the marketing of this construction products. The harmonized European reference standard for ceramic flooring is NP EN 14411:2008 – Ceramic flooring and coverings. Definitions, classifications, characteristics and marking [18]. This standard defines and specifies the requirements and criteria for marking ceramic tiles and when in compliance with the standard, ensure the ability to perform its function at stated performance levels.

Table 1 shows main characteristics of the product. Some values were obtained at Telheiro Alberto Rocha and others were determined in tests carried out at the Construction Materials Laboratory, Algarve University.

Table 1. Main characteristics

Laboratories Test	Water Absorption (1)	Flexural Strength (2)	Compressive Strength (2, 3)
Standard	EN ISO 10545–3	EN ISO 10545–4	EN 772–1
Results	16,4%	Modulus of rupture – 6,1 N/mm2 Breaking load – 1650 N	normalized breaking stress – 38,3 N/mm2
Classification according to NP EN 14411	A III	A III	Not applicable

(1) Tests carried out at the Construction Materials Laboratory of DEC/ISE
(2) Values indicated by Alberto Rocha
(3) Tests carried out on solid bricks

Regarding water absorption, Group III, according to NP EN 14411:2016, refers to tiles with high water absorption, E > 10%.

6 Sustainability

A brief description of the manufacturing process, here presented, is of high relevance, in order to comprehend the low carbon emissions and reduce footprint associated to the fabrication of this natural product, and therefore considered a sustainable product. Also, the vernacular heritage is, in essence, sustainable. It contributes to the local economy, creating jobs; it is also closely linked to the environment around it, using only local materials, and ancient techniques, and it has a very strong social facet, because the vernacular heritage is, according to the Charter of the Vernacular Built Heritage [3], "the fundamental expression of the identity of a community, its relations with the territory and at the same time, the expression of the cultural diversity of the world… is the natural and traditional way in which communities have produced their own habitat".

Here are two aspects of the sustainability of the ceramic industry of Santa Catarina: in relation to the traditional tiles production industry, in the production, when molding the pieces, the ash from the furnace is used, which is mixed with sand, to avoid the wracking of the clay material to the hands and the shape, facilitating the transport of these pieces to the drying area [5]. Another fact that causes a reuse of materials avoiding waste is due to the type of fuel used in the furnaces. Being significant in this parish the production of olive oil, the ceramic industry of Santa Catarina uses the bagasse from the mills as fuel, reducing the waste of materials from other industries.

7 Conclusions

Santa Catarina da Fonte do Bispo has a heritage linked to the traditional ceramic industry that is very rich and that is in danger of disappearing, being including in the "Red List of Crafts in the Algarve" [7]. It is urgent to preserve this heritage and the know-how of the professionals in this industry.

During its life Santa Catarina tiles´ requires no special maintenance. It is a porous material providing an indoor space with good hygrothermal performance. It is a very versatile material that can be applied as a coating material for floors, walls, ceilings and terraces. When applied outside, it has good characteristics, namely, resistant to sunlight even in places of great exposure. It is also resistant to shock and temperature changes. Ceramic tiles are additionally fireproof, unlike wood and products with a synthetic origin. Due to their anti-slip characteristics, they are very useful on ramps, stairs, near a swimming pool or on surfaces frequented by the elderly, children, or people with some type of physical handicap. Finally, the tiles are made with a natural and ecological material. For millennia the food has been cooked and served in clay containers. Clay is also used for medicinal purposes and in various aesthetic treatments, therefore, there is no risk of allergies when applying these materials on building construction.

References

1. Tavira. https://pt.wikipedia.org/wiki/Tavira#/media/Ficheiro:LocalTavira.svg. Accessed 28 Feb 2023
2. Tavira (Concelho de). https://geneall.net/pt/mapa/134/tavira/. Accessed 28 Feb 2023
3. ICOMOS, Carta do Património Vernáculo Construído. Cidade do México (1999)
4. Priberam Dicionário. https://dicionario.priberam.org/telheiro. Accessed 21 Feb 2023
5. Santos, M.: Telheiros. In: Materiais, sistemas e técnicas de construção tradicional. Contributo para o estudo da arquitetura vernácula da região oriental da serra do Caldeirão, 1st ed. CCDR do Algarve, Porto, pp 170–177 (2008)
6. Barros M de P.: Tradição artesanal de Santa Catarina da Fonte do Bispo. Os telheiros. Faro (2009)
7. AA.VV. RED Book. Lista Vermelha das Atividades Artesanais Algarvias. Intrreg España Portugal; Magallanes - ICC; CCDR Algarve (2021)
8. Viegas, F.M., Faro, L.: Geologia e génese do relevo da Rocha da Pena (Algarve, Portugal) e o seu enquadramento educativo (2006)
9. Trindade, M.J., Rocha, F., Dias, M.I., Prudêncio, M.I.: Mineralogy and grain-size distribution of clay-rich rock units of the Algarve basin (South Portugal). Clay Miner. **48**, 59–83 (2013). https://doi.org/10.1180/CLAYMIN.2013.048.1.04
10. Steps & Stones – Geoloquium. https://geoloquium.eu/en/steps-stones/. Accessed 14 Mar 2023
11. Linnemann, U., Pereira, M.F., Jeffries, T.E., et al.: The cadomian orogeny and the opening of the Rheic Ocean: The diacrony of geotectonic processes constrained by LA-ICP-MS U-Pb zircon dating (ossa-morena and saxo-thuringian zones, Iberian and bohemian massifs). Tectonophysics **461**, 21–43 (2008). https://doi.org/10.1016/J.TECTO.2008.05.002

12. Tethys Sea | Definition, Location, & Facts | Britannica. https://www.britannica.com/place/Tethys-Sea. Accessed 14 Mar 2023
13. Feroldi, A., da Silva, E., Gonçalves, M.M.: Drystone walls in the Algarve, Portugal. Characterization and interconnection with the geology and lithology. In: IOP Conference Series: Materials Science and Engineering, vol. 1203 (2021). https://doi.org/10.1088/1757-899x/1203/2/022129
14. Programa Reordenamento e Gestão da Paisagem das Serras de Monchique e Silves (PRG-PSMS) | DGT. https://www.dgterritorio.gov.pt/Programa-Reordenamento-e-Gestao-da-Paisagem-das-Serras-de-Monchique-e-Silves-PRGPSMS. Accessed 15 May 2022
15. Carvalho, J.: RECURSOS MINERAIS NA REGIÃO DO ALGARVE DO ALGARVE Potencial e Ordenamento do Território. Faro (2009)
16. Application and treatment - Terracota do Algarve. https://terracotadoalgarve.com/en/application-and-treatment/. Accessed 14 Mar 2023
17. EUR - Less - Acess to European Union Law, Document 32011R0305 EN - Regulation (EU) No 305/2011 of the European Parliament and of the Council of 9 March 2011 laying down harmonised conditions for the marketing of construction products and repealing Council Directive 89/106/EEC Text with EEA relevance. In: Chapter 13 . Croatia, pp 113–151 (2021)
18. UNE EN 14411:2016 Ceramic tiles - Definition, classification, characteristics, assessment and verification of constancy of performance and marking (2016)

Technical and Sustainability Analysis of Construction Processes and Covering Systems for Outdoor Swimming Pools

Miguel José Oliveira[1] (✉), Fátima Farinha[1], David Marín-García[2], and Elbes Muniz[1]

[1] Institute of Engineering, University of Algarve, Faro, Portugal
mjoliveira@ualg.pt
[2] Higher Technical School of Building Engineering, University de Seville, Seville, Spain

Abstract. Nowadays, outdoor swimming pools are equipments that integrate villas, as well as multifamily buildings or tourist complexes. Architects and engineers have been working on the development of innovative solutions that can reduce the environmental impacts of swimming pools. The selection of the construction process for a swimming pool, even a small one, depends on a series of determining factors, namely, architectural design, soil type, functionalities, aesthetics, durability, budget, and construction time. Regarding the covering system, there are several solutions in the market with quite different costs. The option for one of these solutions depends on the architectural solution, as well as the main expected performance for the cover, for example, for winter hibernation, all year round for fall protection, water temperature maintenance. This article presents a synthesis of the main constructive processes and covering systems for outdoor swimming pools. In addition, it also presents a comparative analysis of economic, technical and sustainability factors for the solutions listed to systematize and facilitate decision-making for those who want to install a swimming pool.

Keywords: Swimming pools · Construction processes · Covering systems · Sustainability Analysis

1 Introduction

The swimming pool (from the Latin Piscina, derived from piscis "fish") is a water tank suitable for swimming, diving and other sports, such as water polo and hydro gymnastics, or simply for recreation. The first swimming pools built for commercial use appeared at the end of the 18th century and beginning of the 19th century. The first registered public swimming pools were in Europe, between the years 1761 and 1781. The adaptation of commercial swimming pools to residential ones comes from several factors, expressly, the desire to obtain a new form of leisure for hygiene habits [1].

In recent times, architects and engineers have been working on the development of innovative solutions that can reduce the environmental impacts of swimming pools. Construction methods, even for small pools, are influenced by several factors, such as

J. F. L. C. Semião et al. (Eds.): INCREaSE 2023, ASST, pp. 234–242, 2023.
https://doi.org/10.1007/978-3-031-44006-9_17

architectural design, soil type, functionality, aesthetics, durability, budget and construction time. In addition to these factors, we highlight the parameters that are required by the equipment to be installed in the pool (example: cover system), as well as the regulatory standards in force for swimming pools.

In Portugal, a wide variety of pool types and sizes can be identified [2]. For example, the Algarve region has areas with a high concentration of swimming pools (Fig. 1).

This concentration of swimming pools favors the development of companies that work in the field of construction and maintenance of swimming pools. It is estimated that around five hundred companies operate in this region of the country [3].

Fig. 1. Aerial view of villas with swimming pools in Carvoeiro, Algarve, source Google Earth

As one of the most popular destinations in Portugal, the Algarve receives millions of tourists annually. This is mainly due to the fact that the climate is attractive and the magnificent beaches recognized internationally. However, even with the small distance to the sea, many houses have a swimming pool for the greater comfort of the residents. The swimming pools in the Algarve are mostly small and generally have water volumes of up to 150 m^3 [3].

Pools are a point of reference for socializing with family and friends. In addition to all this aspect, many homeowners see the pool as an aesthetic element for their home. This means that there are different solutions, both in terms of tank and equipment. Regarding to equipment, covers are an element of great importance both in terms of safety and also a cost-saving measure, due to the possibility of controlling water evaporation and energy consumption. To meet these factors, suppliers have developed different solutions for covering systems.

In view of the development of the swimming pool market, the understanding of construction methods and the elements that make up the set, such as the heating system, filtration process and cover system evolve according to the need that has been presented by the demand of this market, presenting increasingly interesting solutions to be applied in swimming pool construction.

Thus, this article presents, in Sect. 2, a synthesis of the main constructive methods and covering systems available in the market for swimming pools. In Sect. 3 a technical and sustainability analysis for those solutions is made. Finally, the main conclusions are presented in Sect. 4.

2 Construction Processes and Covering Systems

In the following a brief description of the construction process for swimming pool tanks and the main solutions of covering systems are presented.

2.1 Construction Processes

Reinforced Concrete Swimming Pool.
Composed of a set of reinforcements which are surrounded by a formwork (one internal and one external). This formwork, after checking, is concreted and then treated with a waterproofing agent and coated with the defined finish. The main stages of the construction process are: excavation, soil compaction, execution of reinforcement, formwork, concreting, removal of formwork, surface regularization, waterproofing, execution of coatings and filling of the pool.

Shotcrete Swimming Pool.
Shotcrete swimming pools have as their main characteristic the projection of the concrete. In this case, the concrete has a low water/cement ratio, so this concrete can be projected and supported by the reinforcement and the lost formwork, which is usually done in concrete masonry. The main construction stages of shotcrete swimming pools are: excavation, soil compaction, execution of lost formwork, execution of reinforcement, concrete projection, surface regularization, waterproofing, execution of coatings and pool filling.

Masonry Swimming Pool.
Masonry swimming pools, as the definition says, are made up of masonry side walls, but still need a reinforced concrete floor like that of reinforced concrete and shotcrete systems. The main constructive stages of this system are: excavation, soil compaction, execution of the base in reinforced concrete, with the placement of the waiting reinforcement of the masonry (in the corners), execution of the structural masonry, surface regularization, waterproofing, execution of coatings and pool filling.

Fiber Swimming Pool.
The fiber pools have a structure of certain rigidity, with polyester composition and with reinforcement of fiberglass layers. In this model, the designers generally consider a structure in iron profiles, to reinforce the support of the pool, as well as some bands in fiberglass, to increase the weight of the model. In addition to the layers of fiberglass, the structure still receives a layer of vinyl ester resin, which gives the model a certain anti-corrosion durability, in addition to functioning as thermal insulation. Finally, apply

the "gel-coat" modified with "NPG", which is the finish that gives a smooth surface and is resistant to UV rays.

Wooden Swimming Pool.
Wooden pools have different installation methods depending on where they are installed. They can be installed at ground level, partially buried or above ground. The main construction steps are: soil preparation, pool assembly, landfill, water and electricity connections and pool filling.

Swimming Pool with EPS Blocks.
These pools are made up of a self-supporting structure with formwork in EPS, but still need a reinforced concrete floor like that of reinforced concrete and shotcrete systems. The main construction stages are: soil preparation, execution of the reinforced concrete base with the placement of the vertical rods, placement of the EPS blocks, assembly of the EPS blocks with the placement of the horizontal reinforcement, concreting of the blocks, waterproofing, coatings, water and electricity connections and filling the pool.

Swimming Pool with Glass Elements.
Pools with glass elements on the walls are usually made up of a concrete floor with the placement of elements to support the glass, whether in shotcrete, masonry or even a metallic structure. For systems supported on reinforced concrete or masonry walls, a system that functions as a frame for placing the glass is still necessary. The main phases of this constructive method are: excavation, soil compaction, execution of the base in reinforced concrete, placement of the structure that will receive the glass or even execution of the supporting masonry (stage described in the masonry swimming pool model) or the elements in reinforced concrete (step described in the model of swimming pool in reinforced concrete), waterproofing of the base and walls - if applicable, installation of frames (if this model is used) and multilaminate glass and filling the pool.

2.2 Swimming Pool Coverings

Lamellar cover with winder box on the ground.
In this cover model, the structure where the blades are rolled up is exposed, as the name itself refers to "on the ground". The blades that make up the model are generally made of PVC, and the shaft is made of stainless material. These covers can be customized and are suitable for different pool designs.

Submerged lamellar coverage.
Like the lamellar coverings on the ground, the structure is in stainless steel and the slats in PVC. This model has a mechanical bearing kit to facilitate the opening and closing process. In some cases, it can be installed in pools that have already been built.

Telescopic coverage.
This type of cover has a more robust structure, and its seals are made of transparent glass. Telescopic covers can be low or high.

Removable flat cover.

This model is characterized by having flat modules, which are raised or stacked and whose assembly uses casters. In addition to the caster model, this model can also be manual.

Retractable deck cover.
This cover model can be adapted to any pool design. As it has a motorized model, it turns out to be a limited solution, as it requires considerable space to be installed. The deck material can be wood or composites.

Protection covers.
Protection covers are a simple solution that is adaptable to any pool design. As for durability, they have a shorter useful life and require complete replacement. There are solutions with different thicknesses and with improved thermal performance.

3 Technical and Sustainability Analysis

3.1 Technical-Economic and Sustainable Aspects - Comparison of Construction Methods and Cover Systems

In line with the above, with the view to demonstrate a comparative analysis with technical and economic aspects, and points related to sustainability in construction, Table 1 and Table 2 will be presented. This information allows for the comparison of different models most common in the market. However, EPS block construction methods and glass wall elements will not be presented due to their limited presence in the market and the fact that they do not provide enough information to compare with others.

To define the costs of the pool and the cover, the following dimensions were considered for the tank: 5 m long, 10 m long and 1.7 m deep. For the composition of the values presented for the construction methods, according to their construction method, they were consulted through the CYPE Ingenieros S.A database on the website http://www.geradordeprecos.info/. As for the roofs, the values were provided by specialized subcontractors.

To define the estimated useful life of the solutions presented, bibliographies [3–5] were used. It is important to note that the service life is affected by many factors. Among them, we can indicate definitions established in the project, construction, use, maintenance and environmental characteristics. Therefore, the indicated values must be seen as indicative.

To reference some aspects of sustainability, the information source "LiderA" (https://www.lidera.info/) was used. Additionally, the "LíderA" program is a voluntary Portuguese system, whose objective is, in an efficient and integrated way, to evaluate and certify the sustainability of constructions and is used in buildings.

3.2 Comparative Analysis

Reinforced concrete pools, molded in situ as well as shotcrete, can have the advantage of great structural mechanical resistance, longer service life and flexibility in terms of the design. In this case, due to the high structural durability, rehabilitation involves replacing the waterproofing and the final layer.

Table 1. Technical and sustainability analysis – construction methods.

Pool tank		Cost	Service life time	Sustainability aspects LíderA
Built on site I	 Reinforced Concrete	18 979,42 €	Reinforced concrete: 50 years (1) Waterproofing: 16 years (2) Ceramic tiles and bonding mortars: 10 years (3)	P33 - Low lifecycle costs considering structure that can last 50 years;
	 Shotcrete	19 194,50 €	Reinforced concrete: 50 years (1) Waterproofing: 16 years (2) Ceramic tiles and bonding mortars: 10 years (3)	P33 - Low lifecycle costs considering structure that can last 50 years;
	 Masonry	18 898,35 €	Mansory: 56 years (2) Waterproofing: 16 years (2) Ceramic tiles and bonding mortars: 10 years (3)	P33 - Low lifecycle costs considering structure that can last 50 years;
Prefabricated	 Fiber	17 796,88 €	10 years (Information provided by the supplier.)	
	 Wood	18 546,35 €	10 years (Information provided by the supplier.)	P12 - Responsible sourced products and materials

The advantages of fiberglass pools are the speed of their assembly and installation process. Also, this type of pool is easier to clean due to smoother surface. Waterproofing is not necessary since the fiber can perform this function. As for the disadvantages, fiberglass pools have a reduction in the tonality of their material, as it is exposed to the action of chlorine. However, the biggest disadvantage of this model is the pool's lack of versatility in terms of format and dimension, as the dimensions and shapes available on the market are limited.

The main advantages of glass swimming pools are aesthetics. However, the costs of installing such a pool are very high, considering the material and support structure.

Regarding roofs and energy loss, a more suitable model is telescopic roofs. In addition to protecting the pool from energy loss, it favors heating, due to the incidence of sunlight in the tank, giving the greenhouse effect. According to [6] bubble coverage can reduce by up to 20%, vinyl coverage by 40% and inflated coverage by up to 70%. This is because the cover acts as a barrier that prevents direct contact between the pool water and the outside air. According to the Aquacover website, the roofing has some advantages.

In terms of safety, covers are essential to prevent children from falling. In addition, swimming pools can reduce by up to 50% the costs related to pool cleaning maintenance and that they can generate an increase in temperature between 6 and 10 degrees compared to the outside temperature [7].

3.3 Sustainable Solutions

When sustainability is approached, it is known that this involves practices that reduce damage to the environment. That is, rational use of resources becomes implicit [8]. By analyzing the construction models of swimming pools, their use and their types of

Table 2. Technical and sustainability analysis - Covers Systems.

Covers	Cost	Sustainability aspects LíderA
 Lamellar with winder box on the ground	11.836 € (refers to the electric model)	P10 - Decreases water consumption by significantly reducing evaporation losses; P13 - Ensures greater durability of the pool, protects against physical and environmental agents such as solar radiation and biological attack; P18 - Reduces the need to use, for water heating, equipment that works with combustion; P21 - Safety and risk control for children falling.
 Submerged lamellar	14.899 € (refers to the electric model)	P10 - Decreases water consumption by significantly reducing evaporation losses; P13 - Ensures greater durability of the pool, protects against physical and environmental agents such as solar radiation and biological attack; P18 - Reduces the need to use, for water heating, equipment that works with combustion; P21 - Safety and risk control for children falling.
  Telescopic	low: 25.000 € high: 55.500 €	P7 - Passive solar performance; P10 - Decreases water consumption by significantly reducing evaporation losses; P13 - Ensures greater durability of the pool, protects against physical and environmental agents such as solar radiation and biological attack; P18 - Reduces the need to use, for water heating, equipment that works with combustion; P21 - Safety and risk control for children falling.
 Removable flat cover	18.050 €	P10 - decreases water consumption by significantly reducing evaporation losses; P13 - Ensures greater durability of the pool, protects against physical and environmental agents such as solar radiation and biological attack; P18 - Reduces the need to use, for water heating, equipment that works with combustion; P21 - Safety and risk control for children falling.
 Retractable deck	33.531,25 €	P10 - decreases water consumption by significantly reducing evaporation losses; P13 - Ensures greater durability of the pool, protects against physical and environmental agents such as solar radiation and biological attack; P18 - Reduces the need to use, for water heating, equipment that works with combustion P21 - Safety and risk control for children falling; P40 - Innovation. It is a differentiated coverage model that has the possibility of generating a new environment that can be occupied when closed.
 Protective covers	200 € (floating without attachment)	P10 - Decreases water consumption by significantly reducing evaporation losses; P13 - Ensures greater durability of the pool, protects against physical and environmental agents such as solar radiation and biological attack; P18 - Reduces the need to use, for water heating, equipment that works with combustion; P21 - Safety and risk control for children falling.

coverage, we are also directly analyzing energy and water consumption as well as other aspects related to the sustainability of these equipment.

On a positive note, it is possible to highlight some solutions that have already been studied and are already being used in construction [9] carried out a study that highlights four possible methods of reducing energy consumption. This is of great importance because many of the essential systems of a swimming pool, such as the filtration, cleaning and heating system, need this energy for their operation [10].

The solutions addressed for the aforementioned study were:

- Construction of a roof;
- Installation of solar collectors;
- Installation of photovoltaic panels;
- Installation of a geothermal pump, which allows the heat from the earth to be transferred to the water. This model has a low environmental impact, is very efficient and uses little energy for operation;
- Rainwater harvesting (depends on the location in question and it is necessary to assess the frequency of rain and the possible volume that can be used, in addition to the treatment system).

To reduce the energy loss from the pools and the volume of water, we highlight the great importance of installing pool covers. This procedure is widely used as it helps to control the evaporation process, which is one of the main factors responsible for the consumption of both energy and water.

Looking for more efficient solutions for swimming pools, the University of the Algarve is currently participating in the development of the ECOPOOL+++ project. The main objective is to develop a heated outdoor swimming pool that is more efficient in terms of energy and water consumption, which also includes a management model of the various systems linked to the pool [11]. In this project, one of the solutions consists in the development of a thermal insulation system, which will be installed through the interior of the pool tank (Fig. 2), helping to minimize energy losses.

Fig. 2. Application of the insulation system on the walls of the pool

In combination with the insulation model, a pool cover model will be defined, which will be composed of multilayer glass and air panels. This roof model will help reduce evaporation and, therefore, also reduce energy losses. For the heating model, solar panels and the heat collector floor model will be used, which is a sustainable solution as well. To complement this model, some ways of storing this heat produced through reservoirs composed of PCM phase change materials are being analysed [9].

Regarding water efficiency, models for reducing water consumption will be studied and analyzed. These solutions are linked to the control and monitoring model, which is capable of measuring water levels (through sensors) and leaks related to pipes problems (with pressure sensors). Additionally, the use of filter washing water is being studied for secondary uses, such as irrigation.

4 Conclusions

This article presents a synthesis of the main constructive methods and covering systems for outdoor swimming pools as well as a comparative analysis of economic, technical and sustainability factors.

From the various construction techniques presented, it is possible to conclude that the choice depends on several issues such as: architectural design, soil type, functionalities, aesthetics, durability, budget, and construction time.

In addition to the construction methods, the covering systems were characterized, which are also closely linked to the sought objective. Currently, covering systems are increasingly important for maintaining water temperature by reducing energy loss, allowing the use of these equipment for longer periods throughout the year.

Regarding the cost analysis of swimming pool tanks, prices are relatively similar, however the expected lifetime is substantially higher in concrete constructions.

With regard to the sustainability analysis, points were indicated that could be considered in the assessment by the LiderA system. However, this system is not adapted to swimming pools, so it is urgent to develop a methodology applied to swimming pools.

With regard to pool covers, there are large price differences on the market because in terms of materials and durability, the solutions are also very different. Anyway, for safety reasons and for water and energy savings, the use of covers should be generalized.

References

1. Gomes, J.P.: Avaliação Hídrica de Piscinas - Caso de estudo realizado nas Piscinas Municipais de Mangualde. University of Porto, Porto (2015)
2. Sequeira, J.: Piscinas em Portugal - Concepção Arquitetónica das Piscinas Municipais de São João da Madeira. Universidade Lusófona de Humanidades e Tecnologias, Lisboa (2013)
3. Oliveira, M.J., et al.: Common typology, durability and pathologies of the swimming pools in the algarve region - implications in the sustainability. In: Monteiro, J., João Silva, A., Mortal, A., Aníbal, J., Moreira da Silva, M., Oliveira, M., Sousa, N. (eds.) INCREaSE 2019, pp. 893–902. Springer, Cham (2020). https://doi.org/10.1007/978-3-030-30938-1_70
4. NP EN 13670:2010: Execução de estruturas de betão
5. Pinto, A.F.: Construção de uma base de dados de apoio à estimativa da vida útil das construções. University of Porto, Porto (2011)
6. Sveinbjörnsson, H.: O Conceito das Piscinas Aquecidas e Formas de Vida através da Arquitetura. University of Porto, Porto (2017)
7. Aquacover Homepage. http://aquacover.pt/. Accessed 09 Oct 2022
8. Gonçalves, J.M.: Práticas para o Desenvolvimento Sustentável - Conceção e projeto de uma arquitetura otimizada. University of Coimbra, Coimbra (2018)
9. Marinopoulos, I.S., Katsifarakis, K.L.: Optmization of energy and water management of swimming pools. In: International Conference on Sustainable Synergies from Buildings to the Urban Scale, SBE16, Science Direct, Thessaloniki (2017)
10. Zhao, J.: Whole system energy and cost optimization of residential swimming pool pumping systems. University of New South Wales, Sydney (2018)
11. Farinha, F., et al.: ECOPOOL +++ project - innovative heated outdoor swimming pools. In: International Conference Construction, Energy, Environment & Sustainability, Coimbra (2021)

Sustainable Bus Stop for Inclusive and Smart Cities

Manuela Pires Rosa[1], Nelson Sousa[2(✉)], João Rodrigues[3], Rui Cavaleiro[4], and Hugo Lamarão[5]

[1] CINTURS & Institute of Engineering, Universidade Do Algarve, Faro, Portugal
[2] Institute of Engineering, Universidade Do Algarve, Faro, Portugal
nsousa@ualg.pt
[3] ISR-Lisboa@LARSyS & Institute of Engineering, Universidade Do Algarve, Faro, Portugal
[4] Polytechnic Institute of Viana Do Castelo, Viana Do Castelo, Portugal
[5] CEiiA - Centre of Engineering and Product Development, Matosinhos, Portugal

Abstract. In this work it is presented a pilot study of an accessible, smart and sustainable bus stop was developed to be located at Faro International Airport, in Algarve, Portugal. Information and Communication Technologies (ICT) are a fundamental tool for sustainable and inclusive territories, being determinant for tourist destinations. As a result, a communicative stop/shelter module is proposed which will incorporate information to passengers and provide aid to people with reduced mobility, such as, timetable, Wi-Fi, Bluetooth, QR code, NFC technology, digital interactive panel, information panel in real time, points for charging mobile devices via a USB port, presence sensors and lighting intensity control. The operationality of all these equipment demands intensive energy consume. In order to promote the use of renewable energy in urban furniture, a photovoltaic system solution is studied for power-supply this inclusive and smart bus stop. At the existing bus stop of the Faro International Airport, local measurements of radiations were developed with a pyranometer. Several approaches in the integration of photovoltaic energy have been explored, demonstrating that the best solution is not a stand-alone system, but a smaller photovoltaic installation with the backing of the public power grid for periods of low photovoltaic production. The conclusions presented are based on economic analysis and environmental performance, particularly on avoided CO_2 emissions, in order to ensure the sustainability of the recommended solution.

Keywords: Photovoltaic Energy · Sustainable Bus Stop · Smart Shelter · Inclusive Bus Stop

1 Introduction

The 2030 Agenda for Sustainable Development presented by United Nations, in 2015, addresses relevant issues, energy related, to ensure a world where human habitats are safe, resilient and sustainable. One of the sustainable development goals is to "ensure access to affordable, reliable, sustainable and modern energy for all" and specifies the need

J. F. L. C. Semião et al. (Eds.): INCREaSE 2023, ASST, pp. 243–257, 2023.
https://doi.org/10.1007/978-3-031-44006-9_18

to guarantee universal access to energy services, to increase substantially the share of renewable energy in the global energy mix and to double the global rate of improvement in energy efficiency [1].

The Urban Agenda for the European Union, launched in 2016, stimulates growth, livability and innovation in European cities and promotes climate adaptation, energy transition, responsible and innovative public procurement, and sustainable use of land and nature-based solutions. Concerning energy transition, it promotes a long-term structural change in energy systems, i.e. shift to renewable energy and energy efficiency. With this purpose, communities must improve energy efficiency (e.g. in buildings), foster innovative approaches for energy supply (e.g. local systems) and increase the local production of renewable energy [2].

In 2020, the European Union (EU) energy mix, meaning the range of energy sources available, consisted of 34% of oil and petroleum products, 24% of natural gas, 17% of renewables, 13% of nuclear energy and 12% of solid fossil fuels. It imported 57.5% of the energy it consumed, as its own production and stock changes satisfied only 42.5% of its needs. The European Union imports more than half of the energy it consumes each year, with a particularly high level of dependency for crude oil and natural gas. Only one sixth (17%) of the EU-28's gross final energy consumption was from renewable energy sources. The sectors that consume the most energy are: the industry sector (32% of final energy consumption); followed by the transport sector (26%); households (25%); services (12%) and agriculture & forestry (3%). Around 23% of the final energy consumed is in the form of electricity and it comes from different sources: renewable energy sources accounted for 39% of the electricity and overtook for the first time fossil fuels (36%) as the main power source. In addition, 25% of the electricity came from nuclear power plants. Among renewable sources, the highest share of electricity came from wind turbines (14%), hydropower plants (13%), biofuels (6%) and solar power (5%). In 2019, the energy producing industries, fuel combustion by users and the transport sector all had the same share in total greenhouse gas emissions (25.8% each) [3]. Considering the solar potential of EU countries, the solar energy that can be produced is an opportunity to accelerate the current clean energy transition.

In Portugal, the production of renewable energy has been increasing, due to the plentiful supply of sun and wind, the reduction in PV panel installation costs and the imperative to invest in solar technologies. In fact, there is an excellent solar energy availability in Portugal, with global horizontal irradiation between 3.41 kWh/m^2 and 5.08 kWh/m^2 per day [4] and so, more investment on PV panel installations is advised. In 2019, in Portugal, fossil fuels accounted for 76% of total energy supply. Oil products were the largest energy source, accounting for 43% of total energy supply. Road transport accounted for most oil products demand (51%), followed by industry (16%) and oil-based building heating (5%) [5].

A diversity of actions has been implemented to support Portugal's goals for transport decarbonization, such as, road vehicle taxation, promotion of electric vehicles (EV), implementation of EV charging infrastructure, implementation of electrified passenger and freight rail, promotion of public transportation systems and cycle and pedestrian infrastructures.

In order to respond to the challenge of carbon-neutral mobility, high-quality public transportation systems are required, contributing, also, to the quality of life of citizens and livable, inclusive, and safer cities. Concerning public bus lines, the rehabilitation of bus stops has been carried out, replacing older shelters, implementing new ones, modernizing them, contributing for the revitalization of public space. In this process, over the world, some innovation has been emerging in the design of modern bus shelters. In the city of Utrecht the bus shelters' roofs are to be turned into gardens for bees and bumblebees and some of them produce energy obtained from photovoltaic panels [6]. In many other cities, photovoltaic panels are being applied to the bus stops, and other urban furniture, in order to reduce the polluting emissions from fossil fuels. This is the environmental context of the research Project ACCES4ALL - Accessibility for All in Tourism, developed at the Engineering Institute of the Universidade do Algarve, in collaboration with the Polytechnic Institute of Viana do Castelo and CEIIA - Centre for Engineering and Development located in Matosinhos. Its main objective is to develop a pilot study of an accessible, smart and sustainable bus stop to be located at Faro International Airport, in the Algarve, Portugal.

In the present digital age, many tourist destinations are keeping up with advances in technology and today offer unique experiences, real or virtual, to their visitors. Many smart cities, possessing cutting-edge technology and open data, are starting to constitute an example of Smart Tourism Destinations, being highly sought after by millennials. Besides the need for connectivity, the connected tourist also attends to sustainability and social responsibility. This is achieved through the use of Service Design tools, Internet of Things (IoT) and Big Data analysis. Information and Communication Technologies (ICT) are a fundamental tool for sustainability and social inclusion, being determinant for a Sustainable and Inclusive Tourism.

In this project, ICT is considered a key tool for promoting equity in access to information. This inclusive bus stop delivers the energy required by the commuters' commodities, such as lighting and an intelligent system that will be integrated inside the bus stop (smart shelter). To improve the bus stop sustainability, the integration of solar technologies was studied and assessed considering the effective carbon emission and the financial feasibility.

2 Case Study: the Project 'Accessibility for All in Tourism'

2.1 The Research Project

This study is a part of the research project 'Accessibility for All in Tourism' – ACCES4ALL – that, since 2017, has been developed at the Institute of Engineering of University of Algarve, in partnership with the Polytechnic Institute of Viana do Castelo and the Centre of Engineering and Product Development (CEiiA) with the objective of investigating the characteristics of sustainable pedestrian infrastructure and modal interfaces in tourist environments. The main focus of this study was to design an accessible, smart and sustainable bus stop for the Faro International Airport, Portugal, as an example of modal interface, to increase the accessibility of the tourists with disabilities and older adult tourists, promoting a more inclusive tourism.

Within the Institute of Engineering, three departments where involved: civil, electrotechnics and mechanical engineering. The first developed the urbanistic and product co-design of the bus stop, the second the smart dimension of the shelter and the third the photovoltaic design of the shelter's roof.

The final layout of the bus stop resulted from a co-design process where technical know-how was obtained through bibliographic review, expert opinion, collaborative workshops, walking schemes with persons with disabilities and older adult tourists survey results. This paper gives emphasis to the present energetic changing paradigm which promotes renewable energy supply and presents the photovoltaic design of the shelter's roof.

2.2 Smart Design of the Shelter

The ACCES4ALL project aims to achieve a holistic and inclusive perspective of bus stops. The implementation and design of the different concepts that determine a smart bus stop was guided by this objective. To define the ICT architecture behind the Internet of Everything (IoE) concept of a smart bus stop (SBS) it is necessary to acknowledge how many subsystems are needed and how they can interact. Despite the main part being the smart bus stop, the stop itself is only one of the subsystems. The other two are the "HUB" subsystem and the bus subsystem (Fig. 1).Thus, a pilot study of an innovative bus stop was developed by researchers and students of the Higher Institute of Engineering, including the components of the generation of energy, making it sustainable, the integration of information in the bus network to communicate, for example, the number of users waiting at the stop, and the creation of an interactive screen that adapts to the needs of any user through an innovative framework developed within this project. The urbanistic shelter design was developed in a co-design process with multiple players and volunteers.

The integration of an innovative Adaptive User Interface (AUI) system on a screen, installed at the bus stop, allows any user, even a tourist with large luggage, a fluid and customizable interaction that adapts to the user's characteristics in real time. This integrated framework is able to use a simple monocular camera to describe users, in a non-intrusive way, by means of computer vision capabilities and machine learning techniques. This innovative solution is prepared to recognise the human silhouette, pose and face, and from there be able to recognise and classify the person in different ways, such as height, age, gender, emotions, while also classifying their clothes/accessories, such as trousers, skirts, t-shirt, glasses, mobile phones, etc.

With the implementation of this system, it is also possible to predict when a person is with a cane, a characteristic of visual impairment, or accompanied by a guide dog, or in a wheelchair and other walking aid equipment. The fusion of all these technologies allows the creation of an inclusive stop, which adapts to any characteristic of the user who appears in front of the interactive screen. For example, for a visually impaired user, a voice communication system is automatically activated to allow the user accessing all the information. The screen and the LED panel (which indicates bus numbers and the times they take to arrive) has a glass surface with antireflective treatment to avoid glare and make reading possible. These real-time information systems are much appreciated by tourists, and particularly by people with hearing disabilities, as they provide information about the bus you want to catch. This information can also be provided in audio format, for people with visual impairment.

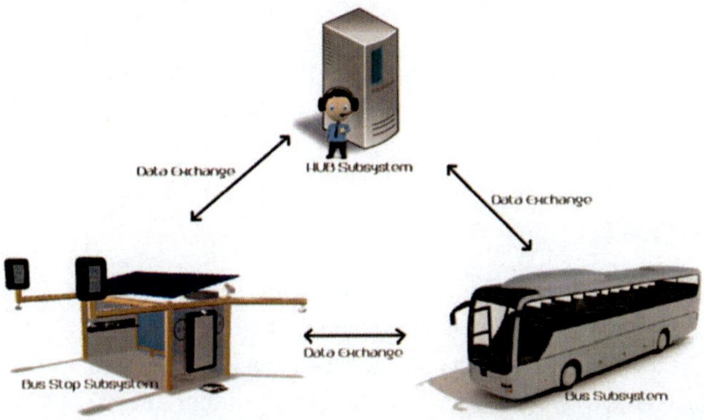

Fig. 1. Global Representation of the model Internet of Everything concept of a Smart Bus Stop [7].

Rain, humidity, temperature, air pollution and noise sensors are available at this type of stop, generating a synergy of useful and relevant information, both for the maintenance and extension of the useful life of solar panels and cell batteries and for sharing meteorological information with users or bus network operators.

There are also cameras installed in the area outside the smart bus stop to manage the number of people waiting for the next bus. Wi-Fi, Bluetooth, QR codes, NFC technology and interactive Braille are provided, allowing users to communicate which route they want to take. When there are no passengers waiting at the bus stop the camera communicates to the subsystem, allowing buses to avoid the stop when unnecessary, optimizing its efficiency.

2.3 Integration of Photovoltaic Production in the Bus Stop

To meet the bus stop energy demand, the photovoltaic technologies are the best option to be integrated into the proposed architecture, as they can be placed in the bus stop roof with minimal visual impact. The evaluation of the solution begins with the study of the solar availability, considering the shading effect by neighbouring buildings followed by the photovoltaic production and solar energy battery accumulation.

2.4 Solar Energy Availability

The amount of electricity produced by photovoltaic panels is influenced by multiple factors, such as the availability of sunlight and the efficiency of the panels themselves. Climate, along with geographical location, plays a significant role in determining the amount of sunlight that arrives on a surface and affects photovoltaic production.

Solar radiation, which reaches its highest value of 1353 W/m^2 outside Earth's atmosphere [8], is not fully absorbed by the Earth's surface due to reflection and absorption in the atmosphere. The latitude, day of year, and time of day affect the path of solar radiation in the atmosphere, and these parameters are used to estimate solar radiation. Other important factors include the inclination and azimuth of the photovoltaic panel with respect to the horizontal surface. Hourly radiation is modeled based on average daily radiation data from a database [9], considering direct radiation, diffuse radiation, and reflected radiation from the surroundings. The simulation results are compared with existing models [9] and local measurements made with a pyranometer to validate their accuracy.

2.5 Shading Effect

The models referenced earlier do not take into account the impact of shading caused by nearby buildings or the topography of the terrain. Shading can have a significant effect on photovoltaic panels, as shadows on one panel cell can affect the entire array of connected panels and the charge controller. This is particularly important to consider at bus stops, which are often located near buildings or trees. It is crucial to account for shading effects in such scenarios to accurately estimate the performance of the photovoltaic system.

Shading caused by obstacles can significantly impact the direct radiation received by photovoltaic panels. When the angle of the sun to the horizontal is less than the angular height of the horizon at a given solar time, the direct radiation is reduced to zero. There are various methods to estimate the shading penalty, such as Solmetric and Solarpathfinder [10, 11]. In this study, a fisheye lens photographic technique was used, and custom software was developed to determine the angular height of the horizon across the azimuth. The most unfavourable location for shading on the bus stop roof was identified by capturing a photograph with a 180-degree amplitude using a Samsung® nx-mini camera with a 1" sensor and a Fujinon® fish-eye 1:1.8/2.7mm lens as depicted in (Fig. 2).

In order to simplify the identification of the skyline, the captured image was processed by converting it from polar to cartesian coordinates, resulting in a panoramic image that preserves the angular relationship. In this image, the base corresponds to 0° and the top corresponds to 90°, as illustrated in (Fig. 3). This transformation allows for easier analysis and identification of the horizon line in the image.

For the bus stop's orientation, which is about 35° east with a panel slope of 20°, it has been observed that shading during the early hours of the morning results in a 7.3% reduction in solar energy availability. The chosen panel slope corresponds to the tilt of the bus stop roof, ensuring better architectural integration of the photovoltaic system. Despite neighbouring shading and azimuth, this slope was found to be the most optimal.

However, under ideal conditions with south-facing orientation and no shading, the best slope would be 35°. When compared to ideal conditions, the orientation penalty for the chosen 20° slope is only 3.8% as presented in (Fig. 4). In fact, the 20° slope benefits solar energy availability during summer months but may result in reduced availability during winter months, which needs to be considered in the integration of the solar system into the bus stop cover.

Fig. 2. Photograph at the bus stop location in the zenith direction with 180° aperture.

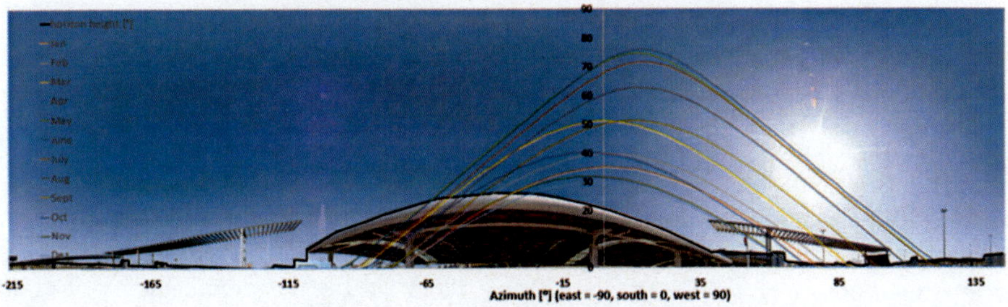

Fig. 3. Panoramic image of the study site, obtained by transforming the image of Fig. 2 and horizon lines and the apparent trajectory of the sun for the average day of each month, considering the legal time.

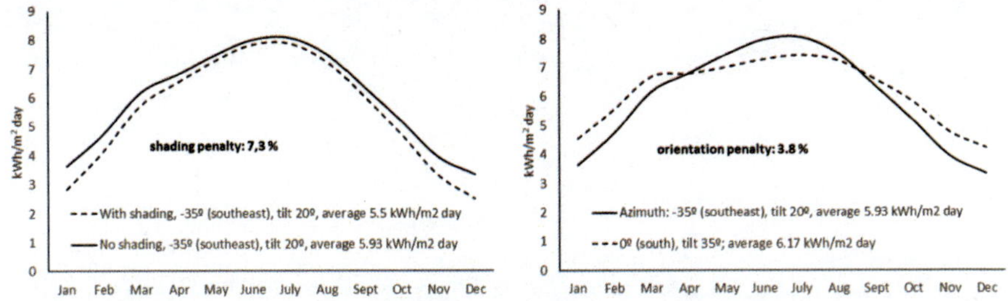

Fig. 4. Penalty of the shade surrounding the bus stop on available solar energy; Penalty of the azimuth of the bus stop and inclination of its coverage in the available solar energy.

2.6 Photovoltaic Production and Energy Storage

Two scenarios were analysed: one where the photovoltaic production area is optimized based on economic considerations, and another where a self-sustaining solution capable of supplying all the electrical energy to the system is sought. For photovoltaic production, the five-parameter and one-diode model [12] was used. This model estimates the electrical power output of the photovoltaic panel based on weather conditions such as solar radiation and environmental temperature.

The model consists of two steps. First, three parameters (m, R_s, R_{sh}) are determined, which correspond to the diode Ideality factor, series resistance, and parallel resistance, respectively. Next, considering the actual temperature and solar radiation conditions, the voltage and current are determined, corresponding to the maximum power point that the charge controller connected to the panels will look for when operating the photovoltaic system.

In the case of this bus stop, the energy consumption or load demand (as shown in Fig. 3) occurs throughout the day, with higher demand at night due to lighting. Batteries are used to store the excess energy generated during the day for use at night, with a daily storage period considered. The efficiency of the battery system, which is approximately 80% for charge and discharge cycles, needs to be taken into account when considering storage.

2.7 Photovoltaic Economic analysis

To determine the optimal configuration of the photovoltaic system, which includes panels, batteries, charge controller, and inverter, the net present value (NPV) is calculated for different system setups. (Fig. 5) displays the results of various simulations for NPV based on different configurations of battery system capacity and number of panels. With a discount rate of 2.5%, the best configuration is found to be using 4 photovoltaic panels with a rated power of 245 W (reference panel) and a battery system capacity of 1000 Wh. The photovoltaic production of this system is shown in (Fig. 6).

Fig. 6. Bus stop load demand and photovoltaic production for a 4-panel power station. Data presented for a middle day of January and June.

Fig. 5. Net Present Value (NPV) depending on the number of panels and the storage capacity of the batteries.

In Fig. 7, 8 and 9 are depicted the production, accumulation, and consumption maps from the electrical grid for the chosen system. Considering that the purchase of electricity is carried out at low voltage, with an undifferentiated tariff throughout the day, the programmed algorithm proposes the use of the energy accumulated at the beginning of each day. This scenario would be different if the tariff were not constant.

Another study carried out was the determination of the maximum power of the electrical and electronic systems of the bus stop capable of being satisfied by an autonomous photovoltaic system. It was considered that the entire roof of the bus stop could be covered by photovoltaic panels, in this case 15 panels. For this scenario, it is necessary to study the most unfavorable month for production, i.e., December. It is verified that the total power of the equipment must not exceed 325W. During the summer months, much of the photovoltaic production potential is not used, and annually this system would remain at 52% of its production capacity. For the autonomous system, due to its nature of being disconnected from the electricity grid, the sale of surplus was not considered. In previous simulations, it was confirmed that the increase in value in the initial investment does not justify the sale of the surplus to the network, at current prices.

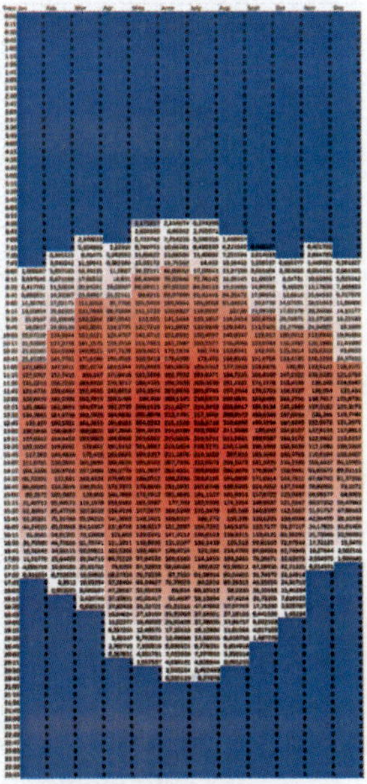

Fig. 7. Photovoltai c production [Wh]

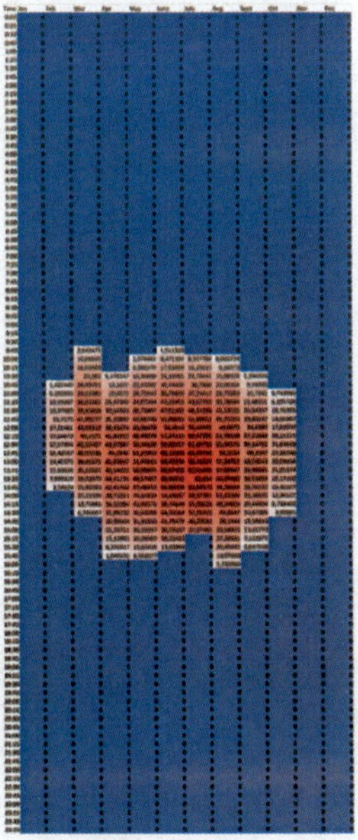

Fig. 8. Battery storage [Wh]

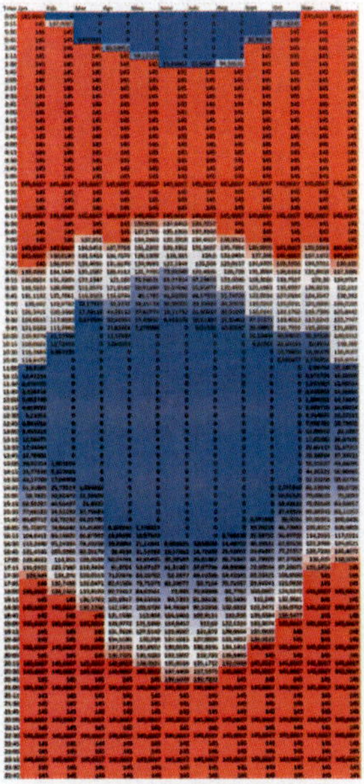

Fig. 9. Electric grid demand [Wh]

2.8 Reduction in Greenhouse Gas Emissions

Looking at the environmental perspective, using photovoltaic electricity can help avoid energy consumption from the public grid, which may have associated CO_2 emissions, depending on the source. Eletricidade de Portugal (EDP), the main electricity supplier in Portugal, reports a value of 263g of CO_2 per kWh for business customers in 2023 [13]. While renewable energy from sources like the sun is considered sustainable, it still requires equipment for capturing it, and studies by various authors have pointed out the equivalent emissions for photovoltaic energy. However, these values tend to vary significantly depending on the authors and study dates. Recent studies suggest an average value of 69 g of CO_2/kWh [14–16], with a range of 60.1 to 87.3 CO_2/kWh.

In real terms, photovoltaic production is not performed at panel full capacity. Depending on the installation, e.g., orientation of the panels, and also on the energy production and utilization thereof. In south-oriented and grid-connected installations with over-sale, photovoltaic production tends to be the maximum. So, considering 69 g of CO_2/kWh for full photovoltaic production over 20 years, the 4-panel solution with southeast orientation and some shading, the non-production equals 11%, so the equivalent CO_2 emission will rise to 77 g/kWh. For the 15-panel solution, assuming the maximum power of the equipment is 325 W, the penalty goes to 268%, so the equivalent CO_2 emission will go to 185 g/kWh.

For batteries, the equivalent emissions over their lifetime are even more variable [17]. CO_2 emission values vary greatly, ranging from 36 to 313 g per kWh of capacity [18], taking an average value of 110 g per kWh. Quanwei et al. [19] presents values between 42 to 140, taking an average value of 91.21 CO_2/kWh, and the justification is mainly based on the origin of the energy used in the manufacturing processes.

Considering the value of 110 kg/kWh, assuming a life cycle of 8 years and adopting the proposed installation, i.e. 4 panels with 245 W peak power coupled with a 1000 Wh battery accumulation system, for a solar fraction of 35%, an equivalent CO_2 reduction of 3434 kg is achieved. The balance of CO_2 emissions for the self-sustaining solution, with 15 panels, cannot be directly compared because in the months of lower solar radiation, solar production does not satisfy the load diagram of the equipment used, forcing a reduction in consumption as shown above.

2.9 Bus Stop Layout for a Tourist Transportation Hub

The proposal of shelter/accessible bus stop stands out from the others existing bus stops by its specific location and wide range of its potential users. In fact, given the characteristics of its location/implantation (access or exit from Faro airport), its area of fruition differs from conventional and current stops, having the freedom to expand its space without any kind of constraints. The specific character of centrality or useful interface and the type of users coming from the various flights allow the development of a differentiating and dynamic design in its function. Its dimensions are different from conventional bus stops which, due to the characteristics of public space in Portugal and its surrounding conditions, usually limit its design, such as the presence of traffic signs, minimum width of pavements, and other functional characteristics of urban design established by norms and standards.

This way, it was intended to develop a proposal for equipment that is close to the concept of "inclusive transportation hub", containing the necessary area to meet the volume of users that every day pass through (and to) the airport, distinguished from the others, not only by its shape and volume, but also by the coupled equipment that will ensure the use and dynamic comfort of its use, transforming it into a more accessible and inclusive stop.

From the different working groups that participated in the co-design process (floor study, study of digital panels, study of photovoltaic panels, study of the type of seating, study of spaces for people with disabilities, etc.), specific information wasinflowing, necessary for the improvement and maximization of the design and elements of this inclusive bus stop. These inputs were included in the conceptual design proposed for the bus stop/shelter under development, serving as a virtual model that seeks to respond effectively to the needs of different users. The final design (Fig. 10 materializes, in a pragmatic way, all the contributions from the different stakeholders.

Fig. 10. Proposed an environmental and inclusive bus stop for a tourist transportation hub

Thus, the module and proposal of the stop/shelter intends to respond to the various types of technical and functional needs that an equipment of this type requires, complying with ergonomic rules and other regulations/laws applicable to this case. From a material and functional point of view, it is intended that, it is prepared to, besides receiving and accommodating waiting passengers, protecting them from the weather/climatic environment, be able to inform them, in real time, of transport timetables in Faro and the Algarve region and, simultaneously, incorporate various support systems to promote social inclusiveness.

The stop/shelter module proposed incorporates various waiting access areas: seats with armrests and backs and a seat with ischemic support, a timetable, QR code about the bus timetables, NFC technology and a digital interactive panel; presence sensors and lighting intensity control; space to accommodate people travelling in wheelchairs (in this case, it was decided to ensure a diameter of 1.80 m instead of the 1.50 m required by law, to make maneuvering easier); alert tactile strip at the edge of the raised waiting platform; at the embarkment area, an alert with neoprene rubber is contrasting colour to their surrounding environment; transparent panels have colour strips; photovoltaic; and telecommunication dispositive. These elements of support and reinforcement of the functional space were planned and came about in order to comply with the stipulations of Decree-Law 163/06 of the 8th of August.

It is foreseen that a great percentage of passengers travel with hand luggage, reason for which the area and dimensioning of several waiting and circulation spaces were enlarged. The dimensions recommended in the plan are: 2.30m x 8.00m as depicted in (Fig. 11).

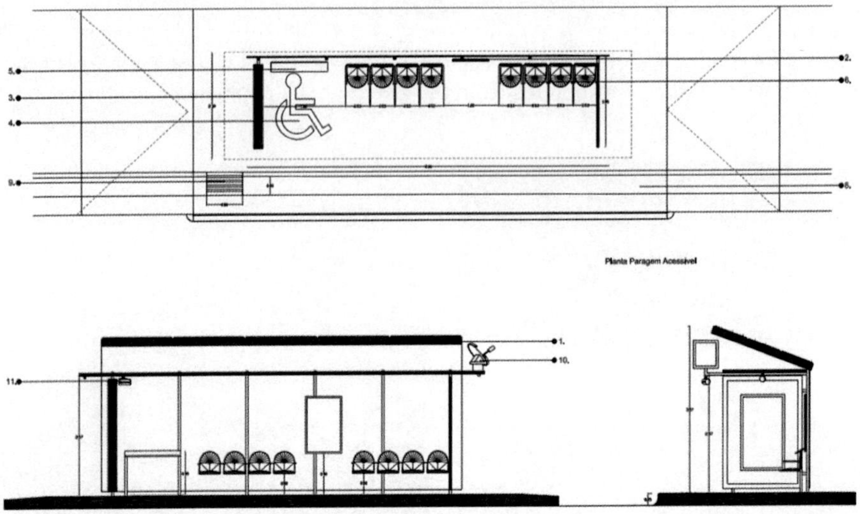

Legend:

1. Photovoltaic panels
2. Timetable, Braille information, QR code and NFC technology
3. Digital interactive panel
4. Space to accommodate wheelchair users
5. Seat with ischemic support
6. Seats with armrests and backs
7. Barrier-free space
8. Alert tactile strip at the edge of the raised waiting platform
9. Alert with neoprene rubber at the embarkment area
10. Telecommunication dispositive, LoRaWAN technology
11. Surveillance camera

Fig. 11. Proposed layout of the bus stop for a tourist transportation hub

The real-time information panel transmits general information also adapted to people with hearing and visual impairment, being foreseen the development of a specific software for this purpose. To improve the visually impaired experience, a tactile pavement line will be applied across the width of the shelter and near the limit of the waiting platform. In this case, it is extremely important to use strong and differentiated colours for better visualisation at a distance and identification of the seating and circulation areas.

In addition to these premises, there is a photovoltaic system on the roof to supply energy for the interactive smart board providing content and services), for free Wi-Fi, the planned LED lighting, as well as two (2) points for charging mobile devices via a USB port, which will be strategically located so as to allow good circulation/mobility of people (with or without luggage).

Shelters should have a minimum average illumination of 50 lx measured at floor level during service hours, and a minimum value of 100 lx at boarding time [20] as persons with vision impairments need more light to overcome gradients safely. The presence of adequate light levels also allows persons with hearing impairment greater visibility for lip and sign reading. The existence of a bus passage sensor is envisaged, which, when the bus stops, triggers an increase in light intensity.

On the side panels there will be additional digital information in the form of an interactive map, and on the back panel there will be updated information and a more active viewing colour strip for people with impaired sight. These panels should have a design which distance from the ground is no more than 10 cm, to allow its detection by visually impaired persons. This shelter/stop module should have a brand/logo, in this case, the number three associated with the letter i - 3i - to convey the conceptual idea of innovative, intelligent, and inclusive mobility equipment.

It is significant to mention that the public space of access to the interior of the stop/shelter should be free and unobstructed from any kind of obstacles, either natural or artificial, the reason why the choice of its location will be fundamental for the success of its function.

Regarding the materials to be used, it was also consider the risk of vandalism, opting for this reason, for a structural system in lacquered aluminum, and side and rear panels in laminated glass and composite panel, respectively. The roof will be in translucent plastic glass and sandwich panel structurally reinforced with transversal aluminum props, to receive and accommodate the required photovoltaic panels necessary to guarantee energy autonomy. For its implementation it will be necessary that the site provides the necessary electrical and digital connections for the correct functioning of this equipment.

3 Discussion and Conclusions

The European Union Solar Energy Strategy demands that solar photovoltaics and solar thermal technologies be rolled-out rapidly, and rewards citizens and businesses with economic and environmental benefits [21].

In many cities, photovoltaic panels are being applied in bus stops and other urban furniture, in order to reduce the polluting emissions from fossil fuels.. According to Aguilera et al. [23], in transport shelters, the solar energy collection system may have photovoltaic modules, the voltage regulator, the inverter and an accumulation system that supplies the LED type lighting (20W LED luminaires are used with a brightness of 1,800 lm at a 90° angle), the bus stop signalling system and the backlit advertising sign on the side of the shelter, and no external power supply is required [21].

However, these energy self-sufficient bus stops require simple elements with low energy dependence. In a context of smart cities, such as London, Paris and Madrid, at many bus stops, there are contactless NFC (Near Field Communication) and QR codes (Quick Response) to provide users with information through their smart mobile phones, as well, free Wi-Fi and dynamic information screens. Technological systems, very energy-intensive, are needed to ensure connectivity between environmental sensors, wireless modules, processors and microcontrollers.

In the ACCES4ALL Project proposals for contemporary shelters were developed, adapted to the current needs of its various users (in transit), especially citizens with impairments and the older adults, using the most sophisticated systems, with integrated and inclusive design, contributing to make this equipment versatile, environmentally friendly and inclusive. A smart bus stop should have four essential features1 (a) added communication, (b) creativity, (c) real-time responses to changes in bus schedules and (d) attractive and eco-friendly designs. One more can be highlighted: (e) uniqueness and inclusion - the adaptation of the smart bus stop to each (unique) individual promoting inclusion and autonomy.

A renewable energy solution is integrated into the intelligent bus stop. The best integration configuration of photovoltaic collectors and storage batteries was studied to ensure the supply of energy with the maximisation of economic gains and the minimisation of equivalent carbon dioxide emissions. For the suggested electronic components, the suggested photovoltaic system can avoid 3434 kg of CO_2 in a 8 year period. The best solution is obtained for a self-sufficient system in the summer months, with grid support the rest of the year.

In the replication of this model, it is important to study the solar potential of the place and consider the impact of shading caused by nearby buildings or the topography of the terrain. Shading can have a significant effect on photovoltaic panels, as shadows on one panel cell can affect the entire array of connected panels and the charge controller. This is particularly important to consider at bus stops and other urban furniture, which are often located near buildings or trees. It is crucial to account for shading effects in such scenarios to accurately estimate the performance of the photovoltaic system.

These sustainable and smart bus stops can encourage greener ways to travel. Promoting photovoltaic panels that guarantee renewable energy in shelters will attract more users and consequently contributing for sustainable mobility.

Acknowledgment. The Project ACCES4ALL – Accessibility for All in Tourism (SAICT-POL/23700/2016) was sponsored by the Portuguese Science and Technology Foundation (FCT) through Portugal 2020, operation number Projeto n.° 023700 | AAC n.° 02/SAICT/2016, and by CCDR Algarve and CCDR Norte, co-funded by European Regional Development Fund (ERDF), through Regional Operational Programme of CRESC Algarve 2020 and Regional Operational Programme of Norte 2020. We also thank VINCI Airports/ANA Aeroportos de Portugal.

References

1. United Nations: Transforming our world: The 2030 agenda for sustainable development, United Nations, New York (2015). https://sustainabledevelopment.un.org/post2015/transformingourworld
2. E: Urban Agenda for EU 'Pact of Amsterdam', Agreed at the Informal Meeting of EU Ministers Responsible for Urban Matters on 30 May 2016 in Amsterdam, Netherlands, Dutch Presidency of the Council of the European Union (2016)
3. EUROSTAT: Shedding light on energy in the EU 2022 Interactive Edition (2022)
4. Global Solar Atlas. https://globalsolaratlas.info. Accessed 21 Mar 2023
5. International Energy Agency Portugal 2021. Energy Policy Review. https://www.iea.org/reports/portugal-2021. Accessed 21 Mar 2023
6. Gawryluka, D., et al.: Green bus stops. In Flórez, M.A., Pilar, C.I., Gawryluk, D. (eds.) urban public spaces: madrid, bialystok, klaipeda. a guide to their functions and meaning, pp. 66–84. Universidad Politécnica de Madrid, Madrid (2022)
7. Viegas, M.: Smart bus stop: people counting in a multi-view camera environment. Master Thesis in Electrical and Electronics Engineering Specialisation in Information Technologies and Telecommunications. Developed under the orientation of Prof. Doutor João M. F. Rodrigues, Universidade do Algarve, Instituto Superior de Engenharia (2019)

8. John, A.D., Beckman, W.A.: Solar Engineering of Thermal Processes. Wiley, Hoboken (2013)
9. European Union. https://re.jrc.ec.europa.eu/pvg_tools/en/tools.html. Accessed 21 Mar 2023
10. Solmetric. Solmetric. http://www.solmetric.com/. Accessed 21 Mar 2023
11. Solarpathfinder. https://www.solarpathfinder.com/PF. Accessed 21 Mar 2023
12. Rui, C.: Uma Introdução às Energias Renováveis. Lisboa: IST (2011)
13. EDP. https://www.edp.pt/origem-energia/. Accessed 21 Mar 2023
14. Fu, Y., Liu, X., Yuan, Z.: Life-cycle assessment of multi-crystalline photovoltaic (PV) systems in China. J. Clean. Prod. **86**, 180–190 (2015)
15. Eskew, J., Ratledge, M., Wallace, M., Gheewala, S.H., Rakkwamsuk, P.: An environmental life cycle assessment of rooftop solar in Bangkok, Thailand. Renew. Energy **123**, 781–792 (2018)
16. Hou, G., et al.: Life cycle assessment of grid-connected photovoltaic power generation from crystalline silicon solar modules in China. Appl. Energy **164**, 882–890 (2016)
17. Dale, H., Lutsey, N.: Effects of battery manufacturing on electric vehicle life-cycle greenhouse gas emissions. The International Council on clean Transportation (2018)
18. Peters, J.F., Baumann, M., Zimmermann, B., Braun, J., Weil, M.: The environmental impact of Li-Ion batteries and the role of key parameters – a review. Renew. Sustain. Energy Rev. **67**, 491–506 (2017)
19. Chen, Q., et al.: Investigating carbon footprint and carbon reduction potential using a cradle-to-cradle LCA approach on lithium-ion batteries for electric vehicles in China. J. Cleaner Prod. **369**, 133342 (2022)
20. Government of Dubai: Dubai Universal Design Code. The Global Alliance on Accessible Technologies and Environments (2017)
21. European Commission: EU Solar Energy Strategy, COM (2022) 221 final Brussels, 18 May 2022 (2022)
22. Aguilera, P., Bach, I., Piña, C., Flórez, M.A., Sepulcre, A., Valiente, M.M.: Transport shelters and bus stops, Madrid. In: Flórez, M.A., Pilar, C.I., Gawryluk, D. (eds.) Urban Public Spaces: Madrid, Bialystok, Klaipeda, a Guide to Their Functions and Meaning, pp. 28–48. Universidad Politécnica de Madrid, Madrid (2022)

The 7D BIM Model Used in the Maintenance of Buildings

Alcinia Zita Sampaio[1,2](✉), Inês Domingos[1,2], and Augusto Gomes[1,2]

[1] Department of Civil Engineering, University of Lisbon, Lisbon, Portugal
{zita.sampaio,ines.domingos,augusto.gomes}@tecnico.ulisboa.pt
[2] Higher Technical Institute, Av. Rovisco Pais, Lisbon, Portugal

Abstract. The Building Information Modelling (BIM) methodology presents as a basic concept the centralization of all information created along the development of a construction design. In the occupation building phase, the maintenance activity has been improved concerning the retrieving of data within the database of the BIM model associated to the building, and the handling of data used in a maintenance perspective. The establishment of adequate maintenance strategy concerning a building requires a deep knowledge regarding the durability of the materials, mainly the degradation perdition of the materials, and naturally, the costs involved. The study illustrate how to act over a set of the most used as finishing of wall facades and roof terraces. In order to to support the estimation of the useful life of the selected materials, two Dynamo scripts were created. The developed approach, allows engineers, in an early design stage, to prepare adequate maintenance plans, based on the knowledge of durability, useful lifetime and the degradation perdition of the materials. The work contributes positively to the implementation of BIM in the maintenance activity, with relevance to the high potential of the Dynamo tool when information management within BIM models is required.

Keywords: BIM · Maintenance Planning · Degradation Prediction · Roofing Material · Wall's Finishing · Dynamo Script · Python Programming Language

1 Introduction

The Construction industry have been progressively incorporate the implementation of the Building Information Modelling (BIM) methodology, supported in the permanent and recent dissemination of its benefits in reports and in the advances of the required technology [1]. This methodology reveals a new and different way of working in the construction sector, when compared to traditional methods based in two-dimensional (2D) technical documents. The BIM work transmits more transparency, better communication and a greater level of collaboration between all partners involved in the development of building projects [2]. When developing a project, using BIM platforms, a BIM model representing a digital three-dimensional (3D) representation of the physical and functional characteristics of a building is created. The model brings together the information produced in the various phases of the life cycle forming a complete database supporting decision makers [3].

J. F. L. C. Semião et al. (Eds.): INCREaSE 2023, ASST, pp. 258–267, 2023.
https://doi.org/10.1007/978-3-031-44006-9_19

The Facility Management (FM) activity, in a building, covers several disciplines in ensuring the functionality of the built environment integrating people, places, processes and technologies [4]. In this practice, BIM integration is still unusual (Sampaio et al., 2016). However, it is recognized by several authors, who emphasize the ability to gather continuously during the various phases of design and construction, all the information necessary for FM, which allows improving the efficiency and productivity for the application of management processes [5]. The ability to store a large amount of information and keep it accessible and updated with the facility of connecting with several FM systems, make BIM solutions very attractive. However, retrieving data from the model requires experience and relevant waste of time [6].

The present study intends to contribute to use the database of the model and its application on the maintenance activity [7]. The text introduces the current practices in BIM-FM and the main advantages and limitations in using BIM platforms. The study describes the development of two Dynamo scripts created to improve the integration of BIM models to FM functionalities.

2 BIM for FM: An Overview

Although available for a few decades, it was only in recent years that BIM and its concepts began to spark more interest in the construction industry, extending their application to the entire life cycle of buildings. The types of building information that might be required in a model to manage the design, construction and operation of built assets can be described as dimensions [8]: 3D model data (3D); time data (4D); costs (5D); sustainability (6D); facility management (7D); safety and accident prevention (8D). So, the 7th dimension comprises the application of BIM methodology in the Facility Management activity. It intends to gather all the information that has been developed and updated throughout the project, representing the as-built model of the building [9].

As the generation of a BIM model is based on the use of parametric objects, each component of construction includes data such as the type of materials applied in the building and its proprieties. The information required in the maintenance activity can ranges from manufacturer data, technical information on the products, warranties, user manuals and expectable service life among others necessary to support these activities [10]. The main challenge is to integrate the accumulated information with the data acquired during the design and construction phases and provide the building management systems with relevant information at the time of operation [11]. In it, the degree of interoperability performed between different software is the main limitation in applying BIM in the operational phase of the building. However, using the native data format of the modeller system, the process of information transfer from BIM model to FM systems can be effectuated without loss of to much information. Additionally, the model can be permanently updated, allowing the facility manager to update the building's operation services [12].

The growing number of researches that address the BIM-FM topic is an indication of the industry's interest in its application. However there are some barriers for its adoption [13]:

- A lack of guidance, protocols and standards for BIM FM;

- Technical expertise needed for the maintenance and updating of the BIM; limited software interoperability;
- Unclear BIM-FM requirements at early project stages which reflects the fragmented nature of the construction industry.

Therefore, it is possible to contribute to the reduction of investment costs and better facilities management services in a building [14].

3 Service Life Prediction

Over time, buildings and their components tend to deteriorate, naturally, and lose some of their original properties. The useful life or period of service of a building corresponds to the period of time during which its properties respond to or exceed the minimum acceptable levels for its operation of intrinsic order to the element, normative or subjective, in a situation of current maintenance [15]. The degradation of a component is a consequence of the natural aging process, associated with exposure to adverse environmental conditions. The evaluation of the degradation level is based on mathematical models, which consider several types of variables, and when properly adjusted, allow estimating the remaining lifetime of the element under analysis.

The awareness of the relevance of buildings' durability for a more sustainable construction led to the establishment of several methodologies for the prediction of a building's service life. The following prediction methods can be considered deterministic, probabilistic (or stochastic) and engineering methods.

The factor method fits into the deterministic methods as a simplified approach to support the prediction of the service life of construction components. This method consists in the prediction of a value for the estimated service life (ESL) of a given construction element, obtain by the product of the reference service life (RSL) value, under standard conditions, by several deterministic factors (Table 1):

Table 1. Deterministic factors required to estimate service life (ESL).

ESL = RSL × factor A × factor B × factor C × factor D × factor E × factor F × factor G		
factor A – quality of components	factor B – design level	factor C – work execution level
factor D – indoor environment	factor E – outdoor environment	factor F – in-use conditions
factor G – maintenance level		

- The RSL value is influenced by data provided by the manufacturer, homologated documentation or thematic bibliography, and by the collection of information verified in similar cases under identical conditions of use;
- The factors should have values between 0.8 and 1.2 where: the unit value is assigned when the factor has no weight for the element under study; a higher value corresponds

to an increase of the ESL; a lower value is reflected in a decrease of the estimated service life.

In the present study, the bibliographical references of Raposo [16], Lopes [17], and Matos [18], were consulted, where three cases of application of the factor method according to the guidelines of the ISO 15686–1 standard were considered. In each case a durability matrix is presented, developed by each author listed in tables, containing the factors applicable to a set of construction components: cladding applied on flat roofs (Table 2); adherent ceramic tiling on façades; ventilated façade tiling. The values of these tables are used in the developed Dynamo script.

Table 2. Factor A of the durability matrix of cladding applied on flat roofs [16]

Factor		Description	Índice
Factor A	A1	Declaration of Conformity and Quality Certificate	
		With EC Declaration of Conformity and Quality Certificate	1,2
		With EC declaration of conformity or Quality Certificate	1,0
	A2	Characteristics of thermal insulation of the cover	
		With ISOLE rating higher than recommended for coverage type	1,2
		With the recommended ISOLE rating for the coverage type	1,0
		With ISOLE rating lower than recommended for coverage type	0,8
	A4	Type of bituminous membrane armor	
		Polyester felt of 250 g/m2 or armed with polyester and fiberglass felts	1,2
		With polyester felt of at least 150 g/m2	1,0
		With polyester felt less than 150 g/m2 or with fiberglass felt	0,8

4 Visual Programming in Dynamo

An adequate performance of buildings is becoming an important aspect to be studied by professionals involved in the construction process, concerned both in the short and long term, in terms of costs and durability, with the implicit need for the maintenance process to be discussed at an early stage of the construction process. In this context, the present study propose to include the factor method for estimating the service life of elements in construction. The study involve the development of two scripts, defined using the visual programming tool Dynamo (Dynamo, 2020): the "Estimate Service Life" script and the "ESL Analysis". The scripts were applied over a case study allowing to verified its efficiency.

4.1 Dynamo Script "Estimate Service Life"

As a data base to be used on development of the script "Estimate Service Life" are the durability matrices of the tables of the referred authors. These were integrated in the script so that the options of each factor can be selected according to the element and material used in order to estimate the respective service life value. The script can run over projects modelled in Revit and applied over the considered constructive elements: roof waterproofing membranes; adherent ceramic tiling on facades; ventilated facade claddings. To script "Estimate Service Life" was created using Dynamo visual programming which main structure is represented in Fig. 1.

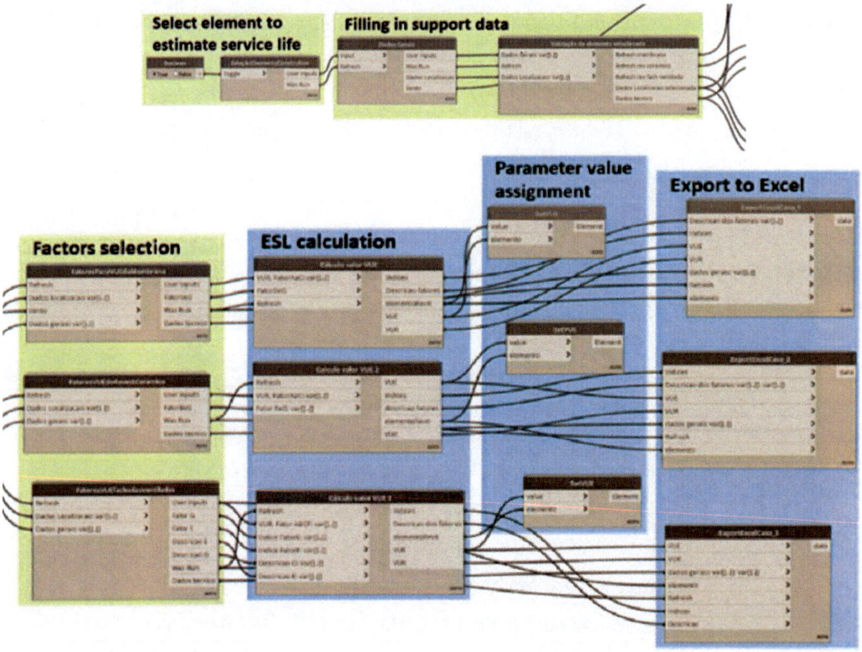

Fig. 1. Dynamo script "Estimate Service Life"

- In order to assist the filling of the factors (in green), a database was created using the DataBase Browser software and the nodes present in the Slignshot package were used to link the database to Dynamo. Depending of the parameters, such as the location of the project, the connection to the database provides data relating to the characteristics of the outdoor environment (summer and winter climate zone, rainfall zone, average annual number of days with frost and solar radiation index) essential to characterize the factor E;
- The groups of nodes (blue) have been programmed to calculate the service life according to the factor method (depending on the indexes selected in each interface); assign the calculated value to the project parameter in Revit; and export information to the selected Excel file.

4.2 Dynamo Script "ESL Analysis"

The "ESL Analysis" script was created as a complement to the previous script, and its use requires that the first script have already been executed. Thus, for the elements of the model to which the ESL is assigned it is possible to make a prognosis of their degradation state for a given year. For this, a linear degradation of the elements was assumed, where a set of colours was assigned according to the range in which the *% of lifespan* is obtained (Fig. 2).

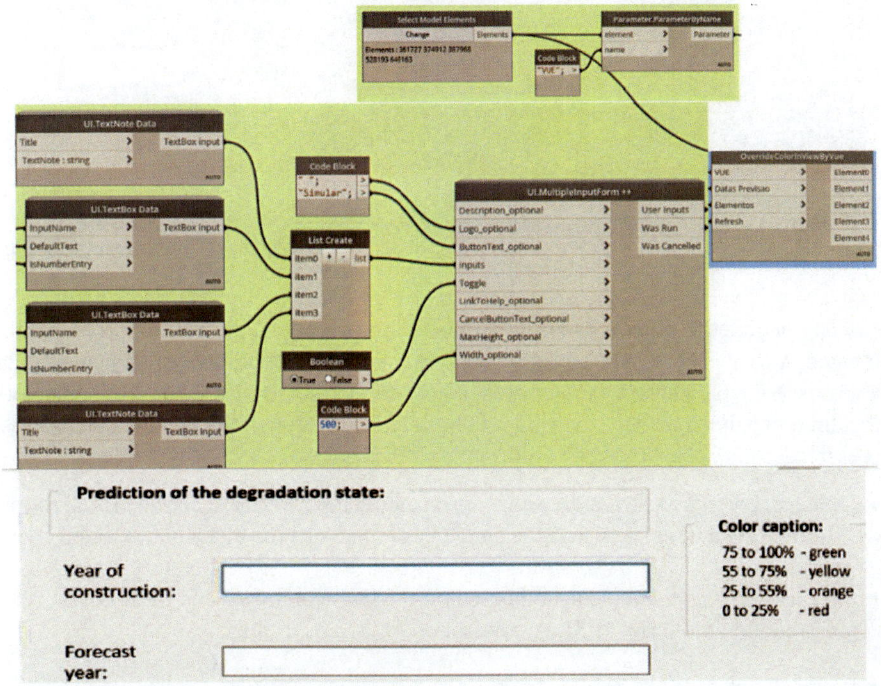

Fig. 2. Dynamo script "ESL analyses" and the assignment of colours.

Depending on the interval in which the calculated percentage is inserted, the element displays the colour assigned to the interval. The green colour corresponds to an initial degradation state, while the red colour is assigned to an advanced degradation state.

5 Case Study

The selected case study corresponds to a building integrated in a private complex, located in the Lisbon area, with 5 elevated floors, 3 floors in semi-basement and roof terrace. It presents two blind warps oriented to West and Northeast. The BIM architectural model was created using the Revit software (Fig. 4):

- The main façade, facing north, consists of a double wall in **brick masonry**, and is covered with adherent tiles in porcelain stoneware;

- The Southeast oriented façade, a solution in **ventilated façade** clad with fibre cement plates;
- On the roof terrace was applied, as a finishing material, the **waterproofing membrane** in APP polymer bitumen self-protected with mineral granulate.

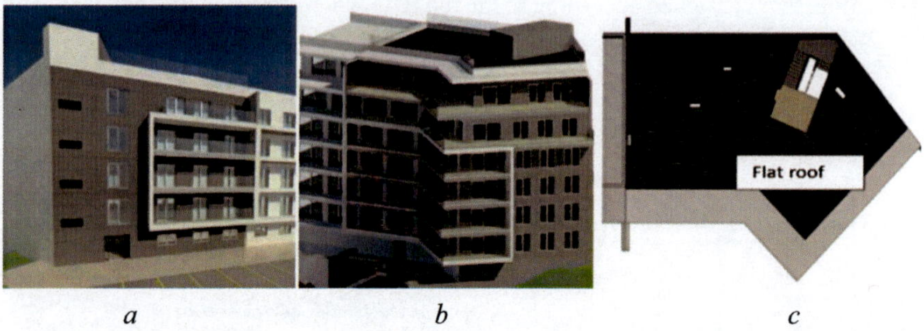

Fig. 3. Main façade with tiles (*a*), rear façade with fibre cement (*b*) and flat roof (*c*).

The parametric objects used in the modelling process of the building were incremented with the necessary parameters related to the ESL of the components under analysis. New parameters were created, associated to the materials, in order to be able to admit, in a dynamic way, the values to be calculated through the two Dynamo scripts presented:

- **ESL**, receives the service life value estimated by the "Estimate Service Life" script;
- **Maintenance Excel**, registers the directory of the Excel file that when selected opens the file;
- **% lifespan**, variable calculate\d based on the ESL parameter and the forecast year, controlled in the script "ESL Analysis".

5.1 Application of "Estimate Service Life" Script

The execution of the script is running through Dynamo Player option included in the upper menu of Revit. The script was successfully executed to estimate the service life of each of the materials identified and the result is visualised in Fig. 4.

A table with the numerical results can be accessed and the aspect of the deterioration is illustrated by the colour assumed by each element. A period of ten years, from year 2020 and 2030, were filled as an example to simulate the year of construction and the year for life forecast respectively.

5.2 Application of "ESL Analysis" Script

To simplify the use of this script, the 3D view of the model was duplicated with an associated filter so that only the objects that contain the ESL parameter are visible. The new view was called "Life Analysis", and in this way the assignment of colours to the objects of the model do not overlap with the colours of the original model. This view

Technician's name:	Inês Domingos
Date:	2020
Element category:	Wall
Element type:	Ventilated wall
Fixing system:	At sight
Maintenance target element:	Fiber cement plates 180x100cm
Year of construction:	2020
Reference service life (RSL):	30
Estimate service life (ESL):	34,6
Total facade area [m²]:	219
Replacement cost [€/m²]:	96
Clean cost [€/m²]:	19,49
Inspection cost [€]:	70

Fig. 4. Visualization of the degree of deterioration and respective values.

is selected to run the script in order to see the life effects in a certain period. When the script is executed, all the elements present in the mentioned view are selected, which include the two facades and the roof under analysis.

When the script is completed, it is possible to observe the elements with a distinct colouring, represented in the view created. The result of the inserted values is the parameter % lifespan (automatically field) calculated for each element in Fig. 4. It can be observed that the wall in porcelain stoneware (coloured in green) presents a greater durability for the year 2030 than the other wall (coloured in yellow) and the waterproofing membrane (coloured in orange) which will be near the replacement year in 2030 if they do not receive maintenance actions.

After, an Excel table is obtain and exported supporting professionals to define or verify maintenance actions or plans. With the Excel data it is possible to explore scenarios of maintenance actions, combined with the element's costs, areas of refurbishing and periodicity of actions to be performed. The planning of the activities can be readjusted to what is necessary according to the state of degradation of the elements during the period in service.

6 Results and Discussion

In the occupation building phase, the BIM methodology have been improving the maintenance activity, namely, in the support of retrieving the required information within the database of the model and in the capacity of promoting an agile handling of data used for maintenance propose. The principal aim of the study was to incorporate into the BIM model, which must be created for each building in analyses, the degradation perdition value concerning some of the components of a building, with an important role in the exterior protection of the building, the facades and roofs.

Using the perdition value, a new strategy of visual impact was introduced supporting experts to improve their studies and easily present results to other partners and building owner, as it is possible to observe with distinct colors applied over the model, the level of degradation of the components for a horizontal period of time. In it two Dynamo scripts were developed.

As the BIM objects did not have the necessary parameters for the application of the factor method, highlighting the need to include, in the digital catalogue of each object the RSL value. This is required to be able to evaluate the durability of the constructive elements in the project to which they are applied, taking into account all the factors to be fulfilled by the technician. The execution of each of the scripts occurred without problems. No errors were reported by Dynamo. However, to apply these scripts to a BIM model it is necessary to create the project parameters: ESL, Maintenance Excel, % lifespan and all the cost's parameters, as they were named when in Dynamo. To get through this point it was necessary to include in the script the creation of each parameter.

7 Conclusion

The use of Dynamo allowed to show the tool's potential in the way it can explore, manipulate and export the data of the BIM model. The developed scripts allowed an analysis of the durability of the elements in the design phase where the change of solutions have less impact on overall costs. These scripts, elaborated for the presented study case, can be easily adapted to other components of the building and to other buildings. Two new parameters must be added to parametric objects representative of other components, of the present model, or of other building cases, and then the describe procedure can applied.

The perspective of the study was directed towards the occupation phase of the building, which since it is the phase of longer duration and where the maintenance activity increases costs that sometimes exceed the initial investment, the use of BIM has the potential to gather all the information to support this phase and reduce the associated costs. Its implementation is not yet recurrent at a business level, particularly in the national context, although as demonstrated in this study, the tools exist to allow it.

References

1. Guzman, G., Ulloa, W.: BIM application in the operation and maintenance management of a sports infrastructure. Tommelein, I.D., Daniel, E. (eds.). Proceedings of the 28th Annual Conference of the International Group for Lean Construction (IGLC28), Berkeley, California, USA (2020). https://doi.org/10.24928/2020/0015, http://iglc.ne/
2. Sampaio, A.Z.: Maturity of BIM implementation in the construction industry: governmental policies. IJETT- Int. J. Eng. Trends Technol. **69**(7), 92–100 (2021). ISSN: 2231–5381,https://doi.org/10.14445/22315381/IJETT-V69I7P214 ©2021 Seventh Sense Research Group®, http://www.ijettjournal.org/archive/ijett-v69i7p214
3. Neuville, R., Pouliot, J., Billen, R.: Identification of the best 3D viewpoint within the BIM model: application to visual tasks related to facility management. Buildings **9**, 167 (2019). https://doi.org/10.3390/buildings9070167
4. Fargnoli, M., Lleshaj, A., Lombardi, M., Sciarretta, N., Di Gravio, G.: A BIM-based PSS approach for the management of maintenance operations of building equipment. Buildings **9**(6), 139 (2019). https://doi.org/10.3390/buildings9060139
5. Bozorgi, P.P., Gao, X., Eastman, C., Self, A.P.: Planning and developing facility management-enabled building information model (FM-enabled BIM). Autom. Constr. **87**, 22–38 (2018). https://doi.org/10.1016/j.autcon.2017.12.004

6. Sampaio, A.Z., Simões, D.G., Berdeja, E.P.: BIM tools used in maintenance of buildings and on conflict detection. In: Sustainable Construction: Building Performance Simulation and Asset and Maintenance Management, vol. 8, pp. 163–183, ch.8 (2016). https://doi.org/10.1007/978-981-10-0651-7_8, http://www.springer.com/gp/book/9789811006500
7. Domingos, I.: The 7D/BIM model used in the maintenance of buildings. Master Thesis in Civil Engineering, University of Lisbon, Lisbon, Portugal (2020)
8. Gao, X., Pishdad-Bozorgi, P.: BIM-enabled facilities operation and maintenance: a review. Adv. Eng. Inform. **39**, 227–247 (2019). https://doi.org/10.1016/j.aei.2019.01.005
9. Gouda, A., Mohamed, M., Abdallah, R., Marzoukc, M.: BIM and semantic web-based maintenance information for existing buildings. Autom. Constr. **116**, 103209 (2020) https://doi.org/10.1016/j.autcon.2020.103209
10. Sampaio, A.Z., Gomes, A.M.: BIM interoperability in structural design, MDPI – civil engineering, special issue: advances in civil engineering. CivilEng **2**(1), 174–192 (2021). https://doi.org/10.3390/civileng2010010
11. Dynamo, The Dynamo Primer (2020). https://primer.dynamobim.org
12. Hovde, P.J., Moser, K.: Performance based methods for service life prediction. In State of the Art Reports, CIB Report: Publication, vol. 294 (2004)
13. Sacks, R., Eastman, C., Lee, G. Teicholz, P.: BIM Handbook: A Guide to Building Information Modeling for Owners, Managers, Architects, Engineers and Contractors, Third Edition Copyright © 2018 by John Wiley & Sons, Inc. New Jersey, USA (2018). https://doi.org/10.1002/9781119287568, ISBN: 9781119287537
14. Sampaio, A.Z., Gomes, A.M., Lite, A.S., Zulueta, P., González-Gaya, C.: Analysis of BIM methodology applied to practical cases in the preservation of heritage buildings. MDPI - Sustainability, Special Issue: Strategies for the Conservation of Industrial Heritage Aligned with the Sustainable Development Goals, vol. 13, p. 3129 (2021). ISSN 2071–1050, https://doi.org/10.3390/su13063129
15. ISO 15686–1. Buildings and constructed assets - Service life planning - Part 1: General Principles. ISO - International Organization for Standardization, Geneva (2000). https://www.iso.org/obp/ui/#iso:std:iso:15686:-5:ed-2:v1:en
16. Raposo, T.: Durability of the building. Estimated life of flat cover coatings. Master thesis in Civil Engineering, University of Porto, Faculty of Engineering, Porto, Portugal (2009)
17. Lopes, C.: Durability in Construction. Estimated life of façade ceramic coatings. Master thesis in Civil Engineering. University of Porto, Faculty of Engineering, Porto, Portugal (2009)
18. Matos, M.: Durability as a design criterion. The factorial method in the Portuguese context. Master thesis in Civil Engineering. University of Porto, Faculty of Engineering, Porto, Portugal (2007)

The 6D BIM Model Applied to Evaluate the Building Energy Performance

A. Zita Sampaio(✉) and Luis Araújo

Higher Technical Institute, Department Civil Engineering, University of Lisbon, Av. Rovisco Pais, Lisbon, Portugal
{zita.sampaio,luis.araujo}@tecnico.ulisboa.pt

Abstract. The Building Information Modeling (BIM) methodology have been applied in all sectors of the construction activity, involving distinct professional working in collaborative way on the development of a global design, building process, and the management and maintenance requirements. The present research analyses the capabilities of BIM on the support of the building management, considering the energetic efficiency related to building exterior walls and floors. The most common exterior walls applied in a current building were selected. A representative parametric object for each wall was defined, associated with relevant parameters related with the energetic simulation, incrementing the wall's and floor's libraries of the BIM software, Revit (Autodesk). All the information introduced in the model, by the insertion of updated values into the new parameters, allows the plugin Green Building Studio (Autodesk), to evaluate the energy consume, supporting a comparative analyses between several alternative façade solutions. The study brings knowledge of applying BIM functionalities in a energetic simulation analyses.

Keywords: BIM · Façade walls · Floors · Physical parameters · Energetic analyses

1 Introduction

The choice of alternative solutions in a construction has a considerable effect in the budget, in the environment and for society. Therefore, it is essential to provide a quick access to the design information in order to perform different alternative solutions. Currently, BIM methodology can be used to get information associated with the multiple stages of design and construction processes, supporting all building lifecycle activities [1].

The sustainability analysis of a project considers, not only the environment aspects, but also a variety of impacts agents [2]. Sustainable development has been growing in prominence with the increase of buildings life-cycle and, consequently, the buildings management and maintenance. The BIM solutions bring an easier way for sustainable development practices by enabling architects and engineers to accurately visualize, simulate, and analyze the project in the preliminary stages, achieving more sustainable projects and reducing constructions time and costs.

J. F. L. C. Semião et al. (Eds.): INCREaSE 2023, ASST, pp. 268–278, 2023.
https://doi.org/10.1007/978-3-031-44006-9_20

Thermal insulation of building components leads to reduce the total thermal energy consumption and to accomplish a significant reduction of CO2 emissions [3]. Special attention must be given to structural systems, concrete beams and lintels which act as thermal bridges providing poor insulation. In addition, consideration should be directed for exposure and durability. A well-insulated home can prevent to 70% of all heat flow through the ceiling, walls, and floors [4].

According to UK Green Building Council, the construction operations account about 40% of global energy consumption and 30% of carbon gases emissions that cause the greenhouse effect. The energy sustainability analysis of a model can be included in the 6D/BIM designation. To obtain the environmental behavior simulation, Revit can make an energetic model simulation, also allowing its transfer, directly or through an IFC file, to other energy analysis software.

It is in the designer's interest to know the tools that can support the decision-making regarding the environmental impacts that the project will create, since one of the factors to be considered in the selection of a certain constructive system is its impact on the building's energy consumption. While granting the required comfort conditions, the building should have the least energy consumption possible.

Hence Revit software has an application, Green Building Studio, capable of performing energy simulations on the 3D/BIM model, retrieving an environmental behavior simulation of the building in different project stages. Autodesk Green Building Studio is a flexible cloud-based service that allows the user to run building performance simulations to optimize energy efficiency and to work toward carbon neutrality earlier in the design process. The plugin can be used as a standalone web service.

This study intends to evaluate the implementation of BIM to support comparative analyses of alternative exterior wall and floors solutions. The study considers the comparative energetic strategy analyses, the materials applied in the composition of façade walls and in floor's finishing and the research of parameters related to energetic characteristics. For this purpose, the following objectives were established [5]:

- Understand BIM´s methodology based in parametric objects and how to enrich the model;
- Identify different architectural solutions based in the use of several type of walls and floors currently applied in the national market;
- Evaluate in an energetic perspective the different solutions using a Revit plugin, the Green Building Studio.

To model and extract all capabilities of BIM, it is necessary that all the parametric objects that are going to be applied in the generation of the required model are available in the BIM tool. The parametric object must contain the physical parameters required by the BIM plugins that are going to be used to obtain different type of analyses [6]. For that new parametric objects of current walls and floors were generated.

2 Methodology

As a methodology several type of facades and floor finishing were first selected, and the respective new parametric objects were created:

- After, specific **parameters** concerning physical proprieties that has impact in the efficiency energy were identified, and included as new parameters into the created wall's and floor's parametric objects;
- By applying distinct solutions, facades and floor, in a case study, an **energetic** analyses **simulation** is then performed;
- The **results** of the energetic consume obtain over each alternative solution allows to achieve important conclusions about the most efficient solution and combination wall + floor solution;
- The methodology followed in the study can be applied over **other** type of construction **components**.

3 Exterior Walls

The designer should be able to quickly access to different options of exterior walls, so the user could decide the right wall to apply [7]. The paper considers the study of the main walls used in Portugal [8]. Therefore, a research throughout the different types of walls has been made. Although it is possible to find different systems, they must respond to environmental and people's demand [6]. They should protect the users from outside: ensure stability in case of fire; safety protection in situation of earthquakes or explosions; garantee comfortable conditions adapted to the environment where they are built. For the present study and in order to define parametric objects of walls, five common facades used in Portugal were selected (Fig. 1):

- Two solutions of double wall brickwork with a layer of air partially filled with a thermal insulator;
- A External Thermal Insulation Composite (ETIC) system, that consist of an outer insulator coated, fixed mechanically or by gluing in brick support panel;
- A ventilated facade insulated on the outside in a brick panel with ceramic tiles spaced from the insulator to allow wall ventilation;
- A Light Gauge Steel Frame (LGSF) system, constituted from interior to exterior, 2 gypsum panels for interior, insulator panels made of rock wool, OSB panel, and for exterior an insulator panel cover with mortar.

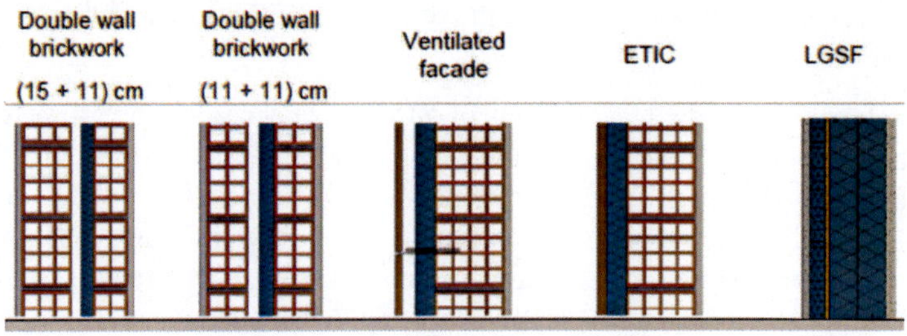

Fig. 1. Selected common facades.

The wall, as a building component, should check a number of requirements. The application of an exterior wall depends of the physical and mechanical properties of each material that composes the different layers of a building element. So to perform multiple analyses and support decisions, it is necessary to select properties that can be associated with parametric objects of exterior walls, to achieve an accurate database accessible to different collaborative parts [7].

Since the internal library of the BIM tool in use, the Revit from Autodesk, don´t contain the main walls used in Portuguese construction it´s necessary to create them. Five types of parametric objects of walls were identified and associated with a set of parameters. A Revit wall system is composed of several layers of different materials, with distinct proprieties (Fig. 2). To achieve an accurate wall it was necessary to identify materials and their properties. Revit contains some of the materials (concrete and mortar) and others were created (OSB panels or gypsum panels).

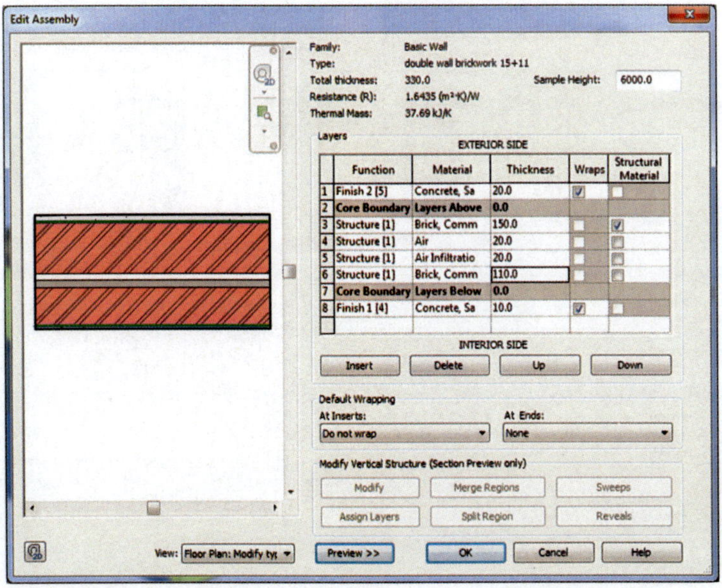

Fig. 2. Parametric object of double wall brickwork.

3.1 Physical Parameters

To generate a BIM model, the initial needs must be clearly defined, which can vary according to the stakeholders and users. The designer needs to visualize details of the project, and a client or a shipyard organizer needs a cost analysis and a project scheduling. Several types of parameters can be added to each BIM model element which enables selecting and providing cost estimates as well as material tracking and ordering. In consequence, several level of detail and level of development (LOD), in their turn can change from one user to another, and can be incremented at different stages of the building project and its occupation [8].

The parameters to be attached to BIM objects should consider studies such as thermal analysis and energy efficiency, construction planning, construction and maintenance costs or provide maintenance plans. In this study where considered, essentially, parameters related to the physical and mechanical properties oriented to the energetic analyses and environmental impact [9]. Thus, the chosen parameters joint with the respective solution are physical parameters such as: thickness, superficial mass, thermal resistance or thermal transmittance; acoustic parameters, sound transmission rate due to aerial transmission or fire rate parameter; environmental parameters such as embodied primary energy. These parameters are listed in Table 1 and must be considered in each wall's parametric object when created.

Table 1. Table of parametric values for each wall type.

	brickwork (15 + 11) cm	brickwork (11 + 11) cm	Ventilated facade	ETIC	LGSF
Type of exterior wall			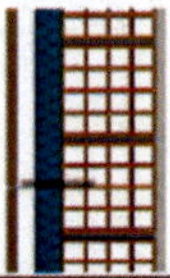		
Physical parameters					
Thickness (m)	0,37	0,33	0,35	0,32	0,235
Superficial mass (kg/m²)	300,84	277,64	255	261,77	86,25
Thermal resistance ((m²K)/W)	3,08	2,95	1,73	2,11	5,43
Thermal transmittance ((m²K)/W)	0,307	0,32	0,525	0,44	0,179
Economical and management parameters					
Construction cost (€/m²)	75,28	73,28	158,72	100,21	111,73
Maintenance cost (€/ano.m²)	2,75	2,75	27,1	11,70	8,94

3.2 Parametric Objects

The 6D and 7D dimensions of BIM model are related with lifecycle-management and sustainability analysis. Buildings are responsible for 40% of energy consumption all over the world, thus the interest to know how the building is going to behave is an information that must be quick access to the designer. A BIM model can be exported to other programs of analysis and simulation such as eQuest, Ecotect, Green Building Studio, to perform several analyses:

- The eQuest is a building energy use analysis tool that provides professionals to perform detailed comparative analysis of building designs and technologies by applying simulation program;

- The simulation of solar energy absorbed by building surfaces is possible to be realized using Ecotect;
- The evaluation of energetic consume can be performed in Green Building Studio.

The properties used by Revit depend of the material, but the main information concerns:

- Identity (description, comments, manufacturer and model mark);
- Graphical properties (color, surface and cutting pattern and textures);
- Physical properties (young´s modulus, poison's ratio, shear modulus, density and thermal expansion coefficient).

The physical proprieties of a composed parametric object like a wall is automatic obtain by Revit, when the set of material layers are chosen and established each thickness. Each material has specific thermal proprieties as shown in Fig. 3 for the brick material.

The use of an adequate BIM rich-data model, with thermal proprieties, allows the environmental and climate analysis, in design initial phases, important when comparing the environmental impact of alternative design cases. A heat analysis can be allowed before and after using thermal insulation, in a wall, and the result can clarify in relation of what solution should be used. By changing the type of wall the analyses values became distinct supporting the best solution for the project.

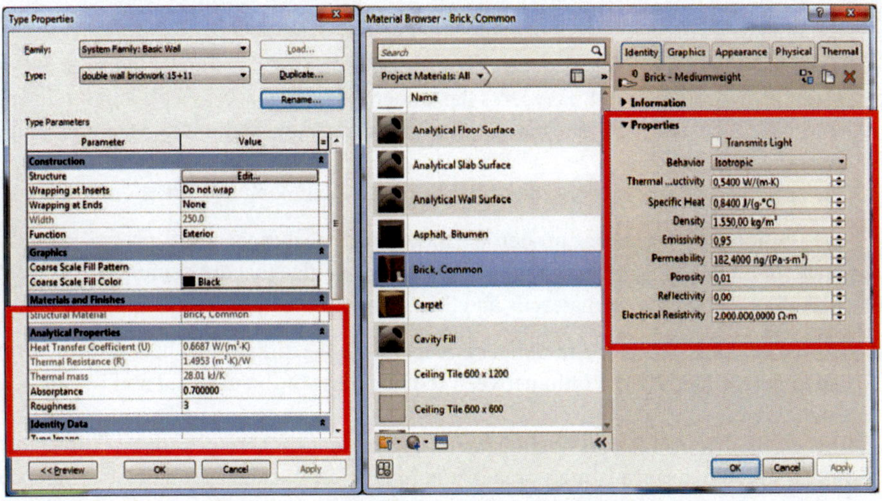

Fig. 3. Physical proprieties of a composed parametric object and thermal proprieties of a material.

4 Floors

In the national market there are a large number of possible materials to be used as flooring finishing, and their selection is an important criterion to be considered by the designer. In the present study, a research about the materials commonly used in frequent building floors was done, oriented to their composition on floors and knowledge of their properties. In the scope of the work, the floor finish classification is made by type of materials (minerals, wood, textile and synthetic). It was considered the most relevant, in relation to the utilization and the traded products in the Portuguese market [10].

4.1 Physical Parameters

In order to support an appropriate selection for a constructive flooring solution in a building, its main characteristics and physical properties were identified. As one of the objectives of the study is to analyze the information that is relevant to the designer to support the decision making on different constructive solutions, this information was materialized in parameters and correspondent values and associated to the parameters of the objects. In the aim of an energetic simulation, the parameters assigned to BIM objects were those associated with thermal and energetic analyzes. They are identical to those selected on the creation of parametric objects of walls.

On the generation of parametric objects relating to floors a set of thermal parameters, linked to each material, allows to characterize the construction solution. The thermal parameters of each solution are the basis of energy analysis that can be performed through the Revit software. So, when creating a BIM model, for the simulation of an energy analysis, the Revit software uses the thermal properties of the materials and of the construction systems. For most materials, except glass, Revit only needs to use two thermal properties [11]:

- The thermal conductivity is a property that characterizes thermally homogeneous materials or products, and represents the amount of heat, that crosses a unit thickness material, when between two flat parallel faces is established a unit of temperature difference;
- The specific heat is the energy required to rise in a temperature unit, a unit of mass.

In addition to the thermal properties and the physical-mechanical properties, it is required for an energy simulation, other thermal properties that are useful for a correct thermal characterization of materials and their constructive systems.

- The thermal resistance of a material corresponds to the difficulty of heat transmission and is determined by the ratio between the thickness of the material and its thermal conductivity;
- The thermal transmission coefficient is the amount of heat per unit of time that runs through a surface of unit area of the immersive element per unit of difference temperature between the environments that the wall separates.

4.2 Floors Objects Library

New parametric objects of floors were created incrementing the library of objects made available in the Revit system. During the definition of constructive solutions representing floors, four groups were considered based on similarity criteria between the flooring materials and the construction system components (constituent materials and their thickness). It was considered in all cases a reinforced concrete support floor. Therefore, the chosen groups are (Fig. 4):

- Natural stone cladding (granite and limestone) or ceramic tiles (porcelain stoneware);
- Floating flooring made of wood, laminate or cork;
- Carpet and vinyl floorings.

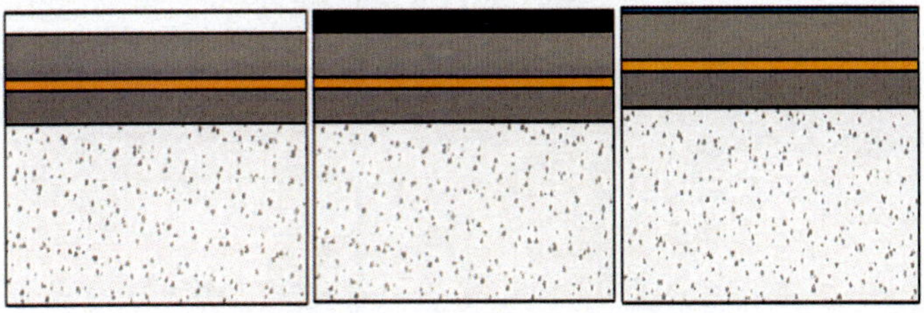

Fig. 4. Limestone, granite and vinyl floorings.

Figure 5 illustrates the floating flooring made of wood constructive solutions, with the respective graphic pattern, type, thickness and function of the material of each layer.

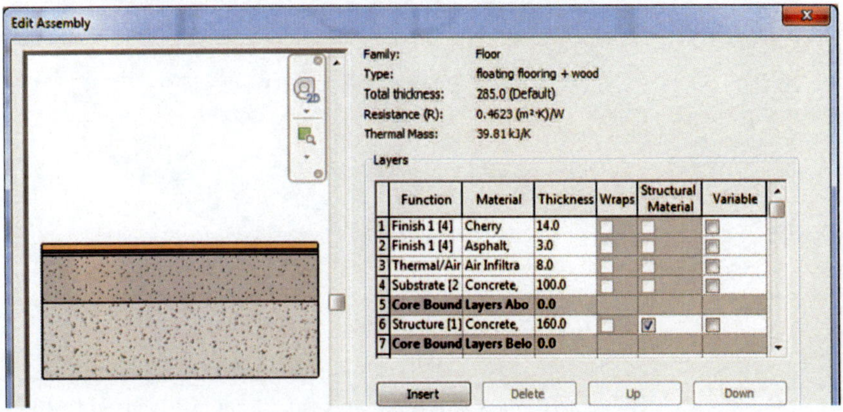

Fig. 5. Generation of a parametric object of a floor.

Each floor solution is characterized by several thermal proprieties obtain from the individual features of the set of materials used in its composition (Fig. 6). When studying the type of floor to apply in a project, Revit allows the user to easily change the type of floor. By selecting an alternative floor option, through another parametric object choose, the result of the process is immediately visualized (Fig. 7). It is possible to observe in a cut 3D perspective, the graphic impact of the floor and the user can consult the proprieties of each floor option.

5 Results and Discussion

In the present work an energy and thermal analysis was carried out. The properties values necessary to the floors constructive solutions were studied and indicated in order to perform such an operation. For an analysis concerning a study phase, the 3D model

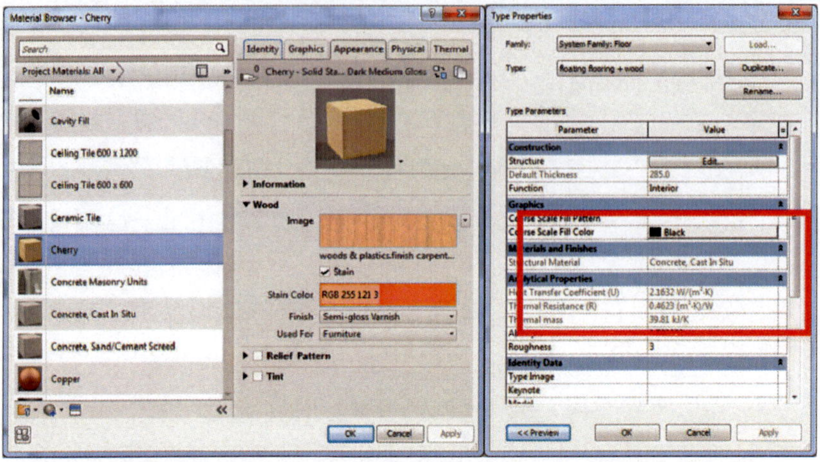

Fig. 6. Selection of a wood material and generation of a parametric object of a floor.

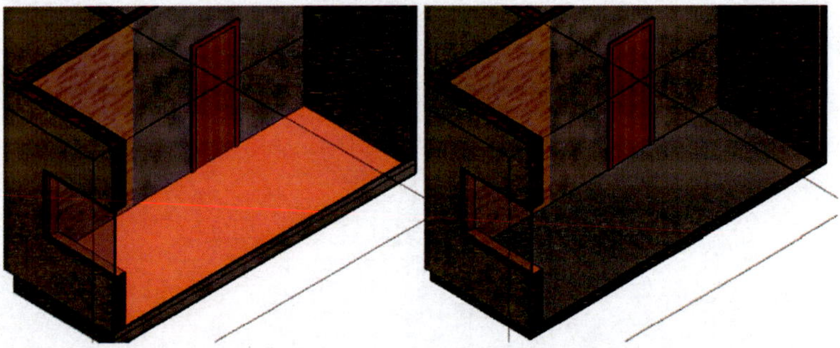

Fig. 7. Changing the type of the applied floor.

can be presented as a conceptual mass, but for analyses with more advanced features, is necessary to give more detail to the constructive elements.

After the required settings are included in the model, it is transferred t for analysis in a cloud, an Autodesk platform. The analysis is performed and a final report is created (Fig. 8).

BIM-enabled design process necessitates innovative methods for design thinking and representation that architects and engineers used to apply and utilize. BIM is a decision-making tool for evaluating environmental factors in conceptual designing, rather than a sole tool for designing. BIM tools allow integrating several settings, in which the target is to improve the environmental and climate-response factors. The thermal parameters of walls and floors objects were used by Revit plugins to obtain results and allowing comparing different solutions. Just the walls and floors were defined with detail, but other elements (light systems or windows) should contain identical thermal parametric to be used in an energetic simulations.

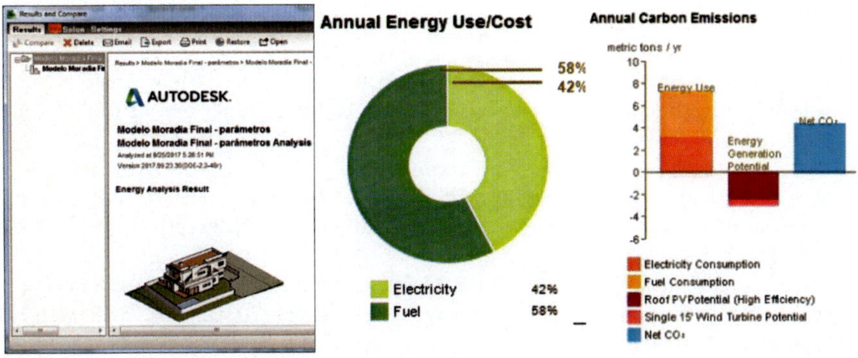

Fig. 8. Energetic analyses report.

Green Building Studio uses a simulator (DOE-2 simulation engine) to estimate the energy use of construction and operating costs, and it is based on the effects and interactions of the building materials, systems, level of use and climate. This application dynamically analyzes the whole building heat and energy levels, and is based on the recommendations. Leadership in Energy and Environmental Design (LEED).

6 Conclusions

BIM would benefit the design process particularly in the area of BIM sustainability application with effective impact during designing. With the increasing global awareness of improving the environmental factors of building performance and the needs of achieving the climate-responsive design, the BIM use associated with environmental analysis of related computer programs appears as the primary key in architectural designing.

The study describes the capacity of BIM to compare solutions supported in energy analyses results, mainly oriented to two principal building components: exterior walls and floors. The characterization of current type of these building components was done and the correspondent parametric objects were created. A set of physical and mechanical proprieties, of materials and construction systems, were identified and selected as of interests to be used in energy simulation tools.

In the study of alternative solutions, decision-making can be grounded using a comparative analysis of the results relating to environmental performance and contribution to the reduction of carbon dioxide emissions allowed by Green Building Studio tool. Environmental building performance can be improved by applying a Revit plugin, Green Building Studio, which gives a wide of sustainability analyses obtained directly from a BIM model. Consequently, architects can easily include environmental design principles in conceptual designing, starting from the proposition and the composition of architectural spaces and forms.

References

1. Carvalho, J.P., Almeida, M., Bragança, L., Mateus, R.: BIM-based energy analysis and sustainability assessment—application to portuguese buildings. Buildings **11**(6), 246 (2021). https://doi.org/10.3390/buildings11060246
2. Edwards, E.E., Lou, E., Bataw, A., Kamaruzzaman, S.N., Johnson, C.: Sustainability-led design: Feasibility of incorporating whole-life cycle energy assessment into BIM for refurbishment projects. J. Build. Eng. **24**, 100697 (2019). https://doi.org/10.1016/j.jobe.2019.01.027
3. El Sayary, A., Omar, O.: Designing a BIM energy-consumption template to calculate and achieve a net-zero-energy house. Sol. Energy **216**, 315–320 (2021). https://doi.org/10.1016/j.solener.2021.01.003
4. Niu, S., Pan, W., Zhao, Y.: A BIM-GIS integrated web-based visualization system for low energy building design. Procedia Eng. **121**, 2184–2192 (2015)
5. Capilla, R., Nava, F., Montes, J., Carrillo, C.: ADDSS: architecture design decision support system tool. In: Proceedings of 23rd IEEE/ACM International Conference on Automated Software Engineering, pp. 487–488 (2008). https://doi.org/10.1109/ASE.2008.84
6. Sampaio, A.Z., Gomes, A.M.: BIM interoperability analyses in structure design. MDPI Civil Eng. **2**, 174–192 (2021). https://doi.org/10.3390/civileng2010010
7. Ren, R., Zhang, J., Dib, H.N.: BIM Interoperability for structure analysis. In: ASCE Construction Research Congress, pp. 470–479 (2018).https://doi.org/10.1061/9780784481264.046
8. Araújo, L.: Characterization of a library of walls supporting BIM methodology, MSc thesis in Construction, IST, University of Lisbon, Portugal (2021)
9. Pereira, V., Santos, J., Leite, F., Escórcio, P.: Using BIM to improve building energy efficiency – A scientometric and systematic review. Energy Build. **250**, 111292 (2021). https://doi.org/10.1016/j.enbuild.2021.111292
10. Sampaio, A.Z., Gomes, A.M., Sánchez-Lite, A., Zulueta, P., González-Gaya, C.: Analysis of BIM Methodology applied to practical cases in the preservation of heritage buildings. Sustainability **13**, 3129 (2021). https://doi.org/10.3390/su13063129
11. Sampaio, A.Z., Fernandes, V., Gomes, A.M.: The use of BIM-based tools to improve collaborative building projects. In: ProjMAN 2022, International Conference on Project MANagement, Lisbon, Portugal, 9–11 November 2022 (2022)

Finite Element Modelling of the Behaviour of Thin-Walled Arch Bridge Made with UHPFRC and Ordinary Concrete

Tomasz Maleska, Damian Beben(✉), and Arkadiusz Mordak

Faculty of Civil Engineering and Architecture, Opole University of Technology, 45-758 Opole, Poland
{t.maleska,d.beben,a.mordak}@po.edu.pl

Abstract. Due to environmental protection, new material solutions are sought that, can replace typical materials used in the construction of culverts and small bridges (steel, reinforced concrete). Ultra-high performance concrete appears to be a more environmentally friendly material as it allows the use of a lower thickness of the structural element than in the case of typical concrete. The paper presents finite element modelling of a thin-walled arch structure made of ordinary concrete and ultra-high performance fibre reinforced concrete (UHPFRC). The arched concrete structure is designed for small bridges and culverts. Four different types of concrete were used, two ordinaries (models I and II) and two UHPFRC (models III and IV). In addition, two arch thicknesses were used (5 cm and 12 cm). Ordinary concrete was not reinforced, and UHPFRC was made with the addition of glass (model III) and steel (model IV) fibres. The obtained results are of a practical nature but still require additional research.

Keywords: arch structure · UHPFRC · ordinary concrete · static load · bridge

1 Introduction

Environmental pollution and the increasing intensity of road and rail transport result in the search for new construction solutions and/or new and better materials that would enable the construction of transport infrastructure facilities (bridges, culverts) of appropriate durability and strength. Most of the bridge structures are built as various types of concrete structures. In the last 40 years, significant development has taken place in the technology of concretes used in bridge construction. One of the milestones in concrete technology was the development and implementation of Ultra-High Performance Concrete (UHPC) as a sustainable material. The beginning of the UHPC development dates back to the 1980s. Moreover, the addition of fibres to UHPC matrices can successfully increase the fracture toughness, tensile strength, ductility and energy absorption capacity of UHPCs [1]. The first use of Ultra-High Performance Fibre Reinforced Concrete (UHPFRC) in Thin Precast Concrete Roof Shell took place at the Shawnessy Light Rail Transit Station in Calgary, Alberta, Canada [2, 3]. Bridge owners have shown a willingness to adopt innovative solutions that can meet the specific challenges related to cost, speed of construction, durability, and service life.

J. F. L. C. Semião et al. (Eds.): INCREaSE 2023, ASST, pp. 279–288, 2023.
https://doi.org/10.1007/978-3-031-44006-9_21

Statistically, small and medium-sized bridges dominate the transport infrastructure, which is why their significance in the transport network is very important. Traditional bridge structures with small and medium spans up to 30 m are usually made of reinforced concrete or steel, or as a steel-concrete composite structure. Recently, bridges have also been made fully composite or hybrid structures with a lightweight concrete slab and an FRP composite girder [4]. There are also known solutions for such bridge structures in the form of reinforced concrete arches (Bebo, Materie, Prefac, OptemARCH, TechSpan, Con-Span systems) or box structures (Box-culvert, Opti-quadro, OptemFRAME systems). Typically, the wall thickness of such structures is min. 0.20 m, but for larger spans, it can even be 0.30–0.40 m. The idea behind this type of solution is to build a structural system of prefabricated elements combined into the entire structure at the construction site. Backfill in such reinforced concrete structures is no longer as important as in the case of soil-steel bridges, where it plays an important role in the transfer of loads [5, 6]. Precast RC culverts and bridges offer advantages such as enhanced quality control, use of higher-strength concrete, lower cost due to mass production, and shorter installation time. In the 1990s there was a significant boom in culvert bridge construction in many countries. Most of these bridges, made of corrugated metal or reinforced concrete, are now approaching 30 years of age and are deteriorating at a high rate. Culvert bridges are subjected to aggressive environments (e.g., exposure to high moisture and treatment with deicing salts). For these structures, combinations of moisture, temperature, and chlorides accelerate the corrosion of the metal pipes or the internal steel reinforcement leading eventually to loss of serviceability.

The last 20 years have seen a rapid increase in the construction of small bridges and culverts made of corrugated sheets, as well as multidirectional research on these objects [5, 7, 8]. They have become an alternative to typical reinforced concrete box culverts or small reinforced concrete bridges made up to a span of 20 m. However, the corrosion resistance aspects of corrugated steel plate culverts and the ever-increasing steel prices on the world markets place these solutions in a less favourable light and have a slow return compared to concrete solutions.

Scientific research on small bridge structures is limited. Most often, tests were carried out for the impact of dynamic loads from passing vehicles and/or fatigue or seismic loads. The aspect of static loads is less frequently discussed. In terms of dynamic research, the main goal of the studies [9–11] was an assessment of the effects caused by passing vehicles. As part of the static tests, the analyses focused mainly on the earth pressure around the tested structure. It is known that the earth pressure in these facilities has a negative effect on the structures. In combination with the external load, it can provide a significant problem for the structure. Chen and Sun [12] studied the influence of soil properties on reinforced concrete culverts under high backfill cover. It was found that this influence is visible and should be taken into account during design. In addition, Ma et al. [13] and Song et al. [14] investigated the influence of negative soil pressure on the concrete structure.

Bearing the above in mind, new material solutions are sought to replace the typical materials used in the construction of culverts and small bridges (steel, reinforced concrete). UHPFRC seems to be such a material. Such concrete usually consists of fine sand, silica fume, high cement content and admixtures (superplasticizers) and dispersed reinforcement in the form of various types of fibres. The advantage of this type of concrete is high homogeneity obtained by reducing the particle size and very good technical parameters, including high concrete deformability, compressive and bending strength. High-strength parameters are obtained thanks to the appropriate selection of ingredients and care treatments in such a way as to improve the properties of the finished material. Examples of the use of this technology are the footbridge in Sherbrooke (Canada) – 1997, and the Gärtnerplatz bridge in Kassel, Germany (2007).

The main goal of the paper is to assess the behaviour of thin-walled arch made from UHPFRC and ordinary concrete under static loads in order to use it in transport infrastructure for the construction of culverts and small bridges. In addition, the work shows the influence of the thickness of the arch made of ordinary concrete and UHPFRC on its behaviour under static load.

2 Methodology of Numerical Study

The computational models of an arch structure made of UHPFRC and ordinary concrete (for comparison purposes) are developed using Finite Element Method (FEM). An important aspect of the modelling stage will be the correct application of the concrete failure model. In these FEM simulations, eight numerical models were made, i.e. four each for the shell thickness of 5 cm and 12 cm, respectively. The arch structure with a segment of a circle with a diameter of 3.0 m and width of 2.3 m (Fig. 1a) was analysed in the three-point-bending scheme.

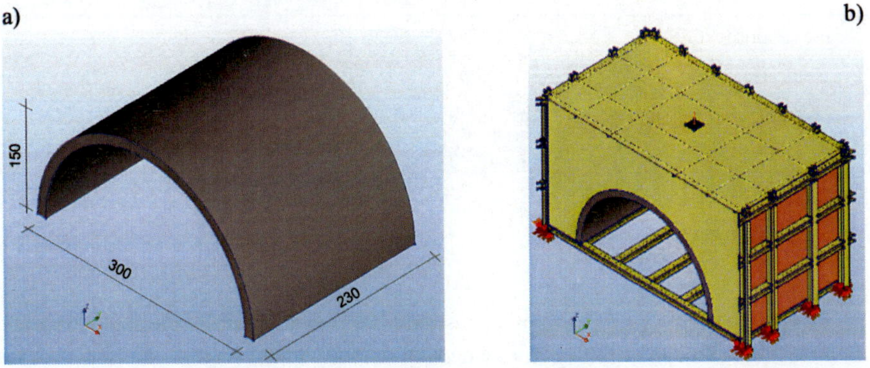

Fig. 1. Numerical model of: a) concrete arch structure, b) test box with arch (dimensions are given in cm).

Ordinary concrete models were modelled in accordance with the standard guidelines (Eurocode 2 [15]) and the recommendations of the DIANA FEA numerical program (model I and II). UHPFRC models were described with a non-linear "Total Strain based crack model" (model III and IV), which will allow the identification of cracks, displacements and strains in the UHPFRC arch. This model is recommended for concrete with glass or steel fibres. Detailed material data for FEM modelling are obtained from laboratory studies and literature [16]. Materials parameters of ordinary concrete and UHPFRC are presented in Table 1. The arch structure was modelled using shell elements.

The backfill was modelled using the Mohr-Coulomb model with the following properties: (i) Young modulus of 100 MPa, (ii) Poisson ratio of 0.2, (iii) density of 2.05 T/m^3, (iv) porosity of 0.95, (v) cohesion of 3 kN/m^2, (vi) friction angle of 30°. The backfill is modelled using solid elements.

In the analysed models, interface elements were used to map the cooperation between the arch and the backfill using the non-linear Coulomb friction model [17, 18]. This model takes into account such parameters as (i) cohesion of 3 kN/m^2, (ii) friction angle of 30°, (iii) dilatation angle of 5° and (iv) rigidity of 100000 kN/m^3 [19]. At this point, it should be mentioned that in some works (e.g. [20]) the influence of the contact layer was omitted, which, in the authors' opinion, does not fully correspond to the actual behaviour of the buried structures.

Boundary conditions for the tested model were set at the base of the test box model (Fig. 1b). Pinned supports were used for the analysis. This type of boundary condition has been used many times in various studies to determine the interaction between the arch structure and the soil.

Table 1. Parameters of ordinary concrete and UHPC used for numerical analysis.

Material parameter	Concrete models (and types)			
	I (C 30/37)	II (C 50/60)	III (UHPC_gf)	IV UHPC_sf)
Young modulus (GPA)	32	37	44	65
Poisson ratio	0.2	0.2	0.2	0.2
Mean uniaxial tensile strength (kN/m^2)	2896.47	4071.63	8600	10130
Mean compressive strength (kN/m^2)	38000	58000	122000	238000
Density (T/m^3)	2.4	2.4	2.301	2.625

Note: gf – glass fibre, sf – steel fibre

FEM simulations were made in the test box (Fig. 1b), in which experimental tests of the concrete arches in question will be carried out in the next stages. The steel box was built of rolled steel sections (HEB 120) and made of S235 steel (according to Eurocode 3 [21]). The steel was modelled with the following properties: (i) Young modulus of 210 GPa, (ii) Poisson ratio of 0.3, (iii) thermal expansion coefficient of 1.2×10^{-5} 1/°C, (v) density of 7.85 T/m^3 and (vi) yield strength of 235 MPa. The walls of the steel box were clad with plywood and modelled as a linear material with the following properties: (i) Young modulus of 10 GPa, (ii) Poisson ratio of 0.35, (iii) density of 0.5 T/m^3.

3 Results from Numerical Analysis

3.1 General Remarks

The numerical program DIANA FEA was used to assess the influence of the thickness and type of concrete on the behaviour of the arch structure under a given static load. The prepared arch structure model was loaded gradually with a point load from 20 kN in 10% steps until reaching 300 kN. The load was applied in 15 loading steps. As part of the numerical analysis, displacements and stresses arising in three main loading steps, i.e. 100 kN, 200 kN, and 300 kN, were analysed. The obtained maximum results are presented in Table 2.

Table 2. Maximal results from numerical analysis of arch made from ordinary concrete (models I and II) and UHPFRC (models III and IV).

Arch of 5 cm thickness				
Force (kN)	Model I^5	Model II^5	Model III^5	Model IV^5
Displacements (mm)				
100	–0.27	–0.25	–0.24	–0.20
200	–0.55	–0.51	–0.49	–0.40
300	–0.83	–0.77	–0.73	–0.61
Stresses (MPa)				
100	-2.39	–2.54	-2.64	-3.11
200	–4.77	–5.07	–5.30	–6.24
300	–7.11	–7.60	–7.96	–9.37
Arch of 12 cm thickness				
	Model I^{12}	Model II^{12}	Model III^{12}	Model IV^{12}
Displacements (mm)				
100	–0.25	–0.24	–0.22	–0.17
200	–0.51	–0.47	–0.44	–0.34
300	-1.11	–0.71	–0.66	–0.51
Stresses (MPa)				
100	-5.77	–6.15	–6.46	–7.40
200	–11.3	–12.2	–12.9	–14.8
300	–19.0	–18.1	–19.4	-22.2

3.2 Displacements

Example maps of the maximum displacements of the arch under the highest applied static load of 300 kN are shown in Fig. 2. In the case of an arch with a thickness of

5 cm (regardless of the type of concrete), it can be seen (Fig. 2ac) that the distribution of displacements is radial from the point of application of the load. On the other hand, for the arch 12 cm thick, it was noticed that the maximum displacements run along the entire width of the arch in the crown (Fig. 2bd). In addition, it can be seen that the shell is flattened and the process of creating a destruction mechanism in the crown is initiated.

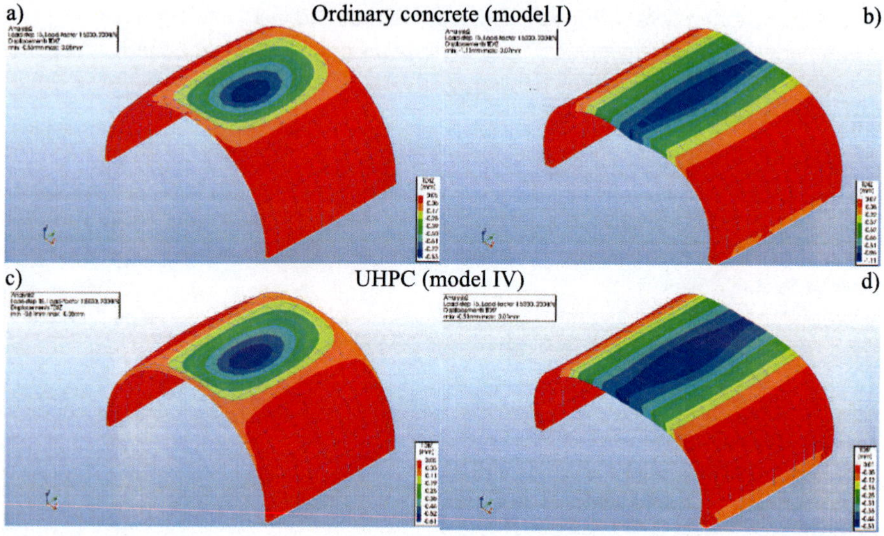

Fig. 2. Distribution of selected maximum displacements in the arch structure under a static load of 300 kN for ordinary concrete and UHPC with a thickness of: a) and c) 5 cm, b) and d) 12 cm.

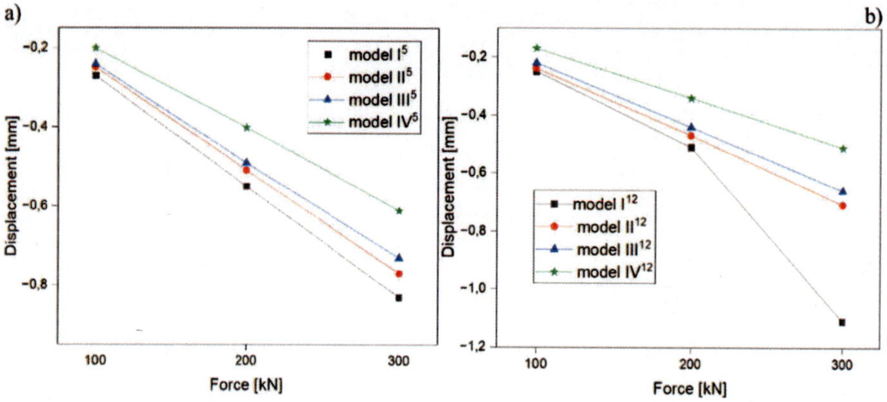

Fig. 3. Maximum vertical displacements for an arch with a thickness of: a) 5 cm and b) 12 cm.

In addition, when analysing the obtained displacement values for the tested models, it can be seen that the class of concrete clearly affects the obtained values, which is consistent with theory and general knowledge. However, it can be noticed that the differences between the 5 cm and 12 cm shells are not too large, especially in models II-IV, where the differences range from 4% to 16%. This is important because the model with a 5 cm shell used as much as 58% less concrete to construct the load-bearing element than in the case of a 12 cm thick arch. As can be seen in Fig. 3, the load of 300 kN caused the largest displacements in the I^5 and I^{12} models. In addition, it was noted that the I^{12} model had larger displacements (1.11 mm) than the I^5 model. This fact may be surprising, but on the other hand, it can be explained by the larger mass of the tested cross-section (arch) of the I^{12} model than the I^5 model. In a thinner shell (model I^5) there is greater flexibility of the structural member and therefore probably a better interaction of concrete and soil. The confirmation of this thesis can be found in UHPFRC concretes, which are characterized by high strength, and due to the applied load, the maximum displacements were smaller than in ordinary concretes.

3.3 Stresses

Figure 4 shows exemplary stress distributions in the arch structure under a load of 300 kN for models I and IV. In the case of a thinner shell (5 cm), it can be seen that the distribution of maximum stresses is more local and appear in the arch crown (Fig. 4ac) than in the case of a thicker shell (Fig. 4bd), where the distribution is more even over the entire arch. In addition, it can be seen that for the thicker shell (Fig. 4bd) the maximum displacements occur in the vicinity of the arch support.

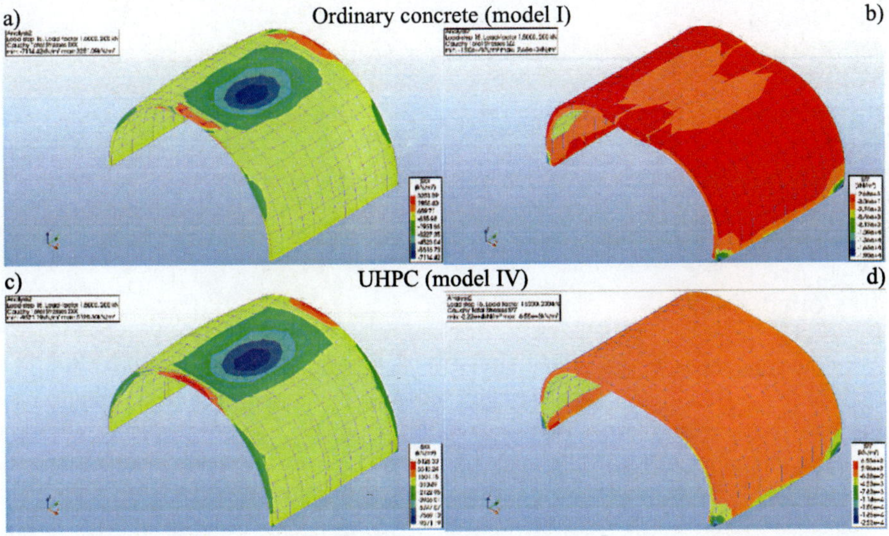

Fig. 4. Distribution of selected maximum stresses in the arch structure under a static load of 300 kN for ordinary concrete and UHPC with a thickness of: a) and c) 5 cm, b) and d) 12 cm.

In addition, it can be seen that in the models with a thicker shell (I^{12}–IV^{12}) higher stresses were noted than in the models with a thinner shell (I^5–IV^5). The highest stresses were obtained in the model where the shell was the thickest and made of UHPFRC concrete with steel fibres (model IV^{12}). In this model, stresses of –22.2 MPa were obtained (for a load of 300 kN). It should be added that the stresses for this shell thickness (12 cm) were obtained in the Szz direction (Figs. 4d and 5b). On the other hand, in models with a 5 cm thick shell (model I^5–IV^5), the greatest stresses were noted in the Sxx direction (Figs. 4ac and 5a). In the case of a 5 cm thick shell, the stresses ranged from about 7 to over 9 MPa (for a load of 300 kN). In the case of a 12 cm shell, the stresses were higher than those of a 5 cm thick shell in the range of 57–63%. Analysing the stress results from Table 2 as well as Figs. 4 and 5, one can also notice a similar trend as in the case of displacements for the tested models I–IV. Greater results in the Sxx (transverse) stress direction in the I^5–IV^5 models may indicate that the arch structure interacts better with the soil than the thicker shell (I^{12}–IV^{12} models). It can therefore be concluded that in this case, negative arching does not have as much impact as in the case of a thicker arch. This is due to the greater flexibility of a thinner shell.

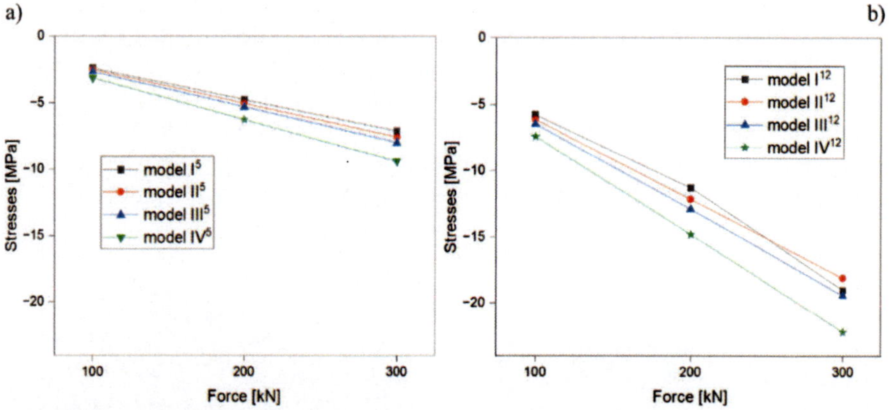

Fig. 5. Maximal stresses for the arch with a thickness of: a) (Sxx) 5 cm b) (Szz) 12 cm.

4 Conclusions

Performed studies have shown that UHPFRC has great application potential in bridge engineering, especially for small bridges and culverts compared to the currently used concretes, so there is a real need to develop a new UHPFRC composition and learn about the behaviour of an arch structure made of this type of concrete. Taking into account the obtained FEM results, the following conclusions can be drawn:

- The model with a thinner shell interacts better with the backfill, through greater flexibility, which results in less impact of negative arching in the buried structures.
- UHPFRC is better at carrying the given load. However, the values obtained for a shell with a thinner thickness (models I^5–IV^5) allow for transferring external loads without destroying the load-bearing structure.
- The use of a thinner shell (5 cm) resulted in maximum stresses in the transverse direction (Sxx), and in models with a 12 cm shell thickness, maximum stresses were obtained in the vertical direction (Szz). In addition, in the thicker shell, the maximum

stresses occur in the vicinity of the supports, and in the thinner shell, in the crown of the arch.

- The use of an arch with less use of concrete (models I^5–IV^5) translates into less weight of the structure. Greater benefits of using UHPFRC would probably become apparent for larger arch spans. Such analyses are planned in subsequent stages. The result is less energy consumption for the construction of the load-bearing structure and lower construction costs.

After analysis, it should be stated that further studies are needed to optimize the consumption of construction materials in buried structures. Experimental studies are planned to confirm the performed numerical calculations. As the analysis showed, the conducted research translates into lower costs and less interference with the natural environment and therefore fits in with the idea of sustainable development.

References

1. Buttignol, T.E.T., Sousa, J.L.A.O., Bittencourt, T.N.: Ultra high-performance fiber-reinforced concrete (UHPFRC): a review of material properties and design procedures. IBRACON Struct. Mater. J. **10**, 957–971 (2017)
2. Nguyen, D.L., Ryu, G.S., Koh, K.T., Kim, D.J.: Size and geometry dependent tensile behavior of ultra-high-performance fiber-reinforced concrete. Compos. B Eng. **58**, 279–292 (2014)
3. Vicenzino, E., Perry, H., Chow, T., Culham, G., Zakariasen, D.: First use of UHPFRC in thin precast concrete roof shell for Canadian LRT station. PCI J. **50**, 50–67 (2005)
4. Siwowski, T., Rajchel, M., Kulpa, M.: Design and field evaluation of a hybrid FRP composite-Lightweight concrete road bridge. Compos. Struct. **230**, 111504 (2019)
5. Beben, D.: Soil-Steel Bridges. Design, Maintenance and Durability. Springer, Cham (2020). https://doi.org/10.1007/978-3-030-34788-8
6. Vaslestad, J.: Soil Structure Interaction of Buried Culverts. Ph.D. Thesis, University of Trondheim, Norway, Trondheim (1990)
7. Maleska, T., Nowacka, J., Beben, D.: Seismic vulnerability of a soil-steel composite tunnel - Norway Tolpinrud railway tunnel case study. Tunn. Undergr. Space Struct. Incorporating Trenchless Technol. Res. **110**, 103808 (2021)
8. Pettersson, L., Sundquist, H.: Design of Soil-Steel Composite Bridges, 5th edn. TRITA-BKN, Royal Institute of Technology, Sweden, Stockholm (2014)
9. Lei, X., Mao, L.: Dynamic response analyses of vehicle and track coupled system on track transition of conventional high speed railway. J. Sound Vib. **271**, 1133–1146 (2004)
10. Li, C., Zhang, W., Wang, X., Pan, B., Zhu, H., Spence, B.F., Jr.: Modeling dynamic responses of a cross-river road shield tunnel under stochastic vehicle loads. Tunn. Undergr. Space Technol. **102**, 103432 (2020)
11. Varandas, J.N., Paixao, A., Fortunato, E.: A study on the dynamic train-track interaction over cut-fill transitions on buried culverts. Comput. Struct. **189**, 49–61 (2017)
12. Chen, B., Sun, L.: The impact of soil properties on the structural integrity of high-fill reinforced concrete culverts. Comput. Geotech. **52**, 46–53 (2013)
13. Ma, Q., Ku, Z., Xiao, H., Hu, B.: Calculation of earth pressure on culvert under lying flexible subgrade. Results Phys. **12**, 535–542 (2019)
14. Song, D., Chen, B., Khan, A.: Analytical solution of the vertical earth pressure on load-shedding culvert under high fill. Comput. Geotech. **120**, 103495 (2020)
15. EN 1992–1-1:2004 Eurocode 2: Design of concrete structures. European Committee for Standardization. Brussels, Belgium (2004)

16. Bońkowski, P., Bobra, P., Zembaty, Z., Jędraszak, B.: Application of rotation rate sensors in modal and vibration analyses of reinforced concrete beams. Sensors **20**(17), 4711 (2020)
17. Zhuang, Y., Cui, X., Hu, S.: Numerical simulation and simplified analytical method to evaluate the displacement of adjacent tunnels caused by excavation. Tunn. Undergr. Space Technol. Incorporating Trenchless Technol. Res. **132**, 104879 (2023)
18. Sedarat, H., Kozak, A., Hashash, Y.M.A., Shamsabadi, A., Krimotat, A.: Contact interface in seismic analysis of circular tunnels. Tunn. Undergr. Space Technol. **24**(4), 482–490 (2009)
19. Maleska, T., Beben, D.: Effect of the soil cover depth on the seismic response in a large-span thin-walled corrugated steel plate bridge. Soil Dyn. Earthq. Eng. **166**, 107744 (2023)
20. Gomes, R.C., Gouveia, F., Torcato, D., Santos, J.: Seismic response of shallow circular tunnels in two-layered ground. Soil Dyn. Earthq. Eng. **75**, 37–43 (2015)
21. EN 1993–1–1:2005 Eurocode 3: Design of steel structures. European Committee for Standardization. Brussels, Belgium (2005)

Partial Replacement of Portland Cement by Stone Cutting Sludge in Mortars – Hygrometric Performance

Fernando G. Branco[1(✉)], José Marcos Ortega[2], Luis Marques[1], and Luis Pereira[1]

[1] ISISE, ARISE, Department of Civil Engineering, University of Coimbra, Coimbra, Portugal
fjbranco@dec.uc.pt

[2] Departamento de Ingeniería Civil, Universidad de Alicante, Alicante, Spain

Abstract. The Construction industry is currently facing major challenges, trying to adapt itself to the growing concern with problems related to concepts of sustainability and reduction of the impact on global warming. The production of Portland cement is associated with the production of high amounts of CO_2, a product that is frequently blamed for the general temperature increase in the planet. In order to try to minimize the impact of this concrete constituent, research has been carried out in order to reduce the amount of Portland clinker present in the concrete. The reduction of the clinker content in concrete can be obtained through its total or partial replacement by alternative materials, namely residues from other industries. This replacement also makes it possible to contribute to sustainability in Construction, allowing the implementation of the circular economy concept, with the reduction of waste, and the amount of natural resources and raw materials used. This communication analyzes and compares the performance of mortars produced with two types of cement (CEM I and CEM III), analyzing the influence of partial replacement of the binder by a residue from the processing industry of natural stone (stone cutting sludge) on the performance of mortars in the presence of water.

Keywords: mortar · cement · rock cutting waste · sludge · hygrometric properties · life4stone

1 Introduction

Production of waste represents one of the main constraints of the ornamental stone industry, both in its extractive and transforming components. The low utilization rate of ornamental rock quarries, with values ranging between 20%–30% for sedimentary limestone, is the main reason for the accumulation of large amounts of waste in landfills, with the well-known environmental consequences. In terms of processing units, it is estimated that waste production can vary between 20%–40%. The largest portion of this waste assumes the form of sludge from cutting and sawing stone (very small-sized particles resulting from the original rock, mixed with water).

J. F. L. C. Semião et al. (Eds.): INCREaSE 2023, ASST, pp. 289–299, 2023.
https://doi.org/10.1007/978-3-031-44006-9_22

One of the main challenges that affect the ornamental stone industry is the valorization of the produced waste: Rock blocks and debris of stone produced in quarries, which are deposited in heaps, and sludge producing in the rock-cutting plants). The solutions found so far consist essentially in waste disposal in landfills, located near the factories or transport of the waste to landfills specifically developed for this purpose, usually abandoned quarry sites. As an alternative, sludge is frequently sent to cement plants. Larger particle-sized waste is also transported to other industries that manufacture calcium carbonate derivatives (lime and other products). However, the percentage of waste reintroduction in the production cycles of other activities still represents a reduced percentage of the produced waste, making it necessary to identify new possibilities for the application of waste.

Concrete is the most used building material in the world. Its constitution contains cement produced from Portland clinker, a material whose manufacturing process gives rise to considerable amounts of CO_2. In order to turn this material more sustainable and environmentally friendly, numerous studies have been carried out to reduce the amount of cement used, partially replacing it with industrial waste with pozzolanic properties [1–6]. Industrial by-products such as blast furnace slag, fly ash or silica fume have been studied and characterized to be used as cement replacements [6–8].

The incorporation of stone dust from stone cutting residues in the manufacture of new materials has also been studied [9, 10]. Stone residues of smaller size (filler) are commonly used in concrete, namely self-compacting concrete [11]. The performance of concrete containing limestone filler as an addition has been studied, with its influence being quantified in terms of workability, reduced porosity and permeability [12]. The synergistic effect of the interaction of limestone filler and fly ash in improving the mechanical properties of cementitious materials has been documented [13]. Studies aiming at integrating limestone filler in alternative cements to Portland clinker have also been carried out [14]. The use of carbonated rock dust residues in products applicable to civil construction has been addressed in laboratory studies, such as the manufacture of brick-type constructive elements, with the addition of cementitious binders in percentages of around 10 to 18% [15], or the addition of vegetable fiber residues, namely cotton, showing promising results [16].

Within the scope of the Life4Stone project, research has been carried out in order to study and characterize solutions that allow the development of composites integrating waste and that can be used in applications in the Civil Construction field, such as the production of pieces of urban furniture.

This communication presents some preliminary results obtained within the scope of this project. Partial replacement of cement makes it possible to reduce the clinker content used in the mortar production, leading to better environmental performance. Two types of cement produced in accordance with the EN 197–1 [17] standard (CEM I and CEM III) were studied, and different series were produced for comparison in which the amount of cement was partially replaced by an equal mass of stone cutting mud. The influence of this replacement on mortar performance was evaluated using three parameters: mechanical strength, water absorption by capillarity and by immersion. The study was carried out over time, and results obtained in mortars up to 180 days are presented. Further research is currently under way on the use of stone cutting sludge as partial replacement of Portland cement in cementitious materials. The research will extend to concrete production, and different types of binders and aggregates will be included as research variables.

2 Materials and Methods

This section characterizes the materials used to produce the tested samples and summarily describes the test methods and laboratory procedures followed.

2.1 Materials

All mortars were manufactured with the same aggregate (river sand) batch. Two different binders (ordinary Portland cement, CEM I 42.5 R and a commercial cement with blast furnace slag CEM III/B 32.5 N/SR [17]) have been tested (see Table 1). Figure 1 presents the particle-size distribution of the used river sand and the stone cutting sludge.

Table 1. Description of the binders used.

Binder (EN 197–1)	Description (percentages in weight)
CEM I 42.5R	Commercial cement consisting of ≥ 95% clinker and ≤ 5% minority components
CEM III/B 32.5 N/SR	Commercial cement consisting of 70% blast furnace slag, 25% clinker and 5% minority components

A mortar mix mass proportion 1:3 (binder:aggregate) was taken as reference. Water/binder ratio (W/C) of 0.5 was used in all tested series. In addition to a reference series produced solely with cement, comparison series were produced in which the amount of cement was partially replaced by an equal mass of stone cutting sludge. The influence of this replacement on mechanical strength, water absorption by capillarity and by immersion was evaluated. The study was carried out over time, with results obtained in mortars up to 180 days being presented.

The different studied series specimens were given the designation "xPC-yy", where "x" designates the type of binder (1 for CEM I and 3 for CEM III) and "yy" represents the amount of cement replaced by sludge, in mass percentage (0, 5, 10, 15 or 20%).

2.2 Methodology

For each series, 160x40x40mm specimens were produced, intended for mechanical characterization and for carrying out water absorption tests. Three specimens were used in each test. After demoulding, the specimens were kept in a climatic chamber at a controlled temperature (20 ± 2 °C) and relative humidity (98 ± 2%) until the date of the tests.

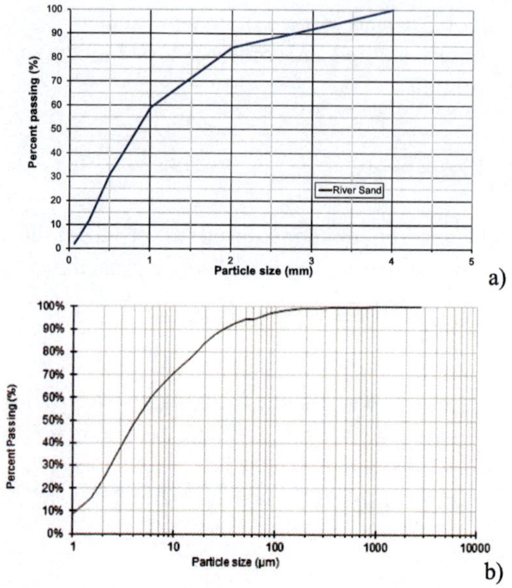

Fig. 1. Particle-size distribution: a) River sand; b) Stone cutting sludge.

Mechanical Strength. Mechanical strength tests (compression and flexural tests) were carried out in accordance with EN 1015–11 [18] standard. In the flexural test, the specimens were placed in the test machine, subjected to a 3-point bending load, and a constant load was applied until failure. The final flexural strength of each sample was calculated using Eq. 1.

$$R_F = 1.5 \times \frac{F \times l}{b \times d^2} \tag{1}$$

where, R_F is the flexural strength (MPa), F is the maximum load applied during the test at the moment of rupture (N), l is the span between supports [mm], b is the width [mm] and d the height [mm} of the test specimen cross-section. The test procedure was repeated for all samples.

The compressive strength tests were carried out using the halves of the test pieces that had previously been subjected to the flexural strength test. The load was applied to a 1600 mm^2 surface, and the result was obtained by applying Eq. 2.

$$R_C = \frac{F}{A} \tag{2}$$

where, R_C is the compressive strength final value [N/mm^2], F is the maximum rupture Load [N] and A is the cross-section of the applied load [mm^2]. The test procedure was repeated for all samples.

Water Absorption by Capillarity. One of the most important ways of ingress of aggressive substances in a construction material is through water absorption [19], especially in the case of mortar coatings and plasters [20, 21]. Therefore, the study of this property in the analyzed lime mortars is relevant. In this research, the coefficient of water absorption by capillarity (Cc coefficient [22]) was determined by performing the test procedure described in the standard EN 1015–18 [22]. Before starting the test, the specimens were dried in a ventilated oven at (60 ± 5 °C) temperature until reaching a constant mass. Distilled water was used in this test. Three different prismatic samples were tested for each mortar type and curing condition studied.

Water Absorption by Immersion after 48 h. This test was performed according to specification E394 [23] of the Portuguese National Laboratory of Civil Engineering (LNEC). It allows obtaining the water absorption after total immersion of the material in water for 48h. For performing this test, the manufactured prismatic samples were used. The preconditioning procedure of the specimens before the test consisted of drying them using an oven at a temperature of (60 ± 5 °C) until they reached a constant mass. Once finished that process, the samples were allowed to cool in a desiccator and they were weighted in order to obtain their dried mass. The test consisted of totally immersing the specimens in a hermetic box filled of water for 48 h. After finishing the 48 h period, the specimens were taken out of the box, cleaned using a damp cloth and then weighted for obtaining the saturated mass of each sample.

The parameter determined with this test is the percentage of water absorption by immersion after 48 h, calculated using the expression

$$W_{48h} = \frac{m_2 - m_1}{m_1} \cdot 100 \tag{3}$$

where W_{48h} is the percentage of water absorption by immersion after 48 h [%], m_2 is saturated mass of the specimen after immersion [g], m_1 is the mass of the dried sample prior to immersion [g]. For each of the mortar series, three specimens were tested at each testing age.

3 Results and Discussion

This section presents and discusses the results obtained in the laboratory tests carried out to assess the parameters under study.

3.1 Mechanical Strength

Laboratory tests to quantify the mechanical strength of mortar mixes under flexural or compressive loads were carried out. Table 3 presents the results obtained in flexural strength tests, while Table 2 presents the results regarding compressive tests.

The results show that, in terms of flexural strength, no clear tendency related to binder replacement is visible. Strength losses of up to about 10% for age 7 days were observed at higher replacement percentages. However, for the age of 28 days, many of the mixtures recovered, with even higher strengths than the reference being observed in several series.

With regard to compressive strength, mixtures produced with CEM III cement are apparently more sensitive to replacement, with this sensitivity especially evident in 28-day results. These mixtures show an approximately linear loss of strength with increasing

Table 2. Average mortar flexural strength (MPa).

Series	7d	28d	90d	180d
1PCREF	4.10	4.97	4.87	5.21
1PC5	3.96	5.11	4.73	4.88
1PC10	4.11	4.69	4.72	5.02
1PC15	3.93	4.83	4.44	4.76
1PC20	3.89	5.80	4.30	4.68
3PCREF	4.71	5.29	6.26	6.98
3PC5	4.53	5.51	6.25	5.50
3PC10	4.56	5.59	6.02	5.01
3PC15	4.33	5.22	6.07	5.06
3PC20	4.36	5.36	5.61	5.23

substitution percentage. The strength loss for 20% cement replacement is about 19% of the reference. In mixtures containing CEM I, it is observed that, for cement replacement up to 15%, the performance improves (up to 4%) when compared with the reference. The strength loss for 20% replacement is only 7% of the reference.

Table 3. Average mortar compressive strength (MPa).

Series	7d	28d	90d	180d
1PCREF	22.70	27.37	30.49	28.10
1PC5	19.97	28.11	27.91	25.84
1PC10	20.84	28.52	28.50	25.36
1PC15	21.34	27.07	27.47	26.56
1PC20	20.24	25.54	26.68	22.81
3PCREF	23.59	32.99	37.74	29.65
3PC5	21.28	31.79	36.85	33.34
3PC10	22.43	29.29	34.13	28.21
3PC15	22.69	28.30	33.69	31.00
3PC20	19.54	26.75	31.24	29.33

It should be noted that, despite the strength reduction observed, mixtures with cement CEM III exhibited better global mechanical performance than mixtures containing CEM I. In fact, for a replacement percentage of 20%, 3PC20 series presents compressive strength of about 26.8MPa, while the reference mix produced with CEM I possessed 27.4MPa strength. One may take into account that CEM III contains only 25% of clinker.

Thus, the amount of clinker in series 3PC mortar is considerably lower than that of series 1PC even when no cement replacement is applied.

It is interesting to notice a general slight reduction of strength observed for 180 days old specimens. This reduction in the mechanical performance was not expected, and may be related with the presence of micro-cracks associated with mortar retraction phenomena along time. Further observation is required to clarify this observation.

3.2 Water Absorption by Capillarity

Water absorption by capillarity tests were performed and the results obtained are shown in Table 4.

Table 4. Water absorption coefficient Cc (x10^$^{-5}$kg.m-^{2}.min$^{-1/2}$).

Series	28d	60d	90d	180d
1PCREF	11.5	12	5	10
1PC5	9.5	10	6	8
1PC10	6.5	7	7	6
1PC15	8.5	8	8	13
1PC20	11.5	12	3	12
3PCREF	4.5	4	6	10
3PC5	5.5	5	7	8
3PC10	6.5	6	2	9
3PC15	4.5	5	4	8
3PC20	9.0	9	7	8

The results obtained revealed a tendency for the absorption coefficient to decrease as the cement replacement percentage increases, up to 10% substitution percentages, in the mixtures produced with CEM I. For mixtures with higher percentages, the coefficient increases again, showing similar values to the reference when 20% replacement occurs. Regarding mixtures containing CEM III, the absorption coefficient increases slightly up to 10% substitution. The value obtained at 15% replacement is similar to that observed in the reference series, increasing again to a replacement percentage of 20%. It should be noted that the values observed in mixtures with CEM III are tendentially lower than those observed in specimens with CEM I for the same replacement percentage.

Over time, the evolution of the capillary absorption coefficient shows a trend to decrease over time. This tendency is more evident in mixtures with lower replacement percentages, up to 10%. Results obtained from 180 days old specimens present a slight absorption increase when compared with 90 days old ones. These results were not expected. One possible explanation may be the existence of a higher number of thin pores for older ages, which would be able to help the capillarity process when compared with a microstructure containing a larger amount of larger pores, characteristic of younger ages. Another possible explanation may be related with the degradation on mechanical properties also registered in 180 days old specimens. In any event, further tests must be carried out, namely mercury intrusion tests that permit acquiring more information on actual the pore dimensions for older ages.

3.3 Water Absorption by Immersion after 48 h

Table 5 presents the results obtained in water absorption by immersion tests. This parameter was quantified for 28 and 180 days old specimens. The results reveal that all tested series have a water absorption greater than 7%. A slight tendency for water absorption reduction along time is noticed. However, there is no visible tendency for variation for this parameter with the percentage of cement replacement.

Table 5. Immersion water absorption in 48h (%).

Series	28d	180d
1PCREF	7.52	6.99
1PC5	7.71	7.32
1PC10	7.60	7.19
1PC15	7.73	7.28
1PC20	7.56	7.32
3PCREF	7.58	6.61
3PC5	7.21	6.19
3PC10	7.60	7.14
3PC15	7.45	6.73
3PC20	7.70	6.87

4 Final Remarks

The present work relates preliminary experimental results obtained within the scope of project Life4Stone regarding partial replacement of cement by rock-cutting sludge waste. This replacement allows reducing the clinker content in mortars, thus contributing to better environmental performance. However, the presence of the sludge in the mix may affect the mortar performance. Thus, this work aimed at assessing the mortar performance changes when the sludge is added to the mixture.

Laboratory tests were carried out to quantify the mortar performance according to three parameters: mechanical strength, water absorption by capillarity and by immersion. Test results were obtained in mortars up to 180 days.

The results show that, in terms of flexural strength, no clear tendency related to binder replacement is visible. Strength losses of up to about 10% for age 7 days were observed at higher replacement percentages. However, above 28 days of age, many of the mixtures recovered, and some series exhibited higher performance than the reference series.

In what concerns compressive strength, the performance exhibited different trends for the two types of cement tested. Mixtures produced with CEM III cement are apparently more sensitive to replacement. The compressive strength for these mortars presents a linear loss with increasing substitution percentage. For instance, when 20% of cement is replaced, the mortar presents compressive strength 19% lower than that of the reference. For mixtures containing CEM I, cement replacement up to 15% leads to performance improvement (up to 4%) when compared with the reference. Above this replacement percentage, strength loss when compared with the reference was observed.

Regarding the absorption coefficient, one may notice, in the mixtures produced with CEM I, a tendency for the reduction of this parameter as the cement replacement percentage increases, up to 10% replacement amount. For mixtures with higher replacement percentages, the coefficient increases again. For 20% replacement, the absorption coefficient shows similar values to the reference. Regarding mixtures containing CEM III, the absorption coefficient increases slightly up to 10% substitution. The value obtained at 15% replacement is similar to that observed in the reference series, increasing again to a replacement percentage of 20%. It should be noted that the values observed in mixtures with CEM III are tendentially lower than those observed in specimens with CEM I for the same replacement percentage. Over time, the evolution of the capillary absorption coefficient shows a trend to decrease over time. This tendency is more evident in mixtures with lower replacement percentages, up to 10%.

The results reveal that all tested series have a water absorption greater than 7%. A slight tendency for water absorption reduction along time is noticed. However, there is no visible tendency for variation for this parameter with the percentage of cement replacement.

The results obtained lead to conclude that a small partial cement replacement (up to 15%) by rock-cutting sludge does not affect mortar hygroscopic performance negatively. The replacement may improve the mechanical strength (as seen in CEM I mortars), and also contributes to reduce the water absorption by capillary.

Acknowledgements. This work was partly financed by FCT / MCTES through national funds (PIDDAC) under the R&D Unit Institute for Sustainability and Innovation in Structural Engineering (ISISE), under reference UIDB / 04029/2020, and under the Associate Laboratory Advanced Production and Intelligent Systems, ARISE, under reference LA/P/0112/2020. This works was also partly funded by FEDER through the COMPETE 2020 program, Portugal 2020, within the scope of project CENTRO-01–0247-FEDER-047067 (LIFE4STONE). Authors would like to thank Cementos Portland Valderrivas, S.A. for providing the cements used in this work.

References

1. Berra, M., Mangialardi, T., Paolini, A.E.: Reuse of woody biomass fly ash in cement-based materials. Constr. Build. Mater. **76**, 286–296 (2015). https://doi.org/10.1016/j.conbuildmat.2014.11.052

2. Biricik, H., Aköz, F., Berktay, I., Tulgar, A.N.: Study of Pozzolanic properties of wheat straw ash. Cem. Concr. Res. **29**(5), 637–643 (1999)
3. Wang, S., Miller, A., Llamazos, E., Fonseca, F., Baxter, L.: Biomass fly ash in concrete: mixture proportioning and mechanical properties. Fuel **87**, 365–371 (2008)
4. Cheah, C.B., Ramli, M.: The implementation of wood waste ash as a partial cement replacement material in the production of structural grade concrete and mortar: an overview. Resour. Conserv. Recycl. **55**, 669–685 (2011)
5. Sata, V., Jaturapitakkul, C., Kiattikomol, K.: Influence of Pozzolan from various by-product materials on mechanical properties of high-strength concrete. Constr. Build. Mater. **21**, 1589–1598 (2007)
6. Teixeira, E., Mateus, R., Camões, A., Bragança, L., Branco, F.G.: Comparative environmental life-cycle analysis of concretes using biomass and coal fly ashes as partial cement replacement material. J. Clean. Prod. **112**(4), 2221–2230 (2016)
7. Ferreira, C., et al.: Comparative study about mechanical properties of strutural standard concrete and concrete with addition of vegetable fibers. Mater. Res. (2017). https://doi.org/10.1590/1980-5373-mr-2016-0905
8. Oliveira, M., Ribeiro, A., Branco, F.G.: Shrinkage of self-compacting concrete. A comparative analysis. J. Build. Eng. **9**, 117–124 (2017). https://doi.org/10.1016/j.jobe.2016.12.002
9. Sua-Iam, G., Makul, N.: Utilization of limestone powder to improve the properties of self-compacting concrete incorporating high volumes of untreated rice husk ash as fine aggregate. Constr. Build. Mater. **38**, 455–464 (2013). https://doi.org/10.1016/j.conbuildmat.2012.08.016
10. Turgut, P.: Cement composites with limestone dust and different grades of wood sawdust. Build. Environ. **42**, 3801–3807 (2007). https://doi.org/10.1016/j.buildenv.2006.11.008
11. Silva, P., Brito, J.: Experimental study of the porosity and microstructure of selfcompacting concrete (SCC) with binary and ternary mixes of fly ash and limestone filler. Constr. Build. Mater. **86**, 101–112 (2015). https://doi.org/10.1016/j.conbuildmat.2015.03.110
12. Courard, L., Michel, F.: Limestone fillers cement based composites: effects of blast furnace slags on fresh and hardened properties. Constr. Build. Mater. **51**, 439–445 (2014). https://doi.org/10.1016/j.conbuildmat.2013.10.076
13. Weerdt, K., Kjellsenb, K.O., Sellevolda, E., Justnesa, H.: Synergy between fly ash and limestone powder in ternary cements. Cem. Concr. Compos. **33**(1), 30–38 (2011). https://doi.org/10.1016/j.cemconcomp.2010.09.006
14. Carrasco, M.F., Menéndez, G., Bonavetti, V., Irassar, E.F.: Strength optimization of tailor-made cement with limestone filler and blast furnace slag. Cem. Concr. Res. **35**, 1324–1331 (2005). https://doi.org/10.1016/j.cemconres.2004.09.023
15. Mageeda, A.A., AbdelHaffez, G.S.: Utilization of limestone dust in brick making. J. Eng. Sci. Assiut. Univ. **40**(3), 913–922 (2012)
16. Algin, H.M., Turgut, P.: Cotton and limestone powder wastes as brick material. Constr. Build. Mater. **22**, 1074–1080 (2008). https://doi.org/10.1016/j.conbuildmat.2007.03.006
17. European Committee for Standardization EN 197–1. Cement - Part 1: Composition, specifications and conformity criteria for common cements; European Committee for Standardization, Brussels, Belgium (2011)
18. European Committee for Standardization EN 1015–11. Methods of test for mortar for masonry Part 11: Determination of flexural and compressive strength of hardened mortar; European Committee for Standardization, Brussels, Belgium (1999)
19. Ortega, J.M.; Sánchez, I.; Climent, M.A.: Influence of environmental conditions on durability of slag cement mortars. In: Proceedings of the 2nd International Conference on Sustainable Construction Materials and Technologies, Ancona, Italy, 28–30 June 2010 (2010)
20. Santos, T., Faria, P., Silva, V.: Can an earth plaster be efficient when applied on different masonries? J. Build. Eng. **23**, 314–323 (2019). https://doi.org/10.1016/j.jobe.2019.02.011

21. Columbu, S., Sitzia, F., Ennas, G.: The ancient Pozzolanic mortars and concretes of Heliocaminus baths in Hadrian's Villa (Tivoli, Italy). Archaeol. Anthropol. Sci. **9**, 523–553 (2017). https://doi.org/10.1007/s12520-016-0385-1
22. European Committee for Standardization EN 1015–18. Methods of Test for Mortar for Masonry—Part 18: Determination of Water Absorption Coefficient Due to Capillary Action of Hardened Mortar; European Committee for Standardization: Brussels, Belgium (2002)
23. Laboratorio Nacional de Engenharia Civil Especificações LNEC E 394. Betões—Determinação da Absorção de Água por Imersão; Laboratorio Nacional de Engenharia Civil Especificações: Lisboa, Portugal (1993)

Evaluation of Different Sustainable Solutions to Enable an Increased Utilization of Outdoor Swimming Pools

A. Inverno[1], M. J. Oliveira[1](✉), J. Monteiro[1], C. Cabrita[1], F. Farinha[1], G. Matias[2], and F. Carmo[3]

[1] Universidade do Algarve, Faro, Portugal
{ainverno,mjolivei}@ualg.pt
[2] Universidade de Coimbra, CERIS/Itecons, Coimbra, Portugal
[3] Cristalpools, Portimão, Portugal

Abstract. Over the last decades the construction of swimming pools for recreation activities has grown exponentially, especially in the countries of southern Europe where tourism is one of the main economic activities. In Portugal, swimming pools are located in hotels and resorts but mostly in villas, where they reach hundreds, or even thousands. They can be considered a leisure equipment, but require considerable amounts of energy to operate. The energy requirements of such pools are especially relevant when, to keep the water temperature comfortable beyond summer months, heating systems are used during long periods of the year. At this level, conventional systems used to heat the water commonly make use of non-renewables forms of energy, which besides imposing high energy costs, cause a negative environmental impact. In this scope, the approach here presented describes the implementation of a sustainable swimming pool. In order to improve the energy of such pools two goals should be considered. The first goal is to suppress non-environment friendly systems and, then, find out how to integrate different solar energy collecting systems, either to heat the water directly or to allow thermal storage for later usage. At this level Phase Change Materials can be used. In order to fully implement it, some relevant issues need to be solved beforehand, as for instance the individual and integrated systems' efficiency, the control methods to operate them and, of course, the investment costs. One second goal is to reduce energy losses, through a pool cover that besides energy saving, allows safety and easy operation. This paper evaluates and compares different solutions that allow heating the water of swimming pools using renewable energy sources, to increase their utilization throughout the year.

Keywords: Sustainable Outdoor Swimming Pools · Solar Energy · Thermal Losses

1 Introduction

Swimming pools are commonly perceived as a luxury equipment. In fact, it could be right, but they can actually serve a variety of purposes beyond leisure. In addition to providing a source of relaxation, swimming pools can also offer benefits in terms of

J. F. L. C. Semião et al. (Eds.): INCREaSE 2023, ASST, pp. 300–309, 2023.
https://doi.org/10.1007/978-3-031-44006-9_23

physical and mental wellness, rehabilitation from injuries, athletic training, and above all they can be seen as an investment for touristic activities. However, due to the potential environmental impact of the construction of swimming pools and their operation, there is often debate about their societal value. In order to mitigate negative environmental impacts, it is essential to develop sustainable solutions in terms of utilization of raw materials, water, and energy. This includes reusing demolition materials whenever possible, implementing water conservation measures through effective monitoring and control, and most importantly, utilizing renewable energy sources to provide the necessary heating for swimming pool's water. It also means using water loss minimization techniques and assessment solutions.

When it comes to the heating swimming pool's water, solar energy is a highly recommended renewable energy source. However, other renewable sources such as geothermal energy can also be used, either directly or through heat pumps. Regardless of the energy source, a heating system of some complexity is always required. The use of direct geothermal energy is the least complex option, as it only requires a circulation pump and a heat exchanger, as well as pipes to access the warm flow from the ground. However, this solution is only feasible if there is a nearby geothermal source, which in most cases do not occur. Additionally, it is important to consider the capital costs of implementing a geothermal solution, as its performance is typically only cost-effective for large projects and may not be suitable for smaller, isolated swimming pools. Therefore, while geothermal energy is a viable option for heating swimming pools, it may not be the most feasible or practical solution in all cases. Ultimately, the choice of a renewable energy source for heating swimming pools will depend on factors such as local availability, project size, and budget constraints. Regardless of the chosen source, it is important to consider the environmental impact and seek out sustainable solutions for swimming pool heating.

The main focus of this paper, done in the scope of the EcoPool+++ project - *Innovative heated pools with reduced thermal losses and integration of SMART energy plus water management systems*, is to evaluate and compare different sustainable solutions that allow heating the water of swimming pools using renewable energy sources, with the focus on an increase of their utilization throughout the year. Among the solar energy heating processes, those envisaged in this project are, the solar collectors, the ground-collector and the swimming pool surface, all being effective only when there is enough solar radiation. It is for this reason that the water surface coverage should also be considered as a system that influences energy consumption, as it contributes to heat losses reduction, and so, should be considered a complement to the other solutions.

The work carried out was focused on finding out how to maximize the solar gains using those systems, either individually or combined, looking for an extension of the usage periods of the pools, especially outside the summer season. As will been seen, the simulations show that is not rationally feasible to use solar energy to get a constant water temperature all year around, because the cost of such solar plant investment would be unbearable. Oppositely, the solution tries to answer the following questions: what should be the size of the solar systems to increase the average water temperature not to be lower than a certain level, for instance 24 °C. And at the same time: what is the amount of the collected solar energy that is wasted?

When those solar heating systems are oversized, usually there are many periods with high water temperatures, that aren't suited by swimming pool users. The oversizing of the systems and the dissipation of the excess, even in low sized systems, should be avoided.

2 State of the Art

There are many studies on the modeling of swimming pools and on the integration of renewable energy sources. They mostly started in the first oil crisis that occurred in the remote year of 1973. Since then the scientific and technical literature have published almost uninterruptedly articles about swimming pool's energy usage and the contribution of renewable energy to its heat balance.

An analysis of an outdoor swimming pool, in Elche, with 50 m^2, in the months of June-September, with flat solar collectors without glass, is done in [1]. One of the works focus is the estimation of the evaporation losses and, for that, they experimented various correlations which results they compare with measured values, using the pool water temperature as the comparative parameter. The simulations were carried out in Trnsys, using a simple mathematical model for the pool, and data generated by Meteonorm for Alicante.

The use of Phase Change Materials (PCM) to accumulate heat which can be delivered to the water later on, is the goal of the approach tried in [2]. The authors simulated an outdoor swimming pool measuring 50 m^2 and 1.65 m deep, in three locations in NE Spain (Catalonia) using local climate data. The bottom and walls of the pool is lined with 3 cm of PCM for heat accumulation and alternatively they also simulate an exchanger with the same amount of PCM as it offers the possibility of controlling the temperature of the water in advance (about 20 h) in relation to the predictable period of use of the pool. The results show that the exchanger solution is better than just lining the interior surfaces of the pool; the higher melting point of PCM, around 28/29 ºC, is preferable to a lower value, as it increases heat exchange with water, which reduces heating times.

In [3], a new simplified numerical model to calculate surface temperature and heat transfer of radiant floor heating and cooling system is proposed and established using the conduction shape factor. Measured data, taken from some references, were used to validate the proposed numerical model and the results showed good agreement either for the surface temperature and heat transfer. The results showed that the surface temperature and heat transfer of radiant floor calculated by the proposed model agreed very well with the numerically simulated data when average water temperature changing from 25 °C to 45 °C for radiant floor heating systems and from 10 °C to 20 °C for radiant floor cooling systems.

Through paper [4] the authors search the construction of a model referring to the heat exchanges between the PCM and the water to calculate the temperatures of both media, the PCM and the water, as well as the solid fraction of the PCM, when heat exchanges occur. Moreover, they also studied the influence of flow on heat transfer performance. This work is a good example for its simplicity despite being applied for a purpose other than heating pools.

The problem of heating a swimming pool is a technical but also economical one [5]. In that work, these issues are applied to a swimming pool to extend its usage to winter time, and so a heat pump is used, combined with a PCM storage tank. An interesting part of that work resides in the use of genetic algorithms as the methodology to find out the optimal combination of source power and heat accumulation volume.

3 The Solar Energy Systems

3.1 A General Description of the Systems

Figure 1 shows the layout of the heating systems to be built in the swimming pool prototype of the Ecopool+++ project. Thermal systems are all based on solar energy. In the center of them is the swimming pool. They were thought to be able to work independently or integrated, what can be done throughout a combination of motorized valves. This topic is the subject for other publications, so they are not described in this communication.

In the following a basic description of the heating systems is made, to support a better understanding of the remaining of the text. Besides the systems which directly transfer solar energy to the swimming pool, like the solar collectors and the ground-collector, the system also includes PCM heat accumulators. PCM allow the accumulation of heat, when in excess, to keep the water at a minimum temperature, being then used to heat up the water, for instance, in the morning or even during a mild sunny day.

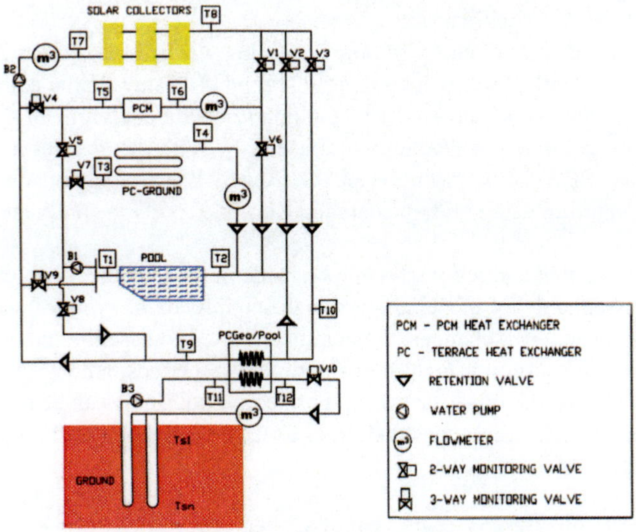

Fig. 1. General layout of the heating systems of the EcoPool+++ project.

The aforementioned systems were modeled using Trnsys 18.0 [6] software for numerical simulation, either using materials already provided by it, called "Types", or specifically built for this project, as for instance for the swimming pool, here named type 251, or the adaptation of one of them, Type 993, to build the type 293, for the ground-collector.

Other "Types" like the circulation pumps, the control valves or the controllers themselves are easily identified. The simulation platform, as shown in Fig. 2, also includes some link equations that are note visible, since they are "inside" the icons with the generic name "equation".

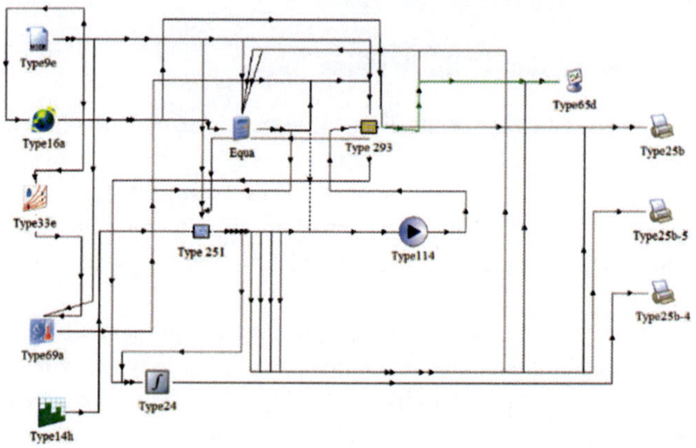

Fig. 2. Trnsys 18.0 studio layout for one of the heating systems.

Numerical Model of the Swimming Pool

The heat balance in an outdoor swimming pool is known. In this regard, we modelled the swimming pool identically to what is done in [1] and [2], using similar equations, solved with a finite difference's method. For the climatic data a dataset from the Portuguese Direção-Geral de Energia e Geologia was used [7]. With that dataset it is possible to generate files in EPW format, which is the one used by Energy Plus software. Most of the simulations were run using climatic data for Faro, the southern continental Portuguese city.

We are aware that corrections need to be made in the numerical model due to many specific factors, mainly for the wind speed in the vicinity of the swimming pool, variable influencing convection exchanges, the parameters used to describe the behavior of the water cover platform during night hours and nonuse periods, among others, affecting convection and evaporations losses. But, those corrections only can be made after monitoring the prototype, which construction is expected to be finalized by the end of the first semester of 2023.

The Sunlight Collection Systems – Solar Collectors

The solar collectors are one of the usual heating systems that can be found both in outdoor or indoor swimming pools, which started to be used a few decades ago. Among their types, the flat one is, maybe, the favorite, which led to its selection in this project. Its performance was simulated using the Trnsys type 1a, for which we only need to characterize the collector, the coordinates of its location and the tilt angle relative to the horizontal plane. The solar collector performance is, thus, computed from its efficiency equation.

For the simulation of the solar collector the following values were used to enable its efficiency equation: $\eta_0 = 0{,}78$ e $a_1 = 4{,}17$ W/(m^2.K), tilted 37 from the horizontal and south oriented, with a water flow of 75 L/m^2, and the solar radiation hitting its surface.

The Sunlight Collection Systems – Ground-Collector

In general, around swimming pools there is space where people lay loungers. Under that portion of soil, or other nearby, water recover piping can be laid to collect sun energy whenever the soil gets sunny.

In order to model it, we used the Trnsys type 993, a routine for modeling radiant heating. Its performance was then verified. We adapted it for outdoor use, and instead of delivering radiated heat, it receives heat from the sun, accumulating it in the mortar of the soil. Because it proved to be effective, we decided to use it.

The equations of this routine discretize the soil in several nodes doing the heat balance around them and solving the equations by the finite difference method. The piping layout as well as its dimensions are important parameters, beside others, which have to be considered to compute the heat gains to the water.

When the mortar temperature is higher than the swimming pool water a circulation pump is activated and the water flows throughout the pipes laid inside the mortar, this way removing the heat.

As will be seen later, the solar-collector system is not as efficient as one would expect.

The Ground Heat Accumulator System

It was expected that the use of solar collectors would lead to high water temperatures, especially in summer. This might sometimes become not pleasant to many of the swimming pool users. When the water reaches high temperatures the water quality control is also more difficult to perform. Even with small solar collector areas this happens, therefore it would be important to evaluate the use of ground heat accumulation, to assure a constant water temperature close to 24 °C.

However, when the solar collectors' area is more convenient to swimming pool heating purposes it seems that the effect of heat accumulations in the ground is small. Another important issue that has to be taken for an economic assessment of this system is the one related with "what to do with the accumulated heat". If the system is to be a simple heat dissipator by means of using a soil accumulator it may not be worth the investment. For this system to be a true energetic system it should provide higher temperatures and energy amounts, either to be "directly" transferred to the swimming pool water, or to allow running a heat pump with the same purpose, or even to use it as a house heating system.

In the following we compare different solutions for water heating and thermal storage. In terms of heat accumulation in the ground, we consider that 50 m long deep holes are used in the soil.

3.2 Some Results of the Simulations

In this section we present the results of the model previously described, computed for the climatic conditions of Faro, in the south of Portugal. As said above, the simulations were initially run individually for the each of the described systems, and then combined.

In all, it can be verified that almost always two identical results can be verified, one for the swimming pool without cover and other with it during the night, i.e. between 20:00 and 8:00 of following day. The influence of the cover in the temperature losses of the swimming pool at the water surface was modelled using some reduction coefficient's affecting the convection and the evaporation phenomena. The results, as shown in Fig. 3, demonstrate an increase of more than 2 °C when the pool surface is covered.

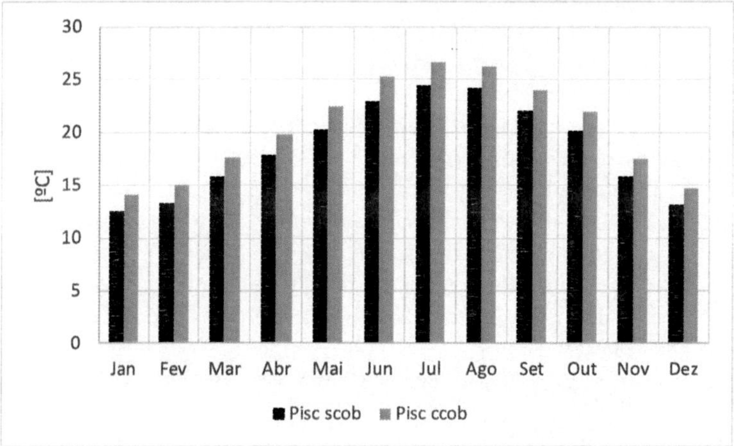

Fig. 3. Monthly mean water temperatures: Pisc scob – without water surface cover and Pisc ccob – with water surface cover.

Despite the consistency of results, we consider some adjustments should be addressed regarding this section of the numerical modeling of the swimming pool when comparing with the testing phase with the prototype fully operational.

Next figure shows, in a similar way, the differences in the monthly water temperature, between the a) a sole swimming pool b) a swimming pool with a 25 m^2 of ground-collector and c) swimming pool with a 30 m^2 of the solar collector (Fig. 4).

While the addition of 25 m^2 of ground-collector (b) does not have a high impact in the water temperature when compared with the results of the sole swimming pool, the impact of having an area of 30 m^2 of solar collectors is considerable. Obviously, the greater the area with collectors, the greater the difference would be, but, at the same time, overproduction in summer months will be responsible for energy loss.

Another metric that matters for the global analysis of the Ecopool+++ is what system is more valuable for getting the lower number of hours in which the swimming pool water temperatures are lower than, for instance, 24 °C. The total number of hours for this condition was calculated, in the period from March to October, after simulating all the systems. These results are shown in Fig. 5. The months outside that interval seems to be less suitable for usage, in outdoor swimming pools. That is the reason for this work.

The numbers near the corners of the defined polygons, in Fig. 5, represent the total number of hours in a month in which the water temperature is lower than 24 °C; thus, the lower lengths are for the summer months. The best results are achieved through the

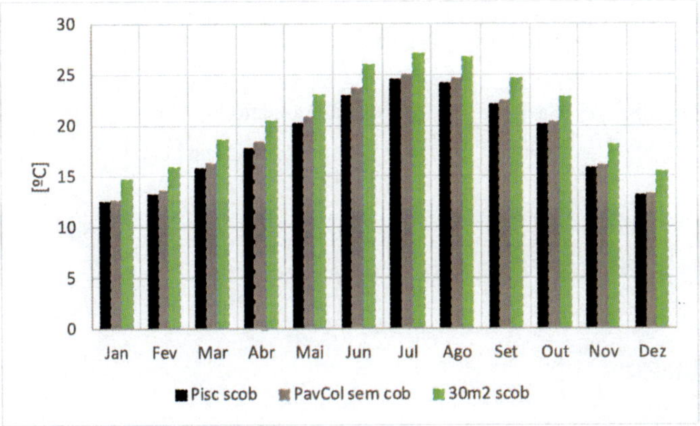

Fig. 4. Monthly mean water temperatures, all case without water surface cover: in black a), gray b) and green c).

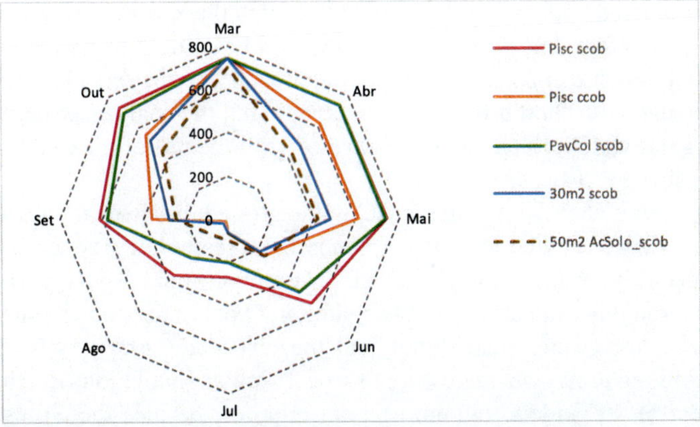

Fig. 5. Variation of the monthly number of hours with water temperature below 24 °C, where: "Pisc scob" corresponds to swimming pool without cover, "Pisc ccob" corresponds to swimming pool with cover, "PavCol scob" corresponds to swimming pool with ground-collector without cover, "30 m^2 scob" corresponds to an installation of 30m^2 of solar collectors without cover, "50 m^2 AcSolo scob" corresponds to the installation of 50 m^2 of soil accumulation without cover.

introduction of solar collectors and also by activating the water surface cover system to the swimming pool during nonuse night hours. All systems can be compared with the simpler case: the sole swimming pool without any other device (Pisc scob).

Table 1 presents the numerical results shown in the previous figures and some additional ones. The designations that have been used until now remain: scob - without cover, ccob - with cover, PavCol – using a Ground-collector; the solar collector areas are quite evident, and finally Alone means the swimming pool with no systems.

Table 1. Number of hours and percentage of total number of hours that the water temperature was greater than 24 °C, between March and October.

For T ≥24 °C	Alone scob	Alone ccob	PavCol scob	PavCol ccob	Solar Collector			
					30 m^2 scob	30 m^2 ccob	50 m^2 scob	50 m^2 ccob
Hours	1152	2570	1490	2886	2861	4140	3044	4286
%	20%	44%	25%	49%	49%	70%	52%	73%

4 Conclusions

As has been quoted in the text, one of the goals of the Ecopool+++ Project is the use of only renewable energy for getting more usable hours in outdoor swimming pools, even knowing that it is a technical and mainly economically issue.

The systems here considered allow collecting solar energy. Even the water surface is a collector of solar radiation, which is a source of energy that on average can be considered predictable. Therefore, this gives rise to two types of question: the first one is the long-term prediction, that can be solved with an approach like the one here developed, the second is related with short term behavior, comprising the analysis of only a few days. Short time solar variations can be compensated with the use of thermal accumulation, for instance using PCM.

The accumulation of "remaining heat" in the ground, has been tried with 50m deep holes in the soil, in one of the simulated systems, but it seems not to be so energy valuable when compared with direct transfer to the swimming pool. Anyway, this evaluation should be thoroughly evaluated with the swimming pool under construction.

As all the results presented throughout the text were computed for the climatic conditions of Faro it may be interesting to repeat similar simulations in other regions.

A project of an outdoor swimming pool requires, besides the issues and results discussed here, an environmental analysis and an economic evaluation.

Acknowledgment. The project ECOPOOL+++ is funded by FEDER, PROJETOS DE I&DT EMPRESAS EM COPROMOÇÃO, Programa Operacional Regional do Algarve, contact n. ALG-01-0247-FEDER-047231. The professional expertise and valuable cooperation of all the partners are gratefully acknowledged.

References

1. Ruiz, E., Martinez, P.J.: Analysis of an open-air swimming pool solar heating system by using an experimentally validated TRNSYS model. Sol. Energy **84**, 116–123 (2010)
2. Zsembinszki, G., Farid, M.M., Cabeza, L.F.: Analysis of implementing phase change materials in open-air swimming pools. Sol. Energy **86**, 567–577 (2012)
3. Wu, X., Zhao, J., Olesen, B.W., Fang, L., Wang, F.: A new simplified model to calculate surface temperature and heat transfer of radiant floor heating and cooling systems. Energy Build. **105**, 285–293 (2015)

4. Wu, S., Fang, G.: Dynamic performances of solar heat storage system with packed bed using myristic acid as phase change material. Energy Build. **43**, 1091–1096 (2011)
5. Li, Y., Ding, Z., Du, Y.: Tecnho-economic optimization of open-air swimming pool heating system with PCM storage tank for winter applications. Renew. Energy Build. **150**, 878–890 (2020)
6. Trnsys 18.0 – Transient Simulation Tool; Doc 04 - Mathematical Reference; Doc 07 - Programmers Guide; Solar Energy Laboratory, University of Wisconsin-Madison
7. Direcção-Geral de Energia e Geologia; Anos Climáticos de Referência; MSExcel, versão 01 (2016). https://www.dgeg.gov.pt/

3D Concrete Printing: Factors Affecting the US and Portugal

Andrew P. McCoy[1], Manuel Vieira[2], Miguel José Oliveira[3](✉), Akhileswar Yanamala[1], and Philip Agee[1]

[1] Virginia Tech, Blacksburg, VA, USA
[2] National Laboratory for Civil Engineering, Lisbon, Portugal
[3] University of Algarve, Faro, Portugal
mjolivei@ualg.pt

Abstract. Housing starts have never recovered from the Great Recession and therefore the lack of inventory has driven up pricing resulting in a crisis of affordable options in the market. Both the United States (US) and Portuguese housing markets are experiencing this crisis driven by high costs of financing, land, materials, labor, and a limited supply chain. Emergent technologies provide one solution to labor, material and lack of production and 3D concrete printing (3DCP) promises solutions, reduces waste, and improves quality. Yet few studies discuss drivers of and barriers to this emerging technology, especially catered to specific markets. This work aims to develop a common set of attributes affecting the diffusion of 3D printing, comparing international market needs and differing factors that could both hinder and bolster the adoption of 3D concrete printing technology. This paper considers only additive manufacturing when referring to 3DCP and analysis is derived from a distillation of various types of media and the lack of empirical work is a limitation for the industry.

Keywords: 3D Printed Concrete · Industrialized Construction · Emergent technologies · Factor Analysis

1 Introduction

The United States (US) housing industry is experiencing an affordability crisis. Housing starts have never recovered from the Great Recession and therefore the lack of inventory has driven up pricing, resulting in a crisis of affordable options at almost all price ranges. Concurrently, the pandemic of 2020 added to the housing crisis as people decided to leave urban environments and move to rural settings. A lack of rural housing caused prices to increase dramatically in markets previously affordable to more people. Furthermore, material and labor pricing has caused the production costs of housing to continue rising [1].

Historically, US affordable housing was possible due to the availability of inexpensive, undesirable land in conjunction with local subsidies that could offset the affordable housing price. Over the last decade, available land has been developed or become desirable, creating a scarcity of options for affordable housing, and driving land prices up.

J. F. L. C. Semião et al. (Eds.): INCREaSE 2023, ASST, pp. 310–324, 2023.
https://doi.org/10.1007/978-3-031-44006-9_24

Simultaneously, federal, state, and local officials have not been able to provide additional subsidies to account for the rising costs of land, materials and production of a home [2].

Prior to the pandemic, the industry was plagued by a lack of trades people who left during the Great Recession and decided not to return once housing demand rose. As NPR reported, the solution to a housing supply shortage should be to build more houses, while the problem is that there aren't enough people who know how to build them [3] As a result, Associated Builders and Contractors (ABC) recently reported the industry average job openings per month in 2022 was the highest level on record and the unemployment rate was the second lowest on record [4].

For Portugal, recent years experienced a sharp drop in construction. Between 2000 and 2020, construction has dropped by about 85% specifically in new homes [5]. Much of this drop occurred during the financial crisis of 2008 with falling house prices and banks holding a large stock of houses, because of outstanding loans, of which they needed to dispose. At the same time, a failing banking system was unable to finance new construction. Construction companies suffered from reduced demand with many disappearing, also resulting in the disappearance of their suppliers and employees, many of whom have not returned to the industry.

Literature [6] suggests a focus on emergent technologies and robotics as approaches toward employee retention or attraction. Emergent technologies also provide solutions to labor, material, and production shortages [7]. Yet no common classification of common diffusion attributes for emerging technology exists to facilitate market adoption. This work therefore focuses on common attributes for one emergent technology: 3D concrete printing (3DCP). 3DCP promises solutions to labor and material costs, saving production time and improving quality [38]. This work aims to develop a framework of attributes affecting the diffusion of 3D printing, incorporating international market similarities and differences that could both hinder and bolster adoption. For this framework, the authors focus on additive manufacturing when referring to 3DCP.

1.1 Background

Before the recession of 2008, the US housing production cycle was already expensive and offered opportunities for cost savings. Before that, producing buildings was expensive. Across the US, the average cost of construction increased by 5 times from 1908 to 1960, by 9 times from 1960 to 1990 and from 1990–2014 the cost of materials alone is 1.8 times higher, and the direct costs of homebuilding have more than doubled. At the same time, we were not as productive as we could be in producing buildings. Productivity in construction has been negative or substantially negative since 1968 while manufacturing productivity has grown 1.7 times more than construction over the same period. Furthermore, the nature of the industry changed a lot just over the last 30 years. For example, the national homebuilding market used to be filled with "small firms that produce only a few homes using their own crews or subcontractors". Now, the list of top-100 companies (in terms of volume) contains a couple dozen publicly traded housing production firms. It is a large industry, fragmented with many stakeholders; vast and far-reaching in terms of its contribution to employment and national spending, including the number of related occupations and economic stakeholders. However, construction is not "one size fits all:" it contains a vast array of firms and employees across a large supply chain before a building is even built. There is also uncertainty and risk that is unique to the production of a building: concentration of suppliers and trades, their supply chains and subcontractor networks. Finally, construction stakeholders typically resist change: each member along the supply chain of producing a building has a different perspective and endorses or vetoes new processes, products, or services. Some of these innovations improve productivity and reduce costs in manufacturing but are seen as posing risk in terms of adoption.

During the last decade, housing productivity has driven growth. The top-100 builder firms produced almost 22% of the total housing starts in 2015. In some metro markets, production builder firms constitute over 30% of housing starts and the trend is above that in many locations. Many small homebuilders and trades contractors were driven out of the market during the recession and there is not a lot of room for them to re-enter now. Costs of producing a new building are often not competitive with existing sales pricing and options and new construction often supports more jobs. We are more diversified and expect more from our housing. Today's firms that produce housing range from affordable housing non-profits to speculative developers of multimillion-dollar luxury mansions; from units with a small footprint to commercial-type residential high rises; from sophisticated financial investments to work-from home offices. Producing buildings remains expensive [7].

The National Association of Homebuilders (NAHB) reports total costs associated with building a single-family dwelling unit (Indirect, Direct and Soft costs) on the rise until the great recession, decreased until 2011 and then rebounded. According to NAHB's two-year breakdown of housing production costs [8]:

- "Finished Lot Costs" and "Total Construction Costs" (or "direct costs"), have increased significantly since 2013 as 54% and 59% respectively and most increases occurred after 2019 as 28% and 32%.
- "Foundation Costs," "Site Work Costs" and "Framing Costs" also increased significantly since 2013 as 84%, 73%, and 71% respectively. Whereas the percentage increases from 2019 are 23%, 58%, and 56%.
- "Interior Finishes" increased at 30% since 2013 but have witnessed a 25% increase since 2019.

Rising house prices have been a recurring theme in Portuguese public discussion as well. Since 2014, the price per square meter has risen by an average of eight percent a year. As of 2012, the number of home purchase and sale transactions more than doubled, which is due to the increase in the number of people wanting to buy a home [5]. The increase in demand for housing by national residents has to do with the decrease in rates. Between 2011 and 2015, the Euribor fell by around two points (Euro Interbank Offered Rate).

Therefore, more people had improved ability and access financing and purchase housing (and more expensive houses) because the monthly fee was lower. With an increase in demand, it was expected that supply would grow in response and an increase in construction would happen to meet the growth in demand. However, this increase in supply did not happen and the Portuguese recovery in real estate sales came years later than expected. The recovery included rising housing prices, which was expected to be offset by a rapid rise in new home construction. Again, that didn't happen. In fact, the sector was still recovering and the operational capacity to suddenly start building again was little, as many workers changed sectors or countries. The Portuguese construction cycle is very long: a lot of time passes from approving the investment to licensing and the end of construction.

In Portugal one of the main factors of increasing construction costs is the higher part of the labor force. The component of labor in the costs of new housing construction has increased during the last decades at a higher rate than materials, namely from the 2008 crisis [11], which with the subsequent massive emigration of construction workers led to a shortage of skilled labor supply.

Another reason for a lack of supply was the transformation of housing into Local Accommodation (AL). Portugal's AL "alojamento local" refers to second homes purchased in Portugal, often for an investment, that are not used in the short term by owners as a primary residence and are therefore rented out to tourists. Since Portugal is one of the most sought-after tourism destinations in Europe, AL generates business opportunities and investors were willing to drive up prices and reduce inventory, making housing even more scarce.

1.2 Housing Affordability

With rising building costs across the world, an innovative, affordable solution is needed. In the US, 29% of all households were "housing cost burdened" in 2019 and half of those households were spending 50% or more of their income on housing alone [9]. The pandemic period has only increased these costs, making housing a burden across the country. An estimated 200,000 affordable rental units are needed in the state to meet demand and for every $1,000 increase to the price of a home, 153,967 families are priced out of homeownership [10]. Recent annual percentage rates for borrowing money will only expand the deficit of affordable housing options.

The right to adequate housing is constitutionally protected in Portugal (see the Constitution of the Portuguese Republic – CRP, article 65.°/1ss). Across Europe, countries are reacting to rising house prices by allowing for a higher level of construction-the big exception is Portugal. Although prices have risen by 44% since 2010, the number of licensed professionals remains approximately the same. In the first quarter of 2018, according to data and the Eurostat house price index, released by the European statistics office, on July 10 of that year, house prices rose by 12.2% in Portugal, against an average of 4.5% in the Euro Zone and 4.7% in the European Union [12].

With house prices and rents rising, the cost of housing can be a burden. This can be measured by the housing cost overburden rate, which shows the share of the population living in a household where total housing costs represent more than 40% of disposable income. In 2021, Portugal had 6.6% (10.4% in EU) of the urban population living in such a household, while the corresponding rate for rural areas was 4.5 (6.2% in EU) [12]. The lower overburden rate compared to EU data may be due to the high level of housing ownership that the Portuguese population has: 78.3% of the population lived in a household owning their home versus 69,9% in EU [13].

It expected that housing cost overburden rate will increase forced by the current inflation rates and the steady growth of housing prices [11], seen during the last decade due to legislation approved in Portugal from 2011, namely the rent law, the golden visa, the urban rehabilitation exception act, licensing simplification, to boost the sectors of tourism and real estate.

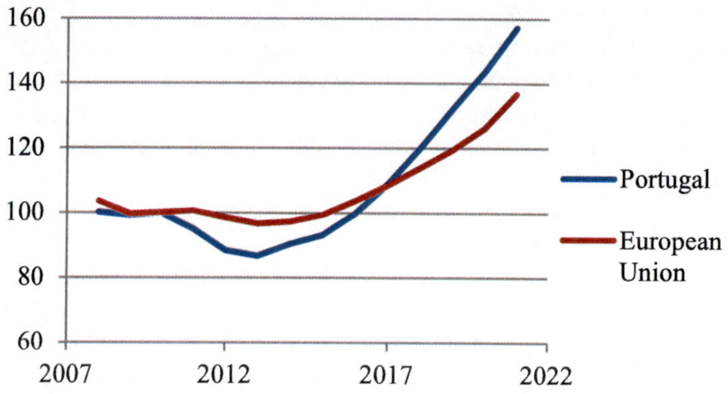

Fig. 1. House price index (2010 = 100) [11]

Meanwhile, Portugal remains at the low end of Europe in terms of social housing stock (Fig. 1). This is one of the conclusions that can be drawn from the recent report by the Housing Europe network [13]. Compared to the total, the relative weight of the public affordable housing stock in Portugal is below 2%, with the European average being around 9% and countries like Denmark, Austria and Nederland presenting shares between 21% and 29%.

Faced with the reality of rising prices and a need for production, the solution is to build more and increase supply. However, there are obstacles to building more, like severe housing costs. Seeking to reduce the costs of materials and labor in any feasible scenario for more housing options, including social housing, is fundamental. Industrialized construction and 3D printed concrete could provide a solution.

1.3 Industrialized Construction

Emergent technologies are playing an increasingly important role in the construction industry, helping to improve efficiency, reduce waste, and enhance safety on construction sites [6, 14–16]. Recent, emergent technologies in construction include Building Information Modeling (BIM); Robotics and Automation; Drones; 3D Printing; Virtual Reality; and Augmented Reality. While these technologies are transforming the construction industry, and it is likely that they will play an increasingly important role in the future, recent efforts have also focused on industrializing construction [35].

Industrialized construction is characterized using standardized, repeatable processes, high levels of automation, and the integration of digital technologies to improve efficiency, reduce waste, and enhance quality control [14]. The benefits of industrialized construction are numerous and include faster construction times, improved accuracy and consistency, enhanced safety, and reduced environmental impact [15]. By producing components in a controlled factory environment, industrialized construction can minimize the amount of waste generated during construction and help to conserve natural resources. Additionally, by using digital technologies to plan and manage construction processes, industrialized construction can help to reduce the risk of errors, minimize rework, and improve the overall quality of the finished structure [15].

Industrialized construction is particularly well-suited to the construction of multi-unit residential buildings, such as apartments, condominiums, and hotels [9, 14, 15]. By using standardized components and repeating the same construction processes over and over, it is possible to significantly reduce construction times and costs, while improving quality control and ensuring that buildings are built to the highest standards [6, 15].

Nevertheless, a key challenge of industrialized construction is compatibility with the on-site environment and successful integration with the overall building design and process [31]. Industrialized construction might also require specialized knowledge and expertise, as well as the use of specialized equipment and tools [17].

1.4 The Promise of 3D Concrete House Printing

Some believe 3D additive manufacturing ("printing") offers a viable, affordable alternative to traditional construction materials and processes in the market. Recently, 3D concrete printing has emerged as a promising type of additive printing for the construction of buildings and houses.

Alternatives to concrete have similar characteristics to concrete (strength, durability, and longevity) while being less carbon intensive and often with an appealing and distinctive appearance [34]. Concrete is by far the highest used construction material and 35%–60% of costs result from formwork materials and installation, adding cost and increasing material waste. Material waste is another huge factor as "80% of the total worldwide waste is generated in the construction industry" [14, 37]. 3D concrete printing promises to eliminate formwork (reducing waste, material, and labor costs), reduce time (labor costs) increase durability and quality and reduce the need for finishes [19, 20, 32].

1.5 3D Concrete Printing (3DCP)

From the different three-dimensional concrete printing (3DCP) technologies, the most promising is the computer-aided process that builds concrete structures by extruding concrete layer by layer through both a digitally controlled mixer and nozzle [18]. Unlike the conventional approach of casting into a mold, 3DCP combines digital technologies and new insights from materials technologies to allow free-form construction without the use of formwork [19]. In fact, development of contour crafting, a method of layered manufacturing process, facilitated 3DCP of large building structures by extruding concrete paste using larger nozzles and higher pressure to account for the different consistencies of concrete [20, 37]. Gantry robots with 3 or 4 Degrees-of-Freedom (DOF) for large projects, and articulated arm robots with 6 or more DOF for smaller objects are most used tools in 3DCP [21].

Broadly stated, benefits of 3DCP include decreased waste, costs, time, and labor, while increasing productivity and improving the quality of construction and also allowing for the highly precise fabrication of complex, geometric, and hollow structures [23, 32, 33, 38]. 3DCP can increase sustainability in construction by reducing the waste of formwork [19] and introducing additives [36]. 3DCP can significantly reduce environmental effects in terms of global warming potential (GWP), acidification potential (AP), eutrophication potential (EP), smog formation potential (SFP), and fossil fuel depletion (FFD) compared to conventional construction methods [24]. Another benefit of 3DCP is its potential to decrease the number of injuries and deaths in the construction sector and upskilling the current workforce [23].

3DCP is envisioned as a disruptor in the construction industry yet all emergent technologies contain characteristics that positively or negatively affect adoption and use. If 3DCP is to successfully provide its touted benefits, a list of the common product attributes affecting diffusion of innovations would be helpful and needs to be developed and shared across possible markets.

2 Research Methods

2.1 A Tale of Two Markets

The degree to which an innovative product exhibits a particular attribute influences the attraction or resistance of residential construction stakeholders to adopt the product. Due to the general lack of information on the adoption and diffusion of 3DCP, the general approach to our international 3DCP attribute development consisted of three stages:

1. *Literature Review.* We collected attributes widely from scientific and popular media regarding 3DCP;
2. *Literature Analysis.* We coded the literature for key words, findings and common statements regarding barriers and enablers of use;
3. *Factor Development and Discussion.* We distilled a set of specific attributes based on commonalities agreed upon by the research team across the Portuguese and US markets.

The US housing market is known for wood construction and well suited for housing production disruption due to its high costs of housing and housing needs for a large portion of the population. Portugal is a country known for concrete and masonry construction and the increasing cost of labor has boosted housing prices and the need for affordable, public housing is high. Therefore, these markets, with their specific conditions, seem appropriate for an emerging technology such as 3DCP.

In developing attributes that affect the adoption of 3DCP, the research team performed an exhaustive *literature review* of available scholarship and collected it into a shared database organized by type of study, industry sector, industrial technology, country of origin and common terminology. The team *analyzed the literature* by identifying (coding) common keywords, terms, phrases and statements across the literature, including digital and social media products. We then developed exclusion criteria based on applicability to the US and Portuguese markets. Our final *development stage* involved combining and distilling a set of common attributes, including definitions, across the team for the framework. While data and analysis of the affordable housing industry has a deep library of information, 3DCP is a new concept in its application, with scientific study and empirical work limited.

3 Analysis and Discussion

3.1 Attributes Affecting the Adoption and Use of 3D Concrete Printing

As with any emerging technology, 3DCP contains specific attributes that could limit or increase its use in the construction industry. As a result of the methods previously discussed, our research team identified the following.

Cost. 3DCP is touted as being very inexpensive in the media, reducing total material, labor, and waste costs below those of standard code-built wood framed buildings in the US. Nevertheless, the experimental side of using 3DCP and the material costs currently do not necessarily make it less expensive than standard construction options. Publicly available, empirical data is required to reduce this barrier for the market.

Schedule. 3DCP could significantly reduce the overall schedule of construction, thereby reducing costs and creating sooner sale or rental revenues. The reduced schedule of 3DCP also reduces overhead costs for the project. Like cost, though, the immediate schedule reductions are not evident due to a lack of standardized systems and objective data by 3DCP builders.

Shrinkage. The high content of cementitious materials and a low water-binder ratio, together with the lack of reinforcement, may increase the risk of cracking by shrinkage. To reduce the effect of shrinkage in 3DPCsome are "using anti-shrink fibers such as those made of glass, polypropylene, polyethylene, basalt or steel" in the mix design. Fibers can clog the printer hose, though, or clump in the extruded concrete, reducing their effect or, worst case, hurting the extrusion quality. Other measures during the curing process to prevent shrinkage include preventing the evaporation of water from the surface of the concrete. This means spraying the surface with water and covering with an anti-evaporation membrane (burlap) or curing compound. Burlap can discolor the concrete or leave a pattern.

Bonding. During the 3DCP process, concrete is printed layer by layer. It is critical that the bond between the layers is strong so that the wall is structurally sound and will not fail. A barrier to the wide adoption of extrusion-based 3DCP can be the weak interlayer bond strength between printed filaments [29]. Therefore, a bonding agent must be placed between printed layers, especially between cold joints when printing is stalled, to improve interlayer bond strength.

Mix Design. Many factors go into a proper 3DCP mix design and these can differ widely based on specific applications, locations, temperatures, and humidity levels. Some common materials found in the mix designs are Portland Cement, natural sand, and silica fume, while there are no standard mix designs that have been tested widely across many applications and objectively studied. Further, a mix design must consider pumpability, extrudability, and buildability. All three of which are required for a successful 3D concrete print.

Nozzle Type. The end of the print head is referred to as the nozzle and is a key component to printing. The nozzle forms the desired shape and size of the next concrete layer. Some of the available nozzles include circular, ellipse, square, and rectangular shapes. According to a review of 3D Concrete Printing Systems and Material Properties, "a circular nozzle gives freedom of printing without having to adjust the printing angle for shape continuity at vertices or changes in angle of the printed element." Nozzle length is also an important factor, as some printer cannot fully reach their intended foundation and a longer nozzle tube adds additional variability to the quality of extruded concrete.

Digitization. Integrating Building Information Modeling (BIM) into the 3DCP process is integral its success, facilitating better quality, detail, and automation. A digital twin of the proposed print reduces questions by local officials and other important stakeholders (owner, designer, general contractor, trades and suppliers). Making sure your model has the appropriate level of detail (LOD) is not the final step, either. You must translate the model into other software needed to print, which adds complexity to an already complex process.

Buildability. Buildability refers to print material that successfully bonds to previous layers as it exits the nozzle via extrusion and subsequently supports the weight of successive layers accumulating on top of it [28].

Pumpability. Pumpability refers to a print mix's (commonly termed "ink") workability constraints: cohesive enough for the pump system, ability to be placed under pressure toward the printing head while retaining its critical *Buildability* properties [26]. On the other hand, a pumpable mix is strictly dependent of the pumping system for which the mix was developed.

Extrudability. Extrudability refers to the print mix material's ability to be extruded continuously and consistently, exiting the nozzle at the printhead unhindered [24, 26].

Value Network. Automated fabrication must be altered to become suitable for large scale products and conventional design approaches, requiring time, cost and the willingness of all stakeholders involved [30]. Certain construction sectors, like residential construction, contain a lower ratio of automated products in comparison with other industries. Therefore, limited materials can currently be used by automated machines in construction. Further, automated machines are often expensive and require construction companies to assume risk and expand beyond their current *value network.*

Finally, factors including the lack of *standards and regulations* may be an obstacle to promote the application and wide acceptance of 3DCP and *seismicity*, or regions with high seismic risk, are less suitable to buildings with structural precast elements requiring additional engineering and structural developments, namely on the connections of those elements.

3.2 Differences in Markets in the US and Portugal

Between 1990–2014 the US cost of materials alone increased 80%, the direct costs of homebuilding more than doubled the industry considerably lagged others in terms of productivity. These market characteristics have only worsened and therefore require disruption that will "flatten the curve" of housing costs and provide quality, efficient and affordable housing options. 3DCP's advantage in the US market is its ability to reduce labor, schedule timing, delays across trades, waste and cost, while also increasing quality and durability.

In Portugal, the need to reduce the costs of materials and labor in any feasible scenario for more housing options requires new solutions as well. Industrialized construction systems like 3DCP expand options for dense housing markets that are in need, like social housing. Concrete is Portugal's predominant construction material, making 3DCP technology accessible in terms of trade knowledge for critical aspects like mix design, placement, and quality. The Portuguese Architecture, Engineering and Construction (AEC) industry would also be prone towards design of 3DCP structures and the ability for supportive trades to work within the details of this type of construction. The ability to upskill Portuguese labor or attract new workers due to the technology itself adds another possible benefit for the Portuguese market.

3.3 Attributes Affecting Cost in the US and Portugal

Having few, current 3DCP applications in Portugal and existing equipment being research lab-based and smaller for printing concrete at laboratory and academic facilities, no data is available on cost impacts of 3DCP technology. Therefore, we discuss the literature, aligning with US experience, and assume that they can also be applied to industry broadly. We apply similar considerations to scheduling.

Some authors state that 3DCP shortens the construction duration and reduces the labor cost and material waste [18, 19]. In particular, the cost of formwork is reduced, because 3DCP technology does not need to use formwork. This technology could also improve the buildability, functional integrity, and accuracy of construction [20]. Nevertheless, preliminary data from US 3DCP housing suggests that the exterior walls can be printed in 5 days for a 1400sf home, which should be like other countries in terms of printing time.

However, cost data remain non-empirical and the media assertion that the cost of construction 3D printing is lower than conventional construction is controversial [18]. The reason is that some scholars have confused attributes between construction 3D printing and traditional construction. The cost calculation of construction 3D printing cannot follow the rules in conventional construction because construction 3D printing is both an industrial product and construction product. Further, most 3DCP projects are experimental prototypes, raising the costs for the initial print. Current plans in the US for production printing of multiple, similar housing types on one property could provide better data on costs for the market to use. To date, US 3DCP construction has not yet been able to produce a less expensive alternative to standard wood construction methods.

When calculating costs of 3DCP, both offsite and onsite factors need to be taken into consideration [6]. Literature suggests that the cost of masonry and formwork could be significantly reduced or eliminated. In addition, depending on the design and quality of the print, other interior and exterior wall system materials and labor could be eliminated, significantly reducing costs. Additional reductions occur through a reduction in general waste and material consumption, labor, and machine cost. So far, 3DCP equipment costs, including transportation and setup costs, increase significantly.

3.4 Additional 3DCP Factor Analysis in the US and Portugal

Besides market characteristics, cost and schedule factors, differences in attributes specific to the 3DCP process could also hinder or accelerate adoption in the US or Portuguese market. A key challenge of industrialized construction and 3DCP is compatibility with the on-site environment (*buildability*) and successful integration with the overall building design and process. The space and vertical dimension of a gantry-type 3DCP requires flat land and larger lot sizes no matter where the printing occurs. If either market requires dense urban infill, 3DCP is not necessarily an option unless a robotic arm printing system can be moved easily around the site. An additional challenge is that industrialized construction might require specialized knowledge and expertise, as well as the use of specialized equipment and tools. While this has been a challenge in the US market, where labor is very tight and disruptive technology is often resisted, the Portuguese already perform much of their housing construction in concrete or masonry, thus making it compatible.

Pumpability and *extrudability* of the mix are the key characteristics of a fresh concrete that governs the printability of this material [18, 19]. While being pumped the mix must segregate in the hose pipe forming a lubrication layer which helps in reducing pump pressure [20, 21]. The printed material must successfully *bond* to previous layers as it exits the nozzle via extrusion and subsequently support the weight of successive layers accumulating on top of it [22]. The conflicting properties in fresh state, deformability to be printable and resistance to withstand its shape and remain stable under the gravitational load of subsequent printed concrete layers without the support by formwork, must be optimized with a suitable mix design [25, 26]. Due to the large size of the United States and varying climate zones, designing for *shrinkage* is a major concern, while these climate differences are less in Portugal. Applications of 3DCP in the US are use bonding agents when necessary and extra care should be taken on high temperature environments due to a decrease of setting time, no matter the market.

However, the performance of the 3DCP mixes is highly dependent of the printing equipment and lack robustness as they are sensitive to small variations of its constituents' content and type. Those two problems must also be solved so this technology may be used widely in any market.

Nozzle type affects the concrete surface finish, affecting its structure, quality and durability, and printers in the US and Europe are experimenting with no nozzle versus "flaps" that smooth the sides. Flaps also limit the tool path, as the printer cannot easily turn without possibly affecting parts of the print. The ability to successfully use flaps is dependent on mix design and geometry of the extruded material in all markets.

Generally, the *mix* ("ink") is reportedly the most important factor in successful 3DCP. In Portugal, the recent shut down of thermoelectric plants interrupted the availability of fly ash, which is suitable to produce flowable mixes with low water content and low risk of bleeding, also improving the interface properties for longer time intervals between consecutive layers [27]. Silica fume has an effect of viscosity modifier, which is suitable for mixes buildability and bleeding resistance. Despite its importance, the use of this addition in Portugal is restrained as it has to be imported, being very expensive, around ten times the cost of cement. Due to the unavailability of those two currently used additions in 3DPC mixes, the development of those kind of materials must go with the study of suitable water reducing and viscosity modifying admixtures.

The growing importance of the digital representation (*digitization*) of building elements, made possible by 'Building Information Modeling' (BIM) methodologies, is introducing very relevant changes in design, construction and operations management. 3DCP, namely off-site at precast plants is perfectly suitable to produce building elements in a digitally managed way. Even though, in Portugal, the implementation of BIM in architecture offices is slow, despite the advantages provided by this technology. Factors such as the impulse of the new generations of technicians that are entering the labor market, the implementation of manufacturers' product libraries, the demands of private clients in the design of their projects, as well as the demand for their use in public tenders, will accelerate the modernization process of architecture firms towards BIM technology. Therefore, the existence of suitable, current options and lack of a *network* that current *values* this technology will hinder its immediate adoption.

In terms of *standards and regulations*, 3DCP design and enforcement in the US largely fits within current codes for concrete and masonry design standards and engineering requirements, making the technology accessible to local officials. While not diffused into the Portuguese market, the same approach could facilitate adoption. Meanwhile, as it is difficult to add reinforcement to 3DCP elements, the Eurocode 2 design requirements for plain and lightly reinforced concrete structures applies to members for which the effect of dynamic actions may be ignored [28, 29].

The *seismic behavior* of 3DCP is important to many regions and previous connections in precast construction systems has been largely recognized as crucial in by the industry and academia. Nevertheless, the complexity of the problem and the variety of inherent issues have made it difficult so far to conceive of appropriate solutions [23, 28–30]. Those issues undermine the proliferation of precast buildings and could for 3DCP.

4 Conclusions

Both the United States (US) and Portuguese housing markets are experiencing a housing affordability crisis driven by high costs of financing, land, materials, labor, and a limited supply chain. Emergent technologies provide solutions while scant work identify factors that could affect adoption and diffusion specific to 3D concrete printing (3DCP) markets internationally.

Currently, 3DCP's hype phase is occurring, where "everything everywhere all at once" seems to be possible. Other technologies were similar: self-compacting concrete (SCC), fiber reinforced concrete (FRC), ultra-high-performance concrete (UHPC), and others, for which the industry has managed to use them where they were best suitable. 3DCP's current stage of knowledge and expectations are experimental and unknown, and for housing, however future applications may reveal its full potential. Until a common set of knowledge is developed, 3DCP's broad benefits exist in the promise of schedule, structure, customization, thermal and acoustic advantages.

3DCP's advantage in the US market is its ability to reduce labor, schedule timing, delays across trades, waste, and cost, while also increasing quality and durability. The Portuguese Architecture, Engineering and Construction (AEC) industry could also be prone towards the benefits of 3DCP structures and the ability for supportive trades to work within the details of this type of construction.

Current cost data remain elusive and non-empirical, and the media assertions of lowered costs are controversial. While preliminary data suggest a reduced schedule, literature and experience show that the cost of masonry and formwork can be eliminated. Depending on the design and quality of the print, other interior and exterior wall system products could also be eliminated, further reducing costs. Additional reductions could be general waste and material consumption, and there also is reduction in human labor cost, material cost, and machine cost. However, transportation and setup equipment costs increase on a 3DCP building.

A proper *Mix Design* is critical. Designing and printing for *shrinkage* is of major concern for the mix design, while climate variability in the US presents more of an issue than in Portugal. *Pumpability* and *extrudability* of the mix are the key characteristics of printability. A final key challenge of 3DCP internationally is compatibility with the on-site environment (*buildability*) and successful integration with the overall building design and process, including *digitization* of the product.

Future work includes the collection of case studies and interviewing those with experience in adopting 3DCP as our next step in refining the attributes and understanding where barriers and accelerators exist.

References

1. https://www.novoco.com/notes-from-novogradac/rising-construction-labor-costs-increase-rents-contributing-affordable-housing-crisis
2. https://www.fanniemae.com/research-and-insights/perspectives/us-housing-shortage#:~:text=The%20pandemic%2Dinduced%20materials%20and,slowdown%20could%20reduce%20demand%20further
3. NPR.org: Planet Money: Why Aren't There Enough Skilled People To Build Houses? Reported by Sarah Gonzalez (2021). https://www.npr.org/2021/07/30/1022642064/planet-money-why-arent-there-enough-skilled-people-to-build-houses
4. https://www.abc.org/News-Media/News-Releases/entryid/19777/construction-workforce-shortage-tops-half-a-million-in-2023-says-abc
5. https://www.gee.gov.pt/pt/indicadores-diarios/ultimos-indicadores/31190-ine-indice-de-custo-de-construcao-habitacao-nova-96. Accessed 13 Mar 2023
6. McCoy, A.P., Yeganeh, A.J.: An overview of emerging construction technologies. In: NAIOP, Commercial Real Estate Development Association and Research, Washington, D.C. (2021)
7. Engineering News Record and ENR.com: Industry Labors to Fill Growing Workplace Gaps by Debra K. Rubin, Scott Van Voorhis, and James Leggate (2021). https://www.enr.com/articles/52797-industry-labors-to-fill-growing-workplace-gaps
8. https://www.nahb.org/-/media/8F04D7F6EAA34DBF8867D7C3385D2977.ashx
9. The National Low Income Housing Coalition. The GAP a Shortage of Affordable Homes (2019). https://reports.nlihc.org/sites/default/files/gap/Gap-Report_2019.pdf
10. https://www.nahb.org/blog/2021/03/New-Priced-Out-Study-Highlights-the-Housing-Affordability-Crisis
11. https://ec.europa.eu/eurostat/cache/digpub/housing/bloc-2b.html?lang=en. Accessed 12 Mar 2022
12. https://ec.europa.eu/eurostat/cache/digpub/housing/bloc-1a.html?lang=en. Accessed 12 Mar 2022
13. Housing Europe: The state of housing in Europe. European Federation of Public, Cooperative and Social Housing (2021). https://www.stateofhousing.eu/The_State_of_Housing_in_the_EU_2021.pdf
14. Hall, D.M., Lessing, J., Whyte, J.: New business models for industrialized construction. In: Bolpagni, M., Gavina, R., Ribeiro, D. (eds.) Industry 4.0 for the Built Environment, vol. 20, pp. 297–314. Springer, Cham (2022). https://doi.org/10.1007/978-3-030-82430-3_13
15. Industrialized Construction: Emerging Methods and Technologies. Computing in Civil Engineering. ascelibrary.org (2019)

16. Qi, B., Razkenari, M., Costin, A., Kibert, C., Fu, M.: A systematic review of emerging technologies in industrialized construction. J. Build. Eng. **39**, 102265 (2021)
17. Template for the Jurnal Teknologi (utm.my)
18. Sanjayan, J.G., Nazari, A., Nematollahi, B.: 3D Concrete Printing Technology: Construction and Building Applications. Butterworth-Heinemann (2019)
19. Mohammad, M., Masad, E., Al-Ghamdi, S.G.: 3D concrete printing sustainability: a comparative life cycle assessment of four construction method scenarios. Buildings **10**(12), 245 (2020)
20. Khoshnevis, B., Hwang, D., Yao, K.T., Yeh, Z.: Mega-scale fabrication by contour crafting. Int. J. Ind. Syst. Eng. **1**(3), 301–320 (2006)
21. Bos, F., Wolfs, R., Ahmed, Z., Salet, T.: Additive manufacturing of concrete in construction: potentials and challenges of 3D concrete printing. Virtual Phys. Prototyp. **11**(3), 209–225 (2016)
22. Khajavi, S.H., et al.: Additive manufacturing in the construction industry: the comparative competitiveness of 3D concrete printing. Appl. Sci. **11**, 3865 (2021). https://doi.org/10.3390/app11093865
23. Bathika, M., et al.: 3D concrete printing for sustainable and economical construction: a comparative study. Autom. Constr. **134**, 104087 (2022). https://doi.org/10.1016/j.autcon.2021.104087
24. Paul, S.C., Yi, W.D.T., Panda, B., Tan, M.J.: Fresh and hardened properties of 3D printable cementitious materials for building and construction. Arch. Civ. Mech. Eng. **18**(1), 311–319 (2018). https://doi.org/10.1016/j.acme.2017.02.008
25. Yang, et al.: EURASIP J. Wirel. Commun. Netw. **2018**, 147 (2018)
26. Le, T.T., Austin, S.A., Lim, S., et al.: Mix design and fresh properties for high-performance printing concrete. Mater. Struct. **45**, 1221–1232 (2012). https://doi.org/10.1617/s11527-012-9828-z
27. Jacobsen, S., et al.: Flow conditions of fresh mortar and concrete in different pipes. Cem. Concr. Res. **39**, 997–1006 (2009). https://doi.org/10.1016/j.cemconres.2009.07.005
28. Zhang, C., et al.: Mix design concepts for 3D printable concrete: a review. Cem. Concr. Compos. **122**, 104155 (2021). https://doi.org/10.1016/j.cemconcomp.2021.104155
29. Nerella, V.N., Hempel, S., Mechtcherine, V.: Effects of layer-interface properties on mechanical performance of concrete elements produced by extrusion-based 3D-printing. Constr. Build. Mater. **205**, 586–601 (2019). https://doi.org/10.1016/j.conbuildmat.2019.01.235
30. SAFECAST: Final Report Summary - SAFECAST (Performance of innovative mechanical connections in precast buildings structures under seismic conditions). European Union (2013). https://cordis.europa.eu/project/id/218417/reporting
31. Zhang, R.-C., Wang, L., Xue, X., Ma, G.-W.: Environmental profile of 3D concrete printing technology in desert areas via life cycle assessment. J. Clean. Prod. **396**, 136412 (2023). https://doi.org/10.1016/j.jclepro.2023.136412
32. Rehman, A.U., Birru, B.M., Kim, J.-H.: Set-on-demand 3D Concrete Printing (3DCP) construction and potential outcome of shotcrete accelerators on its hardened properties. Case Stud. Constr. Mater. **18**, e01955 (2023). https://doi.org/10.1016/j.cscm.2023.e01955
33. Nguyen-Van, V., Li, S., Liu, J., Nguyen, K., Tran, P.: Modelling of 3D concrete printing process: a perspective on material and structural simulations. Addit. Manuf. **61**, 103333 (2023). https://doi.org/10.1016/j.addma.2022.103333
34. Ahmed, G.H.: A review of "3D concrete printing": materials and process characterization, economic considerations and environmental sustainability. J. Build. Eng. **66**, 105863 (2023). https://doi.org/10.1016/j.jobe.2023.105863
35. Al-Tamimi, A.K., Alqamish, H.H., Khaldoune, A., Alhaidary, H., Shirvanimoghaddam, K.: Framework of 3D concrete printing potential and challenges. Buildings **13**, 827 (2023). https://doi.org/10.3390/buildings13030827

36. Dvorkin, L., Konkol, J., Marchuk, V., Huts, A.: Effectiveness of polymer additives in concrete for 3D concrete printing using fly ash. Polymers **14**, 5467 (2022). https://doi.org/10.3390/polym14245467
37. Chen, H., Zhang, D., Chen, P., Li, N., Perrot, A.: A review of the extruder system design for large-scale extrusion-based 3D concrete printing. Materials **16**, 2661 (2023). https://doi.org/10.3390/ma16072661
38. Tay, Y.W.D., Tan, M.J., Wong, T.N.: Tailoring light-weight aggregates for concrete 3D printing applications. Materials **16**, 2822 (2023). https://doi.org/10.3390/ma16072822

Feasibility, Applicability and Incentives in Using Sustainable Materials: Comparative Between Brazil and Portugal

Clauciana Schmidt Bueno de Moraes[1](✉), Stephani Cristine de Souza Lima[1], Cínthia Mara Vital Bonaretto[1], Giulia Malaguti Braghini Marcolini Mártires[1], Natasha Nême Gonçalves de Almeida[1], Manuel Duarte Pinheiro[2], Miguel Pires Amado[2], Rodrigo Prieto Rocha[3], and Leonardo Prudente Torres Gualter[4]

[1] São Paulo State University (UNESP), São Paulo 13506-900, Brazil
clauciana.schmidt@unesp.br
[2] Universidade de Lisboa, Lisbon, Portugal
[3] Marinha do Brasil (CTMSP), São Paulo, Brazil
[4] Universidade de São Paulo (POLI/USP), São Paulo, Brazil

Abstract. The construction industry is a sector whose development generates impacts on the environment, through the generation of waste and the consumption of raw materials and energy. From this fact, it is important to use materials that generate less environmental impact and bear characteristics equivalent and performance equivalent to current materials. The objective of this work was to diagnose the main sustainable materials existing in Brazil and Portugal for use in construction and to apply the methodology proposed in this work, which analyses several indicators of the sustainability of the materials, resulting in a quantitative analysis of the materials surveyed together with a qualitative analysis of legislation and best practices in both countries. The materials analysis methodology consisted of creating a spreadsheet in which grades were assigned to these materials, and, subsequently, a simple arithmetic average was calculated. Still, for the analysis of legislation, the inductive method was adopted. The results consisted of ranking the materials analysed through notes and proposals for incentives in which practices, systems and consolidated agencies in Brazil and Portugal were analysed, adapted and improved for the Brazilian reality and to answered to the gaps in the classic processes of evaluation of sustainable materials used in construction.

Keywords: Sustainable Materials · Sustainability · Sustainable Construction · Public Policies · Tax and Legal Incentives · Indicators

1 Introduction

Among the various human activities that accelerate the increase in environmental impacts, "civil construction is the economic sector that most impacts the environment, especially in the consumption of natural resources and waste generation, with a significant generation of dust and noise pollution in construction sites located within cities" [1–3].

J. F. L. C. Semião et al. (Eds.): INCREaSE 2023, ASST, pp. 325–344, 2023.
https://doi.org/10.1007/978-3-031-44006-9_25

With population growth and the consequent increase in demand in this sector, actions that seek to minimise such environmental impacts and reach the optimisation of processes are essential. From this concern, sustainable construction arises, that is, the search for adequate management of the built environment. Less impactful practices are started, with greater responsibility for socio-environmental issues and lower economic costs [4, 5].

Sustainable construction is "the responsible creation and management of a healthy built environment based on ecological and resource-efficient principles", first cited in Agenda 21 for Sustainable Building [6].

The adoption of more sustainable practices by companies is usually due to pressure from laws and regulations, social pressure (protests and court lawsuits), economic interests related to efficiency and cost reduction, economic interests arising from opportunities to create new lines of goods and services, in addition to image and reputation [7]. "Sustainability is linked to problems of social justice, gender equality and political participation of local actors, configuring the different existing dimensions of life in society, and not only in the environment" [8, 9].

The concept of sustainable construction directs the civil construction sector to rethink. It demonstrates that this sector would be willing to seek sustainable practices, which should start in the design of the project, selection of materials and construction with alternatives that cause less negative impacts on the environment and society [10]. According to [11], "sustainability must be present in all phases of the built environment, such as idealisation; conception; project; construction; use; maintenance; end of life". Therefore, solutions such as sustainable constructions are needed because when using techniques derived from this school, it is possible to increase efficiency, durability, recyclability, performance and use of resources at the construction site, in addition to having positive impacts on the construction site, such as the reduction of noise, debris, materials, maintenance, energy consumption, among other benefits [12].

Due to the growing need for increasing sustainability in the civil construction sector, this work generally aims to study technologies and alternatives in the field of sustainable construction, in particular, the most sustainable materials that can be used in buildings and propose measures and fiscal incentives and economic and the encouragement of sustainable practices and choices in this sector.

Following the quantitative analysis realised on the listed materials, a qualitative analysis of the legislation in force in Portugal and Brazil analyses and documents good practices, tax and legal incentives adopted by both countries in their national territories, and evaluating their particularities.

After listing the main laws, incentives and good practices in both countries, a comparative analysis was realised to explain the best practices used for the sustainability of the civil construction sector, in this way defining proposals that have different levels of coverage (national, state and municipal). They can be adapted and implemented to other Brazilian realities.

Therefore, this work aims to accomplish a quantitative analysis of the selected materials with a qualitative analysis of the laws and to propose new simplified solutions and adaptations in the fields of sustainable materials as well as legislation.

2 Methodology

The first step of this research was the collection of data from the High Quality Environmental (AQUA) and Leadership in Energy & Environmental Design (LEED) certifications. That, followed by an analysis of the recurrent difficulties among the various authors on this theme, originated a short list of parameters to be considered when studying the suitability of the materials used in the work. As a result, a table was created, in which the materials were studied according to several parameters. At the end of this analysis, they received a note to facilitate the study of the most suitable material for the needs in question.

The list of parameters in the analysis worksheet (Table 1) resulted from the difficulties presented mainly by [4, 5, 13–17].

Such parameters were chosen to enable the analysis of materials. The responsibility for selecting the desired materials must analyse each material individually and mark each worksheet line only once in the most indicated answer for each parameter. Each marked response has a different value. Consequently, it is possible to originate a ranking that will be the result of the simple arithmetic average of the scores assigned to each parameter contained in the table, with the values of each parameter being as follows; 5 (five) for excellent, 3 (three) for good, 1 (one) for fair, –1 (less than one) for poor and –3 (less than three) for nonexistent (Table 1).

$$M = \frac{\sum Parameters}{21}$$

Table 1. Spreadsheet for the analysis of selected materials.

Material:					
Parameters	**Excel-lent**	**Good**	**Regu-lar**	**Bad**	**Does not ex-ist**
1. Material durability is reflected in service life and price.					
2. Recycling is more expensive than correct disposal and transport.					
3. High availability.					
4. Low maintenance.					
5. Low cost.					
6. Consumes a lot of water.					
7. Consumes a lot of energy.					
8. It's renewable.					
9. It is reusable and/or reused.					
10. Easy handling.					
11. It works as a thermal and/or acoustic insulator.					
12. Generates little waste.					
13. It needs another material, which can contaminate the construction site.					
14. It requires specialised training.					
15. The material has an environmental seal.					
16. There may be a partnership with NGO's, cooperatives, and non-profit associations.					
17. Can be reused.					
18. It has some hazardous material in the composition.					
19. It has reverse logistics.					
20. It requires more and/or other materials for finishing.					
21. Visually pleasant when installed.					

Score: Excellent (5), Good (3), Regular (1), Bad (-1), Does not exists (-3).

Source: [12, 18].

To help use the table in assigning grades, follow this script, which consists of the ideal answer for each question to allocate the maximum degree, together with small warnings that cause the decrease of the materials in the respective parameters, then follow the script.

1. Material durability is reflected in useful life and price.

To guarantee the maximum score, the price divided by the numerical values of the useful life must be less than or equal to the market values of the equivalent traditional material. For the calculation, use the following formula:

$$Durability = \frac{Useful\ life}{Price}$$

To reduce the score, consider very high prices that do not match the reality of the work and/or the market, as well as the durability and/or useful life of the sustainable material being shorter than that of the traditional material.

2. Recycling is more expensive than correct disposal and transport.

To guarantee the maximum score, it must be verified that the most economical, environmentally correct destination is recycling, with the options of taking it to eco points, sending the waste to Non-Governmental Organizations (NGOs) or partnering with cooperatives, as this reduces the pressure on sanitary landfills and inert. The grade decrease will occur whenever recycling is the highest value correct disposal option, so the more expensive the recycling in relation to other types of correct disposal, the lower the grade.

3. High availability.

To obtain the maximum score, the sustainable material must be easy to acquire, and not need to wait long periods or longer periods than the equivalent traditional material to obtain it. In this way, the grade will be lowered whenever the sustainable material has a longer delivery time than its traditional equivalents, and the greater this difference, the lower the rate.

4. Low maintenance.

The maximum score will always be obtained. The sustainable material preferably needs no or very low maintenance, as it avoids producing more waste. That is, the score will be reduced according to the maintenance required by the material.

5. Low cost.

The maximum score and its decrease are directly related to the price of the material. The cheaper the material to acquire, the higher the score; if it is free, the score will be maximum, and the higher the value to purchase the material, the lower the note.

6. Consumes a lot of water.

The maximum score is related to the low or zero need for water to install the material, together whenever possible with the material's life cycle, to know the amount of water consumption in the production of the material. Therefore, the more water involved in the production and installation of the material, the lower the grade and the lower the water consumption, the higher the grade.

7. Consumes a lot of energy.

The maximum score is related to the low or zero need for energy use to install the material, together whenever possible with the material's life cycle, to know the amount of energy consumption in the production of the material. Therefore, the greater the energy involved in the production and installation of the material, the lower the grade and the lower the energy consumption, the higher the grade. Bearing in mind that the use of renewable energies for production/installation is a factor in which the grade is not reduced, non-renewable energy is a factor that reduces the grade.

8. It's renewable.

The maximum score is related to whether the material is a recycled and/or reused product or to the fact that it is developed with renewable technologies. The score decreases according to the absence of a renewable and/or recycled product produced using non-renewable technologies.

9. *It is reusable and/or reused.*

The maximum score is related to the fact that the material, when finished, can be recycled and/or reused. The score decreases according to the impossibility of providing another destination for the material other than landfill disposal.

10. Easy handling

The maximum score is linked to any function related to the fact that storing and managing the chosen material is simple, not requiring specific machinery or extreme care in which most civil construction workers cannot handle the material. The grade will describe as it does not meet the requirements mentioned above.

11. It works as a thermal and/or acoustic insulator.

The maximum score will be assigned when the material is a thermal and acoustic insulator simultaneously. The factors that describe the grade are whether the material is thermally or acoustically insulating, or neither, and of course, the material's efficiency must be considered.

12. Generates little waste.

The maximum score will be assigned when the material, when used/installed, generates little or no waste. The decrease in the score occurs according to the amount of waste generated and whether this waste needs a different disposal and/or storage.

13. It requires another material, which can contaminate the construction site.

The maximum grade will be assigned according to the lesser need for other installation materials or when the required material is already part of the construction site or the work itself. The score decreases when many materials are needed for the installation, if the extra material has toxicity for the workers and the environment or if the necessary material is of very high value and/or extremely specific.

14. It requires specialised training.

To obtain the maximum score, the selected material should not require any training or course for it to be installed. Any worker in the civil construction area should be able to install it. The grade will decline whenever a paid course or training is required that the company does not provide.

15. The material has an environmental seal.

The maximum score will be assigned whenever the material has one or more recognised environmental seals, nationally or internationally. Remembering that the grade can increase or decrease according to the importance of the seal.

16. There may be a partnership with NGO's, cooperatives, and non-profit associations.

The maximum score will occur whenever the waste is destined for places, institutions and functions other than landfills.

17. Can be reused.

The maximum score will occur whenever the material can be reused at the construction site or another destination where it is not discarded unnecessarily.

18. It has some hazardous material in the composition.

The maximum score will always occur when the material does not have substances or materials that are dangerous to humans and the environment in its composition, and it decreases according to the level of danger of the material in question.

19. It has reverse logistics.

The maximum score will always be assigned when the material has functional reverse logistics, and it decreases according to the difficulty or lack of reverse logistics.

20. Requires more and/or other materials for finishing.

The maximum score will always occur when the analysed material does not require the addition of any other type of material to perform the finishing. Therefore the score will decrease whenever it is necessary to add more materials for the finishing, and the greater the amount needed and the more expensive the added material lower the note.

21. Visually pleasing when installed.

The maximum score will occur whenever, after completing the installation, the sustainable material is visually pleasing and its equivalent traditional material. The decay of the score occurs whenever visually, the sustainable material is presented as inferior to the traditional one.

How to Score Materials

When choosing the best choice to score the analysed parameter, aid in the choice of grades follows.

Excellent: Note chosen whenever the response to the parameter has a very high positive direction. For example, Low Cost: "Excellent", as the market price is a direct competitor of commonly used products, and may even be lower, thus being accessible and competitive to the vast majority of the membership.

Good: Note chosen whenever the response to the parameter has a high positive direction but is not the closest to the ideal. For example, Low Cost: "Good" because the market price is not a direct competitor to commonly used products, but in promotions, it may have lower prices, but not serving consumers from all social strata.

Regular: Note chose whenever the response to the parameter has a sense of neutrality and/or indifference. For example, Low Cost: "Regular", as prices are equivalent to a higher value than the market price, in which many social classes are excluded.

Bad: Note chosen whenever the response to the parameter has a negative direction. For example, Low Cost: "Bad" because the price of the commonly used product is already higher and the environmentally correct product analysed has an even higher price, so a good part of the social classes was excluded.

Does Not Exist: Note chose whenever the response to the parameter has a meaning of not applicable. For example: Low Cost: "Does not exist" because the environmentally correct product is not sold or there is no equivalent environmentally correct product.

After completing the quantitative analysis of the surveyed materials, it is necessary to proceed with a qualitative and comparative analysis of the Brazilian and Portuguese legislations on existing fiscal and legislative incentives for sustainable construction and using more sustainable materials. In this way, it will be possible to determine which good international practices can be replicated or adapted to the Brazilian reality and which good national practices can be improved to encourage and popularise sustainable construction in Brazil linked to a consolidated legal basis. Therefore, it will be a work in which the inductive method will be used, as it will be a mental process that will start from particular facts, such as material analysis tools, national and international incentives, rankings and other information to reach a general conclusion on the subject [19].

3 Rankings

Initially, a diagnosis and previous survey of sustainable materials used in civil construction were made to analyse and rank these materials using the methodology suggested in the previous section 12 types of materials were selected and subdivided into 24 items [16, 20]. They are Concrete; Portland cement; Cement with Construction and Demolition Waste; Cement with blast furnace slag; Cement limestone powder; Aggregate with Construction and Demolition Waste (RCD); Steel aggregate or steelworks slag; Aggregate with ash from sugarcane bagasse; Aggregate with rice husk ash; Aggregate with fly ash; Aggregate with red ceramic; Aggregate with waste tire; Aggregate with glass; Wood; Bamboo; Earth; Soil-cement brick; ISOPET (blocks made of lightweight concrete with recycled styrofoam); Polyethylene Terephthalate (Tubopet); Plaster; Steel Frame and Wood Frame; Phase change materials; Glass; Mineral paints.

Thinking about how to help in the best choice of materials to be used in the work, specific rankings and an overview of the analysed materials were created as follows:

General Ranking
(See Table 2).

Table 2. Table with the general ranking of the analysed materials.

Materials	Score
Bamboo	3.57
Wood	3.19
Soil-cement brick	2.9
Soil	2.71
Steel frame and Wood frame	2.33
Aggregated with rice husk ash	2.14
Aggregated with sugarcane bagasse ash	2.14
Glass	2.14
Phase change materials	2.05
Aggregate with construction and demolition waste	1.95
Portland Cement	1.86
Cement with blast furnace slag	1.76
Cement with construction and demolition waste	1.76
Limestone powder cement	1.76
Aggregate with waste tire	1.67
Mineral paints	1.57
Aggregate with red ceramic	1.29
Aggregate with fly ash	1.29
Aggregate with glass	1.29
Steel aggregate or steel slag	1.29
Plaster	1.29

Source: [12, 18].

Traditional Structural Ranking
(See Table 3).

Table 3. Table with the ranking of traditional structural materials.

Materials	Score
Aggregated with rice husk ash	2.14
Aggregated with sugarcane bagasse ash	2.14
Aggregate with construction and demolition waste	1.95
Portland Cement	1.86
Cement with blast furnace slag	1.76
Cement with construction and demolition waste	1.76
Limestone powder cement	1.76
Aggregate with waste tire	1.67
Aggregate with red ceramic	1.29
Aggregate with fly ash	1.29
Aggregate with glass	1.29
Steel aggregate or steel slag	1.29

Source: [12, 18].

Sustainable Structural Ranking
(See Table 4).

Table 4. Table with the ranking of sustainable structural materials.

Materials	Score
Wood	3.19
Soil-cement brick	2.9
Soil	2.71
Steel frame and Wood frame	2.33

Source: [12, 18].

Finish Ranking
(See Table 5).

Table 5. Table with the ranking of materials focused on finishing.

Materials	Score
Bamboo	3.57
Wood	3.19
Glasses	2.14
Phase change materials	2.05
Mineral paints	1.57
Plaster	1.29

Source: [12, 18].

4 Comparison of Brazilian and Portuguese Legislation with Incentive Proposals

Europe is at the forefront of fiscal and/or tax incentives aimed at sustainable construction, so it is important to gather theoretical and practical information on the use of various tools that encourage the improvement of sustainable construction [21].

The European Union (EU) theme has established standards, laws, incentives, objectives and other tools aimed at sustainability so that its member countries can reduce environmental impacts, improve technologies, reduce the consumption of natural resources and many other benefits. Thus leveraging the entire production chain due to the continental scope that the guidelines cover [21].

To acquire more information and details about government agencies, practices and tools used to improve the environmental issue, with a greater focus on a member country of the EU if necessary, we will analyse Portugal in the sequence [21].

4.1 Sustainable Construction Legislation in Portugal

Portugal and the other members of the EU must follow the guidelines established by the economic bloc, preserving the particularities of each country. However, each country has the autonomy to implement more restrictive measures to benefit the country and the environment.

In this way, we can cite some legislation and actions implemented by Portugal that contribute to the achievement of the objectives proposed by the European Union and the benefit and scope established by the relevant government agencies through prescriptive actions and economic incentives.

Portuguese Environment Agency
Portugal created the Portuguese Environment Agency (APA) intending to manage environmental issues linked to environmental policies with all interested parties in an integrated way, aiming at protecting and enhancing the environment [22].

The APA performs monitoring, planning and evaluation, licensing and inspection functions, being responsible for the following topics: water and coastline, waste, climate change and air, noise, radiological emergencies, environmental impact assessment, circular economy, or environmental education and State Report of the Environment of Portugal [22].

Therefore, this agency is the main environmental regulator in the country; it contains all the environmental policies in force in the country, such as water, climate change, sustainable development, and waste, among others. In this way, the instruments and information systems contribute to the execution and implementation of the objectives, and in this way, it is possible to achieve the stipulated goals. Among these important tools, we have an environmental impact assessment, strategic environmental assessment, and inspection, among other options [22].

However, special emphasis should be given to the tax incentive system, as it establishes the Tax Incentive System for Business R&D (SIFIDE), which aims to increase the competitiveness of companies by reducing the amount charged in tax on corporate income (IRC) through the evaluation and approval of projects that stimulate research and development activities [22].

Environmental Product Declaration Registration System
Aiming to map and reduce the negative environmental impacts caused by products and services used in the sustainable construction chain, the Environmental Product Declarations (DAP) tool was created, which aims to communicate the environmental information of the products/services offered, ensuring the scientific validity of their environmental performance, due to third-party verification [23].

To collaborate, Portugal created the DAP Habitat System, which consists of the national registry of type III environmental declarations, which aims to provide information duly validated in accordance with ISO 14025 and EN 15804 in a publicly accessible database and national and international level [23].

Therefore, with this system, the performance of environmental products and services will be available for consulting by any national or international interested part, linking benefits such as detailed knowledge of the impacts arising from the manufacturing process and possible opportunities for improvement, the possibility of developing ECO-design and optimise solutions aimed at business sustainability [23].

4.2 Sustainable Construction Legislation in Brazilian

After researching international legislation, the same exercise should be carried out for national legislation to better understand the Brazilian context to propose a viable project for the Brazilian reality on the implementation of good international practices, thus requiring greater discussion on the most prominent instruments in the elaboration and application of public policies, namely: payments for environmental services, Green IPTU

(Tax on Property and Urban Territorial Property) and cleaner production (P + L) and eco-efficiency.

Payment for Environmental Services
Dealing with the economic instrument Payment for Environmental Services (PES) which aims to reward everyone who, due to their practices of conservation, protection, management and recovery of ecosystems, maintains or increases the supply of an ecosystem service [24].

Within the ecosystem services category, we can include climate regulation, fertility maintenance and soil erosion control, carbon storage, nutrient cycling, water supply, biodiversity protection, scenic beauty and the maintenance of genetic resources [25].

Therefore, due to the potential for protection and sustainable use of environmental resources linked to the PES, this has become an important tool of Brazilian environmental policy thanks to the possibility of expanding the scale of programs and projects, as well as their replication to other parts of the country and other places in the world and providing a possible improvement in the quality of life of populations that economically depend on the use and exploitation of the environmental resources [24].

The elaboration of public policies is one of the main ways to encourage the existence of PES programs and projects, as through these tools, it is possible to notice that a growing expansion in the number of public policies on this theme has been generated at the regional and local levels.

However, it is necessary to ensure economic and technical feasibility for this purpose to establish a PES policy, as it must be contextualised in the Brazilian legal system in a way that foresees and supports essential elements. The following elements are: its competencies for that purpose (administrative and legislative); whether or not they should be derived from the law; and if they can be developed by normative acts of the public administration, even if there is no law regulating the matter in question.

Cleaner Production (P + L) and Eco-Efficiency
In Brazil, cleaner production, together with eco-efficiency, gained strength with the creation of the Brazilian Business Council for Sustainable Development (CEBDS), which brings together large corporations and aims to promote sustainable development in the business sector. Therefore, even with the great availability of resources available in the territory, the country stands out in these concepts due to the great capacity for innovation and creation present in the scientific business scenario of the country. Due to the great emphasis on eco-efficiency and cleaner production present, the concept is valid of these important instruments [26].

Eco-efficiency aims at greater production using fewer inputs, less toxic and non-biodegradable products, and less negative environmental impacts. In this way, we can say that the basic principles are to reduce the use of raw materials, energy and the dispersion of pollutants; maximise the sustainable use of natural resources, increasing the shelf life of products; as well as the reuse and recycling of residues and by-products [27].

Within the context of eco-efficiency, Cleaner Production (P + L) is linked. It is a preventive concept that deals with an established economic, environmental and technological strategy aimed at sustainable development once it aims to increase efficiency in using raw materials and minimisation or elimination of toxic products [27].

Green IPTU
In Brazil, there is a tax on urban properties, the IPTU, which is a program that aims to encourage actions in favour of the environment and social well-being, such as the preservation of green areas, selective collection, the capture of solar energy, reuse of rainwater, among other activities that private individuals can carry out, therefore encouraging sustainable practices in new or renovated constructions, may result in the application of discounts up to tax exemption [28].

In general, the main measures encouraged by the program to obtain a discount are the installation of rainwater harvesting and/or water reuse systems, the construction of vegetation cover, the presence of permeable areas larger than those required by the municipality, the installation of a solar energy capture system, the planting of trees in front of the residence and the construction with sustainable materials [28].

"Polluter pays" Principle
The "polluter pays" principle is one of the National Solid Waste Plan, Law 12.305/10, which seeks to preserve the environment, as this concept implies a mechanism that aims to attribute to the polluter the costs of the measures adopted by the public authorities to prevent the pollution [29].

This principle has evolved and is also applied in accidental pollution and contamination situations. Therefore, in 1991, the Organization for Economic Cooperation and Development (OECD) defined that it is the responsibility of the polluter to repay the costs of pollution, combating pollution, the damage caused, and the compensation of the victims, that is, he is responsible for all costs always considering the broadest sense of the issue as per [30].

Therefore, the polluter pays principle is a preventive and sanctioning tool to prevent the occurrence of environmental damage and indoctrination of actions by individuals to prevent them from committing acts that may harm the environment [31].

4.3 Comparison of Legislation

During the study and analysis, it was possible to note that the legislation, programs and incentives in Portugal are aimed at improving the production process that makes production more efficient and consumes least resources possible, together with the reduction of costs. Previously mapped environmental impacts. Therefore, we can say that Portugal values global management of the subject, as it covers the extraction of raw materials to the final destination, covering the entire production chain.

Adopting the same analysis for Brazilian legislation, it was possible to notice that Brazil is on the regional scale of projects, laws and incentives due to the short time that sustainable issues have become part of the routine of Brazilians, therefore the theme with the regional scope it is intrinsic to the actions of the private initiative encouraged by the public authorities, the other initiatives still preserve the punctual character of solutions. A summary worksheet on the established conclusions follows (Table 6).

Table 6. Table that presents the main points of Brazilian and Portuguese legislation to encourage the use of sustainable materials and their scope

Country	Methodisation	Functions	Coverage
Portugal	Portuguese Environment Agency	To be the regulatory, supervisory and maintainer of the quality of the Portuguese environment	National
	Product Environmental Declaration Registration System	Map and reduce the negative environmental impacts caused by sustainable construction products and services	Worldwide
Brazil	Payment for environmental services	Reward those who, due to their practices of conservation, protection, management, and recovery of ecosystems, maintain or increase the supply of an ecosystem service	Local
	Cleaner production (P + L) and eco-efficiency	Optimise production processes as much as possible so that production operates using a less resources	Regional
	Green IPTU	Reduction in the amount of the tax according to the reduction in resource use and system performance	Local

Source: [12, 18].

4.4 Possibility of Using Fiscal and/or Tax Incentives

To achieve sustainable national development in Brazil, the public authorities need to develop and adopt fiscal and/or tax incentive policies, as this way encourages the adoption of sustainable measures in a more incisive way [32].

By adopting this attitude, the State starts to encourage companies and sustainable actions through financing and/or subsidies and, consequently, starts to detriment unsustainable actions and activities, as they will lose government incentives and will be taxed through the polluter pays principle [33].

This conclusion derives from the practice of environmental tax incentives in Europe in the 1980s. The dissemination and incorporation of environmental incentive legislation were intensified between 1989 and 1994 when countries such as France, Italy and the United States implemented such practices [32].

Therefore, Brazil can adopt the same position at the federal, state and municipal levels since the Brazilian Constitution of 1988 provides in article 170 the possibility of intervention in the economic and financial order to protect values, principles and diffuse, collective and individual rights. Highlighting item VI which deals specifically with the defence of the environment [32].

In this way, the Constitution supports economic intervention measures aimed at protecting the environment, making incentives aimed at sustainable development legitimate [32].

Examples of Implemented Federal Incentives
Federal taxes may change to encourage sustainable practices, namely the Income Tax, which has already allowed individuals and companies to deduct a fraction or the total amount of the amount spent on forestry and reforestation from their declarations. Another example is Import and Export Taxes, as changing the rates makes it possible to demonstrate a preference for environmentally friendly products during commercial transactions [32].

Examples of Implemented State Incentives
At the State level, we can identify a favourable scenario for sustainable practices since the Tax on Motor Vehicle Ownership in Rio de Janeiro presents different collection values for the type of fuel used and the activity to be performed [32].

Examples of Implemented Municipal Incentives
Municipalities also have initiatives that have legal support and depend only on legislative initiatives. For example, the Urban Territorial Property Tax can be used through temporal progressiveness and the type of use according to the social function of the property. This action is supported by the Statute of Cities, established by Law n. 10.257/2001, which provides in article 7 about the progressiveness of discounts as an instrument of urban policy [32].

4.5 Proposal for the Material Registration System

Portugal developed the DAP Habitat System so that the negative impacts caused to the environment were reduced; it is a system that aims to gather in a single place all the third-party environmental declarations about the life cycle of the material in one place because, in this way, the choice of materials would be carried out with greater environmental awareness based on scientific knowledge.

However, the Brazilian reality does not allow us to establish that the materials used in civil construction have a third-party assessment of the life cycle of the products, as this would result in a greater cost of products since all the prior scientific preparation together with the verification process of a third party could make the competitive trade of the products unfeasible.

Therefore, an intermediate solution appears as the most viable option, so the Brazilian government can establish a system or guideline that implies using the previously mentioned methodology. In this way, it will be possible to rank the materials used by the Brazilian civil construction industry so that better choices can be made.

In addition to smart choices, we can establish parameters for tax discounts. By establishing a metric on the evaluation of materials, we can use this same metric for tax incentives.

Tax and fiscal incentive proposal

There are fiscal and/or economic incentives for good practices in materials and technologies that benefit the environment at the international and national levels. These incentives can occur in smaller spheres, such as municipalities and states, or encompass larger spheres, such as the federal sphere or economic blocks.

Having Brazilian reality and society as a context, the best solutions are intermediate proposals that can be implemented without the need for large financial investments. With the engagement of people and companies, incentives can become more rigid and complex.

In this way, we can base ourselves on the incentive and concept established by the green IPTU and replicate it for the selection of civil construction materials, as previously stated, with the government establishing a system in which the materials used are evaluated and have this evaluation available for consultation by any user, the discounts to be implemented in the taxes inherent to the production process of each material, as this way the price of the productive materials will be even more competitive at the time of purchase by consumers and it will be possible to create greater environmental awareness in people and companies.

Therefore, the proposal is that, according to the note resulting from the application of the material analysis methodology, the discount applied in the production process follows the table below (Table 7):

Table 7. Table showing the relationship between the note received and the discount to be implemented in Brazil and Portugal

Score	Discount
X>4	Up to 85%
2,5 > X ≤ 4	Up to 60%
1,25 > X ≤ 2,5	Up to 45%
0 > X ≤ 1,25	Up to 20%
X ≤ 0	Without discount

Source: [12, 18].

5 Conclusion

With the bibliographical review of the work, it was possible to notice several gaps in the classic processes of evaluation of sustainable materials used in civil construction, together with the fact that these methodologies have a great complexity to be used by the general public.

Therefore, the opportunity arose to develop a new simplified analysis methodology that covers the gaps identified in existing methodologies and critical concepts to analyse the sustainability of materials used in civil construction. In this way, this new methodology allows people with different levels of knowledge on the subject to use the tool and perform the necessary calculations to assign the grade to the studied material and thus make conscious choices about the best materials to be used in civil works.

The materials selected to be analysed in work are used in Brazilian and Portuguese civil works. After being analysed by the proposed methodology, it was possible to conclude that there is still room for improvement in the production process chain along with a margin for innovations and new technologies to reduce the global negative impact of the materials and thus can benefit from not harming the environment.

This benefit would come from the implementation of new tax and legal incentives, as with the analysis of current legislation in Brazil and Portugal, it was possible to identify good national and international practices regarding the materials used in civil construction.

However, Brazil is relatively behind in the environmental issue compared to Portugal because, until now, only specific and regional actions have been carried out, as we can see with the Green IPTU. In contrast to Portugal, projects and products are already being developed with a national vision and even a global system, as we can see with the DAP Habitat System.

This work can assist in practices and actions in the civil construction sector with a view to sustainability linked to the UN Sustainable Development Goals (SDG), 2030 Agenda, exemplifying in this sector: SDG 9 – Industry, innovation and Infrastructure, SDG 11 – Sustainable cities and communities and SDG 12 – responsible consumption and production.

Therefore, it was possible to conclude that by adapting already consolidated methods and systems to the Brazilian reality, it will be possible to develop a greater environmental awareness in society and, in the near future, restrict measures and apply more complex methodologies than the one presented in this work until we match international actions and systems, they may even be shared.

References

1. Agopyan, V., John, V M.: O desafio da sustentabilidade na construção civil. Série Sustentabilidade, vol. 5. 1a edição. Editora Blucher, São Paulo (2011)
2. Dalla Costa, E., Moraes, C.S.B.: Construção civil e certificação ambiental: análise comparativa das certificações LEED (Leadership in Energy and Environmental Design) e AQUA (Alta Qualidade Ambiental). In: FEA/ USP, FGV. Anais do XIV ENGEMA – Encontro Nacional sobre Gestão Empresarial e Meio Ambiente, São Paulo (2012)
3. Dalla Costa, E., Moraes, C.S.B.: Construção Civil e a Certificação Ambiental: Análise Comparativa das Certificações LEED (Leadership in Energy and Environmental Design) e AQUA (Alta Qualidade Ambiental). Revista Engenharia Ambiental (Online) **10**, 160–169 (2013)
4. Moraes, C.S.B., Santesso, C.A., Rocha, R.P., Lima, S.C.S.: Diagnosis and analysis of the economic and socio-environmental viability of technologies and alternatives for sustainable housing construction. In: Monteiro, J., et al. (eds.) INCREaSE 2019, pp. 903–918. INCREaSE 2019. Springer, Cham (2020). https://doi.org/10.1007/978-3-030-30938-1_71
5. Moraes, C.S.B., et al.: Analysis of economic and socio-environmental viability indicators for energy, materials and water management and proposal of technologies and alternatives for sustainable construction in housing. J. Civ. Eng. Archit. **14**, 280–292 (2020)
6. CIB. Agenda 21 on Sustainable Construction. CIB, p. 120 (CIB Report Publication 237), Rotterdam (1999)
7. Gavira, M.O., Moraes, C.S.B., Dadario, A.M.V.: Administração e Gestão Sustentável: Contexto e Ferramentas. Editora Rima, São Carlos (2017)

8. Acselrad, H., Leroy, J.P.: Novas premissas da sustentabilidade democrática (Série Cadernos de Debate Brasil Sustentável e Democrático, 1). FASE, Rio de Janeiro (1999)
9. Sachs, I.: Caminhos para o Desenvolvimento Sustentável. (Org. Paulo Yone Stroch), 2 ed. Garamnond, Rio de Janeiro (2002)
10. CNI; CBIC: Construção verde: Desenvolvimento com sustentabilidade, Brasília (2012)
11. Motta, S.R.F., Aguilar, M.T.P.: Sustentabilidade e Processos de Projetos de Edificações. Gestão & Tecnologia de Projetos, vol. 4, no. 1, pp. 84–119 (maio de 2009)
12. Lima, S.C.S., Moraes, C.S.B.: Análise de Viabilidade Socioambiental e Econômica de Materiais Sustentáveis para aplicação em Residências Unifamiliares. Relatório de Pesquisa. PIBIC, Rio Claro (2018)
13. Bissoli-Dalvi, M., et al.: A necessária simplificação na metodologia de avaliação da sustentabilidade dos materiais: Estudo de caso – ISMAS. Curitiba (2013). Disponível em: http://lpp.ufes.br/sites/lpp.ufes.br/files/field/anexo/artigo378627.pdf. Accessed 15 Apr 2023
14. Santesso, C.A.: Construção sustentável e habitação: diagnóstico de alternativas e tecnologias de energias para a residência unifamiliar. Iniciação Científica (Engenharia Ambiental). FAPESP/Instituto de Geociências e Ciências Exatas, Universidade Estadual Paulista, Rio Claro (2013)
15. Santesso, C.A., Moraes, C.S.B., Santarine, G.A., Goncalves, J.C.: Economia de energia através de estratégias passivas e ativas: um estudo de caso para habitação de interesse social. ESPACIOS (CARACAS), vol. 38, p. 23 (2017)
16. Kwai, L.L., Moraes, C.S.B.: Tecnologias, Conceitos e Propostas do Uso de Materiais para a Construção nas Universidades. In: FEA-USP. Anais do XV ENGEMA – Encontro Internacional de Gestão Empresarial e Meio Ambiente, São Paulo (2013)
17. Silva, V.G.: Indicadores De Sustentabilidade De Edifícios: Estado Da Arte E Desafios Para Desenvolvimento No Brasil. Campinas (2008)
18. Lima, S.C.S., Moraes, C.S.B.: Análise de Viabilidade Socioambiental e Econômica de Materiais Sustentáveis para aplicação em Residências Unifamiliares. In: Anais do XXIX Congresso de Iniciação Científica da Unesp. IGCE/UNESP, Rio Claro (2017)
19. Rodrigues, W.C.: Metodologia Científica. Faetec/ist., Rio de Janeiro (2007)
20. Moraes, C.S.B.: Análise da Eficácia e Viabilidade Socioambiental e Econômica de Tecnologias e Alternativas de Construção Sustentável para Habitação. Relatório de Pesquisa. IGCE/UNESP, Rio Claro (2019)
21. COMISSÃO EUROPEIA. Comissão Europeia. Disponível em: https://ec.europa.eu/info/index_pt. Accessed 11 Apr 2023
22. APA, Agência Portuguesa do Ambiente. Disponível em: https://apambiente.pt/index.php?ref=17&subref=1484. Accessed 11 Apr 2023
23. PORTUGAL. SISTEMA DAPHABITAT SYSTEM. Sistema de registro de declarações ambientais do produto. Disponível em: https://daphabitat.pt. Accessed 15 Apr 2023
24. FUNDAÇÃO GRUPO BOTICÁRIO DE PROTEÇÃO À NATUREZA. Guia para formulação de políticas públicas Estaduais e Municipais de pagamento por serviços ambientais. Fundação Grupo Boticário de Proteção à Natureza, Brasil (2017)
25. Wunder, S. (Coord.), Börner, J., Tito, M.R., Pereira, L.: Pagamentos por Serviços Ambientais: perspectivas para a Amazônia Legal. MMA, (Série Estudos, 10), p. 136, Brasília (2008)
26. Andrade, E.M., Andrade, F.C., Santos, J.M.: Produção mais limpa e ecoeficiência como ferramenta do engenheiro. In: DEPRO/UFS. VIII Simpósio de Engenharia de Produção de Sergipe, pp. 163–172. São Cristóvão, Sergipe (2016)
27. Munhoz, D.: Ecoeficiência - Produção limpa e Ecodesign. In: III Simpósio Mineiro de Química, Belo Horizonte (2008)
28. Feio, L.G.: O IPTU verde e a construção da cidade sustentável. 2018. 119 f. Dissertação (Mestrado) – Universidade Federal do Pará, Instituto de Ciências Jurídicas. Programa de Pós-Graduação em Direito, Belém (2018)

29. Romano, R.T.: Anotações da doutrina sobre os princípios do poluidor-pagador e da prevenção. Revista Jus Navigandi, ano 23, no. 5535. Teresina (2018). Disponível em: https://jus.com.br/artigos/68537. Accessed 30 Nov 2019
30. Barichello, S.E., Araújo, L.E.B.: O princípio do poluidor- pagador e os mecanismos de flexibilização no protocolo de Kyoto. Revista do Direito, pp. 137–149 (2009)
31. Ercolin, A.M.M., et.al: O Princípio do Poluidor-Pagador e sua Aplicabilidade (2015). Disponível em. https://emdeis.jusbrasil.com.br/artigos/311776167/o-principio-do-poluidor-pagador-e-sua-aplicabilidade. Accessed 15 Apr 2023
32. Dantas, M.B., et al.: Mapeamento de incentivos econômicos para a construção sustentável: A indústria da construção brasileira em busca da sustentabilidade. Cbic – Câmara Brasileira da Indústria da Construção, Florianópolis (2015)
33. Claudino, A.: O desenvolvimento sustentado e os pressupostos para uma sociedade eco-sustentada. Revista de Direitos Difusos – Desenvolvimento sustentável. ADCOAS, vol. 1, no. 6, pp. 769–784, abr. 2001, São Paulo (2001)

Sustainable Food Processing and Food Waste Solutions. Food Safety and Quality Engineering

Revisiting Solar Dryers for Small to Medium Production

Bernardo Farrero[1,3](✉), Paulo Bruno Rossi da Silva[2,3], and Luis Frölén Ribeiro[3,4]

[1] Facultad Regional de Córdoba, Universidad Tecnológica Nacional, X5016ZAA Córdoba, Argentina
bernardo.farrero@ipb.pt
[2] Universidade Tecnológica Federal do Paraná, Campus Cornélio Procópio, Cornélio Procópio 86300-000, Brazil
[3] Instituto Politécnico de Bragança, Campus de Santa Apolónia, 5300-253 Bragança, Portugal
[4] Associated Laboratory for Energy, Transports and Aerospace-LAETA/Instituto de Ciência e Inovação em Engenharia Mecânica e Engenharia Industrial-INEGI, Campus da FEUP, 4200-465 Porto, Portugal

Abstract. This article presents an overview of passive solar dryers, an ancient technique that may be revived for food drying capacities below one ton/day. The art of dehydrating food has existed for centuries and was developed to preserve food by extracting its water content, thus delaying its natural deterioration. This article reviews traditional solar dryers and proposes a verification parameter k, parametrising the solar collector's size by the drying capacity. The modelling and parametrisation of the k constant allowed the evaluation of the examples of traditional models from the literature systematically. Results showed that 80% of the evaluated constructed solar dryers might underperform due to lower solar collecting surfaces than those projected by the k parameter.

Keywords: Solar dryer · Food drying · solar dryer parametrisation

1 Introduction

In the Western world, food dryers are often considered craft equipment associated with permaculture when discussing drying equipment. These models are usually one-off designs employing reclaimed or recycled material for a specific site, aiming for low investment and reduced operational costs besides labour. Alternatively, industrialised dryers are available for large productions requiring high investment and energy costs, deemed unsuitable for a small startup to a medium size company.

There is no standardised categorisation regarding the dryer's capacity. For clarity in this article, a producer classification based on daily drying capacity is proposed (24 h). [Table 1]. Small producers are at most 100 kg and medium

J. F. L. C. Semião et al. (Eds.): INCREaSE 2023, ASST, pp. 347–362, 2023.
https://doi.org/10.1007/978-3-031-44006-9_26

Table 1. Producer classification based on the magnitude of the daily mass drying capacity in kg.

0<	Small	$\leq \times 10^2$
$\times 10^2$ <	Medium	$< \times 10^3$
	Industrial	$\geq \times 10^3$

until 1000 kg. Industrial productions exceed a daily capacity of 1000 kg, requiring specific industrial equipment with continuous operation.

With the appropriate designs, instruments and measures, drying can be carried out economically sustainably and industrialized. The drying process becomes more controlled and efficient without incurring high costs for purchasing machines, such as continuous flow dryers. A critical step is to collect data continuously and efficiently communicate to the operator.

It is essential to highlight the two distinct sectors of the food drying industry and the gap between them. This gap creates a space that can be exploited to improve the technical side of food drying, society's belief in it, and how communities can benefit from pre-dried food.

Solar drying technology is worth mentioning, catalogued as dated, unattractive, low-tech, or presently forgotten. However, this technology can be revised and modernized to bridge the gap between low/amateur and industrialized drying.

Food logistics worldwide has been in demand in the last centuries. The transport of certain foods, such as fruits or vegetables, has escalated to levels of crossing continents for final consumer delivery, implying a high energy demand which is directly proportional to the transported weight. The consumption of pre-dried food, either directly (final consumer) or indirectly (food industry), would significantly reduce not only operational costs but would also have a direct impact on the reduction of greenhouse gas emissions. Approximately 91% of the fruits' composition is water; in the case of vegetables, the water content increases to 95% [25]. For example, Spain imported 2011 more than 25.4 million tons of food that travelled an average of 3,827 km [28]. In specific cases, a chickpea travels an average of 7500 km.

The world's leading fruit and vegetable-producing region by far is East Asia, followed by South Asia (Fig. 1 and Fig. 2). Other important producing regions are South America, Southeast Asia and Southern Europe (for fruits), and Southeast Asia (for vegetables). Northern and Western Europe produce relatively little, so they must import much of their consumption [25].

Food drying would not only reduce the energy cost of logistics but also impact the local economy, benefiting the medium farmer by increasing our added value in food without prior processing. In 2018, the world produced 868 million tons of fruits and 1,089 million tons of vegetables [25]. Northern and Western Europe farm relatively minor, so they must import much of their consumption, whereas Asia and Latin America are the largest producers.

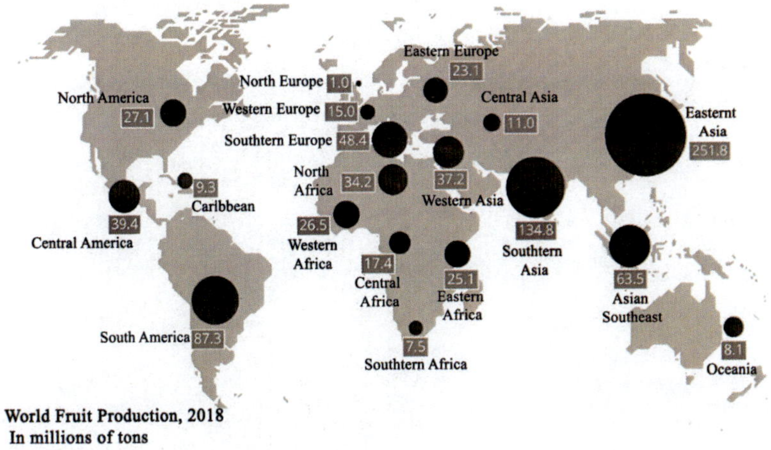

Fig. 1. Fruit production [25].

Worldwide, more than 50% of fruits and vegetables are grown on farms of less than 20 hectares (most of which are family farms). In developing countries, these farms grow the vast majority of horticultural products, more than 80% in most of Asia, sub-Saharan Africa, and China [25]. Since this category (fruits and vegetables) has excellent potential for applying drying techniques, it would directly impact the economy of family farms if it could be carried out efficiently.

Figure 3 shows the world exports of certain foods, most of which can be dried, especially in the main categories.

Problems of distribution and access mean that many people need help to obtain the types or quantities of food they need [25]. A considerable proportion of the harvest is lost or wasted before it reaches consumers' plates. Figure 4

In East and Southeast Asia, losses were greatest during storage (averaging more than 20% loss), processing and packaging. The highest loss levels in sub-Saharan Africa occurred at the farm gate and wholesale markets. Losses and wastage in Central and South Asia tended to be lower, with losses during transport being the highest [25].

Food drying technology is the cheapest method (particularly solar drying), especially suitable for communities with no other preservation possibilities. Applying this technology could significantly reduce storage and transportation losses and transportation costs.

The global dehydrated food market is forecast to register a CAGR (Compound Annual Growth Rate) of 4.8% during the forecast period (2022-*2027) [27]. In 2020 the dried fruit market was expected to reach 4 million tons [29]. EMR estimates the Latin American dried fruit market to reach USD 4.15 billion by 2021. During 2023–2028, the market is anticipated to grow at a compound annual growth rate of 4.77%.

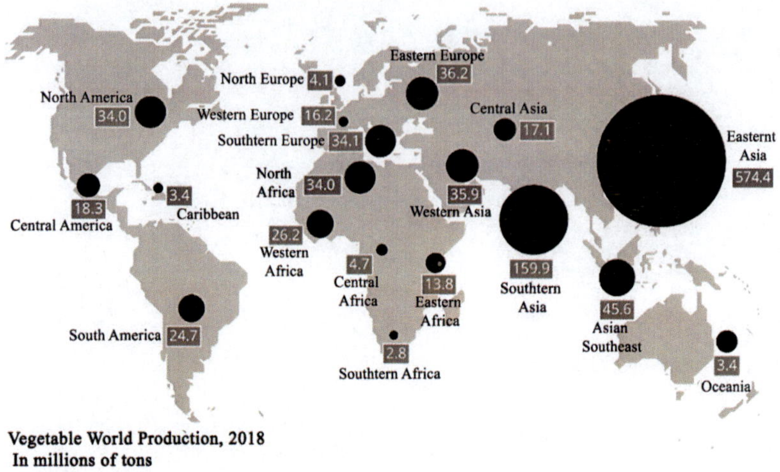

Fig. 2. Vegetables production [25].

Much of the drying technology is applied in North America and Asia. It correlates with Asia's world-leading position in fruit and vegetable production. In 2004, it led the market by more than 50%, Figure 5. On the contrary, North America produces little but dries much of its food, and South America, where fruit production is high but drying is low. The statistics for South America and Africa must be taken with a grain of salt because often, the records of these types of activities are scarce, leading to a distorted perception of the use of these technologies, Fig. 6.

The continuous search for innovation sometimes leads to ignoring low or so-called dated technologies, quickly forgotten or perceived as obsolete. Such is the case of solar drying, a technology that, by some criteria, may not fit into the terms of technology simply because it does not possess electronic complexity or continuous innovation. While it is an ancient technology dating back millennia, it is a technology that, when applied correctly, can currently bring many benefits where simplicity and self-sustainability are a virtue. Revisiting this technology and parametrising the dimensions of principal components with a contemporary lens will lead to even better performance and fulfil the job with an impressively low carbon footprint.

2 Dryers

Food drying is a process by which the product is subjected to certain conditions of temperature and humidity to achieve a transfer of mass and heat. This can be defined as applying heat under controlled conditions to remove a certain amount of water in the food by evaporation. This mechanism works through

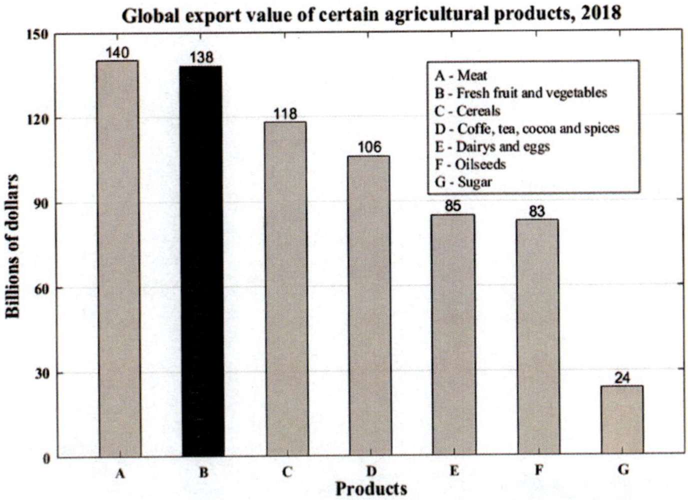

Fig. 3. Production [25]

the difference in humidity between the air in contact with the product and the product's moisture.

"Drying is a physical method of preservation. It consists of reducing, by evaporation, the amount of water contained in the food to prevent or limit microbial growth. By controlling the action of enzymes that chemically deteriorate the food, its shelf life is prolonged" [26].

The critical factors to consider for drying are temperature, relative humidity of the drying agent (air) and the pressure at which this phenomenon occurs.

This process can be carried out in different ways, either in a "natural" way or through dryers. However, there is no standardization of the types of dryers since more than 500 types have been recorded [15]. A classification is accepted according to the technology: Natural (N), Fuel (F), and Solar (S). Table 2 shows the specific parameters of this type of dryer, highlighting particular characteristics that can help one choose according to the needs: investment, drying time, losses, and operating costs.

The investment required to acquire a solar dryer is medium or even low compared to an industrial dryer. This fact already places it in an advantageous position against high-capacity industrial dryers, Table 1. While it is true that the production volume is lower than a fuel dryer, it has to highlight the gap between industrial and home-made dryers for production of up to 1000 kg per day.

As previously presented, one of the primary drying objectives is to reduce product losses due to microbial activity growth, which generates losses of up to 20% in storage, for example. These savings are one of the significant benefits of solar-type dryers. Given the degree of investment and ease of use, reasonably small or even zero losses are obtained in cases where drying is controlled with

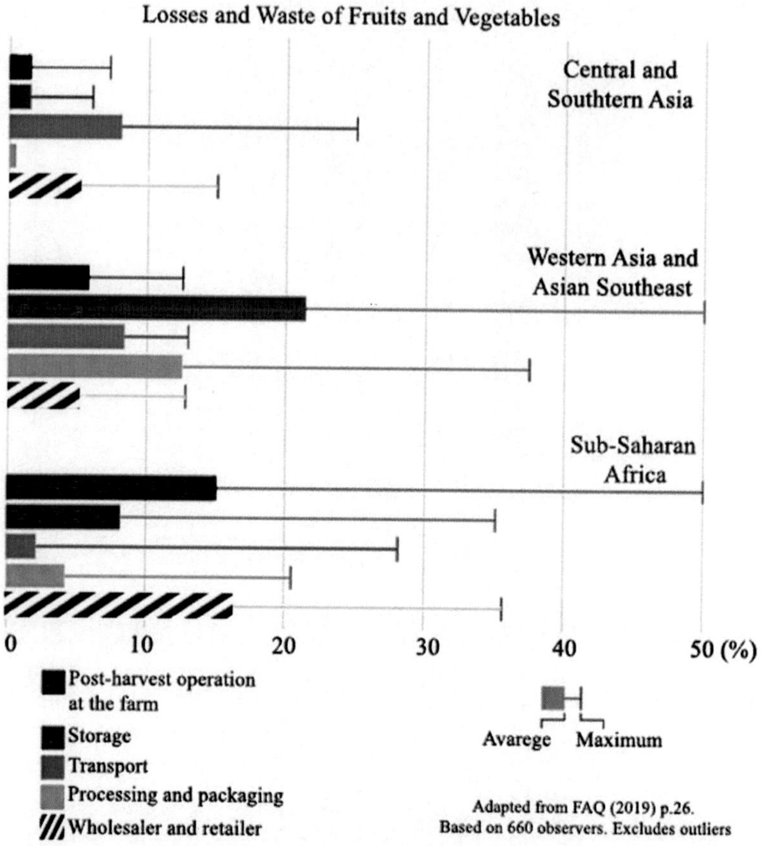

Fig. 4. Losses and waste of fruits and vegetables [25].

the right design and type of solar dryer. This fact positions this dryer category as ideal for a medium sector of food drying producers.

The drying time between the different types of drying is well-marked since natural drying can take days to a week, mainly due to the lack of control over the drying parameters, temperature and humidity. Although drying with fuel accelerates the drying to hours, it can damage the product and expose it to high temperatures generating a loss of nutritional properties. Solar drying can significantly reduce the natural drying time. Drying high-temperature that can damage the quality of the final product can be avoided by monitoring the drying parameters. Moreover, it is acceptable for a medium production volume when the product achieves absolute quality with daily drying rounds. From Table 3, it can be verified that solar dryers are positioned with a specific advantage to make this choice if referring to a start-up or even a medium-sized company. The investigation of this type of dryer and its constituent components is further explored in this document.

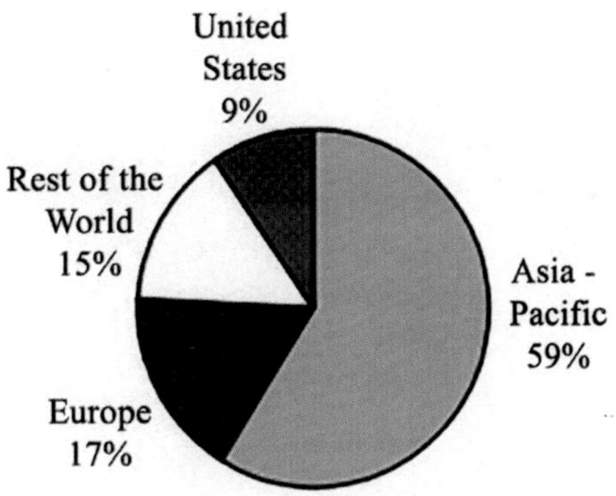

Fig. 5. Distribution of dehydrated market [30].

Table 2. Investment, operational costs, and lost product against the type of dryers: Natural (N); Fuel (F); and Solar (S).

	None	Low	Medium	High
Investment	N	S	S	F
Operating costs	N/S	S	F	F
Lost product	F/S	F/S		N

3 Solar Dryers

Solar technology dryers are equipment used for food preservation by drying using the heat generated by solar radiation [26].

Drying using solar energy is based on obtaining the necessary power to increase the temperature of the circulating air to reduce its relative humidity and thus increase its capacity to absorb moisture from the product. This moisture harnessing is accomplished through one or different components of the dryer.

Being an old technology and not very standardized, the types of solar dryers are vast, and their designs vary in complexity. Although there is no standardization of the types of solar dryers, it has been concluded that the following classification is widely accepted: Direct (D), Indirect (I), and Mixed (M).

The *direct* dryer collects solar energy to increase the air temperature through the roof of the device, i.e., solar radiation not only manages to improve the air temperature but also impacts the product. The ultraviolet radiation degrades the produce, inducing a high rate of losses or low quality of the final product through the loss of food properties, Fig. 7 (a).

The *indirect* dryer collects solar radiation to increase the air temperature. In this case, it is done via an element with an exclusive purpose. The collector is a surface before the drying chamber aiming to capture solar radiation to heat

Fig. 6. Dehydrated food market [27].

Table 3. type of collector relationship between type of dryers/Dryin time.

	Natural	Fuel	Solar
Hours		x	
Days			x
Weeks	x		

the air. The produce is placed in the drying chamber, protected from ultraviolet rays. In such designs, product losses are low or null, Fig. 7 (b).

The *mixed* dryer combines a direct and an indirect dryer. The solar radiation is captured in the collector and the roof of the drying chamber, permitting incoming ultraviolet rays to deteriorate the final product, Fig. 7 (c).

Table 4 shows how the different types of solar dryers are positioned, with the indirect type (I) on the left side of the table, the mixed type (M) predominating on the right side and a transition of the 3 classes in the middle. However, having a low or medium production volume may seem at first sight a disadvantage. It is important to remember that we are looking for a dryer capable of achieving medium production volumes and adaptable to medium-sized producers. The indirect solar dryer is better positioned to occupy this space.

The efficiency of a solar dryer is determined by the amount of heat provided by solar irradiation to the amount of heat delivered to the product to achieve evaporation of the water it contains. The indirect type solar dryer achieves this inefficiently, but this is usually due to design errors or poor construction techniques that can be remedied.

Indirect solar dryers can vary on the prospective market. However, all passive indirect solar dryers always contain the same components: collector, drying chamber, and chimney.

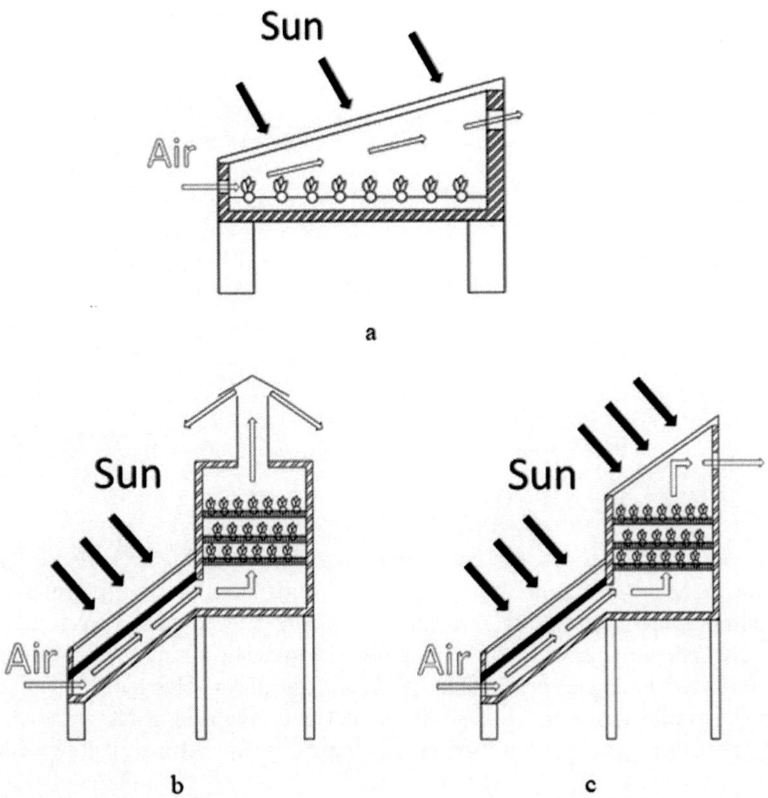

Fig. 7. a: Direct solar dryer, b: Indirect solar dryer, c: Mixed dryer.

3.1 Collector

The collector is the first element in the airflow path, mainly aiming to capture heat from the sun and deliver it to the airflow. It is usually a rectangular hollow parallelepiped, where the upper surface is responsible for absorbing the heat. The air enters one of its lower laterals. It travels through the channel while absorbing heat and increasing its temperature until it leaves it through the opposite extremity, thus decreasing its relative humidity of the air as it expands its capacity to absorb humidity. The collector can have different transparent layers (covers) or thermal insulators on its lateral and inferior surfaces to diminish heat loss by convection.

Table 4. Product damage, the volume of product, and efficiency by type of solar dryer configuration: Direct (D), Indirect (I), Mixed (M).

	Low	Medium	High
Product damage	I	D/M	D/M
Volume of product	I/D	I/D/M	D/M
Efficient	I	D	M

Table 5. Quantity of solar collector layers against efficiency and investment. No layer (0); **one** layer (1); Two layers (2).

	Low	Medium	High
Efficiency	0	1/2	2
Investment	0	1	2

The cover fulfils the objective of reducing the losses of the collector through the greenhouse effect generated between it and the absorber plate. Although transparent plastic covers seem a good solution, they degrade from ultraviolet radiation exposure needing to be regularly replaced. Using a non-transparent cover reduces the collector's efficiency, thus, the dryer. However, it reduces the investment and maintenance costs if the cover is made of plastic. Using two or more covers can generate air tightness promoting the natural airflow, Table 5.

3.2 Chamber

The drying chamber is where the essential part of a solar dryer takes place as it receives the air already dried by the collector. This air circulates the product intending to heat it, evaporate its water content and then transport it to the chimney, conducting a continuous flow.

Although the homogeneity of the flow will depend too much on the final design of the dryer, in general, avoid arranging the trays to generate a parallel flow, as this can lead to uneven drying inside the chamber. This problem can be solved with a cross-arrangement, Table 6.

Table 6. Drying chamber layout and flow homogeneity.

	Serie	Parallel	Cross	Fixe
Even	x		x	x
Uneven		x		

3.3 Chimney

The chimney is the last element playing a vital role in passive dryers once there are no fans to promote the airflow. It works due to the temperature difference between the column of hot air that runs through it and the outside temperature, generating an upward force by pressure difference.

3.4 Instrumentation

The instrumentation of solar dryers to measure and control the operating parameters is often forgotten. However, efficient and controlled drying can be achieved when information about the various elements is available to monitor the process. Accessing these parameters to interpret what happens during drying continuously is crucial. Previous works converge into similar solutions as presented in the Table 7.

The fundamental parameter to measure and control is temperature. It is essential to know the temperature in the interior of the drying chamber for airflow regulation. Also, remember that some foods can not be exposed to high temperatures. It is essential to control that these values are not reached, so all projects generated from solar dryers are made the temperature measurement. Table 7.

The relative humidity is generally measured in Table 7; this allows us to obtain information if the air circulating through the dryer can absorb the moisture generated by the products.

The airflow measurement has a double objective, not only to control this same parameter that allows us to measure if it generates airtightness or not but also to measure the airflow that circulates, enabling calculations indirectly.

On the other hand, there is the measurement of solar irradiation or the weight of the products. Measuring radiation is unnecessary because it can be obtained through meteorological centres. The weight of the load generally serves to access the drying curve of specific products in a particular dryer and at an exact location. It also lets the operator know when the batch can be removed. The drying curve is generally obtained by removing the load and weighing it on an outside scale.

3.5 Parameterization

The collector's efficiency η is given by Eq. 1:

$$\eta = \frac{Q_u}{I \times t \times A_c} \tag{1}$$

where Q_u is the heat from the solar collector to be delivered to the airflow, the airflow temperature increase is due to the heat radiating from the sun I over the collector's surface A_c during drying time t. The total energy E required to evaporate the water contained in the product is given by Eq. 2, [1]:

$$E = \dot{m} \times (h_f - h_i) \times t \tag{2}$$

Table 7. Instrumentation used on reviewed articles.

Reference	Velocity	Temperature	Humidity	solar irradiation	weight
[4]	x	x	x		
[16]		x	x		
[6]		x	x		
[13]	x	x	x		x
[12]	x	x	x	x	x
[10]	x	x	x	x	
[5]	x	x			
[24]		x	x	x	
[2]		x			
[20]	x	x	x		
[18]	x	x		x	
[12]	x	x	x	x	
[23]	x	x			
[22]	x	x	x		
[21]	x	x	x		
[11]		x	x		
[19]		x	x		
[14]	x	x	x		
[17]		x	x		
[7]		x	x		

Being that $\dot{m}$ is the mass flow $[\frac{kg}{s}]$ and $(h_f - h_i)$ is the vaporisation entalpy. t is drying time $[s]$. To achieve complete drying, the heat supplied to the air by the collector must equal the energy required to evaporate the desired amount of water from the product. Introducing Eq. 2 in Eq. 1 and clearing A_c:

$$A_c = \frac{\dot{m} \times (h_f - h_i)}{\eta \times I} \tag{3}$$

where $\dot{m}$ is:

$$\dot{m} = \frac{m_w}{t \times (w_f - w_i)} \tag{4}$$

Introducing Eq. 4 in Eq. 3:

$$A_c = \frac{m_w \times (h_f - h_i)}{\eta \times I \times t \times (w_f - w_i)} \tag{5}$$

where m_w is:

$$m_w = \frac{m_p \times (m_i - m_f)}{(100 - m_f)} \tag{6}$$

Introducing Eq. 6 in Eq. 5:

$$A_c = \frac{m_p \times (m_i - m_f) \times (h_f - h_i)}{\eta \times i \times (w_f - w_i) \times (100 - m_f)} \tag{7}$$

Where i is $I \times t$.

Searching for a ratio between the collector's area and the product's weight to be dried m_p is suitable. By equating Eqs. 7 arrives at:

$$A_c = \frac{k \times m_p}{\eta \times i} \tag{8}$$

The Eq. (8) shows that the collector surface area $[m^2]$ has to be k times greater than the weight product $[kg]$ between solar radiation $[\frac{W}{m^2}]$ and collector efficiency.

Where:

$$k = \frac{(m_i - m_f) \times (h_f - h_i)}{(100 - m_f) \times (w_f - w_i)} \tag{9}$$

k depends on product moisture $(m_i - m_f)$, enthalpy of vaporization $(h_f - h_i)$, absolute humidity $(w_f - w_i)$.

The values of k were calculated from the construction parameters consulted in the bibliography. The articles with a concluded prototype were selected, and the parameters were compiled. Each value of k was calculated using Eq. 9, synthesized and depicted in Table 8. The mean value of k was 2428, and its standard deviation 438. The extreme values, k_{min} and k_{max} fluctuate close to $\pm 18\%$. Thus, the solar collector's surface standardisation by the drying capacity may be explored to evaluate if the proportions were respected within certain values for the constructed solar dryers presented in the literature.

The examples in Table 8 show that the surface of the collector from built solar driers is smaller than required for the quantity of produce in almost all cases. Sometimes, the collector size should be more than double, indicating a systematic underdesign. In such cases, the solar dryer will underperform, not delivering the product as expected. The exception is [9], where the area of the collector exceeds the model by a meagre 6%.

Table 8. Values of k, the calculated surface A_k, and the existing surface A_r as indicated in the references. ϵ is the error for k model as reference.

Ref	k $[\frac{kJ}{kg}]$	A_k $[m^2]$	A_r $[m^2]$	ϵ [%]
[8]	1965	5.68	2.00	–65
[1]	2877	19.99	18.90	–5
[3]	2261	106.81	–	–
[9]	1944	20.00	21.20	6
[6]	2583	59.79	5.80	–90
[10]	2939	4.45	4.00	–10

4 Conclusion

Food drying is a millenary practice. It evolved throughout history as technology advanced, leaving traditional drying techniques in oblivion due to the food industry's need for standardization. Machinery has become more complex and energy-hungry, inhibiting access to medium-sized companies or start-ups to drying technology due to its high acquisition and maintenance costs.

Solar technology for food drying in a medium volume can cover a need found in medium-volume producers, which production volumes do not justify the acquisition of industrial machinery. An indirect solar dryer would adequately fulfil the role of a medium-volume dryer. However, the dryer must be instrumented to monitor the process and ensure the product maintains its nutritional properties to obtain a correct and standard dried product. In other words, it is possible to retrofit an old technology that, although simple, can meet the technical requirements with a more sustainable solution.

The parametrization provided by the k model and the analysis of the different models described in the references highlighted a possible explanation for the underperformance of current traditional solar dryers. The collector's surface tends to be inferior to its intended purposes. In 80% of the examples, the collector surface was inferior to what was required. Furthermore, in 40% of the cases, the surface of the actual solar collector was less than half of what it should be for the type and quantity of the intended food. The dryers, in such cases, will underperform.

References

1. Al-Busoul, M.: Design of fruits solar energy dryer under climatic condition in Jordan. J. Power Energy Eng. **5**, 123–137 (2017). https://doi.org/10.4236/jpee.2017.52007
2. Basumatary, B., et al.: Design, construction and calibration of low cost solar cabinet dryer. Int. J. Environ. Eng. Manag. **4**(4), 351–358 (2013). ISSN 2231–1319
3. Belessiotis, V., Delyannis, E.: Solar drying. Solar Energy **85**(8), 1665–1691 (2011). ISSN 0038-092X, https://doi.org/10.1016/j.solener.2009.10.001, (2010)

4. Mealla-Sánchez, L.E., Morales-Olaciregui, J.D.: Evaluación del comportamiento térmico de secadores solares en régimen pasivo. TecnoLógicas **21**(41), 29–44 (2018)
5. Hernández Cruz, E.: Diseño, construcción y pruebas de un prototipo de secador solar para alimentos. Universidad del Caribe, Cancún, México, Thesis. Departamento de Ciencias Básicas e Ingeniería (2000)
6. Díaz, R.I., Gómez, R.A.J., Lastres Danguillecourt, O., López de Paz, P., Farrera Vázquez, N., Ibáñez Duharte, G.R.: Design, construction and evaluation of a solar dryer for Ataulfo mango. Universidad de Ciencias y Artes de Chiapas-Centro de Investigación y Desarrollo Tecnológico en Energías Renovables. Tuxtla Gutiérrez, Chiapas. Revista Mexicana de Ciencias Agrícolas, Vol. 8 Núm.8 12 de noviembre - 31 de diciembre, 2017, pp. 1719–1732 (2017)
7. Gilago, M.C., Chandramohan, V.P.: Performance evaluation of natural and forced convection indirect type solar dryers during drying ivy gourd: an experimental study. Renew. Energy **182**, 934–945 (2022). https://doi.org/10.1016/j.renene.2021.11.038, ISSN 0960–1481
8. Haro Firmo, N. S.: Secador solar indireto. Instituto Politécnico de Bragança, Universidad Tecnológica Federal de Paraná. Msc Thesis (2020)
9. Forson, F.K., Nazha, M.A.A., Akuffo, F.O., Rajakaruna, H.: Design of mixed-mode natural convection solar crop dryers: application of principles and rules of thumb. Renew. Energy 32(14), 2306–2319 (2007). https://doi.org/10.1016/j.renene.2006.12.003, ISSN 0960–1481
10. Hossain, M.: Design and development of solar dryer for chilli drying. Int. J. Res. (IJR) **2**(1) (2014). ISSN 2348–6848
11. Orellana Iturriaga, J. A.: Automatización y control de un deshidratador solar. Universidad técnica Federico santa María departamento de electrónica Valparaíso - Chile (2017)
12. Mealla-Sánchez, L., Morales-Olaciregui, J.: Evaluación del comportamiento térmico de secadores solares en régimen pasivo. Tecnológicas **21**(41), 29–44 (2018)
13. Montero Puertas, I.: Modelado y construcción de un secadero solar híbrido para residuos biomásicoAvances en Energías Renovables y Medio Ambiente, vol. 17, pp. 03.25-03.32, 2013. Impreso en la Argentina ISSN 2314–1433 - Trabajo selecionado de Actas ASADES2013s.Thesis. Universidad de Extremadura. Escuela de Ingenierías Industriales. Badajoz (2005)
14. Moreno Montiel, G.: Manual de instrumentación para una secadora solar. Universidad autonoma metropolitana Iztapalapa. Seminario de Proyectos ll, Ing. en Energía, UAM-I (2004)
15. Mujumdar, A.: Drying Technology - An Overview. Notes of Industrial Transfer-Processes, National University of Singapor (2011)
16. Roa Marín, G.L., Ortega Torres, D.B.: Diseño y Construcción de un Secador Solar por Convección de Aire Caliente Automatizado de Pequeña Escala, para el Secado de café para la Universidad Nacional de Loja. Universidad Nacional de Loja. Área de la Energía, las Industrias y los Recursos Naturales no Renovables (2011)
17. Rodríguez-Sánchez, M., López-Ramírez, M.E., Cano-Ramirez, J., Ambriz-Colín, F.: Desarrollo de un prototipo de deshidratador solar instrumentado sustentable. Revista de Prototipos Tecnológicos **2–4**, 22–29 (2016)
18. Rodriguez Valencia, H. M.,: Diseño, construcción y caracterización de un colector solar plano de aire. Thesis. Universidad de Santiago de Chile, Facultad de Ciencia, Departamento de Física (2012)
19. Sierra-Canto, X., Mendoza-Pinto, L., Trujillo, M.: Instrumentación virtual para deshidratador solar de alimentos mediante LabVIEW-LINX (2019)

20. Sivipaucar, C., Curo, H., Huancahuari, E., Llantoy, V., Valderrama, A.: Calculation and construction of a solar dryer by natural convection for drying of non-traditional medicinal plants.Centro de Desarrollo e Investigación en Termofluidos CEDIT (2008)
21. Torres Riascos, D.: Instrunmentación y sistema enrgético para una secador de café. corporación universitaria autónoma del cauca facultad de ingeniería ingeniería electrónica (2017)
22. Valenzuela Rojas, C.M.: Diseño de un Secador Solar Mixto de 500 Kg. para reducir la humedad del grano de café en el CCPP Las Pirias – San Ignacio - 2019. Universidad de Cesar Vallejo. Facultad de ingenieria y arquitectura. Thesis. Modelamiento y simulación de sistemas electromecánicos (2021)
23. Vidal Santo, A., Velázquez Camilo, O., Vigueras Zúñiga, M.O.M.I.A., Estela Del, C., Rodríguez, F.: Diseño y Construcción de un Secador Solar como Alternativa Sustentable en el Secado de Frutos en Zonas Rurales. Congreso Internacional de Investigación ISSN.: 1946–5351 ONLINE, 1948–2353 CD ROM, vol. 5, no. 3. chiapas.academiajournals.com (2013)
24. Kokate, Y.D., Baviskar, P.R., Baviskar, K.P., et al.: Design, fabrication and performance analysis of indirect solar dryer, Materials Today: Proceedings (2022). https://doi.org/10.1016/j.matpr.2022.11.439
25. Food and Agriculture Organization Of United Nations. 2020. Frutas y verduras – esenciales en tu dieta. Año Internacional de las Frutas y Verduras, 2021. Documento de antecedentes. Roma (2020). https://doi.org/10.4060/cb2395es. Accessed 10 Mar 2023
26. Instituto Nacional de Tecnología Agropecuaria.https://www.argentina.gob.ar/inta. Accessed 10 Mar 2023
27. MORDOR. https://www.informesdeexpertos.com/informes/mercado-latinoamericano-de-frutas-deshidratadas. Accessed 10 Mar 2023
28. Ministerio de Transicion Ecologica Y el Reto Demográfico. https://www.miteco.gob.es/es/ceneam/carpeta-informativa-del-ceneam/novedades/alimentos-kilometricos.aspx. Accessed 10 Mar 2023
29. Global Industry Analysts. https://www.mundomaritimo.cl/noticias/cuatro-millones-de-toneladas-alcanzaria-el-mercado-mundial-de-frutas-deshidratadas-en-2020. Accessed 10 Mar 2023
30. Data Monitor. http://www.datamonitor.com/. Accessed 10 Mar 2023

Food Waste Assessment in Universidade do Algarve's Canteen

Jaime Aníbal[1,2](✉), Neshum Sapkota[1], and Eduardo Esteves[1,3]

[1] Department of Food Engineering, Institute of Engineering, Universidade do Algarve, Campus da Penha, 8005-139 Faro, Portugal
{janibal,eesteves}@ualg.pt

[2] CIMA – Centre for Marine and Environmental Research & ARNET, Universidade do Algarve, 8005-139 Faro, Portugal

[3] CCMAR, Centro de Ciências do Mar, Universidade do Algarve, Campus de Gambelas, 8005-139 Faro, Portugal

Abstract. Food is discarded at every step of the food chain. Consumers are responsible for around 35% of all food waste. For consumers, food service providers, governments, and international organizations and institutions, it is becoming increasingly essential to tackle this issue. In recent years, just a few studies on food waste, addressing food services in canteens, have been published. The overall objectives of this study were to estimate the amount of food waste in one of the canteens at the Universidade do Algarve (UAlg), to understand the current practices at the university in relation to food waste, and to develop a method to reduce food waste in university canteens, which are used both by students and by administrative and academic staff. This research looked at the food waste at the UAlg's canteen on Campus of Penha, Faro, Portugal, which comprised 483 meals. Food waste per item during the preparation stage varied from 1% (fish) to 78% (collard), with a total wastage rate of 9%, according to the statistics. Based on weighing, 11%, 21%, 24%, and 13% of the given food will be wasted for soup, meals, salad, and dessert, respectively. With additional classification, there were differences in meal types and variation among various meal variants. Because the amount of edible food waste for various types and variety meals ranges from 57% to 91%, UAlg Social Services and the canteen management team should apply every strategy to reduce food waste as soon as feasible.

Keywords: Food waste · Canteen · Universidade do Algarve

1 Introduction

Food waste is a critical issue for global food security and good environmental governance, as it is directly linked to environmental (e.g., energy, climate change, resource availability), economic (e.g., resource efficiency, price volatility, rising costs, consumption, waste management, commodity markets), and social (e.g., health, equality) consequences. Food waste's influence on the environment and food security has become a worldwide issue. According to earlier estimates, one-third to one-half of the world's

J. F. L. C. Semião et al. (Eds.): INCREaSE 2023, ASST, pp. 363–375, 2023.
https://doi.org/10.1007/978-3-031-44006-9_27

food, or around 1.3 billion tons, is wasted [1]. This is linked to overconsumption of natural resources: food waste accounts for 10% of greenhouse gas emissions in developed nations and consumes 550 billion cubic meters of water globally [2]. Despite these figures, the world produces twice as much food as is necessary by nutritional demands per live person. In fact, by wasting less than a fifth of the food we squander, we could feed the world's roughly one billion hungry people. To feed all 9 billion people in 2050, food production will have to rise by 70% if present consumption trends are maintained. These worldwide statistics demonstrate that food waste needs immediate action [3].

Food waste is defined as food that is of good quality and fit for human use but is not consumed because it is discarded before or after it spoils. Food waste related directly to consumption represents approximately 35% of all food wasted in the last phase of the food supply chain [4].

Food waste occurring in consumption can be generally divided into three types of food waste [5]:

- Avoidable food waste: food that was still edible at some stage and was lost for reasons such as leftovers, out of date products.
- Food that some people consume, and others do not eat (e.g., bread crusts) or that can be consumed when a food is cooked in one way, but not in another, can be avoidable (e.g., potato skins).
- Inevitable food waste is a waste that is not edible in normal circumstances (e.g., meat and fish remain (bones), eggshells, tea bags, coffee grounds, hulks, peels) resulting from food or drink preparation. Inevitable food could be reprocessed but cannot be consumed.

Research by eu-fusions.org reported an estimate of 88 million tonnes of food waste in the EU-28 region, this prediction is for 2012 and contains both edible food and food-related inedible fraction. Households (47 $\pm$ 4 million tons) and processing (17 $\pm$ 13 million tons) are the sectors which contribute the most to food waste. In 2006, according to Eurostat, Portugal wasted approximately 635 and 385 tons of food from the stages of production and consumption, respectively.

Food waste from the food services industry is considerable, accounting for around 14% of all food waste in the EU [6], ca. 12 million tons. Most of the food services studies focus on plate waste in school canteens, to assess if students are dismissing valuable nutrition in school lunches. Despite their potentially considerable food waste production, a more comprehensive examination of overall food service losses, particularly in university and corporate cafeterias, has been given little attention [7]. Focusing university research efforts on food waste creation might offer several benefits. Firstly, colleges house people with knowledge and expertise who can readily effect change in their surroundings. Secondly, food waste is generally a significant, if not the most significant, part of the waste percentage generated on college campuses [7]. For example, the canteen of the Institute of Agriculture, from the University of Lisbon (ISA-UL), promoted a reduction in plate waste and increased the recycling attitude of the users (mostly students) by displaying simple posters with educational messages and conducting "guerrilla actions" during the busy lunch hour. Total plate waste was reduced by 15%, resulting in a daily reduction of 17% in meal waste, a 55% reduction in bread waste, and a 42% reduction in meat waste. This study demonstrated the need to raise staff awareness to the problem of food waste. It also showed that staff training on plating techniques, utilizing different sizes of utensils, and enabling customers to choose their own meal quantities will lead to a reduction in food waste [8].

The Universidade do Algarve (UAlg) is a Portuguese public higher education institution located in the Algarve, mainland Portugal's southernmost region. Its headquarters and two of its three campuses are in Faro (Gambelas and Penha campuses), with a third campus in Portimão. The institution has a student body of over 10,000, and a staff (professors, researchers, and employees) of around 1000 members. UAlg's Social Services (SAS) manages many food units, spread through the university campuses: three canteens, two grill restaurants, seven bars, and one restaurant.

Thousands of meals are served every day in the canteens and restaurants, with quality and diversity guaranteed. Every day, the menus provide choices of meat, fish, or vegetarian meals, all of which are planned and prepared under the supervision of a nutritionist. Soup, main course, bread, dessert (sweet or fruit), and drink (water or juice) are all included in the whole meal. The Campus da Penha, with 402 seats, and the Campus de Gambelas, with 594 seats, are the two largest canteens. The academic community can also eat light meals or just have a coffee in the bars and cafeterias during their breaks between classes.

The main objective of this work was to determine the food waste generated in the canteen of Universidade do Algarve's Campus da Penha.

2 Methodology

2.1 Study Plan

The practical part of the work was carried out in the canteen on Campus of Penha, Universidade do Algarve, during April to June 2021. This period resulted from the need to comply with previous time(s) of national lockdown and institutional constraints imposed by the COVID-19 pandemic. Consequently, in-person classes were halted and almost all (expected) canteen users, both students and staff, were unable to use the canteen. As a result, it was impractical to collect "relevant" data on food waste during the preparation of meals or post-consumer stages before the end of March.

During an initial period, spanning approximately two weeks, the practices carried out by the canteen staff for the preparation of the meals and the users' habits and behaviors during lunch time were observed and examined. This allowed for the planning of the sampling times and procedures to collect qualitative and quantitative data, keeping established hygienic and health safety measures and without disturbing the preparation of meals and lunch service.

Thence, the assessment of food waste in the canteens, supported and coordinated with the SAS of UAlg, encompassed the pre-preparation, pre-cooking, serving and post-consumption stages (Fig. 1). In each stage, we determine the (relative) amount per category of food product and type of food waste (see below).

2.2 Food Waste in the Preparation and Pre-cooking Stages

To quantify the food waste during the pre-preparation and pre-cooking stages, the raw food products and ingredients were identified/classified and weighted before and after

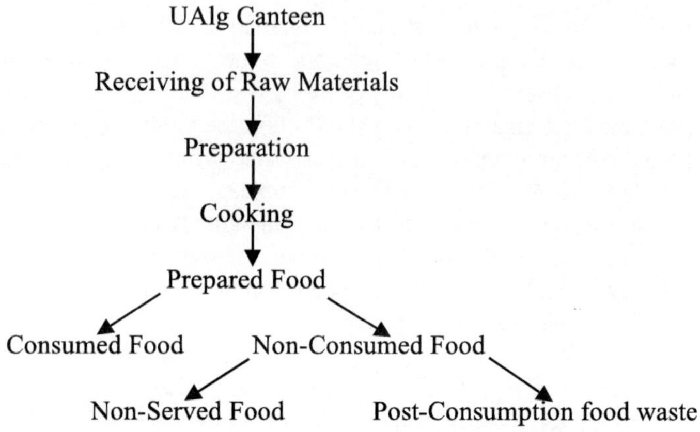

Fig. 1. Simple flowchart of the stages mentioned above.

pre-preparation and pre-cooking procedures. The personnel kindly collaborated by keeping the different types of wastes apart so that it can be weighed. The waste percentage (W, %) per type of raw food products was calculated as:

$$W = \frac{W_W}{W_0} \cdot 100\%$$

where W_W is waste weight (kg) and W_0 (kg) is the initial weight.

2.3 Food Waste Generated by Users of the Canteen

To quantitatively examine the food waste generated by users in the post-consumption stage, a preliminary and then a two-step approach was followed. Initially, starting on March 25, 2021, a random sample of n = 30 items to get the reference values for desserts, bread, and fruits were weighted (±0.1 g). Then, the food waste originated by users (post-consumption stage) in the canteen on Campus da Penha during lunchtime service was assessed for four weeks, from April 28 to May 21, 2021 (n = 16 days). On these occasions, a two-step sampling scheme was implemented.

Because food is provided by the staff to users upon request, served meals, salads, and soup plates were weighted before being delivered to the user in a tray (on average, n = 30 meals, salads and soups were sampled per day; in total, n = 483 meals, salads, and soup plates). This allowed to estimate the (initial) mean weight of the distinct food items served per meal/tray. Since users can only return their trays to one place on the canteen (Fig. 2.a), returned trays were randomly sampled from that place (on average n = 25 trays were assessed per day; in total, n = 483 trays were sampled). Canteen staff helped extensively in managing the trays yet to be sampled and evaluated.

Each tray was then taken to the weighing station (Fig. 2.b), where the type of meal (Full or Economical) and type of main dish (meat, fish/seafood, or vegetarian) were identified, and the contents was separated and weighed (±0.1 g). The type of meals available were: Economical, which includes main dish and salad or dessert; and Full, that includes soup, main dish, salad, and dessert.

After identifying the type of meal and main dish on the tray, isolated meal supplements such as soup, salad, and dessert were measured one by one on their respective plates, while for the main course plate, we visually separated edible and non-edible portions of the meal and weighed it differently on different plates (provided measuring plates), leaving the meal plate completely empty.

(a) (b)

Fig. 2. Structure holding the trays (a) and the weighting station (b).

Edible and Non-edible food waste were identified as: Edible food wastes were defined as any food that was intended to be consumed (including fat), whereas skin, bones, spines, and viscera were classified as non-edible food waste.

The leftovers on the trays were separated into following categories:

- Type of meal (Economic or Full)
- Variety of meal (Vegetarian, Meat, and Fish)
- Total weight of dish with leftover
- Total weight of remaining soup
- Total weight of remaining salad
- Total weight of dessert/fruits leftover
- Total weight of bread leftover
- Edible leftover
- Non-edible leftover

The number of meals served per sampling day together with type of meal and class of user (student, teaching, other staff) were provided by the SAS UAlg (Table 1). During March to June 2021, the average number of served meals increased from 34 meals/day in March (right after the end of lockdown) to 120 meals/day in April and 197 meals/day in May. The canteen is mostly used by students (registered >90% of the meals served).

2.4 Data Analysis

Initially, the "raw" weighted amounts of the different categories of food waste in the pre-preparation, pre-cooking and then in the post-consumer stages were compiled in Excel spreadsheets. Therein, the following weight percentage were determined:

- Pre-preparation/Pre-cooking waste

Table 1. Number of meals served in the Campus da Penha canteen.

Month (2021)	No. Meals	Meals/day*	No. Students	Users % that are students
March	750	34	505	67%
April	2635	120	2406	91%
May	4340	197	4115	95%
June	2102	96	1886	90%

* 22 working days per month.

  - Waste (%) for all raw food products

- Post-consumptions waste categories

  - Overall meal leftover
  - Soup
  - Salad
  - Dessert
  - Edible and Non-edible food waste per type of meal

    - Economical
    - Full

- Edible and Non-edible food waste per type of dish

  - Vegetarian
  - Meat
  - Seafood (mostly fish)

To calculate these values, relatively simple formulas were used; for example, for overall meal leftover W_M (%) we used:

$$W_M = \frac{W_{leftover}}{W_{total}} \cdot 100\%$$

where $W_{leftover}$ is the net weight of the leftovers (g) and W_{total} is the total weight of the dish/soup/salad/dessert. This general formula was used to calculate soup, salad, and dessert leftover percentages. To calculate edible and non-edible food waste per type and variety of meal, Excel = IF() function was used on overall meal leftover to distinguish the abovementioned categories.

Some of the above variables were calculated on a daily basis and may have different averages to reflect the fact that different types of meals may have been served on different sized plates, namely the weights of the main course plate, the meal served, the soup plate, the soup served, the salad plate, the salad served, the dessert plate and the dessert served.

Data is presented as % of total waste in the case of pre-preparation stage and as average values (± standard deviation) or as % (w/w) in the case of post-consumption stage. For the latter, the following summary values (%) were obtained:

- Overall food waste
- Edible and non-edible food waste per type of meal (economical and full)
- Edible and non-edible food waste per type of main dish (vegetarian, meat, and seafood/fish).

3 Results and Discussion

3.1 Food Waste in the Pre-preparation and Pre-cooking Stages

The quantity and percentage of waste per type of raw food products and/or ingredients originated from the preparation of meals (pre-preparation and pre-cooking stages) in the Campus da Penha canteen is summarized in Table 2. The food waste per item type showed a wide range of estimates, ranging from 1% (fish) to 78% ("collard"). Raw products such as turkey or pork were completely used and originated no waste.

According to Rojas-Vargas et al. [9] food waste created during the preparation stage from 5 different educational institutions in great metropolitan area of Costa Rica was 26.10%, which is greater than the average observed herein. Even though the percentage of waste in the Campus da Penha canteen is lower, it should focus its waste reduction plan on the data obtained. The institution has restaurants and cafes with several users and other factors that can act in different ways. Identify what behaviors or situations exist in the canteen kitchens that result in a higher percentage of food waste in the preparation area. For example, the percentage of waste for green leafy vegetables (cabbage and spinach) was much higher than expected.

3.2 Food Waste Generated by Users of the Canteen

Meals Served and Sampled

The number of meals served per type (economical and full) and per type of main dish (meat, seafood, and vegetarian) that were assessed during the period of sampling is shown in Fig. 3. Overall, quite a relevant number of meals was sampled (n = 483) considering the number of meals served during the period of sampling (n = 751), i.e., 64.3%.

Of the meals sampled, most users opted for the economical meal (n = 252 vs. n = 231 for the full meal) and for the main dish including meat (n = 206 vs. n = 173 for seafood and n = 104 for the vegetarian dish) (Fig. 3). Users preferred the lower-cost economical meal, sequentially choosing meat, seafood, and vegetarian options as their main dishes (Fig. 4). On the other hand, the seafood dish was the most popular with the users that opted for eating a full meal, followed closely by the meat and then the vegetarian dishes (Fig. 4).

Table 2. Waste quantity and percentage of raw food products

Variety	Raw food products or ingredients	Net weight (kg)	Waste (g)	Waste (%)
Seafood	Fish	30400	270	1%
	Squids	12000	1300	11%
Meat	Chicken	16000	850	5%
	Turkey	2000	0	0%
	Pork	10300	0	0%
Vegetables	Potato	223000	18450	8%
	Carrot	22000	150	1%
	Spinach	4500	2750	61%
	Broccoli	2250	550	24%
	Pumpkin	80300	5000	6%
	Onion	15000	4500	30%
	Radish	3000	750	25%
	Collard	6000	4700	78%
			Average	9%

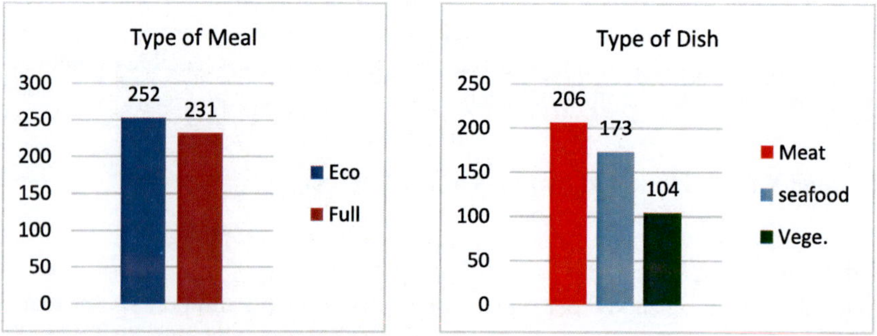

Fig. 3. UAlg Campus da Penha canteen users' preference on type of meal (economical and full) and type of dish (meat, seafood/fish, and vegetarian) during four weeks in April-May 2021.

According to Pinto et al. [8], meat dishes were more popular among University of Lisbon users, followed by fish (seafood) and vegetarian dishes, which is also similar to our findings.

Average Food Waste Generated by Users

Figure 5 depicts the average percentage of waste per item of meal (soup, dish, salad, and dessert) that was originated by users of the canteen on Campus da Penha over the course of the sampling period (four and a half weeks) in April-May 2021. Estimates ranging from 11% (soup) to 24% (salad) were obtained. Thus, salad and the main dish were

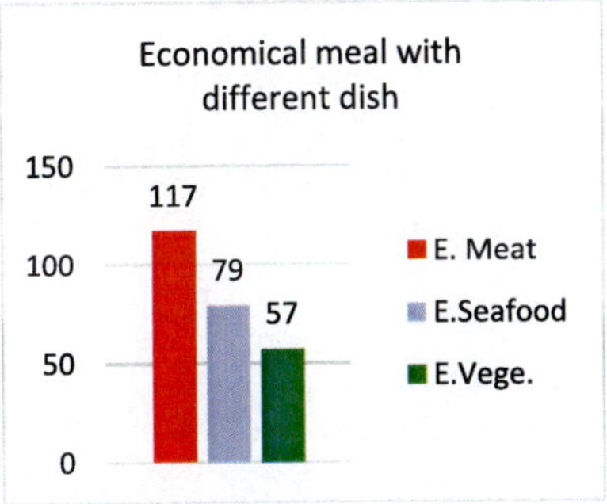

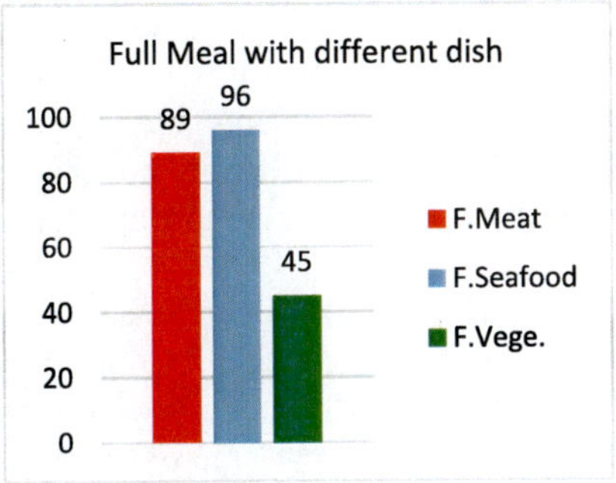

Fig. 4. UAlg Campus da Penha canteen users' preference on different type of dishes per type of meal (economical and full) during 4 weeks in April-May 2021.

the items of the meal with the highest percentage of waste, 24% and 21%, respectively. As with the pre-preparation and pre-cooking stages (see above), vegetables (i.e. salad) showed a higher percentage of waste.

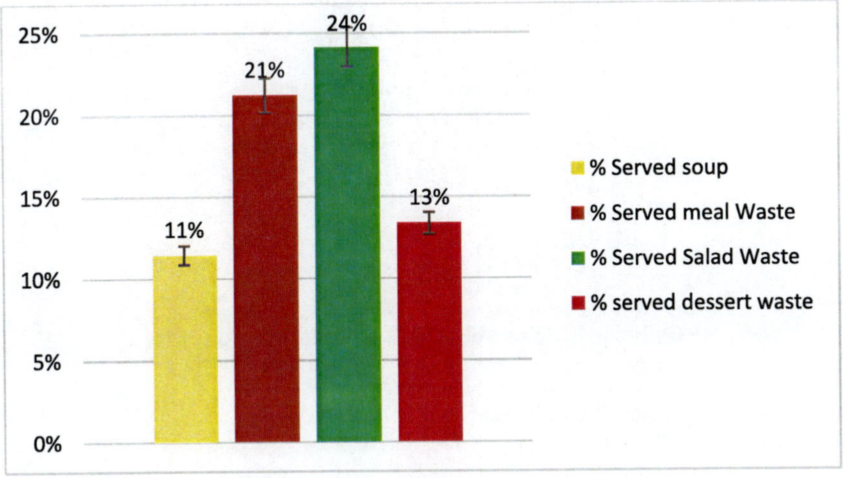

Fig. 5. Percentage waste (%) per item of meals served in the UAlg Campus da Penha canteen during 4 weeks in April-May 2021.

Engström and Carlsson-Kanyama [10], studied food loss in Swedish food service institutions, including two canteens, and reported food waste ranging from 46 g to 115 g per meal, like Pinto et al. [8], who found 41–118 g/meal for the main course plate at the University of Lisbon. Meanwhile, the main course meal waste in our case varied from 8.2 g to 348.3 g, which is quite substantial when compared to the research stated above. However, because all users received the same standard meal composition, the fact that the staff consistently delivered the same amount of meal components may have contributed to the significant plate waste reported in the current study.

The percentage waste of the served soup was 11%, which is comparable to Pinto et al. [8] finding of 8 to 10%. It was suggested that this was due to the soup being overly watery, or that the students just took the soup without inquiry, even though they did not want it. In contrast, we are not able to compare the waste percentages for salad and dessert obtained herein with studies because of unavailability of data.

Edible and Nonedible Food Waste

To examine the food waste originated by users of the Campus da Penha canteen, the edible and nonedible fractions of the food waste were assessed and tentatively quantified. When looking at type of meal, economical vs. full, the waste was determined to be 20% and 22%, respectively (Fig. 6). These estimates demonstrate that the waste originated by users for both types of meals is similar, and they wisely picked the types of meals, demonstrating their understanding of the quantity to consume to avoid hunger and demonstrating good abilities for meal planning.

Moreover, the percentage of edible food waste in the "total" food waste originated from the two types of meals was 62% for economical meals and 63% for full meals. The nonedible portion of the waste for the two types of meals was 15% and 20% of total waste for an economical lunch and a full lunch, respectively (Fig. 6). Because vegetarian meals do not contain any nonedible parts, the nonedible percentage for vegetarian meals was usually 0%.

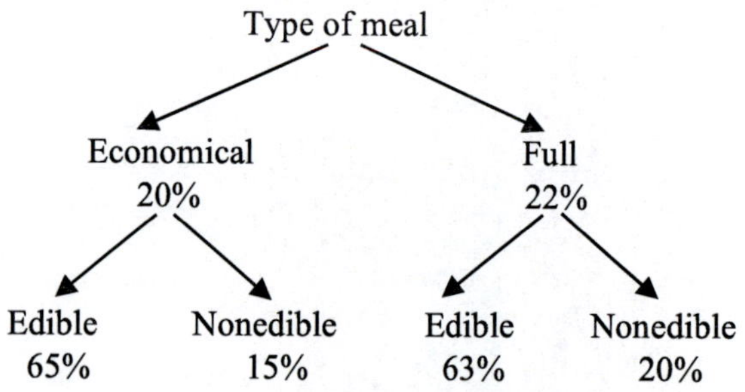

Fig. 6. Percentage waste (%), edible and nonedible, per type of meal in the Campus da Penha canteen during April-May 2021.

Because of the way we categorised our food waste as edible and non-edible, we could not find any other articles or literature for comparison.

When considering the type of dish, food waste from dishes including meat, or seafood or vegetarian dishes was similar, 23%, 20%, and 20%, respectively (Fig. 7). Most of these wastes were classified as edible, 74%, 62%, and 69%, respectively for meat, seafood, and vegetarian dishes. The nonedible waste for seafood dishes was greater than for meat dishes, 29% and 21% of total waste, respectively. Vegetarian dishes had no waste.

There were differences in the percentages of edible and nonedible waste originated from the distinct types of meals (economical and full) and per type of dish (meat, seafood, and vegetarian) (Fig. 8). Generally, economical meals originated less waste (23%, 19%, and 15% of the overall meal in when the main dishes were meat, seafood, and vegetarian,

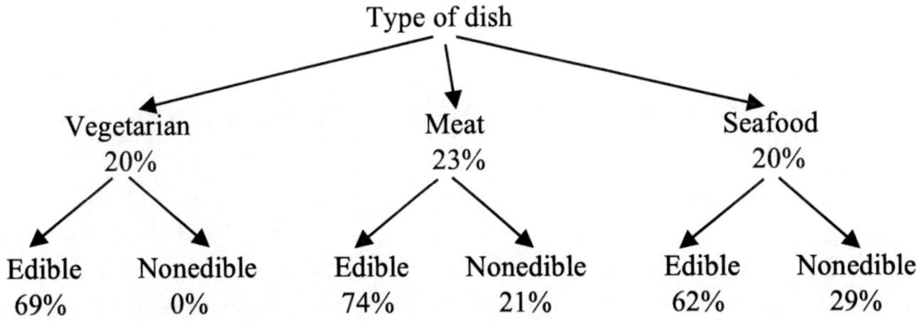

Fig. 7. Percentage waste (%), edible and nonedible, per type of dish in the Campus da Penha canteen during April-May 2021.

respectively) than full meals (24%, 20% and 24%, respectively). In terms of edible meal waste, vegetarian dish waste was found to be greater than seafood and meat dish waste, respectively 93%, 68%, and 62% in economical meals and 91%, 57% and 76% in full meals. Nonedible meal waste for meat dishes is greater than for seafood dishes, 38% vs. 30% respectively. Nonedible meal waste for a full meal on a seafood dish is greater than for a full meal on a meat dish, at 42% and 24%, respectively. No waste (or negligible) waste was found in vegetarian dishes whatever the type of meal.

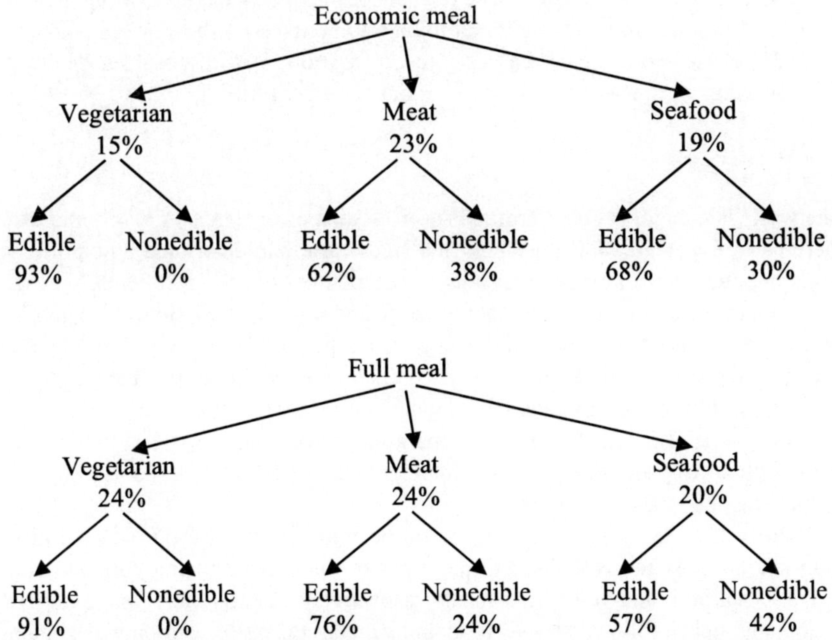

Fig. 8. Percentage waste (%), edible and nonedible, per type of meal and type of dish in the Campus da Penha canteen during April-May 2021.

3.3 Measures to Decrease Food Waste

A variety of best practices and recommendations for food waste minimization have focused on the often overlooked costs of waste in terms of natural resource use and, as a result, the environmental benefits of waste reduction [11, 12]:

- **Reduce.** Because food production has such a large impact on natural resources, and this impact grows as food moves up the food value chain, decreasing food waste is by far the most effective means of reducing natural resource waste. If the supply-demand balance can be better adjusted on the front end, for example, it means not utilizing natural resources to produce food in the first place, reducing pressure on natural resources and allowing them to be used for other reasons;
- **Reuse.** The greatest choice in the case of a food surplus is to keep it in the human food chain. This may entail finding secondary markets or donating it to feed the poor, to preserve its original purpose and save the usage of more resources to generate more food. If the food is unfit for human consumption, the next best alternative is to redirect it for livestock feed, saving resources that would otherwise be needed to manufacture commercial feedstuff;
- **Recycle/Recover.** By-product recycling, anaerobic digestion, composting, incineration with energy recovery, and rendering are the primary recycling and recovery alternatives. All these solutions allow for the recovery of energy or nutrients, which is a considerable benefit over landfill;
- **Landfill.** Landfilling organic waste releases gases like methane (a strong greenhouse gas) and pollutes soil and water, not to mention odor and other societal annoyances. Landfills should only be used as a last resort for food waste disposal, especially given the growing shortage of usable land for human activities.

4 Conclusions

Plate waste is a major problem in food services such as university and school canteens, where part of a meal is left uneaten. This study aimed to determine how much waste occurs in public consumption areas and was carried out in an experimental setting. The research focused on preparation/cooking waste and post-consumption waste, the former being classified as edible food and the latter as non-edible food. It was carried out in the canteen of the Campus de Penha, which serves lunch and dinner to an average of 1200 people from different socio-economic and age groups every day.

Food waste per item during the preparation stage ranged from 1% (fish) to 78% (lard), with an overall waste rate of 9%. Turkey and pork were the only raw material products that generated no waste.

There were 483 meals examined, with the number of economical and full meals being 252 and 231 respectively, and the types of main dishes being 206, 173 and 104 for meat, seafood, and vegetarian meals respectively. Meat variety was preferred for economical meals, followed by seafood and vegetarian, while seafood was popular for full meals, followed by meat and vegetarian.

In general, the average percentage of waste generated by consumers per meal item (soup, main dish, salad, and dessert) ranged from 11% for soup to 21% for main dish and 24% for salad.

The above-mentioned categories were further separated into edible and non-edible fractions for food waste generated by consumers in the canteen. The percentage of edible food waste in total food waste from the two types of meals was 62% for economical meals and 63% for full meals. Similarly, edible waste percentages were 74%, 62%, and 69% for meat, seafood, and vegetarian dishes, respectively. In terms of edible meal waste, vegetarian dish waste was found to be higher than seafood and meat dish waste, with 93%, 68%, and 62% for economical meals and 91%, 57%, and 76% for full meals, respectively. This high proportion of edible waste in every category clearly underlines the importance of the SAS UAlg promoting the procedures/systems already in place and consider implementing alternative solutions that can reduce food waste as quickly as possible in the canteen.

Acknowledgments. Jaime Aníbal would like to acknowledge the support of national funds through Fundação para a Ciência e Tecnologia (FCT), under the project LA/P/0069/2020 granted to the Associate Laboratory ARNET and UID/00350/2020 CIMA. This study also received Portuguese national funds from FCT - Foundation for Science and Technology through projects UIDB/04326/2020, UIDP/04326/2020 and LA/P/0101/2020 (Eduardo Esteves).

References

1. Gustavsson, J., Cederberg, C., Sonesson, U., van Otterdijk, R., Meybeck, A.: Global food losses and food waste: extent, causes and prevention, p. 38. International Congress: Save Food (2011)
2. Stuart, T.: Waste: uncovering the global food scandal. The True Cost of What the Global Food Industry Throws Away. Penguin Books Ltd., London (2009)
3. Fox, T., Fimeche, C.: Global Food: Waste Not, Want Not, pp. 1–14. Institute of Mechanical Engineers, London (2013)
4. Lipinski, B., Hanson, C., Lomax, J., Kitinoja, L., Waite, R., Searchinger, T.: Reducing Food Loss and Waste, p. 40. World Resource Institute, Washington, DC (2013)
5. Quested, T.E., Parry, A.D., Easteal, S., Swannell, R.: Food and drink waste from households in the UK. Nutr. Bull. **36**, 460–467 (2011)
6. Beretta, C., Stoessel, F., Baier, U., Hellweg, S.: Quantifying food losses and the potential for reduction in Switzerland. Waste Manag. **33**(3), 764–773 (2013)
7. Gallardo, A., Edo-Alcón, N., Carlos, M., Renau, M.: The determination of waste generation and composition as an essential tool to improve the waste management plan of a university. Waste Manag. **53**, 3–11 (2016)
8. Pinto, R.S., Pinto, R.M.D.S., Melo, F.F.S., Campos, S.S., Cordovil, C.M.D.S.: A simple awareness campaign to promote food waste reduction in a university canteen. Waste Manag. **76**, 28–38 (2018)
9. Rojas-Vargas, J., et al.: Food loss and waste in food services from educational institutions in Costa Rica. Tecnología en Marcha **34–2**, 187–196 (2021)
10. Engström, R., Carlson-Kanyama, A.: Food losses in food service institutions examples from Sweden. Food Policy **29**, 203–213 (2004)
11. Baldwin, C.J.: Sustainability in the Food Industry. Wiley-Blackwell and the Institute of Food Technologists, Ames, Iowa (2009)
12. Baldwin, C.J.: The 10 Principles of Food Industry Sustainability. Wiley-Blackwell, Chichester, West Sussex (2015)

HACCP Methodology Implementation in a Small Company of Goat Yoghurt Production

Vera Drago[1] and Isabel Ratão[1,2(✉)]

[1] Instituto Superior de Engenharia, Universidade do Algarve, Campus da Penha, 8005-139 Faro, Portugal
{a48678,iratao}@ualg.pt

[2] MED–Mediterranean Institute for Agriculture, Environment and Development & CHANGE–Global Change and Sustainability Institute, Faculty of Sciences and Technology, Universidade do Algarve, Campus de Gambelas, 8005-139 Faro, Portugal

Abstract. Consumers are increasingly interested in food safety, enhancing the demand level and thereafter the rising attention of the authorities and producers. The Hazzard Analysis and Critical Control Points (HACCP) System is of compulsory application since 2006 and can effectively ensure food safety and quality, contributing to expand to new markets, and improve the management level of the manufacturers. The objective of this project was to implement the Hazard Analysis and Critical Control Points system to the goat's yoghurt production process of a small plant. To guarantee the safety of yoghurt production, the biological, chemical, and physical hazards that may exist in its production process were identified. Twenty-eight of the total hazards identified in this study were considered as significant, having been subjected to a Decision Tree to determine the Critical Control Points (CCP). After this analysis, only pasteurization was considered as CCP. Milk and yoghurt storage were considered Control Points (CP). After that, the critical limits, monitoring processes, corrective measures, records, and verifications steps were established. The results of the hazard analysis showed that most of the chemical hazards could be effectively controlled by an effective supplier selection. Relatively to the main microbiological hazards – *Listeria monocytogenes*, *Salmonella* and *E. coli* – specific control measures must be implemented, namely temperature control, particularly in the pasteurization process. Hazard analysis was the opportunity to determine the best choice of these specific control measures.

Keywords: Food safety · HACCP · Goat yoghurt

1 Introduction

Goat milk and its derivatives (yoghurt, cheese and powdered milk) have some benefits for human nutrition. Among them, the possibility of being used by people suffering from cow's milk allergies and gastrointestinal disorders and fulfils the gastronomic needs of connoisseur consumers, which is increasing in many developed countries [1].

J. F. L. C. Semião et al. (Eds.): INCREaSE 2023, ASST, pp. 376–387, 2023.
https://doi.org/10.1007/978-3-031-44006-9_28

Goat yoghurt could contribute to respond to the growing number of individuals with allergies to cow's milk protein. Its production also follows a traditional custom, with the milk being boiled and filtered, followed to powdered goat's milk addition together with lactic ferments (*Streptococcus thermophilus* and *Lactobacillus delbrueckii* subsp. *Bulgaricus*). Fermentation takes place in the packaging itself (glass jars) at a temperature that varies between 38–42 °C.

Goat milk, cheese and yoghurt are widely appreciated around the world and have essential nutritional components for the human diet, but they are also known to be potentially hazardous foods due to various sources of contamination (equipment, personnel, additives and packaging material), during the production process under non-compliant conditions. The implementation of the Food Safety Management System (FSMS), from production to consumption of dairy products, promotes food safety and the prevention of diseases transmitted by these foods [2–4].

In Portugal, particularly in the interior region of the Algarve, goat yoghurt has some importance for the economy of goat farms, since it allows to aggregate added value to the milk, allowing a good increased income for producers [5].

For the development of the yoghurt making activity with quality, which does not pose any danger to the consumer, it is necessary to consider aspects related to licensing and the implementation of a food safety management system [5].

In the last 30 years, Portugal produced an average of 1.2 million tons of dairy products per year. Around 70% of the total volume of milk and dairy products produced by national industry corresponded to milk for consumption. The production of yoghurts, contributed with 7.9%, cheese with 5.1% and other dairy products with an individual contribution less than 2% [6].

According to the National Statistics Institute [7], in Portugal in 2021 around 1169 thousand tons of milk and dairy products were consumed, with the milk being the most consumed product (684 thousand), followed by acidified milk, including yoghurt (213 thousand) and cheese (142 thousand).

Yoghurt is defined by Ordinance n° 742/92 of July 24th [8] as "the coagulated product, obtained by lactic fermentation due to the exclusive action of *Lactobacillus delbrueckii* subsp. *bulgaricus* and *Streptococcus thermophilus* on milk and dairy products (…), the specific flora must be alive and abundant in the final product".

In Portugal, yoghurt consumption increased at an average annual pace of 6.2%, settling in 2021 at 21.5 kg per capita per year [6].

Yoghurts can be of different types depending on the raw material from which they originate. Currently, it is already possible to find yoghurt produced with different types of milk (cow, goat, among others) on the market. These can be natural, sugared, flavoured, solid, shaken, liquid, fat, semi-skimmed or skimmed [9].

Goat yoghurt is very nutritionally rich and is considered an excellent carrier of probiotic cultures, with therapeutic benefits, which have a survival rate above 10^6 cfu/g during the storage period [10].

In the interior region of the Algarve, the production of Algarve goat cheese and yoghurt represents a way to combat desertification and depopulation, reveals the 2014–2020 Diagnostic and Strategy Study: Low Density Territories, published by the Commission of Coordination and Regional Development of the Algarve (CCDR Algarve) in

September 2015, indicating that the creation of goats of this breed should be supported and developed [11].

Dairy products are essential components of the human diet but can be a source of several problems, which can act as causative agents of many diseases, such as microorganisms transmitted by handlers, equipment and water, veterinary drugs, chemical pollutants and toxins. The systematic approach to minimize economic losses and outbreaks of food poisoning at all stages of production of dairy products was positively affected by the development of the Hazard Analysis and Critical Control Points (HACCP) system [3].

Since 2006, Hazard Analysis and Critical Control Points (HACCP) system is mandatory in all the member states of the European Union (Reg 852/2004) [12]. Long before that international organizations such as the World Health Organization, Food and Agriculture Organization, and *Codex Alimentarius* Commission already recommended it [13].

HACCP system is an almost complete process control system that permits to reduce or eliminate the occurrence of foodborne diseases during the processing steps, identifying hazard, establishing control measures, identifying Critical Control Points (CCP), establishing Critical Limits (CL), monitoring processes, indicating corrective measures and doing records and verification [14].

Before HACCP implementation, it is necessary to set the prerequisite program, which could include Standard Operating Procedures (SOP), Good Manufacturing Practices (GMP) and Good Hygienic Practices (GHP).

After the HACCP team formation and the correct description of the product and its intended use, the flow diagram must be draw and a hazard analysis is performed. All the hazards (biological, physical and chemical) are systematically identified and listed, and preventive measures are enumerated [13].

Despite many years of having HACCP implemented, many manufactures still have a negative attitude towards it [15]. HACCP steel continues to be seen, especially by small manufacturers, as a task that is mandatory with the only objective of satisfying the requirement of authorities.

The objective of this project was to implement the Hazard Analysis and Critical Control Points system to the goat's yoghurt production process of a small plant, overcoming some barriers particularly difficult to face.

2 Material and Methods

The complete production process of yoghurt was evaluated including prerequisite program as the structure of each facility, the plant layout, the higienization plan, the technical standards, the staff responsibilities, the safety control measures and the conditions of storage and distribution. All the hazards were identified and preventive measures were addressed. After that, hazards were analysed for Risk Grade (RG) based on severity and probability.

Severity was classified in three levels: Hight, whenever it could evolve death. In this case, an unsafe product could affect the consumers' health in a very drastic way. It could be irreparable, chronical or even lethal. Medium, when it can result in an unsafe product. Pathogenicity is low at the beginning, but can conduct to a significant hazard to the consumer. These effects could be reversed by medical assistance, which could include hospitalization, but not include life risk. Low, for biological agents of low pathogenicity (doesn't cause important effects on consumers health, causing only indisposition and malaise). Eventually could be necessary to see a doctor [16–18].

Probability was established based on the number of occurrences in a year and the organization historic. Hight for more than 8 times a year; medium, between 3 and 8 times a year and low, if it occurs less than 3 times a year [16, 17] (Table 1).

Table 1. Matrix for determination of significant hazards (S) and not significant (NS) (Source: adapted from [1, 19]).

		Severity		
		Low (1)	Medium (2)	High (3)
Probability	Low (1)	1	2	3
	Medium (2)	2	4	6
	High (3)	3	6	9

RG under 3 were covered by Prerequisite Program (PRP). A PRP does not control specific hazards, but it allows the maintenance of an adequate environment to obtain safe products. Only RG equal or superior to 3 were analysed to determine whether they were Critical Control Points (CCP) or not. This analysis was done using the *Codex Alimentarius* Decision Tree [13].

Based on literature and experience, Critical Control Limits were established, as well as monitorization procedures, corrective measures, records, documentation and verifications.

Due to the importance of retardation or even inhibition of microbial growth for food safety, Control Points (CP) and related Control Limits (CL) were defined. The procedures are practically the same as those applied to CCP, except that they constitute essentially preventive points and do not contain elimination criteria.

3 Results and Discussion

3.1 HACCP Team Establishment, Description of the Product and Intended Use (Steps 1–3 of HACCP)

The first step of the work was to define the HACCP team which was recorded in a meeting minute (data not showed). After that, technical specifications were elaborated, including product description and the intended use of the product among other information (data not showed for confidentiality reasons).

3.2 Flow Diagram and *in loco* Confirmation (Steps 4–5 of HACCP)

The complete flow diagram of the yoghurt technical process is presented at Fig. 1, including raw materials reception, storage, weighing, filtration, pasteurization, cooling, lactic ferments addition, packaging, fermentation, labelling, refrigeration/storage, and expedition.

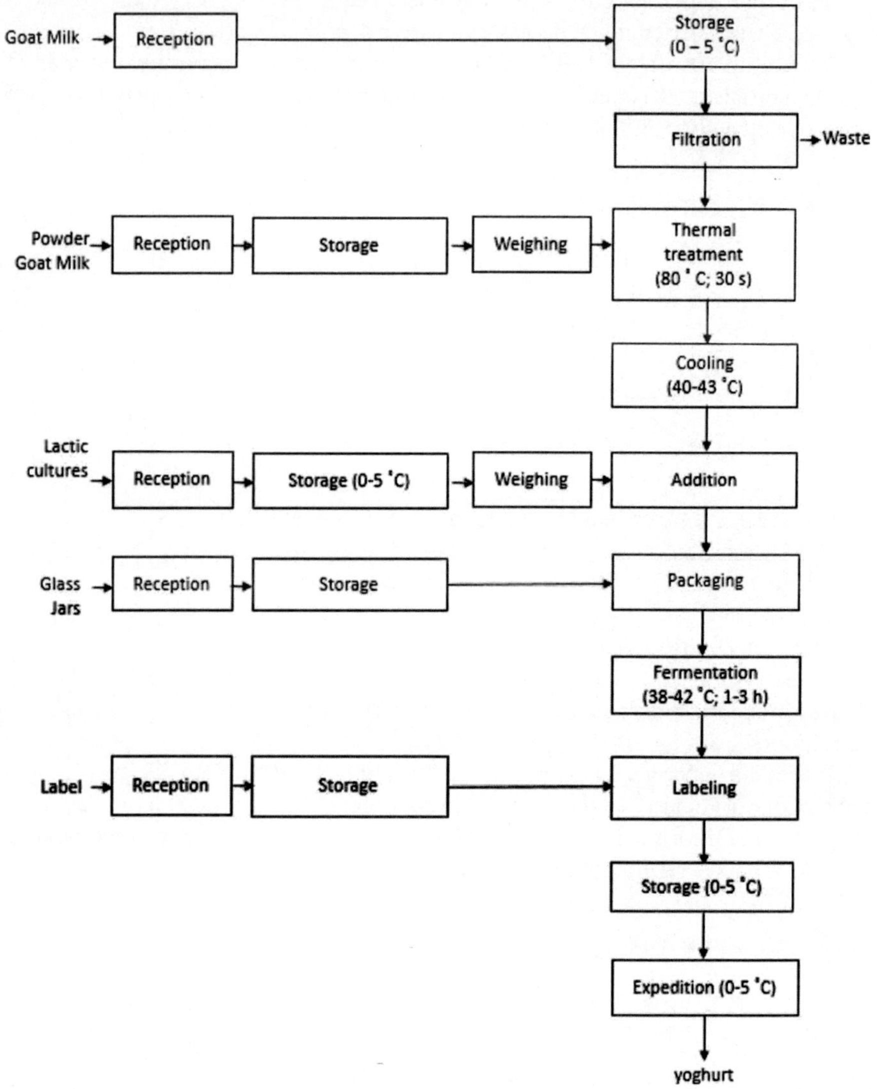

Fig. 1. Flow diagram of the yoghurt technical process.

The flow diagram was verified *in loco* during the course of the operation, through-out the day, being possible to confirm that it is appropriate to the reality.

3.3 Hazard Analysis, Prevention Measures Establishment and CCP Determination (Steps 6–7 of HACCP)

A complete hazard analysis was performed to every step of the process to define any biological, chemical and physical factors that may affect food safety. The severity and the probability were used to determine the Risk Grade and, by consequence, the significance of each hazard. All steps with RG equal or superior to 3 were analysed to determine whether they were CCP or not (Table 2).

Table 2. Example of hazard analysis, preventive measures and management measures establishment. For confidentiality reasons, only part of the analysis will be presented. (Prob-Probability; Sev-Severity; RG-Risk Grade; P- Physical hazard; B-Biological hazard; C-Chemical hazard; NS-Not Significant; S-Significant).

Step	Hazard		Preventive measure	Risk Evaluation		Q1	Q2	Q3	Q4	Management measures
				Prob/Sev	RG					
Milk pasteurization	P	Contamination of milk with hair, waste and pests	Good Hygiene Practices	1/1	1 NS	–	–	–	–	PRP
	B	*Salmonella* spp.; *L. monocytogenes*; *S. aureus*; *E. coli*	Compliance with the binomial time/temperature	1/3	3 S	yes	yes	–	–	CCP
	C	Veterinary drugs[1]; Cleaning and disinfection waste	Good Hygiene Practices	1/3	3 S	yes	No	No	–	Not a CCP

[1]albendazol, amoxicillin, benzylpenicillin, procaine, ceftiofur, chlortetracycline, tetracycline, clenbuterol, cyfluthrin, cyhalothrin, cypermethrin, deltamethrin, dihydrostreptomycin, diminazen, doramectin, eprinomectin, febantel, gentamicin, imidocarb, isometamidium, ivermectin, monensin, neomycin, pirlimycin, spectinomycin, silfadimidine, thiabendazole, trichlorfon, tylosin

Reception of Raw Material and Subsidiary Materials

Milk should meet the requirements of the Commission regulation (EU) n. 37/2010 of December 22nd [20] and Commission regulation (EC) n. 1881/2006 of December 19th [21].

Powder milk should meet the requirements of Reg (EU) 2018/150 of 30 January 2018 regarding to methods for the analysis and quality evaluation of milk and milk products eligible for public intervention and aid for private storage [22].

Lactic cultures should be cultures of *Streptococcus thermophilus* and any *Lactobacillus* species.

Glass jars and labels must comply with the technical specifications integrating the food safety management system.

All suppliers are required to submit quality inspection reports at the delivery time and there will be a random sample and inspection of every batch of raw and subsidiary materials.

Storage of Raw Material and Subsidiary Materials
Milk storage must be done in a cooling tank (0–5 °C). Powder milk storage can be at ambient temperature, as in the case of glass jars and labels, but it must be done in a clean and dry place to avoid fungi growth. Lactic cultures must be stored at refrigeration temperature (<4 °C).

Filtration of Milk
Before use, milk must be filtrated in order to avoid the presence of bristle or some other impurities. Good Hygienic Practices must be met.

Thermal Treatment of Milk
Appropriate thermal treatment will kill most foodborne pathogens. If the temperature is excessively high, it can damage the materials' nutrients and bioactivity, affecting the flavour of the final product. On the other hand, too low temperatures will not be able to eliminate the pathogenic microorganisms. Pasteurization was done at 80 °C for 30 s as it has historically been used in the company where the work took place. At the end of this process, powdered milk diluted in a certain proportion is added and pasteurized again for another 30 s.

Cooling of Milk and Lactic Cultures Addiction
The coagulation temperature influences the action of the lactic cultures; thus, the ideal temperature has to be reached before the lactic ferments could be added. In order to do so, the pasteurized preparation is transferred to a vat and cooled externally with cold water until it reaches a temperature of 40 to 43 °C and the lactic cultures are added and mixed.

Packaging
At this stage the individual jars, pre-sterilized with steam, are filling (weight approx. 250 g) and the lids are placing, so that they remain tight and avoid possible contamination.

Fermentation
The jars are placed inside the yoghurt maker (stove), where they are subjected to the fermentation process for approximately 3 to 4 h at a variable temperature between 38 and 42 °C.

Labelling
The labels are placed on the yoghurt jars. The label consists of a self-adhesive paper disc with all the information required by law (Reg (EU) 1169/2011 from October 25th [23] and National legislation DL n. 62/2017 of June, the 9th [24]).

Storage of the Yoghurt
Yoghurts are stored in a vertical refrigeration cabinet at a temperature of 0 to 5 °C.

Yoghurts are kept under these conditions for a period of 15 days, at maximum. Storage is carried out taking into account the stock rotation procedures - First Expired, First Out (FEFO), where products with a shorter shelf life are stored in the front sections, and products with a longer shelf life in the later sections.

Expedition

Transportation to the selling facilities should be done at refrigerated conditions (0–5 °C) and as fast as possible, in dedicated containers.

3.4 Establishing Critical Control Limits (Step 8 of HACCP)

Based on the result of hazard analysis, the yoghurt production process contains only one CCP (Table 2). The Critical Control Limits (Table 3), which needs to be fulfilled at the Critical Control Points, were established based in the European Regulation [20, 21, 25, 26], *Codex Alimentarius* and previous experience.

Table 3. Example of CCL, monitorization (method, frequency and responsible), corrective actions and records. For confidentiality reasons, only part of the data will be presented. (B-Biological hazard; T-temperature; t-time).

Step	Hazard		CCL	Monitorization			Corrective action	Records
				Method	Frequency	Responsibility		
Milk pasteurization	B	*Salmonella* spp	n = 5 c = 0 Absent in 25 g [25]	ISO 6579 -1:2017 [27]	Every week	Lab technician	reject the batch in case it is already been used for fermentation	analysis bulletin
			T = 80 °C t = 30 s	Control of the t/T binomial	Every Pasteurization	Production technician	Increasing time and temperature	Recording T and pasteurization time
		Listeria monocytogenes	n = 5 c = 0 Absent in 25 g [25]	ISO 11290-1:2017 [28]	Every week	Lab technician	reject the batch in case it is already been used for fermentation	analysis bulletin
			T = 80 °C t = 30 s	Control of the t/T binomial	Every Pasteurization	Production technician	Increasing time and temperature	Recording T and pasteurization time
		Escherichia coli	n = 5 c = 1 m = 100 cfu/g M = 800 cfu/g [25]	ISO 16649-2:2001 [29]	Every week	Lab technician	reject the batch in case it is already been used for fermentation	analysis bulletin
			T = 80 °C t = 30 s	Control of the t/T binomial	Every Pasteurization	Production technician	Increasing time and temperature	Recording T and pasteurization time

CCP - Goat Milk Pasteurization

An inadequate pasteurization process (less time or temperature that is actually needed, 80 °C, 30 s) could allow the pathogens to remain and grow, whereas extra-high temperature or extra-long pasteurization time could negatively affect the product quality.

Control Points (CP)

Due to the importance of maintaining the temperature for delaying or even inhibiting microbial growth, it was considered that special attention should be paid to control the temperature in both goat milk storage and yoghurt storage, having been stablished two Control Points (CP1 and CP2) at those steps (Table 4). The Control Points appear in stages whose control measures do not allow or suggest the definition of critical limits but of control limits. At this points temperature must be controlled every two hours in order to guarantee constant refrigeration conditions.

Table 4. Example of Control Limit (CL), monitorization (method, frequency and responsible), corrective actions and records. For confidentiality reasons, only part of the data will be presented. (B – Biological hazard; T-temperature).

Step	Hazard		CL	Monitorization			Corrective action	Record
				Method	Frequency	Responsibility		
Yoghurt storage	B	*Salmonella* spp.; *L. monocytogenes*; *E. coli*	T < 5 °C	Control of refrigeration T	Every 2 h	Quality technician	Restoration of T	Recording of refrigeration T

The procedures are practically the same as those applied to the CCP, however, being essentially preventive points, they do not contain batch elimination criteria.

3.5 Establishing Monitoring Procedure (Step 9 of HACCP)

A monitoring procedure was established (Table 3) in the HACCP system, to guarantee that the critical limit established for every CCP were continuously fulfilled. This procedure contains the hazard (e.g., *E. coli*); method (e.g., ISO 16649-2:2001) [29]; frequency (e.g., every week) and responsibility (e.g., lab technician).

3.6 Establishing Corrective Measures, Verification, Record and Documentation (Steps 10–12 of HACCP)

Every time a deviation to a CCP occurs, corrective measures (Table 3) should be carried out, in order to correct the problem and to put the production back in control.

The main objective of verification procedures is to verify whether HACCP can effectively control food safety. This verification should be done, at least once a year and every time that changes are made in the product or the process. In this yoghurt factory, internal and external audits were conducted, at least once a year, in order to verify the correct implementation of both the PRP and the HACCP plan, including record keeping verification.

Additionally, random samplings of the products are performed and products were testing as planed in the HACCP plan.

This company have a complaint book and all customer complaints are monthly analysed so that corrective action can be taken.

Every action taken, like CCP monitoring, deviation, its associated corrective measures and verifications are recorded and documentation such as record sheets, hazard analysis and CCP determination were kept during five years.

3.7 Main Difficulties

The main difficulty found during this process was to be able to persuade the employees to change their habits, since they have performed the same tasks for many years, always in the same way and do not understand the need for change. It was necessary a transition period to the introduction of new routines or to change those which already existed.

Sometimes there were also some difficulties to persuade the management team to invest in the maintenance of some areas and equipment of the plant. Nevertheless, the persistence and documentary evidence proved to be very good allies.

3.8 Main Benefits

After the implementation of the HACCP methodology, the company is now able to accomplish the legal requirements for Yoghurt producing, according European regulation [12], and assuring the safety consumption of its products by their customers.

4 Conclusion

Through the analysis of the Food Safety Management System previously implemented in the establishment under study, it can be concluded that it needed some improvements. Therefore, the Prerequisite Program and the HACCP were updated and improved.

The supplier evaluation and selection assume an important role, since the raw milk, the powder milk and the lactic cultures must be carefully evaluated in order to not be a critical control point.

Regarding to HACCP plan, it states that pasteurization is the Critical Control Point for the process. It was created and implemented a record procedure which guarantee the overall process control.

In addition to the CCP, it has been concluded that some steps, namely refrigerated milk storage and refrigerated storage of the yoghurt should be considered as Control Points. At those points, there isn't a rejection criterion, but control measures were implemented in order to assure the product safety.

All the documentation and records generated in the HACCP system can be used to help tracing the origin of eventual problems, thus preventing further production of nonconform products with consequent loses in work force, material, and financial resources.

References

1. Haenlein, G.F.W.: Goat milk in human nutrition. Small Rumin. Res. **51**, 155–163 (2004)
2. El-Hofi, M., El-Tanboly, E., Ismail, A.: Implementation of the hazard analysis critical control point (HACCP) system to white cheese production line. Acta Sci. Pol. Technol. Aliment. **9**(3), 331–342 (2010)
3. Peristeropoulou, M., Fragkaki, A.G., Printzos, N., Laina, I.: Implementation of the hazard analysis critical control point (HACCP) system to a dairy industry: evaluation of benefits and barriers. J. Food Nutri Diete (1), 1–5 (2015)
4. Park, Y.W., Haenlein, G.F.W., Wendorff, W.L.: Handbook of Milk of Non-bovine Mammals. 2nd ed. Wiley Blackwell (2017)
5. DRAAlg., FERN., ANCCRAL., NTLD: Manual de Boas Práticas - Fabrico de Queijo de Cabra no Algarve. Projeto Agro "Obtenção e Valorização do Queijo Curado de Cabra no Algarve" (2005)
6. INE. Estatísticas da Produção e Consumo de Leite 2021 (2022)
7. INE. Consumo humano de leite e produtos lácteos por tipo de leites e produtos lácteos – Anual (2022)
8. Portaria 742/92 of 24 de July 24 do Ministério da Agricultura e do Comércio e do Turismo (1992)
9. Esteves, R., Barbosa, M.: O Iogurte – Para Saber Mais. Coleção E-books APN: No. 28. Associação Portuguesa de Nutricionistas (APN) (2013)
10. Desouky, M.M., El-Gendy, M.H.: Physicochemical characteristics of functional goats' milk yogurt as affected by some milk heat treatments. Int. J. Dairy Sci. **12**(1), 12–27 (2017)
11. IESE. Algarve 2014 – 2020. Diagnóstico e Estratégia. Territórios de Baixa Densidade. Instituto de Estudos Sociais e Económicos. Comissão de Coordenação e Desenvolvimento Regional do Algarve (CCDR Algarve) (2015)
12. Regulation (EC) No 852/2004 of the European Parliament and of the Council of 29 April 2004 on the hygiene of foodstuffs (2004)
13. CAC Recommended International Code of Practice: General Principles of Food Hygiene, CAC/RCP 1-1969, Adopted in 1969. Amended in 1999. Revised in 1997, 2003, 2020. Editorial corrections in 2011 (2020). https://www.fao.org/fao-who-codexalimentarius/sh-proxy/en/?lnk=1&url=https%253A%252F%252Fworkspace.fao.org%252Fsites%252Fcodex%252FStandards%252FCXC%2B1-1969%252FCXC_001e.pdf
14. Meng, L., Yang, Q., Cheng, J.: Study on the application of HACCP system in ice-cream premix powder processing. J. Anhui Agric. Sci. **39**, 6531–6533 (2011)
15. Toropilová, J. Bystrický, P.: Why HACCP might sometimes become weak or even fail. In: International 58th Meat Industry Conference "Meat Safety and Quality: Where it goes?" on Procedia FOOD SCIENCE, vol. 5, pp. 296–299. Elsevier (2015)
16. Batista, P., Venâncio, A.: Sistemas de gestão da segurança alimentar. 1st ed. Guimarães: Forvisão - Consultoria em Formação Integrada, S.A. (2003)
17. Fernandes, E., Silva, M.F.L., Ramalhosa, E.: Sistemas de Gestão, da Segurança Alimentar. Guia para a sua Implementação em Unidades de Restauração. 1st ed. Sílabo, Lisboa (2012)
18. Olofsson, I.: Guidelines for food safety control of artisan cheese-making. Nordic Council of Ministers, Copenhagen (2010)
19. Furtado, V.G.: Sistema de Gestão de Segurança Alimentar Aplicada a uma Adega: Caraterização e Otimização do Processo. Lisboa: Universidade Nova de Lisboa. Dissertação para obtenção de grau de mestre em Tecnologia e Segurança Alimentar (2013)
20. Commission Regulation (EU) No 37/2010 of 22 December, on pharmacologically active substances and their classification regarding maximum residue limits in foodstuffs of animal origin (2010)

21. Commission Regulation (EC) No 1881/2006 of 19 December, setting maximum levels for certain contaminants in foodstuffs (2006)
22. COMMISSION IMPLEMENTING REGULATION (EU) 2018/150 of 30 January, amending Implementing Regulation (EU) 2016/1240 as regards methods for the analysis and quality evaluation of milk and milk products eligible for public intervention and aid for private storage (2018)
23. Regulation (EU) No 1169/2011 of the European Parliament and of the Council of 25 October, on the provision of food information to consumers, amending Regulations (EC) No 1924/2006 and (EC) No 1925/2006 of the European Parliament and of the Council, and repealing Commission Directive 87/250/EEC, Council Directive 90/496/EEC, Commission Directive 1999/10/EC, Directive 2000/13/EC of the European Parliament and of the Council, Commission Directives 2002/67/EC and 2008/5/EC and Commission Regulation (EC) No 608/2004 (2011)
24. Decreto-Lei no. 62/2017, de 9 de junho, estabelece o regime aplicável à composição, rotulagem e comercialização do leite, dos produtos derivados do leite e aos produtos extraídos do leite, transpondo a Diretiva (UE) no. 2015/2203 (2017)
25. Commission Regulation (EC) No 1441/2007 of 5 December, amending Regulation (EC) No 2073/2005 on microbiological criteria for foodstuffs (2007)
26. Commission Regulation (EC) No 2073/2005 of 15 November, on microbiological criteria for foodstuffs (2005)
27. ISO 6579-1:2017. Microbiology of the food chain — Horizontal method for the detection, enumeration and serotyping of *Salmonella* — Part 1: Detection of *Salmonella* spp (2017)
28. ISO 11290-1:2017. Microbiology of the food chain — Horizontal method for the detection and enumeration of *Listeria monocytogenes* and of *Listeria* spp. — Part 1: Detection method (2017)
29. ISO 16649-2:2001. Microbiology of food and animal feeding stuffs -- Horizontal method for the enumeration of beta-glucuronidase-positive *Escherichia coli* - Part 2: Colony-count technique at 44 °C using 5-bromo-4-chloro-3-indolyl beta-D-glucuronide (2001)

Freshness, Spoilage Dynamics and Safety of Deep-Water Pink Shrimp, *Parapenaeus longirostris*, Stored in Ice

Eduardo Esteves[1,2(✉)], João Lagartinho[1], and Jaime Aníbal[1,3]

[1] Department of Food Engineering, Institute of Engineering, Universidade do Algarve, Campus da Penha, 8005-139 Faro, Portugal
{eesteves,janibal}@ualg.pt

[2] CCMAR, Centro de Ciências do Mar, Universidade do Algarve, Campus de Gambelas, 8005-139 Faro, Portugal

[3] CIMA – Centre for Marine and Environmental Research and ARNET, Universidade do Algarve, 8005-139 Faro, Portugal

Abstract. Spoilage dynamics and safety of fresh deep-water pink shrimp, *Parapenaeus longirostris*, are of paramount importance for its storage and commercialization in good sanitary conditions, owing to its high perishability and limited shelf life. This study focused on the analysis of physicochemical indices, microbiota growth and sensory analysis during storage of deep-water pink shrimp in ice at 5 °C. The specimens were kept for 13 d in a perforated plastic box covered with ice that allowed the melted water to drain; new ice was added every 8–12 h. A sensory analysis panel assessed several attributes to develop a species-specific, 14-demerit points Quality Index Method (QIM) scheme. Water content remained at ca. 80% and water activity values ranged 0.98–1.00 during the experiment, confirming the high perishability potential of the samples. Unexpectedly, there were no significant changes in the concentrations of TVBN and TMA during this storage trial. Despite this, the abundances of mesophilic and psychrotrophic bacteria increased consistently from ca. 3.0–3.5 log CFU/g to about 6.9–7.4 log CFU/g. The absence of increase in nitrogen-based compounds despite microbiota proliferation might be related to a leaching process due to melting ice (that needs to be further experimented). Moreover, Enterobacteriaceae were not present in any of the samples analyzed, indicating good sanitary and hygienic conditions during handling and storage and safety of samples. Data suggests that deep-water pink shrimp should not be stored in ice at 5 °C more than 7 days before consumption. Food safety is critical to food security and public health, thus relevant to United Nations' SDG 1 (No poverty) and SDG 2 (Zero hunger).

Keywords: Deep-water pink shrimp · freshness evaluation · quality index method · shelf life

J. F. L. C. Semião et al. (Eds.): INCREaSE 2023, ASST, pp. 388–402, 2023.
https://doi.org/10.1007/978-3-031-44006-9_29

1 Introduction

Deep-water pink shrimp, *Parapenaeus longirostris* (Lucas, 1846) is a decapod crustacean that is distributed along the east and west coasts of the North Atlantic as well as in the Mediterranean Sea commonly at depths of 100 to 400 m [1–3]. The species is among the most fished shrimps/prawns worldwide [4], particularly in countries bordering the Mediterranean Sea, mainly Spain and Italy, wherein it has very high commercial value (e.g. in Portugal, landing prices of ca. 12–14 €/kg compared to 1.65 €/kg for all fishes; [5]). Together with red shrimp *Aristeus antennatus* and Norway lobster *Nephrops norvegicus*, pink shrimp are the most valuable species caught by the Portuguese crustacean (demersal) trawl fleet and are widely appreciated by Portuguese consumers [6].

Crustacean shellfish are highly perishable, commonly spoiling faster than other muscle foods because of their composition [7]. Rosa and Nunes [6] determined the proximate composition, amino acid, and fatty acid profiles, lipid classes and cholesterol and glycogen contents of pink shrimp. The nutritional composition of *P. longirostris*, 75 to 79% water content, 17 to 22% protein and 0.3 to 0.6% total fat, particularly the high content in free amino acids and amines [7], in addition to the fact specimens die just after being captured (usually by trawlers), limit shelf-life. Tissue enzymes and microbial activities are primary proxies of spoilage and limited shelf-life of shrimps [8], that predominantly occurs in the head (cephalothorax) wherein most visceral enzymes and ca. 75% of the microbial load is found. Microbial growth, namely of Gram-negative psychrotolerant bacteria [9] such as *Pseudomonas* spp. and *Shewanella* spp., have been found responsible for the spoilage of the shrimps that are not appropriately iced and stored in proper conditions [10–12]. Several studies (cf. [13]) addressed the microbial spoilage of several species of shrimps. Moreover, enzyme-induced oxidation in phenolic precursors influenced by environmental factors (sunlight, temperature, etc.) triggers the development of irresolvable black pigments (melanin) or "black spots" – melanosis– in the internal surface of shell [7]. This process starts in the cephalothorax and progresses thru the pleon (abdomen) and the pereiopods, to the telson (tail) during ice storage [14]. Even though black spot formation is considered harmless for consumers, it markedly reduces product's market value and acceptability. To prevent/delay melanogenesis, sodium metabisulphite is widely used as preservative.

In the European Union, Regulation 2074/2005 [15] altered by Regulation 1022/2008 [16], establishes the control measures and limit values for certain categories of fishery (fish) products as well as the analysis methods to be used when "unprocessed fishery products [are] regarded as unfit for human consumption [after] organoleptic assessment has raised doubts as to their freshness". The limits imposed therein for TVBN range from 25 to 35 mg N/100 g but do not refer to crustaceans (or other seafood other than fish). Officially, «organoleptic assessment» of freshness (sensu Regulation 2406/96 [17]) is realized thru ratings established in Annex I (Part E) by reference to appraisal criteria specific to each of them. Alternatively, the Quality Index Method (QIM) was developed in the late 1970s-early 1980s at the Tasmanian Food Research Unit of the Commonwealth Scientific and Industrial Research Organization (CSIRO) of Australia. The QIM is an attribute scoring methodology that considers the evaluation of typically 10 to 15 attributes, that change during storage of a particular species using short scales, usually from a minimum of 0 to maxima of 1 to 3. The sum of the scores provides an overall index that is (cor)related with the (remaining) shelf life of fish and other seafood [18]. Recently, Esteves and Aníbal [19] carried out a meta-analysis to summarize the ca. 70 QIM schemes developed since the inception of the method in the early 1980s and to attest the species-specificity of QIM schemes. Therein, authors refer only eight studies developing QIM schemes for other crustaceans. Howgate [20] pointed out that the recent popularity of the QIM (cf. QIM Eurofish website, www.qim.eurofish.com) applies just to measurements of freshness in the raw state.

Food safety is critical to food security and public health, and therefore relevant to several United Nations' SDGs (https://sdgs.un.org/goals), namely SDG 1 (No poverty) and SDG 2 (Zero hunger).

The aim of the present study was to investigate the QIM to determine the quality and freshness of deep-water pink shrimp and to correlate the results with microbiological and physicochemical changes that occurred during the period of storage in ice at 5 °C.

2 Materials and Methods

2.1 Specimens Acquisition, Preparation, and Sampling

The specimens were purchased in a local fish market in Faro (Algarve, South Portugal) but they were fished off Algarve (south Portugal) and landed that same day at the nearby port of Vila Real de Santo António. Immediately after purchase, they were transported to the laboratory in isothermal insulated boxes with ice (−1 to 3 °C). Shrimps were kept covered with ice for up to 13 days in a perforated plastic box that allowed the melted water to drain; new ice was added every 8 to 12 h. The box was stored in a walk-in refrigerator at 5 ± 1 °C. Samples (n = 3 or 10 g) were regularly taken to determine physicochemical and microbiological quality parameters as well as for sensory analysis.

2.2 Physico- and Biochemical Parameters

Water content (in %) and water activity (aW, in the range 0 to 1) were determined using an infrared moisture analyzer (Mettler Toledo LP 16, Spain) and an aW-meter (Rotronic Hygrolab, USA), respectively. For the determination of the TVBN and TMA content, the microdiffusion method proposed by Conway and Byrne [21] and described in the Portuguese standard NP 2930 [22] was used. All solutions used were prepared with distilled water and the reagents were of analytical grade. TVBN and TMA contents are presented as mg N/100 g.

2.3 Microbiological Determinations

Sample preparation for the enumeration of microorganisms was carried out following international standard [23]. In short, samples (10 g) of shrimp's muscle were aseptically placed into sterile Stomacher® bags containing 90 ml of peptone water with NaCl (0.85% w/v) (Merck, Darmstadt, Germany) and homogenized for 2 min (Stomacher® 400, Seward Ltd., London, UK). Aliquots of 1 ml were poured in Petri dishes according to serial decimal dilutions before addition of appropriate media. For the enumeration of mesophilic aerobic and psychrotrophic bacteria, plate count agar (PCA, Scharlau, Germany) was incubated at 30 °C for 72 h ± 3 h [24] and at 6.5 °C for 10 days [25], respectively. Enterobactericeae were enumerated after inoculation of 1 ml aliquots into 10 ml of molten (at 45 °C) violet-red bile glucose agar (VRBGA, Scharlau, Germany). After settling, a 10 ml overlay of molten media was added, and plates were incubated at 37 °C for 24 h ± 1 h [26]. All plates were examined visually for typical colony types and morphological characteristics associated with each medium. Microbiological data, i.e. the number of colony forming units per unit mass were log-transformed prior to analysis, as log(CFU/g).

2.4 Sensory Analysis

A panel of 8 individuals, co-opted from the faculty, staff and graduate students of the Departamento de Engenharia Alimentar, Instituto Superior de Engenharia da Universidade do Algarve, and experienced in the sensory assessment of seafood products, was involved in this part of the work. Initially, a provisional QIM table (Table 1), adapted from previous published literature [27–30], was prepared to assess five sensory attributes (appearance/color, melanosis, adhesion of head-body and exoskeleton-muscle, and odor) each scored from 0 to 3 (demerit) points to give a total of 14 points. The QIM scheme was applied by the assessors daily and under standardized lab conditions [31].

Table 1. QIM scheme used to assess pink shrimp stored in ice.

Parameter	Description	Score
Specimens' color	Strong pink to orange	0
	Light pink to orange	1
	Translucent pink to orange with yellowish tones	2
	Whitish with yellowish and greenish tones	3
Melanosis* (dark spots)	Absence of melanosis	0
	Some melanosis in the head**	1
	Intense melanosis in the head**	2
	Melanosis in the head** and body***	3
Adhesion of the shell§ to the flesh	Weak (easily peeled from the flesh)	0
	Average	1
	Strong (hardly peeled from the flesh)	2

(continued)

Table 1. *(continued)*

Parameter	Description	Score
Adhesion of the head** to body***	Strong	0
	Average	1
	Weak	2
	Head** separated from body***	3
Odor	Fresh, smooth aroma of seaweeds	0
	Weak aroma of sea air	1
	Weak ammoniacal	2
	Strong ammoniacal, putrid	3
Quality index		14

* Appearance and extent of melanosis. ** Cephalothorax. *** Abdomen. § Exoskeleton, incl. Carapace (head) and segments (abdomen).

2.5 Statistical Analysis

Results are given as mean ± standard deviation (descriptive statistics) or parameter estimate ± standard error (regression models).

Changes in TVB-N and TMA content over storage time were tentatively modelled using linear, exponential and spline models. Appropriate growth models were fitted to the observed changes in the abundance of microorganisms during storage time. The exponential and Baranyi & Roberts models [32] were fitted via nonlinear regression analysis (using the Levenberg-Marquardt algorithm thru the function nls in R package nlsMicrobio [33]) and their goodness-of-fit assessed using the residual deviance and pseudo-R^2 [34]. The (linear) relationships between the averages of scores given to sensory attributes in each sampling occasion and storage time and between the sum of demerit points (QI) and storage time were modelled using linear regression. Following Sveinsdottir et al. [35] and Rossini et al. [36], the QIM scheme was validated by an estimate of the uncertainty of prediction of days in ice from the QI using partial least-squares (PLS) regression with leave-one-out (LOO) cross-validated predictions R package pls [37]. The root mean-square error (RMSE) of prediction provided a prediction error in original units (days) and bias was calculated as the averaged difference between predicted and QI values for all samples in the validation set. The standard error of performance (SEP) was used to estimate the precision of the results, corrected for bias [35].

All statistical procedures were carried out at the 0.05 level of significance and using R statistical software [38].

3 Results and Discussion

The water content and water activity of deep-water pink shrimp stored in ice remained relatively constant throughout the experiment at levels of 0.76 to 0.83 and 0.94 to 1.00 (Fig. 1). These values are in line with published data for other penaeid shrimps (e.g. [39,

40]). At these levels of water content and particularly of a_W, growth of microorganisms and rate of chemical reactions responsible for seafood spoilage are extremely promoted [41].

The initial concentrations of TVBN and TMA in shrimp's muscle averaged 22 mg N/100 g and 3 mg N/100 g, respectively (Fig. 2). These are biochemical indicators of fish and seafood freshness, and values are well within reported TVBN and TMA contents for other crustacean species studied under a diverse set of conditions. Laghmari and Marrakchi [27] obtained values of 31.1–31.6 mg N/100 g for TVBN and 0.67 mg N/100 g for TMA in fresh *P. longirostris*. In another study by Huidobro et al. [42], TVBN and TMA showed levels of ca. 18 mg N/100 g and 0.23–0.47 mg N/100 g, respectively, after 24 h storage in flaked/liquid ice. In samples of fresh *P. brasiliensis* and *P. paulensis*, Moura et al. [43] measured TVBN values ranging from 27.6 to 73.0 mg N/100g. In whole shrimp *Pandalus borealis*, the TVB-N and TMA values were 33.5 mg/100 g and 0.5 mg N/100 g at the beginning of storage trial, in that study 2 days after catch [30]. Shamshad et al. [44] found 16.7 mg N/100 g for TVBN in fresh specimens of *P. marguiensis*. For that species, Fatima et al. [45] reported initial contents in TVB-N and TMA of 11.5 mg N/100 g and 0.01 mg N/100 g, respectively. Kirschnick et al. [46] obtained initial values of TVBN in fresh prawns *Macrobrachium rosenbergii* of 18.75 mg N/100 g. The initial TVBN content of Pacific white shrimp *Litopenaeus vannamei* found by Nirmal & Benjakul [47] was 7.2–7.4 mg N/100 g. Kirschnick and Viegas [48] reported values of TVBN content of 10.8 mg N/100 g for fresh freshwater shrimp *Macrobrachium rosenbergii*. Akintola and Bakare [49] measured the TVBN content in fresh *M. vollenhovenii* and obtained 16.7 mg N/100 g. Moreover, the low concentration of TMA compared to TVBN found herein substantiate the hypothesis that the main fraction of total volatile bases in pink shrimp might be ammonia as suggested by Huidobro et al. [42].

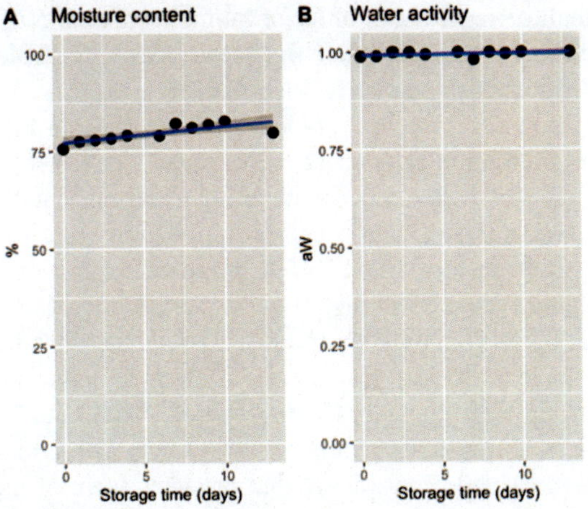

Fig. 1. Changes in (A) moisture content and (B) water activity measured in deep-water pink shrimp stored in ice (at 5 °C).

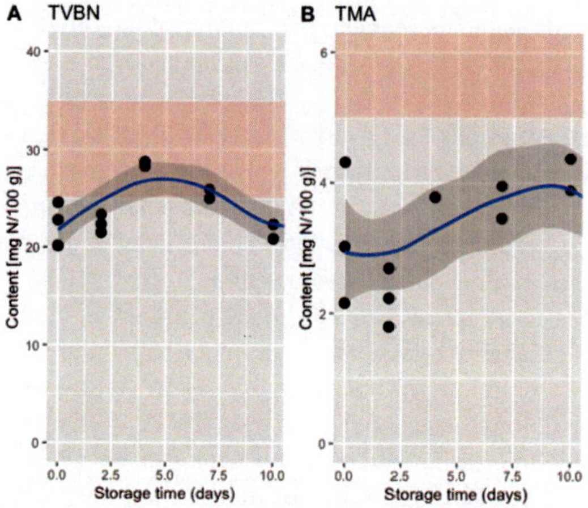

Fig. 2. Changes in TVBN and TMA observed in deep-water pink shrimp stored in ice (at 5 °C). Smooth splines fitted to the data (TVBN: smoothing parameter = 0.102, equivalent df = 5.89, Penalized Criterion (RSS): 0.1557 and GCV: 2.030; TMA: smoothing parameter = 0.139, equivalent df = 5.80, Penalized Criterion (RSS): 0.0548 and GCV: 0.399).

Unexpectedly (or not), the concentrations of TVBN and TMA in shrimp's muscle did not increase during storage in ice (Fig. 1). Similar results were found in several previous studies [42, 49–54], suggesting that a leaching phenomenon during storage of shrimp in melting ice is taking place. Results of a complementary experiment [55] seem to support this hypothesis. In addition, several authors [e.g. [52, 56–58] question the use and/or validity of TVB-N (and TMA) contents as indicators of freshness in species of fish, mollusks and crustaceans. Notwithstanding, several others have used it effectively and TVB-N is the regulated parameter within the European Union to assess if "unprocessed fishery products shall be regarded as unfit for human consumption where organoleptic assessment has raised doubts as to their freshness" [15, 16]. A literature review and meta-analysis could contribute to clarify this issue.

Initial TVC and counts of psychrotrophic bacteria in pink shrimp were about 3.5 log CFU/g (Fig. 3) and no Enterobacteriaceae were found. The latter clearly suggests that correct handling practices were carried onboard and during landing [59]. In addition, the relatively low levels of TVC and psychrotrophes imply that shrimps were of high microbiological quality. In fact, ICMSF [60] proposed a limit of 7 log (CFU/g) that was reached only after 7 days of storage (Fig. 3). At the end of our storage trial, TVC and psychrotrophic bacteria in pink shrimp *P. longirostris* reached abundances of 8.45 log(CFU/g) and 6.78 log (CFU/g), respectively. Relatively lower initial counts of total bacteria in pink shrimp were found by Huidobro et al. [42], about 2.4–2.7 log (CFU/g), that increased to 2.5–3.5 log(CFU/g) after 4 days in ice. Those authors suggest that a correct handling was carried onboard and that products were of high microbiological quality even if Enterobacteriaceae were found (2.7–3.0 log(CFU/g)). On the day of catch counts of total microorganisms in shrimp stored at 0 °C of around 4.5–4.7 log (CFU/g) have been reported [61–63]. Moreover, Kirschnik and Viegas [48] counted ca. 1.44 log(CFU/g) in shrimp *M. rosenbergii* at start of their storage trial. After 10 days of storage in ice, *M. rosenbergii* muscle presented abundances of psychrotrophes of 3.5–4.0 log(CFU/g). In contrast, Oliveira [64] observed quite higher initial abundances of psychrothropes, ca. 5 log (CFU/g), in *L. vannamei* but a growth pattern, similar to that observed herein, that exceeded 7 log (CFU/g) on day 8. At the onset of another experiment [30], the microbial quality of the shrimp *Pandalus borealis* (in terms of total bacteria) was marginal, i.e. ca. 5.3 log (CFU/g). Then, TVC increased steadily during storage in ice at 1.5 °C and brine mixed with flake ice at –1.5 °C to 8.48 log(CFU/g) at the end of storage period.

Baranyi and Roberts models [32] fitted well ($R^2 > 0.99$) the changes in abundances of TVC and psychrotrophic bacteria during storage trials (Fig. 3): the 4-parameter (complete) model to the former and the 3-parameter (no asymptote) model to the later. Besides, estimated lag periods (1.446 ± 0.498 d and 1.540 ± 0.309 d) and maximum (specific) growth rates μmax (1.088 ± 0.064 d-1 and 1.052 ± 0.067 d-1) were similar (Table 2).

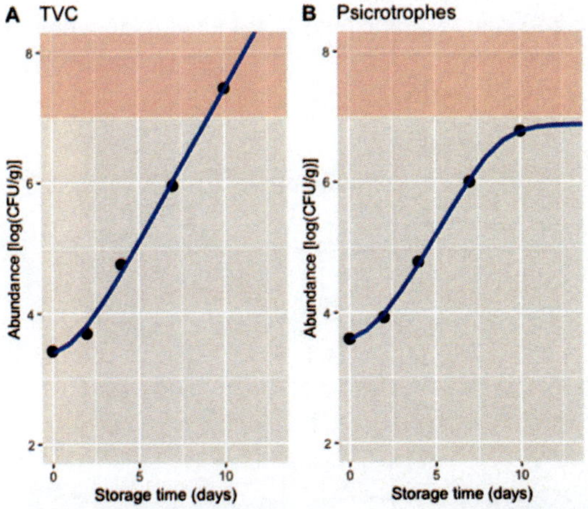

Fig. 3. Changes in log(CFU/g) abundance of TVC (A) and psychotropic bacteria (B) observed in deep-water pink shrimp stored in ice (at 5 °C). Curves represent fitted Baranyi and Roberts models, respectively (A) the 3-parameter/no asymptote model and (B) the 4-parameter/complete model. See main text for further details on fitted models' parameters and statistics.

In terms of sensory analysis, the (demerit) scores for all the attributes considered in the QIM scheme increased linearly and were highly correlated with storage time (Fig. 4), with R^2 ranging 0.64 for shell adhesion to 0.96 for melanosis. Similar patterns were observed by Khodanazary [65] for shrimp *Maetapenaeus affinis* stored in ice, by Gonçalves et al. [66] for spiny lobster and by Lee et al. [67] for black tiger shrimp (*P. monodon*). Herein, on day 8 of the storage trial the samples were rejected by panelists due to their unacceptable sensory attributes. On that occasion, the specimens of *P. longirostris* studied herein were translucent pink to orange/yellowish, had intense melanosis in the head, the shell was hardly peeled from the flesh, often the head was already separated from the body, and emanated an ammoniacal odor. In other studies, in comparable conditions, the rejection of specimens based on sensory attributes varied between 3.7 days for shrimp *P. borealis* in flake ice [30] or 4 days (50% rejection) for

Table 2. Parameters (estimate ± SE) and goodness of fit of the Baranyi & Roberts [32] models fitted to TVC and psychrotrophic bacteria counts (log(CFU/g)).

	Model	Parameter	Estimate	SE	p-value	R^2	RSE
TVC	3-parameter/no asymptote Baranyi & Roberts model	$\log(N_0)$	3.395	0.131	0.0015	0.997	0.135
		Lag (d)	1.446	0.498	0.1009		
		μ_{max} (d^{-1})	1.088	0.064	0.0035		
Psychrotrophes	4-parameter/complete Baranyi & Roberts model	$\log(N_0)$	3.588	0.064	0.0114	0.994	0.066
		Lag (d)	1.540	0.309	0.1262		
		μ_{max} (d^{-1})	1.052	0.067	0.0404		
		$\log(N_{max})$	6.877	0.099	0.0091		

shrimp *P. longirostris* in ice [50] and 9 days for shrimp *M. affinis* stored in ice [65] or 10 days for spiny lobster [66].

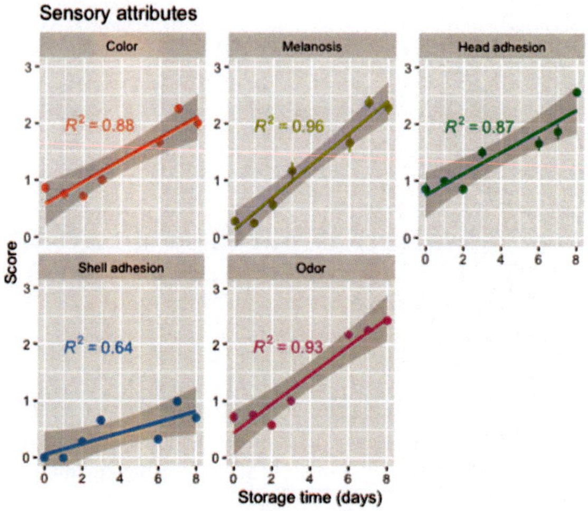

Fig. 4. Changes in the average score per sensory attribute assessed in *P. longirostris* during storage at 5 °C.

The sum of (demerit) points, *i.e.* the quality index (QI), also increased linearly with storage time (Fig. 5). The loss of freshness can be described by the equation QI = 1.96 + 1.01·days. Considering the time to rejection of 8 days, the estimated QI = 10.04 is substantially lower the maximum QI of 14 (Table 1).

The PLSR model (Fig. 6) indicates a good adjustment of the QIM to the experimental data with RMSE of 0.107 (and a mean square of errors (MSE) of 0.011) These are rather lower than the errors obtained by Khodanazary [65] for shrimp *M. affinis* stored in ice, 1.153, or Gonçalves et al. [66] for spiny lobster *Panulirus argus* in ice, 2.745.

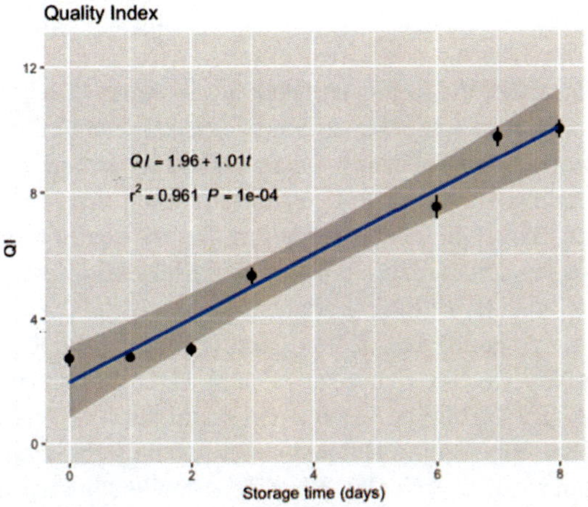

Fig. 5. Average quality index (QI) (±SD) per storage day versus storage time in *P. longirostris* during storage at 5 °C.

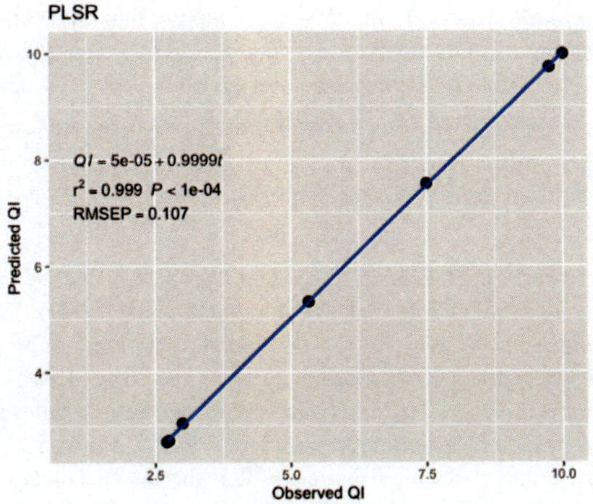

Fig. 6. Plot of partial least squares regression (PLSR)-predicted QI vs. observed QI for *P. longirostris* during storage at 5 °C. RMSEP-residual mean squared error of prediction of the PLSR.

4 Conclusions

The Quality Index Method developed here for pink shrimp showed a high linear correlation between the QI and the storage time on ice ($QI = 1.96 + 1.0 \times$ days on ice; $r^2 = 0.961$, $p < 10^{-4}$). The PLSR model's mean square of errors indicate very good accuracy. Thus, it was possible to predict freshness and remaining shelf-life for consumption.

The study showed that despite the growth of microbiota, TVB-N (total volatile basic nitrogen) and TMA (trimethylamine) contents did not increase during storage. The absence of increase in nitrogen-based compounds despite microbiota proliferation might be related to a leaching process due to melting ice (that needs to be further experimented).

On day 10 of storage, TVC (total viable count) exceeded 7 log(CFU/g), which is the limit for acceptable microbiological quality for seafood in many countries. Furthermore, on day 8, the samples were rejected by panelists due to their unacceptable sensory attributes. These findings suggest that deep-water pink shrimp should not be stored in ice at 5°C for more than 7 days. It is important for the seafood industry to be aware of these findings and to adjust their storage and transportation practices accordingly to ensure the safety and quality of their products.

Food safety is an essential part of food security and public health, thus relevant to UN SDGs, particularly SDG 1 (No poverty) and SDG 2 (Zero hunger). In addition, maintaining food safety during the storage and distribution of seafood products whilst reducing food waste is an ongoing challenge that is within the scope of SDG 12 (Food waste and management of chemicals). Without an understanding of and control over food safety issues, progress may be hampered.

Acknowledgments. This study also received Portuguese national funds from FCT - Foundation for Science and Technology through projects UIDB/04326/2020, UIDP/04326/2020 and LA/P/0101/2020 (Eduardo Esteves). Jaime Aníbal would like to acknowledge the support of national funds through Fundação para a Ciência e Tecnologia (FCT), under the project LA/P/0069/2020 granted to the Associate Laboratory ARNET and UID/00350/2020 CIMA.

References

1. Fisher, W., Schneider, M., Bauchot, M.L.: Identification des espèces pour les besoins de la pêche. Mediterranée et mer Noire. Fishes FAO. Volume I–II (1987)
2. *Parapenaeus longirostris* (Lucas, 1846). Deep-water rose shrimp. https://www.sealifebase.se/summary/Parapenaeus-longirostris.html. Accessed 23 Nov 2022
3. FAO 2023. *Parapenaeus longirostris* Lucas,1846. Fisheries and Aquaculture Division, Rome. https://www.fao.org/fishery/en/aqspecies/2598/en. Accessed 12 Mar 2023
4. Gillett, R. Global Study of shrimp fisheries. FAO Fisheries Technical Paper 475 (2008)
5. DGPA. Recursos da Pesca 2021. Direção-Geral de Recursos Naturais, Segurança e de Serviços Marítimos, Lisboa (2022)
6. Rosa, R., Nunes, M.L.: Nutritional quality of red shrimp, Aristeus antennatus (Risso), pink shrimp, *Parapenaeus longirostris* (Lucas), and Norway lobster, *Nephrops norvegicus* (Linnaeus). J. Sci. Food Agric. **84**, 89–94 (2004). https://doi.org/10.1002/jsfa.1619
7. Küçükgülmez, A., Baykal, M., Kadak, A.E., Cukurova, M.C.: Quality changes in crustaceans during and after processing. In: Genç, I.Y., Esteves, E., Diler, A. (eds.) Handbook of Seafood: Quality and Safety Maintenance and Application, pp. 127–137. Nova Science Publishers, Inc., New York (2016)
8. Ashie, I.N.A., Smith, J.P., Simpson, B.K., Haard, N.F.: Spoilage and shelf-life extension of fresh fish and shellfish. Crit. Rev. Food Sci. Nutr. **36**(1–2), 87–121 (1996)
9. Gram, L., Dalgaard, P.: Fish spoilage bacteria–problems and solutions. Curr. Opin. Biotechnol. **13**(3), 262–266 (2002)
10. Flick, G.J., Granata, L.A.: Biological safety of fresh and processed shellfish. SRAC Publication, 4901 (2010)

11. Huq, K.A., Rahaman, S.M.B., Islam, B., Shaha, B.K., Ghosh, A.K., Sayeed, A.B.: Quality aspects of frozen shrimp product in processing industry: a case study in Khulna, Bangladesh. Bangladesh Res. Publ. J. **3**(2), 945962 (2009)
12. Hossain, A., Mandal, S.C., Rahman, M.S., Rahman, M., Hasan, M.: Microbiological quality of processed frozen black tiger shrimps in fish processing plant. World J. Fish Marine Sci. **2**(2), 124–128 (2010)
13. Lagartinho, J.: Estudo da deterioração da gamba *Parapenaeus longirostris* (Lucas, 1846). MSc dissertation, Departamento de Engenharia Alimentar, Instituto Superior de Engenharia, Universidade do Algarve, Faro (2010)
14. Gonçalves, A.C., Lopez-Caballero, M.E., Nunes, M.L.: Quality changes of deepwater pink shrimp (*Parapenaeus longirostris*) packed in modified atmosphere. J. Food Sci. **68**(8), 2586–2590 (2003)
15. EU. Regulamento (CE) 2074/2005 da Comissão. Jornal Oficial da União Europeia L **338**, 27–59 (2005)
16. EU. Regulamento (CE) 1022/2008 da Comissão. Jornal Oficial da União Europeia L **277**, 18–20 (2008)
17. EC. Regulation (EC) 2406/96 of the Commission. Official Journal of the European Communities, **334**(1), 1–15 (1996)
18. Esteves, E.: Relating sensory and instrumental analyses of well-known and emerging fish and seafood products. In: Genç, I.Y., Esteves, E., Diler, A. (eds.) Handbook of Seafood: Quality and Safety Maintenance and Applications, pp. 31–64. Nova Science Publishers, Inc., New York (2016)
19. Esteves, E., Aníbal, J.: Sensory evaluation of seafood freshness using the quality index method: a meta-analysis. Int. J. Food Microbiol. **337**, 108934 (2021). https://doi.org/10.1016/j.ijfoodmicro.2020.108934
20. Howgate, P.: A history of the development of sensory methods for the evaluation of freshness of fish. J. Aquatic Food Prod. Technol. **24**(5), 516–532 (2015). https://doi.org/10.1080/10498850.2013.783897
21. Conway, E.J., Byrne, A.: An absorption apparatus for the micro-determination of certain volatile substances. I. The micro-determination of ammonia. Biochem. J. **27**, 419–429 (1933)
22. IPQ. NP 2930. Produtos da pesca e da aquicultura. Determinação do teor de azoto básico volátil total (ABVT). Instituto Português da Qualidade, Lisboa (2009)
23. ISO. ISO 6887-3: Microbiology of the food chain—Preparation of test samples, initial suspension, and decimal dilutions for microbiological examination—Part 3: Specific rules for the preparation of fish and fishery products. International Organization for Standardization, Geneva (2003)
24. IPQ. NP 4405. Microbiologia Alimentar: Regras gerais para contagem de microrganismos: Contagem de colónias a 30 °C. Instituto Português de Qualidade, Lisboa (2002)
25. ISO. ISO 17410: Microbiology of Food and Animal Feeding Stuffs – Horizontal Method for the Enumeration of Psychrotrophic Microorganisms. International Organization for Standardization, Genève (2001)
26. IPQ. NP 4137. Microbiologia Alimentar: Regras gerais para determinação de Enterobacteriaceae sem revitalização. Técnicas do número mais provável e de contagem de colónias. Instituto Português de Qualidade, Lisboa (1991)
27. Laghmari, H., Marrakchi, A.: Appréciation organoleptique et physico-chimique de la crevette rose Parapenaeus longirostris (Lucas, 1846) conservée sous glace et à température ambiante. Rev. Méd. Vétérinaire **156**(4), 221–226 (2005)
28. Martinsdóttir, E., Sveinsdóttir, K., Luten, J., Schelvis-Smit, R., Hyldig, G.: Sensory evaluation of fish freshness. QIM Eurofish, Ijmuden (2001)

29. Oliveira, V.M., Freitas, M.Q., São-Clemente, S.C., Mársico, E.T.: Método do Índice de Qualidade (MIQ) desenvolvido para camarão (Litopenaeus vannamei) cultivado. Rev. Ciências Vida **29**(1), 60–71 (2009)
30. Zeng, Q.Z., Thorarinsdottir, K.A., Olafsdottir, G.: Quality changes of shrimp (Pandalus borealis) stored under different cooling conditions. J. Food Sci. **70**(7), 459–466 (2005)
31. IPQ. NP EN ISO 8589. Sensory analysis. General guidance for the design of test rooms (ISO 8589:2007/Amd 1:2013). Instituto Português da Qualidade, Lisboa (2014)
32. Baranyi, J., Roberts, T.A.: A dynamic approach to predicting bacterial growth in food. Int. J. Food Microbiol. **23**, 277–294 (1994)
33. Baty, F., Delignette-Muller, M.L.: nlsMicrobio: data sets and nonlinear regression models dedicated to predictive microbiology. R package version 0.0-1 (2013)
34. Ritz, C., Streibig, J.C.: Nonlinear Regression with R. Springer, Heidelberg (2008). https://doi.org/10.1007/978-0-387-09616-2
35. Sveinsdóttir, K., Hyldig, G., Martinsdóttir, E., Jørgensen, B., Kristbergsson, K.: Quality index method (QIM) scheme developed for farmed Atlantic salmon (Salmo salar). Food Qual. Prefer. **14**, 237–245 (2003)
36. Rossini, K., Verdu, S., Cariou, V., Qannari, E.M., Fogliatto, F.S.: PLS discriminant analysis applied to conventional sensory profiling data. Food Qual. Prefer. **23**, 18–24 (2012)
37. Mevik, B.-H., Wehrens, R., Liland, K.H.: PLS: partial least squares and principal component regression. R package version 2.6-0 (2016)
38. R Core Team. R: A language and environment for statistical computing. R Foundation for Statistical Computing, Vienna, Austria (2022). https://www.R-project.org/
39. Cui, H., Xue, C., Xue, Y., Su, W., Li. Z., Cong, H.: Development of shelf-stable, ready-to-eat (RTE) shrimps (*Litopenaeus vannamei*) using water activity lowering agent by response surface methodology. J. Food Sci. Technol. **50**, 1137–1143 (2013)
40. Cadun, A., Cakli, S., Kisla, D.: A study of marination of deepwater pink shrimp (*Parapenaeus longirostris*, Lucas, 1846) and its shelf life. Food Chem. **90**, 53–59 (2005). https://doi.org/10.1016/j.foodchem.2004.03.024
41. Hall, G.M.: Fish Processing – Sustainability and New Opportunities. Wiley-Blackwell Publishing Ltd. (2011)
42. Huidobro, A., Lopez-Caballero, M., Mendes, R.: Onboard processing of deepwater pink shrimp (Parapenaeus longirostris) with liquid ice: effect on quality. Eur. Food Res. Technol. **214**, 469–475 (2002)
43. Moura, A.F.P., Mayer, M.D.B., Landgraf, M., Tenuta-Filho, A.: Qualidade química e microbiológica de camarão-rosa comercializado em São Paulo. Braz. J. Pharm. Sci. **5**(2), 203–208 (2003)
44. Shamshad, S.I., Kher-Un-Nisa, Riaz, M., Zuberi, R., Qadri, R.B.: Shelf life of shrimp (Penaeus merguiensis) stored at different temperatures. J. Food Sci. **55**(5), 1201–1205 (1990)
45. Fatima, R., Khan, M.A., Qadri, R.B.: Shelf life of shrimp *Penaeus merquiensis* stored in Ice (0 °C) and partially frozen (−3 °C). J. Sci. Food Agric. **42**, 235–247 (1988)
46. Kirschnik, P.G., Viegas, E.M.M., Valenti, W.C., Oliveir, C.A.F.: Shelf-life of tail meat of the giant river prawn, *Macrobrachium rosenbergii*, stored on ice. J. Aquat. Food Prod. Technol. **15**(2), 57–71 (2006)
47. Nirmal, N.P., Benjakul, S.: Effect of ferulic acid on inhibition of polyphenoloxidase and quality changes of pacific white shrimp (*Litopenaeus vannamei*) during iced storage. J. Food Chem. **116**, 323–331 (2009)
48. Kirschnik, P.G., Viegas, E.M.M.: Alterações na qualidade do camarão de água doce *Macrobrachium rosenbergii* durante estocagem em gelo. J. Ciência Tecnol. Alimentos **24**(3), 407–412 (2004)

49. Akintola, S.L., Bakare, S.B.: Effects of ice storage on the biochemical composition of *Macrobrachium vollenhovenii* (Herklots, 1857). J. Fish. Aquatic Sci. **8**, 213–217 (2012). https://doi.org/10.3923/jfas.2013.213.217
50. Mendes, R., Huidobro, A., Lopez-Caballero, M.E.: Indole levels in deepwater pink shrimp (Parapenaeus longirostris) from the Portuguese coast. Effects of temperature abuse. Eur. Food Res. Technol. **214**, 125–130 (2002)
51. Salvi, D.: Study of Quality deterioration of shrimp by chemical, microbial and electronic nose analysis. AIT thesis, Bangkok, Thailand (2005)
52. Castro, P., Padrón, J.C.P., Cansino, M.J.C., Velázquez, E.S., Larriva, R.M.: Total volatile base nitrogen and its use to assess freshness in European sea bass stored in ice. J. Food Control **17**, 245–248 (2006)
53. Grigorakis, K., Alexis, M., Gialamas, I., Nikolopoulou, D.: Sensory, microbiological, and chemical spoilage of cultured common sea bass (Dicentrarchus labrax) stored in ice: a seasonal differentiation. Eur. Food Res. Technol. **219**, 584–587 (2004)
54. Hernández, M.D., López, M.B., Álvarez, A., Ferrandini, E., García, B.G., Garrido, M.D.: Sensory, physical, chemical and microbiological changes in aquacultured meagre (Argyrosomus regius) fillets during ice storage. Food Chem. **114**, 237–245 (2009)
55. Aníbal, J., Lagartinho, J., Esteves, E.: Is total volatile basic nitrogen (TVB-N) a reliable spoilage indicator for the deep-water shrimp (*Parapenaeus longirostris*)? In: International Meeting on Marine Resources (IMMR'12), Escola Superior de Turismo e Tecnologia do Mar, Instituto Politécnico de Leiria, Peniche, Portugal (2012)
56. Anastasio, A.: Correlations between pH, total volatile basic nitrogen, trimethylamine and sensory evaluation in fresh fish slices. Arch. Lebensmittel Hygiene **50**(3), 63–66 (1999)
57. Kyrana, V.R., Lougovois, V.P.: Sensory, chemical and microbiological assessment of farm-raised European sea bass (Dicentrarchus labrax) stored in melting ice. Int. J. Food Sci. Technol. **37**, 319–328 (2002)
58. Papadopoulos, V., Chouliara, I., Badeka, A., Savvaidis, I.N., Kontominas, M.G.: Effect of gutting on microbiological, chemical, and sensory properties of aquacultured sea bass (Dicentrarchus labrax) stored in ice. Food Microbiol. **20**(4), 411–420 (2003)
59. Kornacki, J.L., Johnson, J.L.: Enterobacteriaceae, coliforms, and *Escherichia coli* as Quality and safety Indicators. In: Downes, F.P., Ito, K. (eds.) Compendium of Methods for the Microbiological Examination, pp. 68–81. American Public Health Association (2001)
60. ICMSF. Microorganisms in Foods. 2. Sampling for Microbiological Analysis: Principles and Specific Applications, 2nd edn. University of Toronto Press, Toronto (1986)
61. Ho, M.L., Cheng, H.H., Jiang, S.T.: Effect of modified ice storage on the shelf-life of shrimp. Bull. Jpn. Soc. Fish. Oceanogr. **52**, 479–488 (1986)
62. Lannelongue, M., Finne, G., Hanna, M.O., Nickelson, R., Vanderzant, C.: Storage characteristics of brown shrimp (Penaeus aztecus) stored in retail packages containing CO_2-enriched atmospheres. J. Food Sci. **47**, 911–923 (1982)
63. Layrisse, M.E., Matches, J.R.: Microbiological and chemical changes of spotted shrimp (*Pandalus platyceros*) stored under modified atmospheres. J. Food Protect. **47**(6), 453–457 (1984)
64. Oliveira, V.M.: Estudo da qualidade do camarão branco do Pacífico (*Litopenaeus vannamei*) inteiro e descabeçado estocado em gelo. Ph.D. thesis, Universidade Federal Fluminense (2005)
65. Khodanazary, A.: Freshness assessment of shrimp Metapenaeus affinis by quality index method and estimation of its shelf life. Int. J. Food Properties **22**(1), 309–319 (2019)

66. Gonçalves, A.A., de Lima, J.T.A.X., de Paula, F.E.R.: Development of quality index method (QIM) scheme for spiny lobster (*Panulirus Argus*, Latreille, 1804) stored in ice. Food Control **47**, 237–245 (2015)
67. Tam, L.N., Khue, D.N., Thanh, N.B., Thi, T.T.V.: Towards improved quality benchmarking and shelf life evaluation of black tiger shrimp (Penaeus monodon). Food Chem. **235**, 220–226 (2017)

How to Reduce the Ethanal or Acetaldehyde in *Arbutus unedo* L. Fruit Distillate

Vera Francisco[1] and Ludovina Galego[1,2](✉)

[1] Instituto Superior de Engenharia, Universidade do Algarve, Campus da Penha, 8005-139 Faro, Portugal
lgalego@ualg.pt

[2] MED - Mediterranean Institute for Agriculture, Environment and Development and CHANGE – Global Change and Sustainability Institute, Évora, Portugal

Abstract. Ethanal or acetaldehyde is a natural compound of fermentations. The Portuguese legislation imposes, to the Arbutus unedo L. fruit distillate a maximum limit of 40 g/hl of pure alcohol (p.a.) to this compound. Although the main causes of its formation are known, different fruits fermentation have detailed specifications that can enhance its formation. The aim of this study was to investigate possible processes to reduce this compound before distillation (part I) and after distillation (part II). In this study to reduce ethanal or acetaldehyde before distillation, nine fermentations were planned. Three fermentations using 25% fruits harvested from the ground (A), three fermentations using 25% arbutus pricked from birds (B) and three fermentations where only whole harvested fruits from the tree where used (C). To reduce acetaldehyde after distillation 3 batches (D, E and F) from the same producer, all with acetaldehyde excess were used. The samples were placed in a bath at 35 °C with 50 r.p.m agitation, during two hours. As main results, from part I, when the fermentations use only fruits whole harvested picked from the tree (C), the ethanal or acetaldehyde content is 28.6 ± 3.8 g/hl p.a., in the other case the compound exceed significantly the law limits. After distillation, part II, in the first half hour ethanal or acetaldehyde present a reduction of 12.8 ± 4.9 g/hl p.a.in all batch (D, E and F). Ethanol has a reduction 1.2 ± 0.2% vol. in all the samples, every half hour. Ethyl acetate is also reduced, methanol decreases in the first half hour, higher alcohols and acidity tend to increase over time.

Keywords: Ethanal or acetaldehyde · Arbutus unedo L. fruit distillate · Ethyl acetate · Methanol · FID-GC

1 Introduction

Arbutus unedo L. fruit distillate also known as strawberry tree distillate is a traditional homemade distillate produced in several Mediterranean countries with different relevance. In Greece [1] and in Portugal [2–4] this distillate is mostly produced by several small-scale industries, in Spain [5] or Italy [6] only scientific experiments are known. This distillate in Portugal is produced all over the country, currently using fruits from new orchards with clones adapted Arbutus plants [7, 8] and modern distillation systems

J. F. L. C. Semião et al. (Eds.): INCREaSE 2023, ASST, pp. 403–408, 2023.
https://doi.org/10.1007/978-3-031-44006-9_30

[9], but it is in the Algarve mountain area, where always has been great economic importance, which motivated early quality studies [6] and local quality control mechanisms [10–12]. Ethanal or acetaldehyde is a natural compound of many fermentations [13, 14]. The Portuguese legislation [15] imposes a maximum limit of 40 g/hl p.a. to this compound, because some toxicity is associated [16]. Acetaldehyde is a water-soluble compound with 20.2 °C boiling point [17]. Theoretically, the formation of this compound occurs during the conversion of sugar into ethanol by yeast [13, 14, 16] or oxidation of ethanol when exposed to oxygen [16]. Although the main causes of its formation are known, different fruits fermentation have detailed specifications that can enhance its formation. The aim of this study was to investigate possible processes to reduce this compound in *Arbutus unedo* L. fruit distillate.

2 Experimental

2.1 Material

Part I - Acetaldehyde reduction before distillation, was performed using nine fermentations. Three fermentations using 25% of fruits harvested from the ground – Experience A -, three fermentations using 25% of fruits pricked from birds - Experience B - and three fermentations where only harvested fruits from the tree are used – Experience C. All the fermentation took place in the same period under the same conditions. Distillation take place in traditional copper alembic.

Part II - Acetaldehyde reduction after distillation. Three Arbutus distillates batch (D, E and F) with different content of acetaldehyde were used to test the reduction in a thermal bath at 35 °C ± 2 °C with stirring a 50 r.p.m (OVAN Model BS127 E Spain). From each batch half-liter samples were taken and placed in beakers with 1 L of capacity. When the temperature inside the goblet stabilized at 33.5 °C. 5 ml samples were collected every 30 min for two hours.

2.2 Chemicals

Ethanol, methanol, and methyl-2-butanol, 4-methyl-2-pentanol, (Merck, Darmstadt, Germany), 3-methyl-1-butanol, copper, phenol red indicator (Panreac, Madrid, Spain), acetaldehyde, ethyl acetate, 2-methyl-1-propanol, (Fluka,New Ulm, Switzerland).

2.3 Volatile Quantification Compounds

The quantification of volatile compounds was done using a GC (PerkinElmer Clarus 400) equipped with a FID (Waltham, USA) detector. A BP 20 capillary column (30 m; 0.32 mm I.D.; 1.0 μm film thickness; SGE, Australia) was used with the following oven temperature program: 5 min at 40 °C, 5 °C min^{-1} increase until 210 °C. The injector was set to 250 °C, the injections were made in the split mode (30:1), the injection volume was 1 μl and the detector was set to 270 °C. The quantification was done using the internal standard 4-methyl-2-pentanol [9].

2.4 Total Acidity

The total acidity was measured by a potentiometric titration or titration using phenol red as indicator according Portuguese official standard for spirits NP 2139 (1977) [18].

2.5 Statistical Analysis

All analytical measurements were run at least in triplicate. Measurements were averaged and the results are given as mean ± standard deviation. The F-Test two-samples for variances were used. Differences were considered significant at $p < 0.05$ level, using Microsoft office 2019 excel software.

3 Results and discussion

Part I - Table 1 shows the average concentration of acetaldehyde and other quality important substances (ethyl acetate, methanol, higher alcohols sum and total acidity) to the nine batch Arbutus unedo L fruits fermentations: three using 25% of fruits harvested from the ground - **A**, three using 25% of fruits pricked from birds - **B** and three where only harvested fruits from the tree are used – C.

Table 1. Contents (g/hl p.a.) of mean compounds analysed from the three experiences A, B e C

Compounds (g/hl p.a)	A	B	C
Acetaldehyde	$46.1^{\alpha} \pm 4.4$	$62.2^{\beta} \pm 4.9$	$28.6^{\gamma} \pm 3.8$
Ethyl acetate	$173.7^{\alpha} \pm 5.5$	$140.6^{\beta} \pm 6.4$	$139.9^{\beta} \pm 4.8$
Methanol	$803.6^{\alpha} \pm 10.7$	$853.6^{\beta} \pm 9.8$	$856.1^{\beta} \pm 12.9$
Higher alcohols sum	$154.0^{\alpha} \pm 8.1$	$163.0^{\beta} \pm 9.6$	$153.3^{\alpha} \pm 6.6$
Total acidy*	$63.5^{\alpha} \pm 1.8$	$56.9^{\beta} \pm 1.9$	$47.9^{\gamma} \pm 1.1$

The results correspond to the mean and standard deviation of 3 replicates. In each line the same symbol means that there is no significant difference with a significance level of 95% ($p < 0.05$) (n=3). *Express in acetic acid.

Fruits that have fallen to the ground are overripe, many are already fermenting and are generally crushed during transport. The same is happens for the pricked fruits by the birds. In both cases fermentation to begin in contact with the air, what leads to the oxidation of ethanol, when exposed to oxygen and this practice increase the acetaldehyde formation [16]. In the case of fruits pricked by the birds, it is even more serious, because normally these fruits are on the tree, for a few days fermenting in contact with the air, until they are harvested. There is significant variation between all experiments, relatively to acetaldehyde. The same happens with total acidity, which is also very dependent on the oxygen contact. Greater oxygen contacts higher acidity content [16].

Part II - In the tests carried out to reduce acetaldehyde after distillation. Samples in the thermal bath were collected at 0 min, 30 min, 60 min, 100 min, 130 min and 160 min. The samples were analysed until it was verified that the acetaldehyde content was below the minimum limit accepted by legislation. The acidity was only analysed at the beginning and at the end (160 min in the thermal bath).

Figures 1, 2, 3, 4, 5 and 6 show the variation tendency for some quality important compounds. Figure 1 show the acetaldehyde decrease with the time for samples E. In the others experiences (D, F) the behaviour is similar. A 12.8 $\pm$ 6.9 g/hl p.a decrease on ethanal or acetaldehyde was observed, on average, in all samples in the first half hour, with this methodology. However, it is important to know the behave of other important quality compounds with the thermal treatment.

Ethanol also have a decrease with the time, 1.2 $\pm$ 0.2% vol. Every half hour, spent in the thermal bath (Fig. 2). The ethyl acetate also has a reduction more than 15% for every half hour in the thermal bath (Fig. 3). Methanol has a less regularly tendency to decrease, on average near 5% in all the time in the thermal bath (Fig. 5 and 6). Higher alcohols (Fig. 4) and total acidity (table 2) have a contrary tendency.

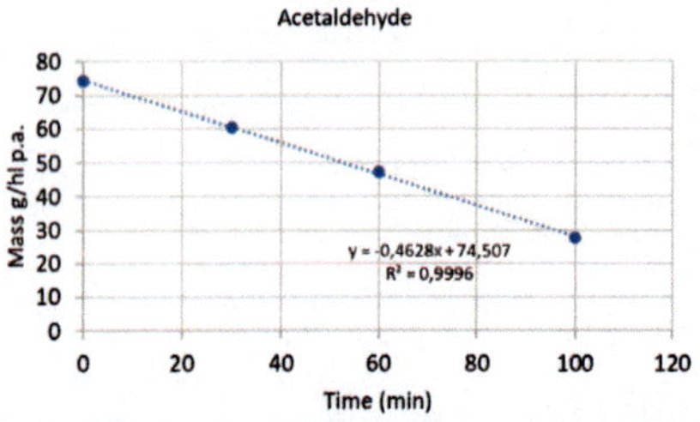

Fig. 1. Acetaldehyde decrease with the time, experience E

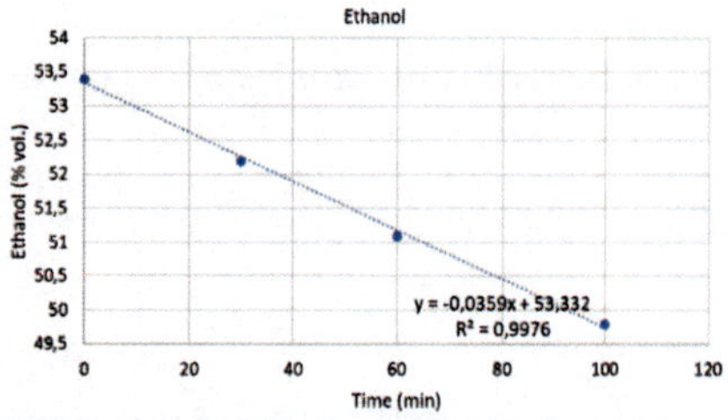

Fig. 2. Ethanol reduction decrease with the time, experience D

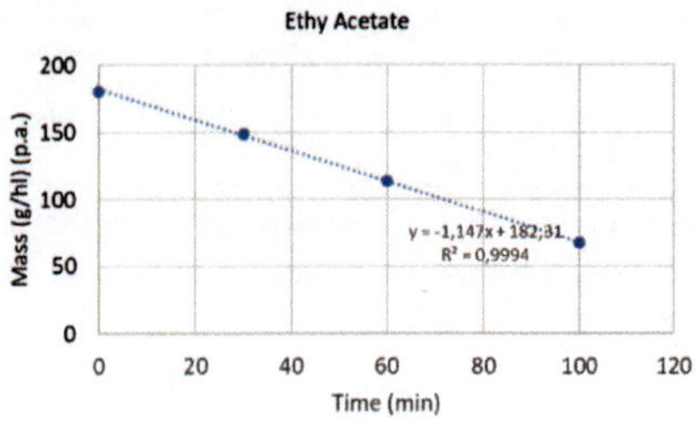

Fig. 3. Ethyl acetate decrease with the time into the thermal bath, experience E

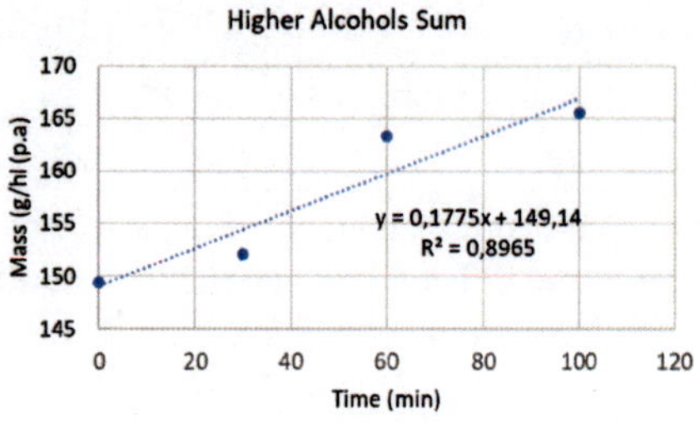

Fig. 4. Higher Alcohols sum tendency with the time into the thermal bath, experience D

In all experiments there is an increase in acidity, as can be seen in the table 2. Because ethanol is reduced and all compounds are quantified relative to 100% alcohol, heavier compounds tend to increase, like higher alcohols and acids.

The distillate heating increases significantly the total acidity content, in any of the experiments done.

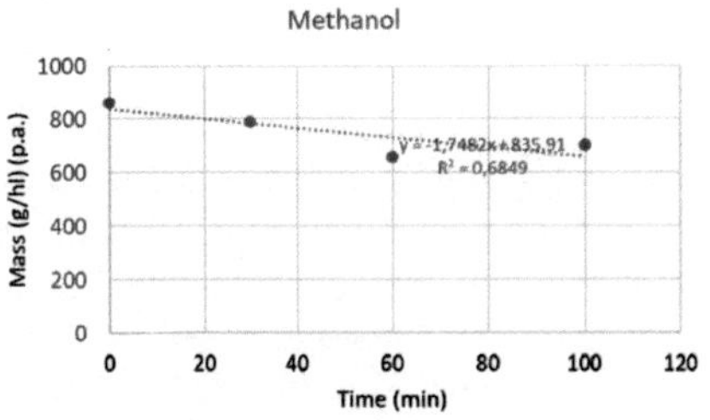

Fig. 5. Methanol tendency with the time into the thermal bath, experience F

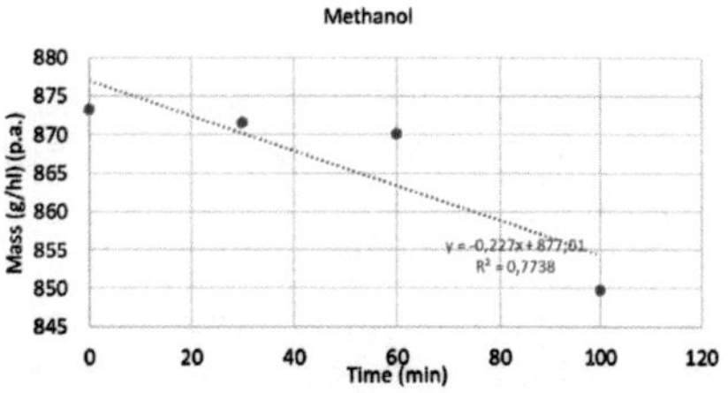

Fig. 6. Methanol tendency with the time into the thermal bath, experience E

Table 2. Total acidity experiences determination in the initial and in the final

Experience	Initial total acidity* (g/hl a.p.)	Final total acidity* (g/hl a.p.)
D	$32.0^{\alpha} \pm 0.8$	$43.7^{\beta} \pm 0.9$
E	$45.6^{\alpha} \pm 1.2$	$54.9^{\beta} \pm 1.5$
F	$42.3^{\alpha} \pm 1.6$	$52.8^{\beta} \pm 0.8$

The results correspond to the mean and standard deviation of 3 replicates. In each line the same symbol means that there is no significant difference with a significance level of 95% ($p < 0.05$) ($n = 3$). *Express in acetic acid

4 Conclusions

The knowledge of all the steps in the production of any quality product are fundamental. In the *Arbutus unedo* L. fruit distillate, it is important do not use fruits that has fallen to the ground or pricked by the birds, because these fruits contribute to increase the ethanal or acetaldehyde content in the final distillate, may exceed the legislation limits. However, as the desired optimal harvesting conditions are not always possible, it is necessary to find loss mitigation mechanisms. The present study showed that it is possible to recover distillates batches with ethanal or acetaldehyde excess. In order to recover a distillate batch with ethanal or acetaldehyde in excess it is necessary to place the distillate in a 35 °C thermal bath, with a 50 r.p.m. agitation. The acetaldehyde content decreases (in the first half hour, 12.8 ± 6.9 g/hl p.a.), there is also loss of alcohol, ethyl acetate and methanol. Higher alcohols and total acidity tend to increase. Batches distillates with high initial levels of total acidity and/or the higher alcohols may have limitation on the application of this thermal methodology. Batches distillates with higher ethyl acetate and/or methanol levels have a double quality improvement if this thermal methodology is used. Likewise, an alcohol content reduction is always desired, if does not fall below the legislation limits.

Acknowledgments. The authors are grateful for the samples and information about the production process given producer association APAGARBE from Monchique Algarve – Portugal and Herdade da Tamanqueira – Odemira – Alentejo – Portugal.

References

1. Soufleros, E.H., Mygdaline, S.A., Natskoulis, P.: Production process and characterization of the traditional Greek fruit distillate "Koumaro" by aromatic and mineral composition. J. Food Compos. Anal. **18**, 699–716 (2005)
2. Santo, D.E., Galego, L., Gonçalves, T., Quintas, C.: Yeast diversity in the Mediterranean strawberry tree (*Arbutus unedo* L.) fruits' fermentations. Food Res. Int. **47**, 45–50 (2012)
3. Galego, L.R., Almeida, V.R.: Aguardentes de frutos e licores do Algarve História, técnicas de produção e legislação, Edições Colibri, Lisboa (2007). ISBN 987-972-772-722-3
4. Versini, G., Seeber, R., Dalla Serra, A., Sferlazzo, G., Carvalho, B., Reniero, F.: Aroma compounds of arbutus distillate. In: Charalambous (ed.) Food Flavors Generation, Analysis and Process Influence, pp. 1779–1790. Elsevier Science, London (1995)
5. González, E.A., Agrasar, A.T., Castro, L.M.P., Fernández, I.O., Guerra, N.P.: Solid-state fermentation of red raspberry (*Rubus ideaus* L.) and arbutus berry (*Arbutus unedo*, L.) and characterization of their distillates. Food Res Int. **44**, 1419–1426 (2011)
6. Versini, G., Moser, S., Franco, M.A., Manca, G.: Characterisation of strawberry tree distillate (*Arbutus unedo* L.) produced in Sardinia. J. Commod. Sci. Technol. Qual. **50**, 197–206 (2011)
7. Baleiras-Couto, M.M, Caldeira, I., Gomes, F., Botelho, G., Duarte, F.L.: Saccharomyces cerevisiae diversity in Arbutus unedo L. Fermentations in association with the volatile and sensory similarities of the distillates. Foods, **11**, 1916 (2022)
8. Caldeira, I., Gomes, F., Mira, H., Botelho, G.: Distillates composition obtained of fermented Arbutus unedo L. fruit from different seedlings and clonal plants. An. Agric. Sci. **64**, 21–28 (2019)
9. Galego, L., Francisco, V., Ratão, I.: *Arbutus unedo* L. fruit distillate from tradition to innovation. In: Mortal, A., et al. (eds.) INCREaSE, pp. 151–157. Springer, Cham (2018). https://doi.org/10.1007/978-3-319-70272-8_10
10. Galego, L.R.: Otimização de parâmetros para a produção de aguardente de medronho, dissertação apresentada na Faculdade de Ciências e Tecnologia da Universidade Nova de Lisboa para a obtenção do grau de Mestre em Tecnologia Alimentar/Qualidade (1995)
11. Galego, L.R., Martins, M.A.N., Almeida, V.R., Versini, G.: Valorization of arbutus destillate. Euro Food Chem VIII Atas, Austria, pp. 341–344 (1995)
12. Galego, L.R., Bento, F., Almeida, V.R., Martins, A.N., Jesus J.P., Vilas Boas, L.: Variação do teor em cobre ao longo de uma destilação experimental para obtenção de aguardente de medronho. In: Atas do XII encontro Luso-Galego de Química, Porto, p. 5.02 (1998)
13. Xu, X., Niu, C., Liu, C., Li, Q.: Unraveling the mechanisms for low-level acetaldehyde production during alcoholic fermentation in *Saccharomyces pastorianus* lager yeast. J. Agric. Food Chem. **67**, 2020–2027 (2019)
14. Aguera, E., Sire, Y., Mouret, J.-R., Sablayrolles, J.-M., Farines, V.: Comprehensive study of the evolution of the gas−liquid partitioning of acetaldehyde during wine alcoholic fermentation. J. Agric. Food Chem. **66**, 6170–6178 (2018)
15. Decreto-Lei nº 238/2000 de 26 de setembro, diário da República I série A, Imprensa Nacional, **223**, 5145–5147 (2000)
16. Tsakiris, A., Kallithraka, S. & and Kourkoutas, Y. Grape brandy production, composition and sensory evaluation. J. Sci Food Agric. **94**, 404–414 (2014)
17. Anjos, C.A.R.: Influence of the process in the generation of acetaldehyde and residual levels in poly (ethylene terephthalate) (PET) packaging and drinks – review article. BioEng. Campinas **1**(3), 277–290 (2007)
18. NP 2139. Determination of total acidity content. IPQ, Lisbon (1987)

Effect of Edible Chitosan/Clove Oil Films and High Pressure Processing on the Quality of Trout Fillets

I. Albertos[1](✉), I. Jaime[2], A. B. Martin-Diana[3], M. J. Castro-Alija[1], and D. Rico[3]

[1] Recognized Research Group: Assessment and Multidisciplinary Intervention in Health Care and Sustainable Lifestyles of the University of Valladolid, Valladolid, Spain
irene.albertos@uva.es

[2] Department of Biotechnology and Food Science, University of Burgos, Plaza Misael Bañuelos s/n, 09001 Burgos, Spain

[3] Agricultural Technological Institute of Castile and León (ITACyL), Government of Castile and León, Ctra. de Burgos Km. 119, Finca Zamadueñas, 47071 Valladolid, Spain

Abstract. It has been explored the potential utilisation of combined high-pressure processing (HPP) and chitosan edible films with or without clove essential oil (CEO) in order to improve quality (oxidative, sensorial) of rainbow trout fillets at 4 °C over 22 days in comparison with untreated samples or cooked and HPP alone.

The optimal CEO concentration was determined previously through total phenol and antioxidant capacity determination. Films up 20 gkg^{-1} of CEO did not show significantly higher phenolics compounds and antioxidant capacity. Both traditional cooking and HPP reduced the spoilage as reflected from TVBN results when compared to the control. HPP had less impact on lipid oxidation (TBARS, hexanal concentration and rancid odour) and produced lower colour changes (L* and a* parameters) than cooking. However, some significant changes occurred in pressurised samples. In this sense, application of chitosan films to pressurised samples reduced lipid oxidation and colour alteration. The use of 20 gKg^{-1} CEO-chitosan films showed a further improvement in oxidative parameters compared to that obtained with chitosan alone. Nevertheless, the remaining odour from the CEO can be a limitation in their application in fresh fish but it could be added to marinated seafood.

Keywords: Edible films · clove essential oil · high-pressure processing · trout · chitosan

1 Introduction

Preservation technologies are specially required to maintain high perishable food such fatty fish. Rainbow trout (*Oncorhynchus mykiss*) contained high content of polyunsaturated fatty acids (PUFAs) making it highly susceptible to oxidation and development of unpleasant off-flavors (Baron et al., 2007).

J. F. L. C. Semião et al. (Eds.): INCREaSE 2023, ASST, pp. 409–423, 2023.
https://doi.org/10.1007/978-3-031-44006-9_31

The combination of preservative factors or hurdle strategy, which has been empirically used traditionally in foods, is currently intelligently selected and intentionally applied in novel food products (Leistner, 2000). Hurdle technologies exploit synergistic interactions between preservation treatments with the aim of minimising their impact on sensory and nutritive properties of the food (Ross et al., 2003), being of fundamental importance in the preservation of intermediate-moisture and high-moisture foods (Leistner et al., 2000).

The use of edible coating and films allows extending the shelf life of many food products. Chitosan has been widely used as hydrocolloid to constitute edible films due to its biodegradability, biocompatibility, antimicrobial and antioxidant activities and nontoxicity (Ruiz-Navajas et al., 2013). Moreover essential oils such clove essential oil (CEO) can be applied to the formulation of edible films, providing additional antioxidant and/or antimicrobial properties (Gómez-Estaca et al., 2010). Nevertheless, no information on the antioxidant capacity of chitosan film containing CEO has been found. The application of CEO through film pretended to lessen their incorporation. In this way, their disadvantages such as intense aroma and high cost will be reduced (Sánchez-Gonzalez et al., 2011). In addition, edible films with essential oils added produce a gradual liberation of the additive to the food along its shelf life (Campos et al., 2011).

Non-thermal technologies have only recently received considerable attention (Ross et al, 2003). High hydrostatic processing (HHP) is widely applied as a minimal processing technique to prolong the shelf life of additive-free foods (Ojagh et al., 2011). It is well documented that HHP processing may induce lipid oxidation (Cheftel and Culioli, 1997; Medina-Meza et al., 2014). This effect depends on factors like pressure level, fish species and subsequent storage conditions. However, very little information exists on HHP effect on seafood oxidation, especially lipid oxidation (Yagiz et al., 2007). Relatively few studies on the oxidative effects of the HPP on fish have been conducted (Gómez-Estaca et al., 2010; Ojagh et al., 2011: Núñez-Flores et al., 2013; Günlu et al., 2014). A protective effect on lipid oxidation in the shelf life of chilled trout submitted to HPP was found by Günlu et al. (2014) by the use of chitosan films. There are no previous reports on the combination of HPP and edible films formulated with chitosan and essential oil, or HPP and edible films including CEO.

Although our previous study (Albertos et al., 2015) yielded interesting results in the combination of hurdle technologies (chitosan films, with CEO and HPP) on the shelf-life of trout. The effects of these novel treatments on other quality parameters such as lipid oxidation or sensorial acceptance were unknown.

Therefore, the aim of the present study were to (1) determinate the optimal concentration of CEO in chitosan films in antioxidant in vitro assays and (2) study the quality and sensorial aspects of chilled trout by using the combination of chitosan or chitosan with CEO films and HPP, and compared with the treatments alone (film or HPP) or a cooked and unprocessed samples.

2 Material and methods

2.1 Formulation of the films

The films were prepared by aqueous solution of 15 gkg $^{-1}$ of chitosan (Sigma Aldrich Chemical Co Steinheim, Germany) and 10 gkg^{-1} acetic acid (E-260 aditio, Panreac Química, Barcelona, Spain). The mixture was stirred at 40 °C in an enzymatic digester (GDE, Velp Scientifica, Italy) for 2 h to obtain an homogenous solution. Afterwards, 5 gKg^{-1} of glycerol of biopolymer was added as plasticiser (E-422 aditio, Panreac). Food grade clove oil (Panreac) was incorporated in a proportion of 5, 10, 15, 20 and 50 gKg^{-1}, respect to the filmogenic solution, together with 2 gkg^{-1} of biopolymer of Tween 80 (E-433 aditio, Panreac) as emulsifying agent. The clove-added film forming solutions were homogenised at 13.400 xg (Ultraturrax IKA T25 digital, IKA-Werke, Germany) keeping in ice bath to avoid overheating. Later, solutions were degassed using an ultrasound bath (Selecta SA, Spain) for 30 min and kept overnight at room temperature in darkness. The films were obtained by casting 20 mL in 90 mm-diameter Petri dishes and dried at 42 °C in a force-air incubator (Biosan ES-20, Biogen Científica SL, Madrid, Spain) for 15 h. Prior to analyses, the films were conditioned in desiccators over a saturated solution of KBr (Panreac) (58% relative humidity).

2.2 Optimization of the Antioxidant Properties of the Films

Antioxidant properties of films with different concentration of clove oil were tested (0, 5, 10, 15, 20 and 50 gKg^{-1}).

Extract Preparation: A 200 mg piece of conditioned film was placed in a polypropylene tube containing 30 mL of methanol (Panreac), the lid tightly closed and the content stirred at 300 xg for 90 min at room temperature. After centrifugation at 3214 xg for 10 min the supernatant was collected for antioxidant capacity and total phenol determinations. Measurements were carried out in triplicate.

Total Phenols (TP): Total phenols were measured using the Folin-Ciocalteu method (Slinkard and Singleton, 1977). Results were expressed as mg Gallic Acid (GAE) per g^{-1} sample using a calibration curve with gallic acid (Sigma Aldrich) as standard (9.8 μM to 70 μM).

ORAC (Oxygen Radical Absorbance Capacity): The procedure was based on Ou et al (2001). Results were expressed as μmoles Trolox Equivalents (TE) g^{-1} sample.

TEAC (Trolox Equivalent Antioxidant Capacity): The analysis was carried according Re et al. (1999). Results were expressed as μmoles TE g-1 sample.

2.3 Combined Treatment of Trout Fillets with Film and HPP

Preparation and Treatment of Fish Samples: Gutted aquaculture rainbow trouts (*Oncorhynchus mykiss*) were provided by aquaculture farm IPEASA (Villa de Fuentidueña, Segovia, Spain). After skinning and filleting, they were cut into pieces of 5.7 × 2.5 mm (weighing approximately 10 g). Fish samples were randomly allocated into 7 batches: fresh fish (Fresh), chitosan film (Film), chitosan film with 20 g kg^{-1} of clove oil (Clove), cooked (Cook), high pressure processing (HPP), chitosan film and high pressure treatment combination (HF) and chitosan film with 20 g kg^{-1} of clove oil and high pressure treatment combination (HC). Pieces of the samples for film treatments (film, clove, HF and HC) were individually wrapped with 90 mm-diameter films with the help of tweezers in hygienic conditions. Afterwards, all samples (fresh, film, clove, HPP, HF and HC) were placed in co-extruded polyamide/polyethylene (30/130 μm thickness) flexible bags with oxygen permeability of 30 mL mm^{-2} day^{-1} bar^{-1} and water vapour transmission of 1.4 g m^{-2} day^{-1} (Industrias Pargón, Salamanca, Spain), and vacuum sealed. A slightly different procedure for cooked samples was followed: within the bags the samples were placed in a convection oven at 90 °C for 15 min, blast chilled to 4 °C and vacuum sealed. After vacuum packaging, high pressure processed samples (HPP, HC and HF) were treated in a high pressure unit Wave 6000/135 (NC Hyperbaric, Burgos, Spain) for 10 min at 300 MPa at 12 °C. All treatments were stored at 4 °C until analysis at days 0, 6, 9, 15 and 22. The assay was made twice. All analyses were performed in triplicate.

Total Volatile Basic Nitrogen (TVBN): TVBN was determined according to the method described by EC Regulation 2005/2074 (EC, 2005). Results were expressed as mg TVBN/100 g muscle.

Thiobarbituric Acid Reactive Substances (TBARS): Samples were analysed using the methodology described by Tarladgis et al. (1960). Analyses were performed in triplicate and results were expressed as μg malondialdehyde (MDA) g^{-1} muscle.

Hexanal: Hexanal content in fish samples was determined by Headspace Gas Chromatography/Mass spectrometry (GC/MS) using Solid Phase Dynamic Extraction (SPDE). Three grams of fish samples were placed in 20 mL vials with 100 uL of cyclopentanona as internal standard. Headspace volatile were sampled by SPDE with 90% dimethylsiloxane and 10% active carbon, under the following conditions: preincubation time: 1 min; incubation temperature: 70 °C; syringe temperature: 75 °C; number of extraction strokes: 50; extraction fill/eject speed: 40 μL s^{-1}; pre-desorption time: 30 s; desorption gas volume: 500 μL; desorption flow speed: 15 μL s^{-1}.

GC analysis were accomplished by HP 6890N Series instrument (Agilent Technologies) equipped with HP-5MS capillary column (60 m × 0.32 mm, 0.25 um film thickness) and a 5973i mass spectrometer detector (Agilent Technologies) working with the following temperature programme: 40 °C to 240 °C at 3 °C per min; injector and detector temperatures: 250 °C; carrier gas helium at 1 ml/min in mode splitless.

Identification of the compounds was based by their retention time and the comparison of their MS with the library spectra present in Wiley and National Institute of Standards and Technology (NIST) databases. All determinations were performed in triplicate.

Colour: Colour parameters, lightness (L*), redness (a*) and yellowness (b*), were measured using a spectrophotometer (Minolta CM-2002, Osaka, Japan). The illuminant was D65 and standard observer 10°. Measurements were taken on samples packaged with transparent plastic bags at six different points.

Sensorial Analysis: A trained panel consisting of 12 panellists were recruited from the department of Biotechnology and Food Science of the Burgos University, Spain for their previous experience in sensory analysis. All the tests were carried out in a sensory laboratory equipped with individual booths (ISO 8589:1988). The samples were coded and randomly presented to panellists without cooking (in raw). The following descriptors were studied: colour intensity, off-colours, surface defects, drip loss, fishy odour intensity and off-odours. The panellists were asked to evaluate these descriptors in comparison with control fish, ranking the samples between −4 (lower values than control) to 4 (higher values than control), scoring 0 if samples do not differ from control (Fresh).

2.4 Statistical Analysis

The statistical analyses were performed using Statgraphics Centurion XVI. One-way analyses of variance were carried out. Differences between pairs of means were assessed on the basis of confidence intervals using Fisher LSD (Least Significant Difference) with a level of significance of $P < 0.05$.

3 Results

3.1 Antioxidant Properties of Films

Total phenols and antioxidant capacity (ORAC and TEAC) of chitosan film were shown in Table 1.

Table 1. Total phenol and Antioxidant capacity of chitosan films with different concentration of CEO.

CEO (g Kg^{-1})	Total Phenols (mg GAE g^{-1})	ORAC(μmoles TE g^{-1})	TEAC (μmoles TE g^{-1})
0	0.44 ± 0.02^a	1.84 ± 0.10^a	0.02 ± 0.00^a
5	0.83 ± 0.09^b	3.29 ± 0.43^b	0.08 ± 0.01^b
10	1.21 ± 0.11^{bc}	4.33 ± 0.76^b	0.12 ± 0.00^c
15	1.73 ± 0.15^c	6.65 ± 0.26^c	0.19 ± 0.01^d
20	1.98 ± 0.28^d	8.28 ± 0.34^d	0.23 ± 0.01^e
50	2.06 ± 0.29^d	9.76 ± 0.53^e	0.24 ± 0.01^e

Values (mean ± standard deviation, n = 3) followed by different lowercase letter are significantly different ($P < 0.05$).

Total phenolic content and antioxidant capacity of edible films increased as the CEO concentration did it, up to 20 g kg^{-1}. Although no previous studies on the antioxidant activity of chitosan films with CEO were found, similar behaviour has been observed for chitosan films formulated with other essential oils such as *Zataria multiflora* Boiss and *Thymus* spp. Essential oils (Moradi et al., 2011; Ruiz-Navajas et al., 2013). A further significant increase in antioxidant capacity ($P < 0.05$) was observed from 20 and 50 g kg^{-1} concentrations only if ORAC was determined. A possible explanation was proposed by Moradi et al. (2011). Films with high amount of CEO (50 g kg^{-1}) caused an increase in the hydrophobicity. This fact could reduce the migrations of the phenolics from the film.

Phenols are very important constituents of CEO, being eugenol the majority constituent. Phenols have scavenging ability due to their hydroxyl groups (Gülçin et al., 2004). The antioxidant activity of CEO according to Jirovetz et al. (2006) may be based on various mechanisms such as scavenging radicals and chelation.

It has been found good correlation between total phenols and antioxidant activity in chitosan films with CEO with the different methods carried out to evaluate antioxidant capacity. All correlations were above $r^2 = 0.96$.

3.2 Combined Treatment of Trout Fillets with Film and HPP

Total Volatile Basic Nitrogen (TVBN): Initially TVBN values ranged from 11.90 ± 0.99 to 7.00 ± 0.08 mg TVBN $100g^{-1}$. These values were in accordance with data reported in the scientific literature, which were typically between 5 and 20 mg TVBN $100g^{-1}$ (Rezaei and Hosseni 2010). An increase in TVBN values was observed during storage in all the treatments. Throughout the storage period, all treatments were more efficient compared to untreated samples (Fresh). Nevertheless, any treatment did not exceed the legal limit established by European Union of 25–35 mg of TVBN in 100 g^{-1} of muscle (EC, 2005). Chytiri et al. (2004) reported TVBN values of 18.31 mg N_2 $100g^{-1}$ for filleted trout samples at the time of sensory rejection. Based on above it may conclude that chitosan film prevented the deterioration of trout fillets respect to raw fish (fish), cooked and HPP samples. In fact, at the end of storage, HPP samples showed 18.90 mg N_2 $100g^{-1}$, while TVBN value in HPP treatment samples covered by chitosan film was 15.4 mg N_2 $100g^{-1}$. Souza et al. (2010) and Doğan and Izci (2016) also found a reduction in NVBT values in chitosan coated samples compared to control in salmon and smoked trout fillets, respectively. These differences were attributed to the antimicrobial activity of chitosan, reducing the bacterial population and consequently their capacity for oxidative deamination. Günlu *et al.* (2014) described a significant synergistic effect in the use of HPP and chitosan films different from our results. Nevertheless, they used a higher storage time, specifically 44 days.

Thiobarbituric Acid Reactive Substances (TBARS): Lipid oxidation was studied using the TBARS method to monitor levels of secondary oxidation products formed (Table 2). During the storage, TBARS values were under the limit beyond which fish would normally develop an undesirable odour (1–2 μg MDA g^{-1}) (Souza et al. 2010). This fact can be attributed to the presence of carotenoids, which can act as a strong endogenous antioxidant system in rainbow trout (Ojagh et al. 2011). The initial TBA values of trout fillets varied a lot in literature ranging from 0.07 μg MDA g^{-1} (Hamzeh and Rezaei 2012) to 10.43 μg MDA g^{-1} (Chytiri et al. 2004; Mexis et al. 2009). These differences could also depend on the filleting process, which may affect rancidity levels of fish samples, probably as a result of exposure of the fish lipids to atmospheric oxygen accelerating oxidation (Chytiri et al. 2004).

Cooking treatment induced the highest TBARS increase at the beginning of storage. Nevertheless, TBARS levels remained constant along storage time. Similar results were found by Ojagh et al. (2011). These increment in the TBARS values in cooked samples probably occurred due to the high temperature that promotes lipid peroxidation, increasing MDA levels (Weber et al. 2008). Specifically, the mechanism was proposed firstly by Frankel (1991). High temperatures have sufficient energy to break covalent bonds in the acyl backbone to form a variety of lipid alkyl radicals and these radicals react to form lipid oxidation products.

In pressurised samples (HPP), TBARS levels increased from day 15 of storage. There were not significant differences between these samples and cooked ones after the period previously mentioned. This trend also was found by Gómez-Estaca et al. (2007), Ojagh et al. (2011) and Günlu et al. (2014). The pro-oxidative effect of HPP (Cheftel and Culioli 1997; Gómez-Estaca et al. 2009; Yagiz et al. 2007) could be counterbalanced because HPP induces changes in fish muscle, increasing asthaxantine availability (Yagiz et al. 2007). Furthermore, Medina-Meza et al. (2014) reviewed the effect of HPP in the degradation of lipid fraction of food. Pressure treatment below 300 MPa had a little effect on lipid oxidation, being linearly increased above this value.

Coating with chitosan edible film enriched with CEO lowered oxidation rate in the same way as HC. Use of natural plant extracts in lipid oxidation prevention has been widely described in literature. In fact, Jirovetz et al. (2006) showed that CEO had a protective role in oxidative processes through TBARS determination. Nevertheless, there are few reports based on edible films formulated with plant extracts. In addition, most of them are focused on physical, chemical and/or antimicrobial properties, not on their antioxidant effect (Gómez-Estaca et al. 2007). Andevari and Rezaei (2011) also demonstrated the major efficacy of coating enriched with other essential oil, such as cinnamon, compared to gelatine coating alone on the oxidation of rainbow trout over chilled storage.

Unlike the case of the batches with CEO incorporated, chitosan films (Film) reduced lipid oxidation at lesser extent. According our results, films exerted a protective role in oxidation compared to cooked, HPP and control (Fresh) samples at the end of storage time (day 22). Similar results were found by Souza et al. (2010) and Doğan and Izci (2016). Apart from antioxidant activity of chitosan, chitosan films reduced the diffusion of oxygen and the availability of oxygen to the fish surface.

Nevertheless, not synergistic effect was observed in HF respect to films. Our findings differed from Günlu et al. (2014), where HF had significant lower TBARS values than chitosan covered samples (Film).

Table 2. Thiobarbituric acid reactive substances (μg malondialdehyde per gram of muscle) of trout fillets subjected to different processing treatments and stored at 4 °C for 22 days.

Treatment/Day	0	6	15	22
Fish	$_{A}0.059 \pm 0.007a$	$_{B}0.252 \pm 0.002b$	$_{AB}0.136 \pm 0.003b$	$_{C}0.403 \pm 0.004^{c}$
Film	$_{A}0.042 \pm 0.007a$	$_{A}0.031 \pm 0.002a$	$_{A}0.031 \pm 0.020a$	$_{B}0.272 \pm 0.023^{b}$
Clove	$_{A}0.015 \pm 0.009a$	$_{A}0.020 \pm 0.007a$	$_{A}0.025 \pm 0.011a$	$_{B}0.062 \pm 0.007^{a}$
Cook	$_{A}0.463 \pm 0.098b$	$_{A}0.674 \pm 0.100c$	$_{A}0.703 \pm 0.065c$	$_{A}0.476 \pm 0.011^{cd}$
HPP	$_{A}0.037 \pm 0.008a$	$_{B}0.050 \pm 0.003a$	$_{D}0.730 \pm 0.007c$	$_{C}0.572 \pm 0.006^{d}$
HF	$_{A}0.056 \pm 0.004a$	$_{A}0.053 \pm 0.010a$	$_{A}0.084 \pm 0.006ab$	$_{B}0.256 \pm 0.04^{b}$
HC	$_{A}0.014 \pm 0.002a$	$_{AB}0.022 \pm 0.009a$	$_{A}0.012 \pm 0.001a$	$_{B}0.038 \pm 0.006^{a}$

Values (mean ± standard deviation, n = 3) followed by different uppercase letter in same row are significantly different ($P < 0.05$). Values (mean ± standard deviation, n = 3) followed by different lowercase letter in same column are significantly different ($P < 0.05$).

Hexanal: Initiation of lipid oxidation in fish is generally associated with the polyunsaturated fatty acids in phospholipids of muscle cell membranes (Decker and Xu 1998). Hexanal has been identified as a good marker for early lipid oxidation and correlated with rancid off-flavour (Timm-Heinrich et al. 2013). Hexanal concentration increased significantly over storage time as shown in Fig. 1. It was remarkably the different pattern in hexanal concentration of control samples (Fresh), which reached the highest value at day 9, decreasing from this date. Among treatments, cooking increased significantly hexanal concentration respect to the rest of them. Moreira et al. (2011) also obtained the highest hexanal concentration in Senegalese sole muscle submitted to cooking.

HPP also induced oxidation, but to a lesser extent than cooking. Similar results were found by Beltran et al. (2003), where hexanal values of minced chicken cooked (90 °C for 15 min) and pressurized (300 MPa for 30 min at 20 °C) were compared. The mechanism of pressure-induced oxidation is related to the effect on the cell membrane. The pattern of hexanal generation was very similar that to the observed in TBARS values. Nevertheless, hexanal was less sensitive to monitor lipid oxidation compared to TBARS, differing from Bragnolo et al. (2005). No significant differences between coated (Film and Clove) and uncoated (Fresh) samples were found. This is consistent with the results from Kim et al. (2012) where the volatile profiles of coated with defatted mustard meal film and uncoated smoked salmon were not distinguishable.

	6	9	15
Fresh	Aa	Cc	Ba
Film	Aa	Aa	Ba
Clove	Aa	Ba	Ca
Cook	Ac	Aa	Bb
HPP	Ab	ABb	Ba
HF	Aa	Aa	Ba

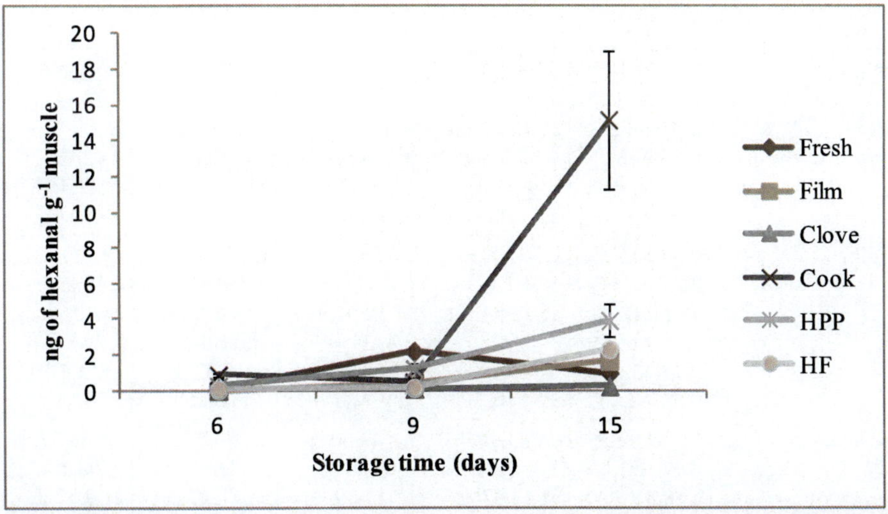

Fig.1. Hexanal concentration (ng/g) in rainbow trout fillets subjected to different processing treatment and stored at 4 °C for 22 days. Values (mean ± standard deviation, n = 3) followed by different uppercase letter in same row are significantly different ($P < 0.05$). Values (mean ± standard deviation, n = 3) followed by different lowercase letter in same column are significantly different ($P < 0.05$).

Colour: Table 3 showed the evolution of colour parameters along 22 days of storage. Lightness significantly increased in cooked and HPP treated samples. This increment in lightness (L*) as a result of cooking is well documented (Ojagh et al. 2011; Picouet et al. 2011; Rodriguez-Turienzo et al. 2011). HPP also induced an increase in the opacity of fish muscle with a loss of its raw appearance (Gómez-Estaca et al. 2009; Yagiz et al. 2007; Ojagh et al. 2011). Matser et al. (2000) reported lightness (L*) increment at pressures beyond 150 MPa. Films did not exert a protective role in changes produced by HPP. Chitosan film (Film) did not induce changes compared to uncoated (Fresh) samples in L* parameter over storage.

Cooked samples presented the most pronounced reduction in redness (a*). These results were in agreement with Ojagh et al. (2011). The reduction in a* can be explained by the thermal degradation of carotenoids (mainly astaxanthin and canthaxantin). On the other hand, carotenoids are bound to some myofibrillar proteins and thus, the degree of protein denaturation may also influence colour changes (Rodriguez-Turienzo et al. 2011). No significant changes were found among HPP and films samples. Similarly, yellowness (b*) did not suffer modifications with the different treatments.

Table 3. Colour parameters of rainbow trout fillets subjected to different processing treatment and stored at 4°C for 22 days.

Treatment/Day	0	6	9	15	22
Fresh	L* 35.61 ± 0.47 Aa a*10.38 ± 0.66 Aab b* 9.58 ± 1.70 Aa	L* 38.32 ± 4.27 Aa a* 12.72 ± 4.17 ABb b*14.30 ± 3.45 Bab	L* 37.60 ± 4.71Aab a*10.44 ± 3.33Aab b*11.73 ± 3.04ABab	L* 37.19 ± 5.90Aab a*15.95 ± 2.47Bcd b*19.18 ± 3.18Cd	L* 36.46 ± 3.28Aa a*12.02 ± 2.45Ab b*11.44 ± 2.45ABa
Film	L* 32.15 ± 2.29 Aa a* 11.47 ± 0.76 Ab b* 9.43 ± 0.62 Aa	L* 35.3 ± 3.01 Aa a* 13.54 ± 3.52 Ab b* 13.18 ± 3.84 Aa	L* 36.92 ± 6.63Aa a* 12.97 ± 5.4Ab b* 13.39 ± 2.45Aab	L* 33.36 ± 2.75Aa a* 17.50 ± 1.57Bd b*18.67 ± 4.11Bcd	L* 36.77 ± 2.12Aa a* 10.97 ± 2.29Aab b*13.41 ± 3.05Aa
Clove	L* 42.02 ± 4.17 Ab a* 8.81 ± 1.06 Aa b*12.70 ± 6.11 Aab	L* 43.96 ± 2.31 ABb a* 13.97 ± 3.71 Bb b*17.27 ± 3.96 Bb	L* 42.60 ± 4.86Ab a* 11.39 ± 1.67ABab b*10.54 ± 1.04Aa	L* 39.80 ± 5.33Ab a* 11.27 ± 2.37ABb b*11.36 ± 1.13Aab	L* 48.57 ± 2.74Bb a* 9.52 ± 0.69Aab b*11.27 ± 3.07Aa
Cook	L* 62.95 ± 3.58 Ad a*14.61 ± 2.92 Bc b*16.75 ± 3.37 Bb	L* 60.60 ± 3.53 Ad a*6.53 ± 1.55 Aa b*12.03 ± 3.02 ABa	L* 61.27 ± 5.67Ad a*6.89 ± 4.12Aa b*13.36 ± 6.8ABab	L* 60.56 ± 5.49Ad a*6..71 ± 0.68Aa b*8.08 ± 0.30Aa	L* 61.47 ± 5.86Ac a* 8.54 ± 0.77Aa b*14.19 ± 2.63ABa
HPP	L*52.99 ± 4.64Ac a*15.69 ± 1.51 ABc b*14.09 ± 3.09 Aab	L*58.4 ± 2.06Ad a*15.59 ± 0.51Cb b*12.89 ± 1.22Aa	L*56.38 ± 1.94Acd a*14.02 ± 4..36BCb b*15.05 ± 3.57Ab	L*54.76 ± 3.63Acd a*10.46 ± 3.34Ab b*12.51 ± 4.45Aab	L*56.44 ± 1.67Ad a*11.77 ± 1.80ABb b*11.64 ± 1.57Aa

(continued)

Table 3. (*continued*)

Treatment/Day	0	6	9	15	22
HF	L*53.46 ± 1.90ABc a*10.96 ± 1..64ABab b*16.28 ± 2.59Ab	L*54.07 ± 2.73ABc a*9.03 ± 2.60Aa b*14.42 ± 2.88Aab	L*54.43 ± 2.87Bc a*10.21 ± 3.66ABab b*15.66 ± 3.22Ab	L*52.25 ± 2.07ABc a*12.71 ± 2.99Bbc b*14.19 ± 2.83Abc	L*50.48 ± 5.05Ab a*10.19 ± 2.90ABab b*12.96 ± 1.78Aa
HC	L*53.43 ± 1.03ABc a*10.96 ± 1..64ABab b*16.28 ± 2.59Ab	L*54.07 ± 2.73ABc a*9.03 ± 2.60Aa b*14.42 ± 2.88Aab	L*55.10 ± 2.77Bc a*10.21 ± 3.66ABab b*15.66 ± 3.22Ab	L*52.25 ± 2.07ABc a*12.71 ± 2.99Bbc b*14.19 ± 2.83Abc	L*50.48 ± 5.05Ab a*10.19 ± 2.90ABab b*12.96 ± 1.78Aa

Values (mean ± standard deviation, n = 3) followed by different lowercase letter in same row are significantly different ($P < 0.05$). Values (mean ± standard deviation, n = 3) followed by different uppercase letter in same column are significantly different ($P < 0.05$).

Sensorial Analysis: Sensory analysis is the best available method to understand changes caused by lipid oxidation which are perceived by humans. Figure 2a, 2b and 2c show the mean values for the sensory characteristics of trout fillets (fishy odour, strange odour, drip loss) at 0, 6, 9, 15 and 22 days of chilled storage at 4 ºC. Sensory analysis tried to determinate if differences between treated samples (Clove, Film, HPP, HC, HF, Cook) and control (Fresh) occurred along storage.

Fishy odour gradually increased in all samples (Fig. 2a), as the storage progressed until day 15. From this point, a decrease took place. Among treatments, fishy odour at the end of storage was qualified as rancid in samples submitted to cooking (Cook). In fact, these samples also had the highest hexanal values. Volatile organic compounds are responsible for the overall aroma of fish products. Some of them contribute to positive sensory attributes but other produce "off-flavours". In relation to HPP, lipid rancid odour has been hardly detected. Although TBARS values in HPP samples were exceeded the organoleptic detectable thresholds of off-flavour in muscle foods from day 15 (Gray *et al.* 1996). The combination of films and HPP did not contribute to improve the fishy odour.

Results (Fig. 2b) showed that strange odour was detected by panellist in samples covered with CEO (Clove). Other essential oils such thymol, applied as coating, have also a strong odour and they did not produce unfavourable changes according to sensory analysis (Chamanara et al. 2012). In this work, apart from the application way (coating instead of film), the essential oil concentration used was the half than CEO contained in the films in the present study. Nevertheless, chitosan films made no unfavourable change in odour. Odour was described as herbal.

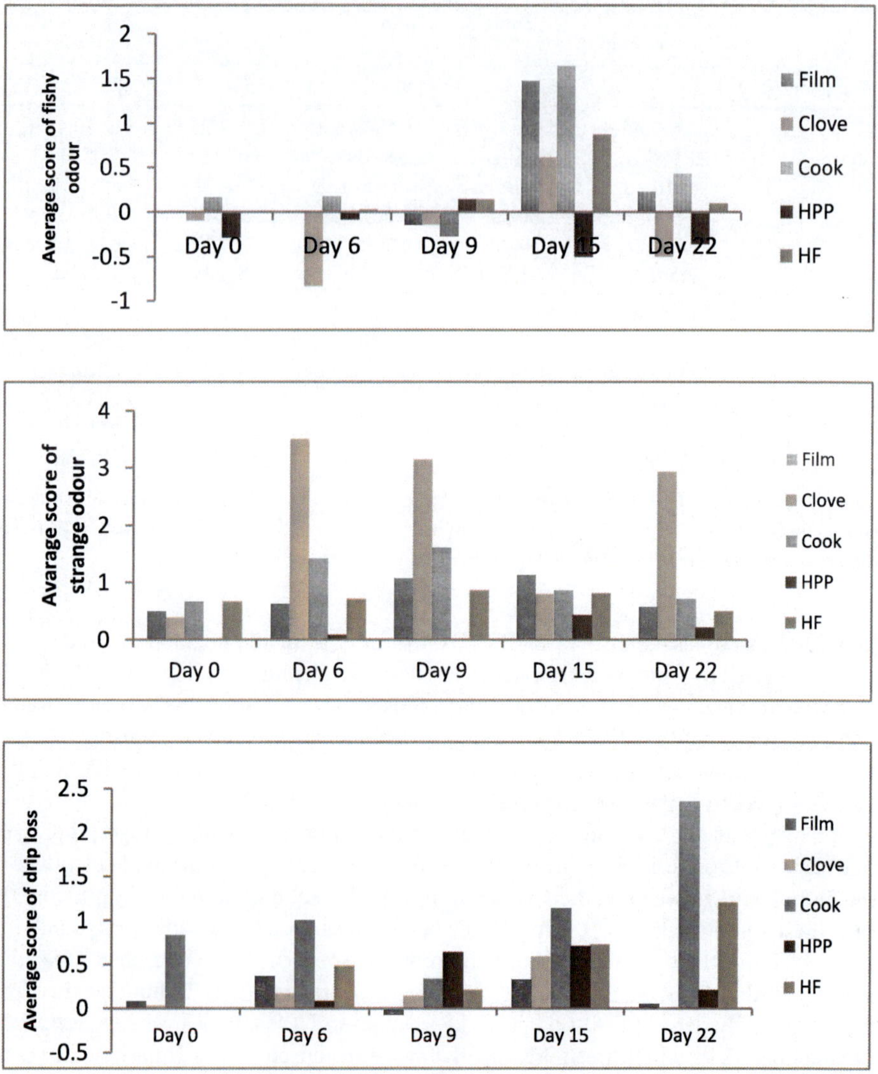

Fig. 2. (a) Average score of fishy odour in rainbow trout fillets subjected to different processing treatment compared to untreated (Fresh) and stored at 4 ºC for 22 days. (b) Average score of strange odour in rainbow trout fillets subjected to different processing treatment compared to untreated (Fresh) and stored at 4 ºC for 22 days. (c) Average score of drip loss in rainbow trout fillets subjected to different processing treatment compared to untreated (Fresh) and stored at 4 ºC for 22 days.

Panellist perceived that Cook samples had significant more weight loss (Fig. 2c) and increased over storage. HPP also had effect on drip loss at lesser extent to cooking. The use of films (with or without CEO) protected from drip loss through the water absorption.

Concerning to colour (data not shown), sensory analyses were in accordance with instrumental measurement. Cooking treatment causes the most negative changes among samples. Samples submitted to cooking presented white areas typical of cooked appearance. In fact, some panellist identified cooking samples. From 15 days of storage whiteness are turning into yellowness. Cooking samples presented porous surface with some cracks in their surface, surface defects are increasing along storage. HPP also caused colour changes, at lesser extent to thermal treated. Pink appearance was gained due to HPP. As to increasing the storage time, appeared brown spots at the edge of the HPP samples. The use of chitosan films (Film) did not adversely affect the colour.

Considering sensory analysis chitosan films are suitable to increase the shelf life in vacuum refrigerated trout.

4 Conclusions

Phenolics compounds and antioxidant capacity increased with increasing CEO concentration, until 20 g kg^{-1}. Testing for quality and sensory trout preservation, both treatments (HPP, Cook) induced lipid oxidation and changes in organoleptic markers (colour parameters and sensorial analysis). Nevertheless they were effective reducing TVBN compared to control (Fresh). The combination of HPP and Film would be a promising technology to reduce the negative effects of HPP such as lipid oxidation. The inclusion of CEO reduced significantly the lipid oxidation according TBARS measurements compared to chitosan films. Although the use of CEO in films did not induced colour changes. In sensorial test, CEO caused strange odour.

In conclusion, the use of chitosan films with or without CEO is shown to be a suitable treatment for trout fillets, reducing the negative effects of thermal and high pressure treatment. The odour inconvenience of CEO can be solved by means of adding it to fish products where this odour will be accepted such as marinade.

References

Albertos, I., Rico, D., Diez, A.M., Gonzalez-Arnáiz, L., Garcçia-Casas, M.J., Jaime, I.: Effect of edible chitosan/clove oil films and high-pressure processing on the microbiological shelf life of trout fillets. J. Sci. Food Agric. **95**(14), 2858–2865 (2015)

Andevari, G.T., Rezaei, M.: Effect of gelatin coating incorporated with cinnamon oil on the quality of fresh rainbow trout in cold storage. Int. J. Food Sci. Technol. **46**, 2305–2311 (2011)

AOAC Procedure 984.25. Association of official analytical chemists, official methods of analysis, 14th edn. AOAC, Washington (1990)

AOAC Official methods of analysis of AOAC international, 16th edn. The Association of Analytical Communities, Arlington (1995)

Baron, C.P., Kasgard, I.V., Jessen, F., Jacobsen, C.: Protein and lipid oxidation during frozen storage of rainbow trout (Oncorhynchus mykiss). J. Agric. Food Chem. **55**, 8118–8125 (2007)

Beltran, E., Pla, R., Yuste, J., Mor-Mur, M.: Lipid oxidation of pressurized and cooked chicken: role of sodium chloride and mechanical processing on TBARS and hexanal values. Meat Sci. **64**(1), 19–25 (2003)

Bragagnolo, N., Danielsen, B., Skibsted, L.H.: Effect of rosemary on lipid oxidation in pressure-processed, minced chicken breast during refrigerated storage and subsequent heat treatment. Eur. Food Res. Technol. **221**(5), 610–615 (2005)

Campos, C.A., Gerchenson, L.N., Flores, S.K.: Development of edible films and coating with antimicrobial activity. Food Bioprocess Technol. 4, 849–875 (2011)

Cheftel, J.C., Culioli, J.: Effects of high pressure on meat: a review. Meat Sci. **46**, 211–236 (1997)

Chamanara, V., Shabsnpour, B., Gorgin, S., Khomeiri, M.: An investigation on characterisitic of rainbow trout coated using chitosan assisted with thyme essential oil. Int. J. Biol. Macromol. **50**, 540–544 (2012)

Chytiri, S., Chouliara, I., Savvaidis, I.N., Kontominas, M.G.: Microbiological, chemical and sensory assessment of iced whole and filleted aquacultured rainbow trout. Food Microbiol. **21**(2), 157–165 (2004)

Decker, E.A., Xu, Z.: Minimizing rancidity in muscle foods. Trends Food Sci. Technol. **9**, 241 (1998)

Doğan, G., Izci, L.: Effects on quality properties of smoked rainbow trout (Oncorhynchus mykiss) fillets of chitosan films enriched with essential oils. J. Food Process. Preserv. **41** (2016)

EC. Commission Regulation (EC) No 2074/2005. Official Journal of the European Union, 22.12.2005, L338/27 (2005)

Frankel, E.N.: Recent advances in lipid oxidation. J. Sci. Food Agric. **54**, 495–511 (1991)

Gómez-Estaca, J., López-Caballero, M.E., Gómez-Guillén, M.C., de Lacey, A.L., Montero, P.: High pressure technology as a tool to obtain high quality carpaccio and carpaccio-like products from fish. Innov. Food Sci. Emerg. Technol. **10**, 148–154 (2009)

Gómez-Estaca, J., de Lacey, A.L., López-Caballero, M.E., Gómez-Guillén, M.C., Montero, P.: Biodegradable gelatin-chitosan films incorporated with essential oils as antimicrobial agents for fish preservation. Food Microbiol. **27**, 889–896 (2010)

Gray, J.I., Gomaa, E.A., Buckley, D.J.: Oxidative quality and shelf life of meats. Meat Sci. **43**, 111–123 (1996)

Gülçin, I., Güngör, I., Beydemir, S., Elmastas, M., Küfrevioglu, I.: Comparison of antioxidant activity of clove (Eugenia caryophylata Thunb) buds and lavender (Lavandula stoechas L). Food Chem. **84**, 393–400 (2004)

Günlü, A., Sipahioğlu, S., Alpas, H.: The effect of chitosan-based edible film and high hydrostatic pressure process on the microbiological and chemical quality of rainbow trout (oncorhynchus mykiss walbaum) fillets during cold storage (4±1°C). High Pressure Res. **34**(1), 110–121 (2014)

Jirovetz, L., Buchbauer, G., Stoilova, I., Stoyanova, A., Krastanov, A., Schmidt, E.: Chemical composition and antioxidant properties of clove leaf essential oil. J. Agric. Food Chem. **54**, 6303–6307 (2006)

Kanatt, S., Chander, R., Sharma, A.: Chitosan and mint mixture: a new preservative for meat and meat products. Food Chem. **107**(2), 845–852 (2008)

Kim, I., Yang, H., Noh, B., Chung, S., Min, S.C.: Development of a defatted mustard meal-based composite film and its application to smoked salmon to retard lipid oxidation. Food Chem. **133**(4), 1501–1509 (2012)

Leistner, L.: Basic aspects of food preservation by hurdle technology. Int. J. Food Microbiol. **55**, 181–186 (2000)

Matser, A.M., Stegeman, D., Kals, J., Bartels, P.V.: Effects of high pressure on colour and texture of fish. High Pressure Res. **19**(1–6), 109–115 (2000)

Medina-Meza, I.G., Barnaba, C., Barbosa-Cánovas, G.V.: Effects of high pressure processing on lipid oxidation: a review. Innov. Food Sci. Emerg. Technol. **22**, 1–10 (2014)

Mexis, S.F., Chouliara, E., Kontominas, M.G.: Combined effect of an oxygen absorber and oregano essential oil on the shelf life extension of the rainbow trout fillets stored at 4 °C. Food Microbiol. **26**, 598–605 (2009)

Moradi, M., Tajik, H., Rohani, S.M.R., Oromiehie, A.R.: Effectiveness of Zataria multiflora Boiss essential oil and grape seed extract impregnated chitosan film on ready-to-eat mortadella-type sausages during refrigerated storage. J. Sci. Food Agric. **91**, 2850–2857 (2011)

Moreira, N., Valente, L.M. P., Castro-Cunha, M., Cunha, L.M., Guedes de Pinho, P.: Effect of storage time and heat processing on the volatile profile of Senegalese sole (Solea senegalensis Kaup, 1858) muscle. Food Chem. **138**, 214–219 (2011)

Núñez-Flores, R., Castro, A.X., López-Caballero, M.E., Montero, P., Gómez-Guillén, M.C.: Functional stability of gelatin-lignosulphonate films and their feasibility to preserve sardine fillets during chilled storage in combination with high pressure treatment. Innovative Food Sci. Emerg. Technol. **19**, 95–103 (2013)

Ojagh, S.M., Núñez-Flores, R., López-Caballero, M.E., Montero, M.P., Gómez-Guillén, M.C.: Lessening of high-pressure-induced changes in Atlantic salmon muscle by combined use of a fish gelatine-lignin film. Food Chem. **125**, 595–606 (2011)

Ou, B., Hampsch-Woodill, M., Prior, R.L.: Development and Validation of an improved oxygen radical absorbance capacity assay using fluorescein as the fluorescent probe. J. Agric. Food Chem. **49**, 4619–4626 (2001)

Picouet, P.A., Cofan-Carbo, S., Vilaseca, H., Carboné-Balbe, L., Castells, P.: Stability of sous-vide cooked salmon loins processed by high pressure. Innov. Food Sci. Emerg. Technol. **12**, 26–31 (2011)

Re, R., Pellegrini, N., Proteggente, A., Pannala, A., Yang, M., Rice-Evans, C.: Antioxidant activity applying an improved abts radical catión decolorization assay. Free Radical Biol. Med. **26**, 1231–1237 (1999)

Rezaei, M., Hosseini, S.F.: Quality assessment of farmed rainbow trout (Orcorchynchus mykiss) during chilled storage. J. Food Sci. **73**, 93–96 (2010)

Rodriguez-Turienzo, L., Cobos, A., Moreno, V., Caride, A., Vieites, J.M., Diaz, O.: Whey protein-based coatings on frozen Atlantic salmon (Salmo salar): influence of the plasticiser and the moment of coating on quality preservation. Food Chem. **128**(1), 187–194 (2011)

Ross, A.I.V., Griffiths, M.W., Mittal, G.S., Deeth, H.C.: Combining nonthermal technologies to control foodborne microorganism. Rev. Int. J. Food Microbiol. **89**, 125–138 (2003)

Ruiz-Navajas, Y., Viuda-Martos, M., Sendra, E., Pérez-Álvarez, J.A., Fernández-López, J.: In Vitro antibacterial and antioxidant properties of chitosan edible films incorporated with Thymus moderi or Thymus piperella essential oils. Food Control **30**, 386–392 (2013)

Sanchez-Gonzalez, L., Vargas, M., González-Martinez, C., Chiralt, A., Cháfer, M.: Use of essential oils in bioactive edible coatings. Food Eng. Rev. **3**, 1–16 (2011)

Souza, B.W., et al.: Effect of chitosan-based coatings on the shelf life of salmon (Salmo salar). J. Agric. Food Chem. **58**, 11456–11462 (2010)

Slinkard, K., Singleton, V.L.: Total phenol analyses: automation and comparison with manual methods. Am. J. Enol. Vitic. **28**, 49–55 (1977)

Tarladgis, B.G., Watts, B.M., Younathan, M.T., Dugan, L.R., Jr.: A distillation method for the quantitative determination of malonaldehyde in rancid foods. J. Am. Oil Chem. Soc. **37**, 44–48 (1960)

Timm-Heinrich, M., Eymard, S., Baron, C.P., Nielsen, H.H., Jacobsen, C.: Oxidative changes during ice storage of rainbow trout (oncorhynchus mykiss) fed different ratios of marine and vegetable feed ingredients. Food Chem. **136**(3–4), 1220–1230 (2013)

Yagiz, Y., Kristinsson, H.G., Balaban, M.O., Marshall, M.R.: Effect of high pressure treatment on the quality of rainbow trout (Oncorhynchus mykiss) and Mahi Mahi (Coryphaena hippurus). J. Food Sci. **72**, 509–515 (2007)

Author Index

J. F. L. C. Semião et al. (Eds.): INCREaSE 2023, ASST, pp. 425–426, 2023.
https://doi.org/10.1007/978-3-031-44006-9